Research Topics in Aerospace

Series Editor

Rolf Henke, Mitglied des Vorstands, DLR, Cologne, Nordrhein-Westfalen, Germany

DLR is Germany's national centre for space and aeronautics research. Its extensive research and development work in aeronautics, space, transportation and energy is integrated into national and international cooperative ventures. Within this series important findings in the different technical disciplines in the field of space and aeronautics, as well as interdisciplinary projects and more general topics are addressed. This demonstrates DLR's outstanding research competences and capabilities to the worldwide scientific community and supports and motivates international research activities in exploring Earth and the universe, while also focusing on environmentally-friendly technologies and promoting mobility, communication and security.

More information about this series at http://www.springer.com/series/8625

Johann C. Dauer

Editor

Automated Low-Altitude Air Delivery

Towards Autonomous Cargo Transportation with Drones

Springer

Deutsches Zentrum
DLR für Luft- und Raumfahrt
German Aerospace Center

Editor
Johann C. Dauer
Institute of Flight Systems
German Aerospace Center
Braunschweig, Germany

ISSN 2194-8240 ISSN 2194-8259 (electronic)
Research Topics in Aerospace
ISBN 978-3-030-83146-2 ISBN 978-3-030-83144-8 (eBook)
https://doi.org/10.1007/978-3-030-83144-8

This Springer imprint is published by the registered company Springer Nature Switzerland AG
The registered company address is: Gewerbestrasse 11, 6330 Cham, Switzerland

Foreword

Automated Low-Altitude Air Delivery: The title alone is unwieldy, and now there is an entire book about it. If DLR had published a book about each and every one of its projects, it would have produced over 1,000 different books throughout its history since 1907. So why publish a book on this topic and what makes it so special?

The answers are simple: There are the technical challenges, then the social aspects, and finally a component which I would like to call "Zukunftssehnsucht"—a yearning for the future.

Let us start with the story from a popular-science point of view: Unmanned aircraft have been around for a long time, and another special branch of aviation is also closely connected with them: the "air taxis". These have been popular since the "Jetsons" cartoons, are featured regularly in magazines such as "Popular Mechanics" and appear time and again in sketches of visions of the future. It took two elements from outside of civil transportation aviation to provide these topics with a new impetus. Firstly, consumer electronics have demonstrated that by using inexpensive sensors, controllers and software, stable flight attitudes can be achieved even for simple devices. Secondly, military aviation has also developed application possibilities outside of armed missions, such as monitoring/observation and heavy-goods transport.

In the 2010s, a new aspect was added: Once the delivery of goods via Internet orders had been extensively automated, it would only be a matter of time before both large Internet companies and start-ups began to address the "last mile". The first video from the "Prime Air" segment of a major Internet provider in October 2015 became legendary, and since then, the topic has been discussed intensively—from the fast transport of organs to the "pizza helicopter".

This all provided the ingredients which led to the status in 2020: Automatic freight transport is "within reach", and even air taxis are "en vogue" again. Freight transport begins at just a few grams for cameras. For a major research facility such as DLR in general, and for the transport class described in this book in particular, payloads in the range of several hundred kilos are addressed; the smaller ones are still left to existing companies, and the larger ones fall into the category of cargo aircraft.

The start-ups mentioned exert a great deal of pressure on the air transport system, which, in my opinion, is very good for it. For too long, passenger and freight transport by air has been visible only in one way, which has led to a certain saturation of the mind. It is now necessary to think about new concepts and to question the existing, classic modes. The overall system is facing new, overarching challenges—not only for the vehicle but also for the airspace.

In the case of air taxis, new and interesting alliances have formed alongside the start-ups (but some of them have already dispersed again—an indication of the dynamics of the system): Airbus and Audi, Boeing and Porsche, Hyundai and Uber. It is apparent that, for one thing, completely new types of aircraft are seeking their place in the system, and that, for another, new business models will also characterize this "new aviation".

DLR has always had unmanned aircraft in operation, and institutes such as Flight Systems, Robotics and Sensor Systems Technology are scientifically engaged in this field. As society's interest in these vehicles grew, it became obvious that we would have to address this topic more intensively.

DLR's aeronautics research strategy aims to ensure an evaluation capability throughout the entire aeronautics system; we want to be able to provide meaningful information on components, vehicles, processes and effects, with proven, reliable accuracy, via models, simulation and experiments. In order to sharpen our tools for this, we defined six guiding concepts in 2013, whereby individual topics from various institutes are raised to a higher level of integration. Until 2019, one of these guiding concepts was "Unmanned Freight Transport", which corresponds to the era of the ALAADy project. In the meantime, we have expanded this guiding concept to "Urban Air Mobility", in order to take into account the current social relevance of the topic. The topics addressed by ALAADy will, of course, continue to be taken into consideration.

Three new DLR facilities will contribute to the success of the ALAADy program in the future:

- In 2019, DLR founded its "National Experimental Test Center for Unmanned Aircraft Systems" in Cochstedt near Magdeburg. The Center covers an extensive area of land and enables research into all practical aspects, including those relating to future infrastructures. This is where the ALAADy demonstrator made its maiden flight, and this is where we will continue to test this type of vehicle in the future—whether developed at DLR or by partners.
- Furthermore, in 2020, DLR founded its "Small Aircraft Technology" facility in Aachen, which will be set up during the ALAADy trials. The topics planned for this facility, such as the design, production, propulsion, maintenance of small aircraft in the form of the ALAADy demonstrator or air taxis, will pursue the continuation and implementation of the results obtained through ALAADy.

- And finally, at the Institute of "System Architectures in Aeronautics", founded in 2017, the entire system can be analyzed. Not only the vehicle but also its operation in a given system over the entire life cycle is relevant. For this reason, new architectures must be developed which deviate significantly from the classic airport-airspace concept of today's transport aircraft.

In addition to these three completely new facilities, the scientific focus of the ALAADy project in recent years has been achieved through many "classical", experienced and proven aviation institutes and their expertise, above all the institutes for "Flight Systems" and "Flight Guidance". One institute alone could address solely the individual aspects, e.g., the vehicle, the airspace or the logistics in its field of activity. By combining the capabilities of all the institutes through DLR's aeronautics program, comprehensive solutions are, however, created which support the industry in terms of products and the politicians in their decisions.

One goal of the ALAADy project was the construction of a demonstrator, a low-flying freighter with a payload capacity of several hundred kilos. Despite the fact that a product such as the "ALAADy Demonstrator" cannot be bought, it was, in addition to considerations concerning airspace and logistics, imperative to acquire a freely available product from the world of manned aviation and to convert it into an unmanned aircraft. The learning effect in such a case is particularly strong: For every single component, it is necessary to consider what must be available and with what reliability. All sensors and actuators, existing and additional, which are required as a result of the now absent pilot, must be subjected to a critical examination in order to be utilized in unmanned operations. All this is carried out on almost unknown terrain, as the corresponding approval regulations do not yet exist. As a consequence, however, this procedure is future-oriented for DLR as well: Other bodies can also benefit from such a future "construction manual". As a matter of fact, the institute performing the project received numerous enquiries during the development phase, and a business model can certainly be expected in the coming years.

Equally challenging was and is the investigation of the movement of such an unmanned device through the general airspace. This concerns the immediate surroundings in the sense of the requirement for pilots to avoid collisions: "see and avoid", which must now become "sense and avoid". Accordingly, the question as to the rules under which large numbers of unmanned devices should operate, both in their interaction with each other and in the choice of suitable flight paths, taking into account all other traffic participants such as rescue helicopters or general aviation, is of paramount importance. Two major DLR projects have hereby provided valuable contributions and have attracted worldwide attention: In the project "Unmanned Freight Operations UFO", transport via large Airbus-class aircraft was investigated, resulting in the derivation of rules for smaller vehicles for ALAADy. And one of DLR's most significant projects on the topic of "Urban Air Mobility", the "City ATM" project, extends the airspace integration of ALAADy to include operation over urban areas.

As one of the very few research centers in the world, DLR is able to analyze, simulate and optimize both the individual vehicles and their utilization in a holistic manner. For selected points, universities or other research centers in Europe also achieve success, but only the breadth and depth of DLR's expertise can result in usable concepts for low-altitude cargo delivery, i.e., ALAADy.

While this preface is being written, the "ALAADy Demonstrator" is undergoing flight testing at the new DLR test center in Cochstedt. Simultaneously, the planning for further demonstrators—and, consequently, other projects within DLR and also with partners in Germany and Europe—is being carried out in the fascinating field of unmanned aircraft in all their diversity.

At the same time, work is being undertaken in Europe and internationally on certification rules, which DLR will help to define. In addition to new regulations for the vehicle and its movement in the airspace, new certification rules are also required for innovative—often electric—propulsion systems. In the case of the ALAADy Demonstrator, the propulsion system is still "classic", but we will also use this knowledge and share it with the authorities in order to forge ahead in the broader context of unmanned small aircraft.

I hope that this book about ALAADy will be a rewarding experience for all readers. The authors and supporters deserve my sincere thanks; it takes a tremendous amount of work and commitment not only to deliver a project report but also to scientifically analyze this topic from the very beginning. ALAADy was "only" a project in DLR aviation research, but it is also a synonym for our working principle: interdisciplinary, holistic, comprehensive, with thematic breadth, high standards and scientific depth in the support of research, industry and politics.

August 2020   Rolf Henke

Preface

This book presents the condensed results of the project Automated Low Altitude Air Delivery (ALAADy) on cargo delivery with unmanned aircraft developed at the German Aerospace Center (DLR). We summarize the findings of the years 2016 to 2019. Payload capacities of one metric ton are considered. At the same time, it is investigated if the operational risks for such comparably large aircraft can be reduced to a point where a design for the operation within the EASA specific category is justified. The risks such unmanned aircraft pose to the environment are minimized by flying below general air traffic and exclusively over sparsely populated areas. Beside contributing to and overcoming many technical challenges, this work serves two major goals. First, the EASA specific category is challenged to identify possible limitations in size and operational conditions. Second, a safety benchmark is created to serve as a test for the Acceptable Means of Compliance for this operational category.

The book is divided into four parts. Part I establishes the basis of the project, which provides a reference for the following parts of the book. These fundamentals include applying the methods to consider the operational risk in the development of new unmanned aircraft. Additionally, the potential use-cases and economic perspectives are discussed. Part II of the book examines the aspects of the aircraft configuration. The focus is on the impact of the operational-risk-based methods on the aircraft itself. Part III is dedicated to the components and methods to enable safe autonomy. The focus includes airspace integration, detect and avoid, onboard system architecture, route planning, runtime monitoring and datalinks. PART IV of the book presents the verification and validation of the study. It summarizes different forms of simulations and several technology demonstrations.

The first chapter of the book introduces the goals of the ALAADy program in detail and spans the various research questions that arise. The following chapters are each dedicated to answer individual questions based on each technical discipline. The final chapter of the book brings together the results from the different research activities and provides an overall perspective of the findings.

The concepts presented in this book will be further explored in the years to come. For updates on the developments, the reader is referred to the website www. dlr.de/ft/alaady or to contact any of the authors of this book directly.

The institutes of the DLR actively contributing to this book are:

- Institute of Flight Systems, Braunschweig
- Institute of Flight Guidance, Braunschweig
- Institute of Air Transport and Airport Research, Braunschweig and Köln
- Institute of Communication and Navigation, Oberpfaffenhofen
- Institute of Engineering Thermodynamics, Stuttgart
- Institute of Aeroelasticity, Göttingen
- Air Transportation Systems, Hamburg

Contents

Part II: Unmanned Aircraft Configuration

Performance-Based Preliminary Design and Selection of Aircraft Configurations for Unmanned Cargo Operations 107

Yasim Julian Hasan and Falk Sachs

Configurational Aspects and Vehicle Specific Investigations for Future Unmanned Cargo Aircraft 133

Falk Sachs

Part III: System Components and Safe Autonomy

Cargo Drone Airspace Integration in Very Low Level Altitude 247
Niklas Peinecke and Thorsten Mühlhausen

Part IV: Verification, Validation and Discussion

A Multi-disciplinary Scenario Simulation for Low-Altitude Unmanned Air Delivery

Simon Schopferer, Alexander Donkels, Sebastian Schirmer,
and Johann C. Dauer

Capacity and Workload Effects of Integrating a Cargo Drone in the Airport Approach

Thorsten Mühlhausen and Niklas Peinecke

Part I: Low Risk Operations of Unmanned Cargo Aircraft

Automated Cargo Delivery in Low Altitudes: Concepts and Research Questions of an Operational-Risk-Based Approach

Johann C. Dauer and Jörg S. Dittrich

Abstract The project Automated Low Altitude Air Delivery–ALAADy is a study on the design of a heavy-lift transportation drone in the context of certification based on operational risks. This chapter is the opening of a book providing the results of the initial phase of the project within the years 2016 to 2019. Conceptually, in ALAADy, unmanned aircraft systems (UAS) are envisioned to provide a payload capacity of around one metric ton. Research questions on how the newly established EASA Specific Category of drone operation and its acceptable means of compliance, the Specific Operations Risk Assessment (SORA), influences the UAS are introduced and motivated in this chapter. Applying the SORA imposes limits to the operation to ensure safety. The project focus is investigating the impact on the UAS design if the operational limitations are maximized by restricting the flight over sparsely populated areas and below common air traffic only, while exploring large-scale UAS at the same time. The aspects considered include UAS design, aircraft configuration, system architecture, datalinks, safe autonomy, and airspace integration. The impact on the use-cases of such a UAS as well as the effect on the economical perspective are also addressed. Finally, this chapter introduces the need to design and implement scaled technology demonstrators and presents the overall project's perspective to experimental analysis. One demonstrator is based on a manned micro light gyrocopter converted into a cargo drone. It enables payload transportation of around 150–200 kg and is applied to validate the conceptual part of the ALAADy project beyond simulation.

Keywords UAS · Drone · Cargo delivery · UCA · Heavy-lift cargo drone · Autonomy · Aircraft design · Concept study · Technology demonstration · Automated low altitude air delivery · ALAADy · Safety assessment · Safe drone operation · Safe autonomy · Scaled demonstrator

J. C. Dauer (✉) · J. S. Dittrich
German Aerospace Center (DLR), Institute of Flight Systems, Lilienthalplatz 7,
38108 Braunschweig, Germany
e-mail: johann.dauer@dlr.de

J. S. Dittrich
e-mail: joerg.dittrich@dlr.de

© Deutsches Zentrum für Luft- und Raumfahrt e. V. (DLR) 2022
J. C. Dauer (ed.), *Automated Low-Altitude Air Delivery*, Research Topics in Aerospace,
https://doi.org/10.1007/978-3-030-83144-8_1

1   Introduction to Automated Low Altitude Air Delivery

Drones with a payload capacity of a few kilograms are currently on the verge to be put into regular operations to transport goods in many specific applications. There are well-known concepts considering last-mile delivery for high demand parcel delivery or humanitarian transportation of emergency goods like medical supplies or equipment. A recent enabling factor is the development of new regulatory frameworks for civil certification of drones and concepts of assessing the safety of operation.

The certification of unmanned aircraft and the proof of operational safety is currently one of the core issues in unmanned aviation. A new process, the *Specific Operations Risk Assessment* (SORA) not only considers the drone as a product itself, but includes the mission for which it is intended and its operation (Murzilli 2019). Its intention is to establish a level of operational safety equal to that of manned aviation. With the SORA, the safety of the operation is holistically assessed, which includes the UAS itself, operator and crew qualification, and mission parameters. Afterwards, the degree and depth of certification requirements for each of the above is determined. In other words, if there are no people on board and the aircraft and mission pose little risk to the environment, the effort for the certification can be reduced.

This process is a key that has long been missing in civil unmanned aviation to assess the safety of a more complex operation, such as regular beyond visual line of sight (BVLOS) flights while maintaining the economic benefit of unmanned systems. In recent years, the DLR (The German Aerospace Center) has carried out a project to develop transport concepts further and assess the applicability of the newly defined safety concept. The project *Automated Low Altitude Air Delivery* (ALAADy) envisions the use of drones with payload capacities of one metric ton. The initial phase of the project was carried out between 2016 and 2019 and is the source of this book's content. Is it possible to operate an aircraft with this payload capacity within the new safety framework? If a technological solution exists, what does that imply for the economic value of such an unmanned aircraft system (UAS)? The present book is an extract of the individual work packages of this first project phase. Several DLR institutes were involved in the project and incorporated their perspectives and expertise into a overall systems concept study.

When the project started in 2016, early stage projects did focus especially on the last-mile delivery problem. These included, for example, Alphabet's daughter Wing Aviation LLC that lowers the payload without landing (Shannon 2019), Amazon Prime Air (Sudbury 2016), and the DHL Parcelcopter (Schütt 2017). At the time JD.com, Rakuten Inc. and Drone Delivery Canada (DDC) advanced as well, cf. Perera (2019). An important application is the delivery of medical goods and equipment. A pioneering example is the launching system designed by Zipline for dropping blood reserves (Ackerman 2018). An early heavy lift drone is the unmanned version of the Kaman K-Max (Sauvageau 2011) with external slung load. The increasing importance of this topic also became apparent as the IATA World Cargo Symposium, 2016 in Berlin hosted its first cargo drone related track. However, except for the K-Max, all these examples are all small UAS.

In parallel to the ALAADy project, the technology matured to an extent that the market share of start-ups, which provide solutions for delivery drones, increased significantly. Companies like Wingcopter and Matternet are just two examples that already partner up with major logistics companies on their way towards regular operations. DHL has further developed and validated its concept of the transport drone in various boundary conditions (Poljak 2020). Applications vary from parcel delivery, over medication to medical equipment like defibrillators (AED). Antwork Ltd. works on urban delivery with small UAS (Antwork 2015). All applications have in common that the additional energy consumption by airborne delivery is justified either by the time criticality of the delivery or the lack of ground bound infrastructure for alternative means of transportation.

Heavy lift unmanned aircraft are currently progressing as well. Examples are the companies Sabrewing (De Reyes 2020), Elroy Air with a patented aircraft configuration (Merrill 2019), the amphibious Singular Aircraft FlyOx (Singular Aircraft 2014), the Volodrone from Volocopter GmbH and the drones of Wings for Aid (Wings for Aid 2014). One example of converted light manned aircraft is the Star UAV Co., Ltd. AT200 (Star UAS 2015). The development in the area of urban air mobility is currently extraordinarily active and attracts significant media attention. Focus is not only cargo but also transportation of passengers. Startups like Volocopter GmbH, Lilium GmbH, eHang but also established aviation companies like Airbus are investing in these often called air-taxis (cf. Polaczyk 2019). In summary, during the course of the project, three development strands emerged that were actively being worked on: increasing the payload capacity and hence the size of unmanned aircraft, urbanizing of the operations, and enabling transportation of passengers.

While each topic involves its own challenges in the context of safe design and operation, the research project ALAADy evolves around the aspect of increasing the aircraft size and the implications to safe and economical operation. It explores the possibilities to limit investment costs into the UAS technology but rendering the overall operation safe by managing the operational risks. This approach explicitly avoids operation over populated urban environments and passenger transport. Comparatively large payloads and taking advantage of the fact that no person is on board has the potential to create cost-efficient transport solutions for time-critical goods in areas where little infrastructure is available. At the same time, the project is flanked by a large number of research activities paving the way for transportation with drones as part of our everyday life in a foreseeable future.

The questions of air space integration and how the airspace can be extended to support operation with drones are key examples of this research. The Unmanned Aircraft System Traffic Management (UTM, cf. Kopardekar 2014) crystallizes in the United States and its European equivalent called U-space emerges (SESAR Joint Undertaking 2019). Additionally, the results of the Joint Authorities for Rule-making on Unmanned Systems (JARUS) provide essential input to ALAADy, which addresses the challenges from a regulatory perspective in order to establish unmanned systems in regular operation. What initially started in the project as a vague assurance approach based on operational risks, found its concretization in the Specific Operations Risk Assessment (SORA) (Murzilli 2019) and was developed by the JARUS

Working Group (WG) 6. The first version of this document was published in 2017 and the second version is currently available. This approach has been adopted by EASA as an Acceptable Means of Compliance (AMC) for UAS operation in the so-called Specific Category in Europe, see EASA (2019a). The results of the ALAADy project were fed back into the JARUS WG6 during the SORA development and served as a benchmark scenario on which the methodology was tested.

It is safe to assume that this new type of safety assessment has a significant impact on existing development cycles of aircraft. Here, the term development cycle is understood as the iterative process of assessing market requirements and designing a technically feasible solution until an effective and efficient product is created. As a result of this development cycle, this product can be implemented with current technological solutions and meets market requirements to ensure a sufficient profit margin at its business models. Hence, the development of a new cargo UAS is not only a technical question.

In general, a major driver of development costs in aviation is the effort of certification. The certification therefore has significant impact on the design of the product. Changes in the development cycle that arise from the assurance specific to the operational risk are examined in this project and understood more precisely. This development cycle is started by examples of what appear to be sensible top level aircraft requirements (TLARs). These TLARs define the specific point of operational risks from which the investigations of this work evolve. At the beginning of the project, it was not validated whether these TLARs actually allow a market-relevant implementation. Thus, the economical aspects were analysed based on these TLARs and the results are included in this work. In contrast to actual product development, the research character of the project does not allow the TLAR to be iterated by the market assessment; instead, the project focusses on many derived questions of the system design resulting from the TLAR specification. This approach distinguishes the research focus of the project from a concrete product design.

This book provides results of one of the very first assessments of the assurance method based on operational risks and also discusses the method's impact on UAS technology. This approach allows us to illuminate specific scenarios in a timely manner and to avoid too abstract concept studies. Our goal is to provide independent and objective assessment of the impact of the new SORA concept without being influenced by economic interests that are connected with actually designing a product for a specific market.

The following sections present the considered overarching research questions and core assumptions made for the project, which are applied in the remaining chapters. The nextsection examines the project's hypothesis in more detail, which reveals that the new methods of certification will result in significant changes in the design of a UAS. In comparison to traditionally established aircraft designs, new optimal combinations of aircraft configurations, system design, and aspired missions might be the outcome of such cycle. The subsequent section introduces the TLARs as the entry point to the project and presents the resulting assessment point of SORA. The concluding section of this chapter gives an overview on the different parts of the book including the fundamentals of the study, aircraft design, safe autonomy related questions, and validation. It guides the reader to the chapters of most interest and explains the relations of the different topics discussed.

2 Research Questions and Assumptions of the Study

The following sections outline the bigger picture of questions which are answered by the contributions of this book. Two approaches are emerging in order to achieve the necessary level of safety for the operation of comparably large scale UAS:

1. Increase the safety of the UAS operation by investing in the reliability of the involved technical systems in order to achieve comparable failure rates to manned aircraft.
2. Restrict the operations to limit the involved safety risks, for example in the form of flight routes in low-level altitudes or by avoiding densely populated areas.

The first of the possibilities means significant development, production, and educational efforts, which clearly casts doubts on the economic advantages compared to manned air freight transportation. The proof of feasibility for such a reliable unmanned aircraft still has to be provided. Nevertheless, this approach will play a major role, especially for the developments of urban air mobility and for aircraft transporting passengers. A pragmatic solution for certain aircraft can be the Special Condition SC-VTOL in the near future (EASA 2019b). However, this SC also does not include certification requirements for different levels of autonomy and its related software components.

The development of the second approach primarily focuses on a near-term implementation, since it bases directly on the current state of technology. The safety-technical problem is solved by a combination of concepts for safe operation and the development of suitable technical safety mechanisms. Thus, this approach has the potential to allow a manifold of different use-cases that require low-cost UAS without compromising safety. The ALAADy project examines a challenging setting of this concept. On the one hand, the size of the aircraft imposes an increased risk factor and on the other hand, environmental conditions are chosen so that the safety risks are limited by flying over sparsely populated areas and in the very low level airspace. It thus represents the most exciting combination of the aircraft's risk classification and the largest possible number of operational mitigations. Can a sensible realization of a transport UAS be created for this initial setting? Is it possible to implement UAS more cost-efficient than manned aviation? If so, which applications benefit from this concept? Figure 1 shows the various aspects that are examined within the scope of the study.

Fig. 1 ALAADy–World: artist impression of the aspects of the study to assess the transportation over sparsely populated areas in very low level altitude – ground handling, datalinks, airspace integration, aircraft configurations, geofencing, risk-based motion planning, safety assessment, and ground control station. Art by Boris Bolm, © DLR

2.1 A Modified Development Cycle

The design of a new aircraft concept should always consider the result of an extensive market analysis. The goal is to find an optimum between the costs of development and operation versus the product capabilities and thus maximized market potential. The introduction of SORA creates interesting new possibilities in this context. The risk involved in an operation is carefully examined and precisely those measures which render the operation safe are implemented. An increasing range of intended use of the UAS leads to more mission variants, which as consequence increase the variations of risks to be considered. This idea creates interplay between the effort required for development, assurance and certification, and production on the one hand and the universality of the intended use on the other hand. These considerations can be simplified in an iterative process as presented in Fig. 2.

A market assessment determines a set of requirements for the UAS. One of the strengths of unmanned aircraft is that the configuration can be selected from a broad set of alternatives due to the absence of a pilot. Aircraft candidates are chosen in a preliminary design. A concept of operations is laid out for their applications, which in turn represents the basis for risk assessment according to SORA. This risk assessment defines the safety objectives that include the development process, qualifications, and aircraft capabilities. Accordingly, the system architecture can be designed, defining the required component redundancies and a selection of technologies. Based on the architecture, the UAS design is carried out, which in turn forms the basis for estimating the costs. Here, cost models are used, which include acquisition or development costs including the previously determined certification measures as well as operating costs. Such a cost model closes the circle and checks the extent to which the

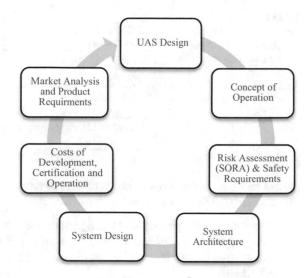

Fig. 2 Development cycle of a UAS including the interplay resulting from the operational risk based approach

technical solution meets the requirements of the market. For product development, the cycle is iterated until an optimum is found.

One of the challenges of this process is the selection of the appropriate level of detail in order to be able to run through this cycle efficiently. The entry into this design circle can differ. It can start from a promising product idea or, alternatively, from an identified market demand. For the project of this book, the point of entry is the set of top level aircraft requirements (TLARs), which are located on the transition from market analysis to UAS design. The design circle will be run through once during the project to illuminate as many unique aspects of this concept as possible. As an example, it can be expected that aircraft configurations will differ from those that have been established in manned aviation over the past decades.

2.2 Top Level Aircraft Requirements

The TLARs represent the most basic form of the requirements, which determine the payload, range, and speed for which the UAS is designed. Their definition allows many design questions to be addressed in parallel. Therefore, the aircraft config-uration can be expedited while the risk assessment and market suitability can be examined at the same time. The focus of the research project is the investigation of the aspects affected by the safety assessment based on operational risks. Due to the novelty of this approach, the goal of designing a market-optimal UAS is a second step subsequent to the initial phase of the ALAADy project.

The definition of the TLAR thus enables an efficient and parallel start of many design tasks which allows an early concrete examination of the methods for the different research questions. For this purpose, TLARs are defined at the start of the

Table 1 Top level Aircraft
Requirements (TLARs) -
Initial assumption of the
project and entry point into
the development cycle of
ALAADy

Cruise speed	200 km/h; 110 kts
Cruise altitude	below common air traffic
Range	600 km; 325 NM
Payload capacity	1 metric ton
Cargo size	length: 3.0 m; height: 1.3 m; width: 1.3 m/5 m^3
Take-off and landing distance	<400 m; <440 yd (STOL capability)

project and remain unchanged for the project to ensure consistency of the findings.
The chosen TLARs are motivated by three aspects:

1. In contrast to the previously existing regulations on drones, the EASA specific category and the SORA does not define an upper weight limit. The limits of the new methods are therefore challenged with a heavy but realistic weight requirement.
2. The desired size of the aircraft indicates an interesting transition point for engine technologies including piston engines, turboprops, and future alternative drive-train concepts, examples of which are electrical, hybrid-electric, and fuel cells. It can therefore be assumed that this aircraft size poses interesting trade-offs to be assessed.
3. The targeted size represents a gap between the different aircraft classes that DLR researches. It is therefore worthwhile to investigate whether this gap poses an interesting niche for unmanned aviation.

The TLARs of this project are compiled in Table 1. The ability to take-off and land on short runways comes from the initial assumption that this type of aircraft will potentially be used for the transport of goods for humanitarian aid. The Short Take-off and Landing (STOL) ability therefore allows the operation of landing sites with little infrastructure and of reduced size. The dimension of the cargo area follows a related argument that simple transport equipment such as Euro-pallets should be enabled. Figure 4 shows artistic variants of such UAS configurations. Reference configurations from manned aviation are Cessna 208 Cargomaster, Cessna 406, Dornier 228, and Bell 412.

2.3 An Approach to UAS Safety Based on Operational Risks

Taking the operations into account to provide safety evidence started as a general idea in the initial phase of ALAADy. Simultaneously, the SORA was published, which represents a concrete implementation of this idea implementing a qualitative risk assessment. The SORA assesses the overall risk for third parties posed by the specific operation of a UAS by analyzing the critical event *loss of control* of operation. The whole argumentation revolves around a bow-tie analysis of this event. To assess the impact, two risk defining aspects are distinguished. First, the *ground risk* considers the damage caused by a crashing aircraft to the environment on the ground - to people and critical infrastructure. The size and weight of the aircraft on the one hand and the population density of the area overflown on the other hand are decisive for risk classification. The second risk aspect is the *air risk*, which depends heavily on the airspace structure in which the drone is operated.

Both risk aspects are divided into classes and their combination results in a total risk representing metric, the SAIL (Safety Assurance and Integrity Level). Depending on the determined SAIL, specifications are made to ensure safe operation. The bow-tie argumentation is then completed by defining the measures that avoid loss of control of the UAS operation. These measures are called *operational safety objectives* and scale in robustness with the SAIL.

The intention of the project is to sound out the limits of applicability of this concept. For this reason, a combination of a comparatively large aircraft and a low-risk operating environment is selected. The operation above sparsely populated areas and below regular air traffic allows minimized impact of a loss of control event to be assumed. The imminent question is, whether the SORA provides a satisfactory list of safety objectives rendering this operation safe, and whether this set of measures maintains advantages compared to a classical full certification approach.

During the project three versions of SORA were considered. First, an early draft version, then the officially published version 1.0, and finally version 2.0 were consulted. It was interesting to observe the evolution of the risk classes concerned with the ALAADy concept. While the risk classification was low at the beginning of the project indicating a significant reduction in certification effort, the classes increased as SORA matured. One of many reasons for the working group creating the SORA to increase the risk classification of cases similar to ALAADy, was the use of the project results as a borderline test scenario. The project thus contributed to the overall integrity of the SORA concept.

In the particular case of low altitude flight above sparsely populated areas, the understanding of safety differs to manned aviation: Normally, all conceivable measures are taken to avoid a crash and additionally, this effort must be proven. In this project, however, the conscious termination of the flight can be considered as an adequate risk mitigating measure. As the aircraft only flies in very low altitudes, it is very unlikely to encounter other aircraft in case of a safe termination, as there are likely no aircraft below, because the very low airspace is sparsely used and even prohibited for most manned aircraft. Additionally, flying over sparsely populated areas also renders it very unlikely that people and infrastructure are affected by a precautionary emergency landing. In other words, importance to prove the reliability of the aircraft is reduced in favor to show a very remote chance that in case of a safe termination people on the ground and in the air are harmed and no critical infrastructure is damaged. The reliability of the drone in this safety understanding is nevertheless connected to economic and environmental aspects.

This argumentation simplifies the assurance of safety. The development processes of safety-relevant hardware and software, for example according to the ARP4754 (SAE 2010) and DO-178b/c (RTCA 2011), multiply the development costs compared to a purely functional implementation. A large part of the development costs is therefore invested in the assurance of the reliability of the aircraft system. These processes have matured over decades of aviation. However, the interplay of the individual steps of this process is complex. It is not possible to directly deduce which level of safety is achieved by adding or omitting certain process steps. Furthermore, these approaches are not easily extended towards new technologies, especially in software development, such as machine learning cf. Cluzeau (2020). From this context, the need for cost reduction and research into alternative approaches to certification becomes evident. The investigations of this book as well as SORA itself focus on these aspects of safety. Their influence on applicable use-cases and system design is analyzed. Questions of environmental impact and social acceptance of this approach is equivalently important to study and is subject of future research.

From the perspective of the overall system, a large number of design questions arise. Which system architectures are possible and necessary for such a safety approach? How must the low level airspace be extended to ensure the shared operation of manned and unmanned systems? What does the approach based on operational risks mean for system components such as data links, control station, and propulsion systems? The question of aircraft configuration is of particular importance: What are the safety characteristics of the alternatives in relation to the operational risk and what are the associated methods for safe termination? These questions are all examined within this book.

The following section presents the structure of the book and the assignment of questions to the respective chapters.

3 Focus of the Study and Structure of the Book

The book is structured into four main parts: first, the fundaments of the study are outlined. Here, details on the SORA and the application to the ALAADy concept are discussed. The interval of feasible safety and integrity levels are determined and their consequences for the operations of the conceptual transport UAS are analyzed. A second perspective is presented which surveys the most appropriate use-cases and their market potential. This part of the book provides a context for the remaining sections wherever concretization is required to limit the number of conceptual variations.

The second part of the book focuses on the aircraft configuration. A preliminary design study selects promising aircraft configurations from a large set of alternatives. This study includes safety characteristics of the configuration itself and structural considerations. The engine design considers new and alternative propulsion concepts and is presented afterwards. Additionally, it is investigated how ground and cargo handling can benefit from the automation of a UAS as well as the specifics of the ALAADy transportation concept.

The third part of the book looks at different aspects of safe autonomy. From the perspective of the UAS, external topics such as airspace integration and detect and avoid (DAA) are addressed. In addition, various components are considered on which the risk-based approach can have a conceptual impact. These components include risk-based path and motion planning. The question of how the system can be structured with a suitable architecture is discussed. The architecture shall ensure sufficient reliability of the UAS with respect to the assigned SAIL. Based on this discussion, a special aspect of runtime monitoring plays an important role, which we refer to as safe operation monitoring (SOM). During flight, this runtime monitor observes the safety of the operation with respect to the SORA assessment and might have the potential to reduce the overall software assurance effort. The last part of the book is concerned with validation of the concepts. Two methods are used to evaluate the findings: First, simulations help to validate and discuss parts of early technology readiness and the development of technology demonstration for aspects of higher maturity. The following sections introduce the research questions in more detail and shall guide the reader to the chapters of most interest.

3.1 Operational, Use-Cases and Market Potential

Under which conditions can air cargo transportation with UAS be carried out from an economic and operational perspective? The operational limitations imposed by the SORA concept also impose limits to the applicable civil use-cases. Nikodem et al. (2021) apply the SORA on this transportation concept. It is outlined how the different steps of the SORA are interpreted to assess the ALAADy operation. On the one hand, this chapter provides guidance on how to use the SORA to assess a concept of operations. On the other hand, the consequences of the different SAIL classes to the operation are outlined and discussed.

The proceeding work commences with identifying potential use-cases in which hold a favorable relationship between the criticality of the delivery time, alternative ground-based infrastructure, and the value of the goods to be transported. Thus, possible applications of the ALAADy drones can be the following:

- humanitarian logistics, that are providing people in disaster areas (in the event of floods, earthquakes, hurricanes, ship accidents, etc.) with relief goods (water, shelter, food, medicines, etc.) and supplying emergency services with equipment
- delivery of time critical goods, parts between the production, and assembly sites of suppliers and manufacturers (e.g. just-in-time production in automobile production)
- spare part logistics, that is transport of urgently needed spare parts from regional stores/distribution centers to the place of need, e.g. in case of downtime of capital-intensive production facilities

Fig. 3 Artist impression of different use-cases of the proposed low altitude air delivery: time critical inter plant logistics or medical equipment, replacement parts, and humanitarian goods. Art by Boris Bolm, © DLR

Figure 3 illustrates a vision of the use of such drones. In order to assess possible use-cases, first, general considerations and market reviews are performed in Pak (2021). To assess cost and yield models for a disruptive technology is a challenging task. This challenge is tackled by interviews with possible users of such a transportation system as well as reviewing statistical data on the use of current transportation infrastructure. The results of these research steps and concrete examples are provided. Based on the identified use-cases, parametric cost models are derived by Liebhardt and Pertz (2021). New aspects of this work are the integration of the potentially unproven reliability of the UAS that are operated in low risk classes. Both, cost models for highly time critical replacement parts as well as for humanitarian aid are presented and discussed.

3.2 Conceptual Aircraft Design

The second part of the book begins with the generation of three reference configurations. The aim is twofold: first, a basis is created for the detailed analysis in order to generate sufficient concreteness, similarly to the identification of the use-cases.

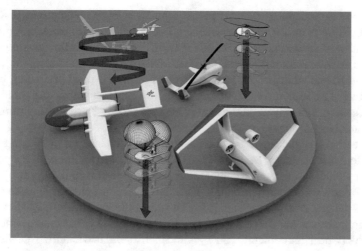

Fig. 4 Artist impressions of alternative aircraft configurations and potential safety concepts. Art by Boris Bolm, © DLR

Using concrete aircraft and missions helps the more technical research aspects to be performed with higher value. The second goal is to investigate to what extent aircraft configurations differ from established configurations due to the operations related risk-based approach. Starting points are the TLARs from Sect. 2.2. Different conceivable, even unusual, configurations are investigated and compared. From this set of alternatives, promising candidates are selected for a detailed preliminary design. Aspects such as efficiency, accessibility of the payload, technical simplicity, and size are taken into account. For these configurations, expected flight performances are calculated. Both fixed-wing and rotary wing are analyzed and a detailed sizing is performed. Finally, three promising candidates are selected. Figure 4 shows an artistic interpretation of the selected configurations, an autogyro and two fixed-wing aircraft. Hasan and Sachs (2021) considers these aspects in detail.

Sachs (2021) goes into more detail about the three configurations. Here, the safety features are examined and different forms of safe flight termination are identified. The natural characteristics of the aircraft, such as the vertical autorotation of the gyrocopter, are examined, but also variants with parachute are compared. Especially, the gyrocopter is a remarkable choice. Due to its natural vertical autorotation and cost efficiency, it provides promising aspects in respect to the project requirements. Since it is a comparatively unexplored configuration, especially when equipped with additional wings, a first scaled technology demonstrator is built. This demonstrator aims at testing the aircraft configuration in wind tunnel and in flight tests as cost-efficiently as possible. As shown in Sachs (2021), the testing led to crucial knowledge about the configuration, which would not have been discovered if purely theoretical approaches were used.

Hecken et al. (2021) also look at further details of these three configurations by examining a structural design. Using the Certification Specifications CS-23 (EASA

2003a) and CS-27 (EASA 2003b) maneuvers and gust loads are determined. Requirements resulting from the safety assessments create load cases for safe termination that are also taken into account. In this way, the requirements for the structure are compiled. A finite element method (FEM) is used to perform a seizing of the structure and to investigate which of the three configurations is the most promising from the perspective of a structural realization. Information on structure thickness and masses are obtained in this way. Additionally, Aptsiauri (2021) investigates future use of electric and hybrid electric propulsion systems.

The section concludes with a conceptual study of the integration of these aircraft into the logistics chains by Meincke (2021). The feasibility of the ALAADy concept is not only an economic and technical question with regard to the UAS, but is also decided by how well it can be integrated into existing supply chains. Thus, the chapter discusses the ground handling of the aircraft and possible realizations. Aspects of the ground infrastructure and its automation are examined and illustrated in simulations. Additionally, possible container concepts and their automated handling are discussed.

3.3   Safe Autonomy in the Context of SORA

Complex assistance systems are already in use to support crews in piloting manned aircraft. In manned aviation, effects of many system failures can be compensated by trained pilots. If an assistance system fails, it is up to the pilot in command (PIC) to understand its nature and to counteract the failure. This fallback solution can be taken into account for the assurance requirements of technical components. However, this approach cannot be transferred to the support systems for unmanned aircraft directly. The PIC, who controls and monitors one or more aircraft from the ground, depends heavily on these support systems due to the physical separation from the aircraft. Therefore, especially software components tend to have a high safety criticality in the context of unmanned aviation.

Assistance systems involve implementation of complex software components. However, the verification of such software development is a complex and expensive process; see Do-178C (RTCA 2011). For many modern software functions such as deep machine learning, verification processes are not fully understood and conceptually feasible only for a small subset of possible methods, see Cluzeau (2020). The aim of this part of the book is therefore to examine questions of system architectures, essential system components and autonomy functions in relation to SORA. It is strived to find solutions that allow cost savings through the new verification concepts and still guarantee safe operation.

In Fig. 5, the aspects discussed in the third part of the book are put in relation to each other from an operational risk perspective. It illustrates the complexity of the interactions between the aspects that define safety and autonomy. The ground risk in the figure is defined by the number of people potentially overflown. This number is affected by the methodology of path planning represented in the concept of operations. For high degrees of autonomy, runtime monitoring can be used to

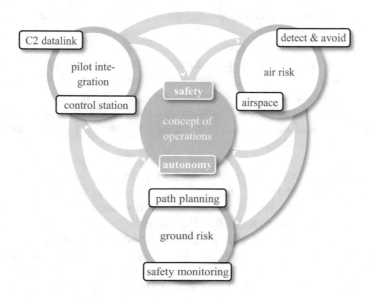

Fig. 5 The level of autonomy and its safety concept defines the concept of operations from a mission perspective. The figure shows the impact on and interplay with the safety defining risks in air and on ground. The third major aspect connected to the level of autonomy is the pilot involvement

supervise the safety status. In ALAADy, these two aspects are specifically considered. On the one hand, a path and motion planning concept is explored which explicitly considers the ground risk. On the other hand, a runtime monitoring based on a formal language is analyzed which has the potential to significantly reduce software development costs. During flight, this monitor supervises the operational constraints defined in the concept of operation.

Depending on the level of autonomy, the impact of the pilot's role on safety varies. This variation changes the requirements for the control station and the command and control (C2) data link, which enables communication between the aircraft and control station. It allows the pilot to monitor the conditions in flight. The issue of air risk is defined by the characteristics of the airspace and the UAS integration. Closely related are the DAA functions and their safety characteristics. However, the topology of the airspace also has a direct influence on the planning of flight routes. The routes and a surveillance by a safe operation monitor can increase or decrease the criticality of the data link reliability. These strong interactions must be taken into account when designing and discussing these components. The procedures and technologies presented in this part of the book contribute to this overarching design questions. The chapters are structured as follows.

First, Peinecke (2021) deals with airspace integration. A large part of the constraints is caused by the design of the airspaces in which the UAS are operated. The proposed concept is illustrated using the challenging example of dense airspace structures, such as those found in Europe and especially in Germany. The approach is set and discussed in relation to the European concept of air traffic management system for unmanned aircraft, U-space (SESAR Joint Undertaking 2019).

Rothe and Nikodem (2021) dedicate a chapter to possible onboard system architectures for the ALAADy concept and examine the extent, to which the system complexity differs due to classification into different SAIL of SORA. For this purpose, the *Operational Safety Objectives* are examined in terms of their influence on the system architecture. The identified architecture variants are compared relatively in terms of development effort in order to estimate whether and which SAIL could yield significant cost savings.

Challenges regarding the design and development of the ground station are outlined in the following chapter (Friedrich et al. 2021). Based on a prototypical development of a ground station, mission and state visualization concepts for a pilot on ground are presented. The communication between aircraft and ground station is then examined by Schalk and Backer (2021). Communication via satellite and cellular networks are assessed and put into context of the operation risk-based approach. In particular, flying over sparsely populated areas and in low altitudes pose challenges to the data link due to obstacles in the line-of-sight to ground-based relay stations. The influence of the coverage of the network is investigated by means of simulation. For this purpose, transmission masts and their switching behavior and the reception characteristics of antennas are modelled. Topological maps are used to estimate the line-of-sight and the reliability of the data link availability is discussed, based on concrete trajectories resulting from the use-cases of the first part of the book.

Detect and avoid (DAA) is one of the key questions for establishing regular operation of unmanned aircraft. The technological maturity of the techniques is constantly increasing. However, how does the operation in very low airspaces influence the requirements for DAA? Do technologies already exist that meet these requirements? The chapter (Schalk and Peinecke 2021) deals with cooperative aspects of these questions.

In order to automatically generate routes that take operational risks into account, various steps are necessary. First, the operational risks must be modelled for the ground risk. The necessary information needs to be stored in geospatial databases. Schopferer and Donkels (2021) propose a solution for this problem first and then describe a method capable of planning paths on these databases that correspond to the operational concepts of ALAADy. Trajectories are planned with the help of sampling-based methods. Aspects of the realization in a software prototype are presented and computational factors such as sampling density and trajectory smoothing are discussed by means of concrete mission examples.

Schirmer and Torens (2021) assess the use of runtime monitoring in the operational risk context. Runtime monitoring has the potential to reduce verification effort by directly linking to the risk-based approach. We call this particular form of monitoring Safe Operation Monitoring (SOM). The basic idea is that the correctness of an implementation no longer needs to be proven extensively if a reliable system monitors the UAS during operation and initiates countermeasures as soon as the permitted operating conditions are threatened. Ideally, the verification of the entire system then falls back to that of the runtime monitor and the countermeasure.

This concept is illustrated using the example of geofencing, a comparatively well understood approach of defined operating limits. Challenges of the representation of the geometrical operating limits are presented and it is explored to what extent formal languages can be used to specify such operating limits. The appeal of such formal languages is that general proofs can be provided for a defined set of language elements, for which tools can be developed and qualified. Such a formal language is chosen so that it mimics expressing the higher-level requirements. By this means, it provides an abstraction of programming languages in the direction of specification. Thus, the developer can directly code the specification instead of performing classical programming. Such formal languages thus facilitate a correct implementation of the desired requirements. The proof of a runtime monitor based on such formal languages can be adapted to the desired application with little effort if qualified tools are available. In this case, the implementation has then already been achieved by qualifying the tools. Ideally, only the correctness of the formal expressions in relation to the requirement remains to be proven.

3.4 Validation Using Simulation and Technology Demonstration

The so far presented chapters cover the fundamentals of the study, possibilities of applications of the concept in a market survey, possible aircraft configurations, and ALAADy specific aspects of safe autonomy. The remaining part of the book is dedicated to validation. Due to the research nature of the project ALAADy, two validation tools are used: simulation and technology demonstration. Some chapters have already taken up results of the validation and discussed them on the spot. This section thus focuses on which tools have been used for the overall systems perspective and how they have been integrated.

Schopferer et al. (2021) is dedicated to the implementation of a simulation infrastructure. The simulation framework is able to integrate modules of this study and tackle multidisciplinary system analysis. The chapter shows how variations of the analyzed system components can be exchanged and studied similarly to parameter studies. A system simulation is presented, which connects the functions of autonomy developed in Part III of the book with the aircraft concepts from Part II. Since the respective chapters have discussed the results of the simulation in place, the chapter introduces the general capabilities of the so called scenario simulation.

Mühlhausen and Peinecke (2021) deal with the question of a possible integration of the ALAADy aircraft into the regular operation of commercial airports. The influence of the UAS on the arrival traffic flows of the airport is estimated, especially, since their flight performance differs significantly from commercial aircraft. This analysis is performed based on a traffic simulation with human-in-the-loop to control the air traffic in the sector of the exemplary chosen Dusseldorf Airport.

How can the interplay of all these aspects discussed so far be validated beyond simulation? Simulations always rely on models of the analyzed components. These models are designed for the target of the considered study. In particular, the complex and interdisciplinary interactions cannot be fully represented in simulations. However, this interplay is critical to understand the system properties. Additionally, concept studies a prone to overlook important aspects when assessing new technology without precedent. Lorenz et al. (2021) therefore present the development of a technology demonstrator called the *ALAADy Demonstrator*, which is a physical realization of a transport UAS. It is designed to test the concepts derived in the previous sections of the book. The development of a technology demonstrator enforces the consistent break of circular dependencies, as shown in Fig. 5, and the selection of one feasible combination of system components. The Concept of Operations, although not as comprehensive as the transport mission of the concept study, had to be written in full and negotiated with the aviation authority. The realization of such a technology demonstrator therefore represents the ultimate test for the developed concepts.

The development of the technology demonstrator started in the first year of ALAADy and is based on a micros-light gyrocopter. The manned gyrocopter is converted into a transport UAS, see Fig. 6. In contrast to the gyrocopter configuration presented in Sect. 3.2, the ALAADy Demonstrator does not have additional wings. However, even without wings, the gyrocopter is an ideal aircraft for technology demonstration. The aircraft itself is extremely cost-effective and has unique safety features which make it stand out for the assessment of operational safety: The main rotor of the gyrocopter is constantly in autorotation as it is propelled by the airflow. For example, it is not affected by the failure of an engine. For this reason,

Fig. 6 System prototype: technology demonstrator based on a micro light manned gyrocopter, converted into a UAS and designed to demonstrate aspects of the study by real world implementation and in flight tests, © DLR

the original hypothesis of the project is, that the gyrocopter can be landed safely and with low descent speed from almost all flight situations and many failure conditions. Thus, no parachute would be necessary as a contingency measure. At the same time, the containment of such a descent is comparably easy to ensure compared to fixed-wing aircraft. The gyrocopter descends almost vertically, which can even be supported by a contained helix trajectory. Hence, the gyrocopter has benefits with respect to necessary safety buffers.

Lorenz et al. (2021) presents many details on the realization of such a demonstrator from mechanical modification of the aircraft, avionics and software design, development of experimental operation, and their limitations. The demonstrator is also the key to the next phases of the project. During the initial project phase, automated flights under visual line of sight conditions have taken place. Enabling real transport missions is the next goal of this technology demonstrator and thus the further validation of the conceptual findings presented in this book.

The book is rounded off with Dauer (2021), in which the findings of the individual chapters are summarized and put into relation. Future work and open questions are summarized and an overall conclusion of the project is presented.

Acknowledgements The authors would like to thank all colleagues and friends who have made this project possible, actively participated with contributions, discussions and feedback. The complete list of people involved exceeds the room of this book and goes far beyond the list auf authors actually writing the different chapters. Nevertheless, we would like to express our thanks to a small number of key persons providing the environment to challenge a new concept that at the time of initiation of this project did not even had a name – the SORA. We would like to thank Prof. Dr.-Ing. Stefan Levedag for setting the context of the flight systems related research, which is one of the core aspects of this study. We also express many thanks to Dr.-Ing. Ingmar Ehrenpfordt who continuously supported the project and enabled many of the interesting research questions that have not been predicted when ALAADy started. Many thanks also to Lorenzo Murzilli and all our friends involved in the JARUS Work Group 6 for accepting our input and considering the ALAADy concepts as a benchmark for the SORA process.

References

Ackerman E, Strickland E (2018) Medical delivery drones take flight in east Africa. IEEE Spectr 55(1):34–35. https://doi.org/10.1109/MSPEC.2018.8241731

Antwork (2015) Antwork Ltd. https://www.antwork.link, Accessed 13 June 2020

Aptsiauri G (2021) Concepts of full-electric and hybrid-electric propulsion and operation risk motivated integrity monitoring for future unmanned cargo aircraft. In: Dauer JC (ed) Automated Low-Altitude Air Delivery - Towards Autonomous Cargo Transportation with Drones. Springer, New York

Dauer JC (2021) Unmanned aircraft for transportation in low-level altitudes: a systems perspective on design and operation. In: Dauer JC (ed) Automated Low-Altitude Air Delivery - Towards Autonomous Cargo Transportation with Drones. Springer, New York

De Reyes E (2020) Can cargo drones solve air freight's logjams? a drone startup says its big vertical-takeoff flier would be quick to land, load, and take off again. IEEE Spectr 57(6):30–35. https://doi.org/10.1109/MSPEC.2020.9099930

Cluzeau JM, van Dijk L, et al (2020) Concepts of Design Assurancefor Neural Networks (CoDANN). Public Report Extract, Innovation Network - AI Task Force, EASA

EASA (2003a) CS-23: Certification Specifications for Normal, Utility, Aerobatic, and Commuter Category Aeroplanes. https://www.easa.europa.eu/document-library/certification-specifications/cs-23-initial-issue, Accessed 9 June 2020

EASA (2003b) CS-27: Certification Specifications for Small Rotorcraft. https://www.easa.europa.eu/document-library/certification-specifications/cs-27-initial-issue, Accessed 9 June 2020

EASA (2019a) Acceptable Means of Compliance (AMC) and Guidance Material (GM) to Commission Implementing Regulation (EU) 2019/947, Annex I to ED Decision 2019/021/R. https://www.easa.europa.eu/document-library/acceptable-means-of-compliance-and-guidance-materials/amc-gm-commission, Accessed 9 June 2020

EASA (2019b) SC-VTOL: Special Condition - Vertical Take-Off and Landing (VTOL) Aircraft. https://www.easa.europa.eu/document-library/product-certification-consultations/special-condition-vtol, Accessed 14 June 2020

Friedrich M, Peinecke N, Geister D (2021) Human machine interface aspects of the ground control station for unmanned air transport. In: Dauer JC (ed) Automated Low-Altitude Air Delivery - Towards Autonomous Cargo Transportation with Drones. Springer, New York

Hasan YJ, Sachs F (2021) Performance-based preliminary design and selection of aircraft configurations for unmanned cargo operations. In: Dauer JC (ed) Automated Low-Altitude Air Delivery - Towards Autonomous Cargo Transportation with Drones. Springer, New York

Hecken T, Cumnuantip S (2021) Structural design of heavy-lift unmanned cargo drones in low altitudes. In: Dauer JC (ed) Automated Low-Altitude Air Delivery - Towards Autonomous Cargo Transportation with Drones. Springer, New York

Kopardekar PH (2014) Unmanned aerial system (UAS) traffic management (UTM): enabling low-altitude airspace and UAS operations. NASA Technical Report

Liebhardt B, Pertz J (2021) Automated cargo delivery in low altitudes: business cases and operating models. In: Dauer JC (ed) Automated Low-Altitude Air Delivery - Towards Autonomous Cargo Transportation with Drones. Springer, New York

Lorenz S, Benders S, Goormann L, Bornscheuer T, Laubner M, Pruter I, Dauer JC (2021) Design and flight testing of a gyrocopter drone technology demonstrator. In: Dauer JC (ed) Automated Low-Altitude Air Delivery - Towards Autonomous Cargo Transportation with Drones. Springer, New York

Meincke P (2021) Cargo Handling, Transport and Logistics Processes in the Context of Drone Operation. In: Dauer JC (ed) Automated Low-Altitude Air Delivery - Towards Autonomous Cargo Transportation with Drones. Springer, New York

Merrill D, Weekes T et al (2019) Unmanned Cargo Delivery Aircraft. US Design Patent D843,889. Accessed 26 Mar 2019

Mühlhausen T, Peinecke N (2021) Capacity and workload effects of integrating a cargo drone in the airport approach. In: Dauer JC (ed) Automated Low-Altitude Air Delivery - Towards Autonomous Cargo Transportation with Drones. Springer, New York

Murzilli L et al. (2019) JARUS guidelines on specific operations risk assessment (SORA). In: Joint Authorities for Rulemaking on Unmanned Systems. JARUS. http://jarus-rpas.org/sites/jarus-rpas.org/files/jar_doc_06_jarus_sora_v2.0.pdf. Accessed 7 June 2020.

Nikodem F, Rothe D, Dittrich J (2021) Operations risk based concept for specific cargo drone operation in low altitudes. In: Dauer JC (ed) Automated Low-Altitude Air Delivery - Towards Autonomous Cargo Transportation with Drones. Springer, New York

Pak H (2021) Use-cases for heavy lift unmanned cargo aircraft. In: Dauer JC (ed) Automated Low-Altitude Air Delivery - Towards Autonomous Cargo Transportation with Drones. Springer, New York

Peinecke N, Mühlhausen T (2021) Cargo drone airspace integration in very low level altitude. In: Dauer JC (ed) Automated Low-Altitude Air Delivery - Towards Autonomous Cargo Transportation with Drones. Springer, New York

Perera S, Dawande M, Janakiraman G, Mookerjee V (2019) Retail deliveries by drones: how will logistics networks change? forthcoming, production and operations management. SSRN: https://ssrn.com/abstract=3476824, https://doi.org/10.2139/ssrn.3476824

Polaczyk N, Trombino E, Wei P, Mitici M (2019) A review of current technology and research in urban on-demand air mobility applications. In 8th Biennial Autonomous VTOL Technical Meeting and 6thAnnual Electric VTOL Symposium, February 2019. Vertical Flight Society, pp 333–343

Poljak M, Šterbenc A (2020) Use of drones in clinical microbiology and infectious diseases: current status, challenges and barriers. Clin Microbiol Infect 26(4):425–430. https://doi.org/10.1016/j.cmi.2019.09.014

Rothe D, Nikodem F (2021) System architectures and its development efforts based on different risk classifications. In: Dauer JC (ed) Automated Low-Altitude Air Delivery - Towards Autonomous Cargo Transportation with Drones. Springer, New York

RTCA (2011) DO-178C: Software Considerations in Airborne Systems and Equipment Certification. RTCA, Inc., Washington, USA

Sachs F (2021) Configurational aspects and vehicle specific investigations for future unmanned cargo aircraft. In: Dauer JC (ed) Automated Low-Altitude Air Delivery - Towards Autonomous Cargo Transportation with Drones. Springer, New York

SAE (2010) ARP4754A: Guidelines for Development of Civil Aircraft and Systems. Aerospace Recommended Practice, S-18 Aircraft and Sys Dev and Safety Assessment Committee. SAE International. https://doi.org/10.4271/ARP4754A

Sauvageau, B (2011) K-MAX cargo unmanned aerial system. kaman aerospace. https://ndiastorage.blob.core.usgovcloudapi.net/ndia/2011/targets/WednesdaySauvageau.pdf, Accessed 7 June 2020

Schalk LM, Becker D (2021) Data link concept for unmanned aircraft in the context of operational risk. In: Dauer JC (ed) Automated Low-Altitude Air Delivery - Towards Autonomous Cargo Transportation with Drones. Springer, New York

Schalk LM, Peinecke N (2021) Detect and avoid for unmanned aircraft in very low level airspace. In: Dauer JC (ed) Automated Low-Altitude Air Delivery - Towards Autonomous Cargo Transportation with Drones. Springer, New York

Schirmer S, Torens C (2021) Safe operation monitoring for specific category unmanned aircraft. In: Dauer JC (ed) Automated Low-Altitude Air Delivery - Towards Autonomous Cargo Transportation with Drones. Springer, New York

Schopferer S, Donkels A (2021) Trajectory risk modelling and planning for unmanned cargo aircraft. In: Dauer JC (ed) Automated Low-Altitude Air Delivery - Towards Autonomous Cargo Transportation with Drones. Springer, New York

Schopferer S, Donkels A, Schirmer S, Dauer JC (2021) A multi-disciplinary scenario simulation for low-altitude unmanned air delivery. In: Dauer JC (ed) Automated Low-Altitude Air Delivery - Towards Autonomous Cargo Transportation with Drones. Springer, New York

SESAR Joint Undertaking (2019) SESAR Concept of Operations for U-space, vol 1, 2 & 3. https://www.sesarju.eu/node/3411, Accessed 9 Sept 2019

Schütt M, Hartmann P, Holsten J, Moormann D (2017) Mission control concept for parceldelivery operations based on a tiltwing aircraft system. In: 4th CEAS Specialist Conference on Guidance, Navigation & Control, Warsaw, Poland

Singular Aircraft (2014) Singular Aircraft S.L. http://singularaircraft.com/, Accessed 10 June 2020

Shannon T, Li Z (2019) Apparatuses for Releasing a Payload from an Aerial Tether. US Patent D10,315,764 B2, Accessed 11 June 2019

Star UAS (2015) Star UAV Co., Ltd., Longstar, Langxing UAV http://www.staruas.com, Accessed 11 June 2020.

Sudbury AW, Hutchingson EB (2016) A cost analysis of amazon prime air (drone delivery). J Econ Educ 16(1):1–12

Wings for Aid (2014) The Wings for Aid Foundation. https://www.wingsforaid.org, Accessed 07 June 2020

Operations Risk Based Concept for Specific Cargo Drone Operation in Low Altitudes

Florian Nikodem, Daniel Rothe, and Jörg S. Dittrich

Abstract The Operations Risk Based Concept for specific UAS evolved around three new categories of UAS operation recently defined by the European Aviation Safety Agency (EASA). This paper gives an overview of these categories and how they are embedded into general aviation regulations. After regulative basics are established the focus is on the *Specific Category* that features a holistic approach based on operational risks as a core element for risk assessment. This Specific Operations Risk Assessment (SORA) is applied to a low altitude, large cargo drone concept developed by the German Aerospace Center (DLR) within a project called Automated Low Altitude Air Delivery (ALAADy). It is shown that the operation of such a UAS is ideally done over unpopulated or sparsely populated environments and in low frequented airspaces. In addition to the application, special particularities of the SORA methodology are discussed. The chapter concludes with an outlook on the expected future development of UAS regulation in Europe.

Keywords Unmanned aircraft system · Risk assessment · SORA · Specific category · UAS operation

1 Introduction

In recent years the use of Unmanned Aerial Systems (UAS) increased to a point where aviation safety agencies all over the world had to answer the question on how to integrate the expected amount and variety of different types of UAS into the civil airspace. Major factors that had to be considered regarding the airspace integration of

F. Nikodem (✉) · D. Rothe · J. S. Dittrich
German Aerospace Center (DLR), Institute of Flight Systems, Lilienthalplatz 7, 38108 Braunschweig, Germany
e-mail: florian.nikodem@dlr.de

D. Rothe
e-mail: daniel.rothe@dlr.de

J. S. Dittrich
e-mail: joerg.dittrich@dlr.de

© Deutsches Zentrum für Luft- und Raumfahrt e. V. (DLR) 2022
J. C. Dauer (ed.), *Automated Low-Altitude Air Delivery*, Research Topics in Aerospace,
https://doi.org/10.1007/978-3-030-83144-8_2

UAS were the different operational scenarios conducted with various UAS types that posed a broad range of safety risk. The key to the aforementioned question is to find an appropriate solution for achievable safety, comparable to manned aviation, that addresses both large scale UAS used in operational scenarios that pose greater risk as well as small, lightweight UAS that are meant to be used by children, and everything in between. The solution shall ensure safety to the society, especially people and safety critical infrastructure, as well as be open enough to allow the corresponding UAS industry to evolve further.

Existing civil aviation regulations for manned aviation show that regulation based on risk is not uncommon for aviation authorities. The requirements to attain a type certificate for a large transportation aircraft are fairly different from the requirements of a small single piloted aircraft or even rotorcraft. Manned aviation regulation already has different regulations for a broad spectrum of different aircraft types. However, in contrast to UAS, manned aircraft pose at least a significant risk to the pilot that is considered throughout all certification specifications.

The DLR's ALAADy project, short for Automated Low Altitude Air Delivery, focuses on the idea of a rather large unmanned cargo drone, operating at low altitudes that do not necessarily need an aircraft certification. ALAADy introduces three concepts for an UAS that is able to carry a payload of one metric ton over a distance of up to 600 km. The ALAADy cargo drone concepts are designed to operate within a new category of UAS operations, called 'specific category', introduced by the EASA. Throughout the project, and focused in this chapter, is the application of the new EASA (European Aviation Safety Agency 2017) driven approach of UAS regulation including a holistic risk assessment method called Specific Operations Risk Assessment (SORA) (Joint Authorities for Rulemaking of Unmanned Systems JARUS 2017).

This paper analyses the regulatory framework of UAS with focus on the specific category and applies the required safety assessment to the ALAADy cargo drone concepts. The goal of this chapter is to show the general suitability of the specific category for the operational concept and the UAS concepts developed in ALAADy.

2   Manned Aviation Regulatory Framework in Europe

As stated above the EASA needed to develop an approach that differs from the known regulatory framework of manned aviation to regulate the full spectrum of unmanned aviation. To allow a better understanding of the similarities and differences between manned and unmanned regulations, a short overview of the manned aviation regulatory framework is provided.

In the late 1960s, the need for uniform and international standards to regulate increasing aviation activities was recognized. Therefore, the European Joint Airworthiness Authorities was founded in 1970, later renamed in the well-known Joint Aviation Authorities (JAA) and in 2002 replaced by the European Aviation Safety Agency (EASA). The EASA performs certification, regulation and standardization as well as investigation and monitoring. The foundation of today's European civil aviation was the *Regulation (EC) No 1592/2002* of the European Parliament and of the Council on common rules in the field of civil aviation and establishing the EASA (The European Parliament and the Council of the European Union 2002). Its current and valid version is the *Regulation (EU) 2018/1139* of 4th July 2018 (The European Parliament and the Council 2018).

Regarding the initial airworthiness of any civil aircraft the *Commission Regulation (EU) No 748/2012* and its latest amendment *Commission Delegated Regulation (EU) 2019/897* are of importance (The European Commission 2019). A crucial part of the initial airworthiness process for an aircraft manufacturer is to attain a type certificate (TC) for the new aircraft type. Necessary requirements and design standards are given by the EASA in so called *certification specifications* (CS) for different aircraft types. The most prominent examples are the CS-23 for *Normal Category Airplanes*, the CS-25 for large aircraft and the CS-27 as well as CS-29 for small and large rotorcraft.

One major point of interest in each CS is the paragraph 1309 respectively 2510 in CS-23 in which equipment, systems and installation requirements are stated. Part of this paragraph is a categorization of failure cases and failure probabilities. The failure cases typically range from *no safety effect* to *catastrophic*. To give a better idea of the failure categorization, the CS-25 for large aircraft shall serve as an example (European Aviation Safety Agency 2019):

> "No Safety Effect – Failure Conditions that would have no effect on safety; for example, Failure Conditions that would not affect the operational capability of the aeroplane or increase crew workload."

> "Catastrophic – Failure Conditions, which would result in multiple fatalities, usually with loss of the aeroplane."

The failure probabilities range from probable to extremely improbable:

> "Probable Failure Conditions are those anticipated to occur one or more times during the entire operational life of each aeroplane."

> "Extremely Improbable Failure Conditions are those so unlikely that they are not anticipated to occur during the entire operational life of all aeroplanes of one type."

To help to fulfill these certification specifications the EASA released acceptable means of compliance (AMC) and guidance material (GM). These documents are meant to show manufacturers how to comply with the requirements stated in the certification specifications. In general they give examples on how to comply with the paragraphs of the CS and show examples for suitable methods and processes. They also refer to wide spread and well established industrial standards such as the Aerospace Recommended Practice (ARP) of the Society of Automotive Engineers (SAE):

- ARP 4754A Guidelines for Development of Civil Aircraft and Systems
- ARP 4761 Guidelines and Methods for Conducting the Safety Assessment Process on Civil Airborne Systems and Equipment

Other well-known and often applied standards are developed by the Radio Technical Commission for Aeronautics (RTCA):

- DO-178C Software Considerations in Airborne Systems and Equipment Certification
- DO-254 Design Assurance Guidance for Airborne Electronic Hardware

Similar to the commission regulation on initial airworthiness and environmental certification there exist several other regulations on all other aspects of manned civil aviation, such as continuing airworthiness and aircrew operations. It is important to keep in mind the broad regulatory landscape that has grown over the past decades. This is especially essential when discussing new regulatory approaches for unmanned aviation.

3 Regulatory Framework of Unmanned Aircraft Systems

In the last decade, the use of UAS of all types and sizes has increased continuously and is expected to grow even further (European Aviation Safety Agency 2017) (Single European Sky ATM Research (SESAR) Joint Undertaking (SJU) 2016). Therefore, the EASA needed to find a new uniform regulation for UAS.

Since *Regulation (EC) No 216/2008,* the directive of UAS up to a maximum take-off weight (MTOW) of 150 kg was the responsibility of each Member State of the EU. The airworthiness certification of UAS with a MTOW of more than 150 kg was covered by an EASA policy statement based on Annex I Part 21 of the known commission regulation. However, the increasing number of UAS with less than 150 kg MTOW prompted the EASA to take action and develop an international regulatory standardization. With release of a *Notice of a Proposed Amendment (NPA)* as an introduction of a regulatory framework for the operation of UAS in 2017 the EASA proposed three categories of UAS operation; *Open, Specific* and *Certified*. This NPA followed an EASA Opinion document released in 2015 where these categories were firstly announced (European Aviation Safety Agency 2015). The category *Open* shall cover operations with very low risk to people and critical infrastructure. Toy drones and very light weight drones for private use as well as model aircraft are UAS that are typically expected in this category. The other end of the operational spectrum is addressed by the category *Certified*. UAS and operations that pose a risk equivalent to manned aviation have to meet safety standards and processes that are based on and equivalent to civil manned aviation regulations. Certainly, the existing certification specifications are not enough to fully cover UAS certification. In contrast to manned aircraft, unmanned systems rely on control stations and especially require detect and avoid (DAA) systems to participate in civil airspace. Also, UAS configurations are yet not covered that typically do not exist in traditional manned aviation. Multi-rotor UAS in their application as the commonly called "Air Taxi" would be an example for such a configuration. Other UAS that typically participate in the certified category are very large aircraft, designed to transport payload, or large military or civilian aircraft, such as the Global Hawk.

However, since dedicated certification specifications for UAS of the certified category still do not exist, the certification specifications for civil manned aircraft apply. It is recommended by the Joint Authorities for Rulemaking for Unmanned Systems (JARUS) to use CS-23 or CS-25 depending on the MTOW of the UAS as basis (Joint Authorities for Rulemaking of Unmanned Systems 2015a) (Joint Authorities for Rulemaking of Unmanned Systems 2015b). The aspects not covered by the CS, as mentioned above, are supposed be discussed in close contact with the national aviation authorities.

The category *Specific* is designed to handle all UAS and operation types that are in between the open category and the certified category regarding their risk to other people and critical infrastructure. To support this, the regulation of this category is build on UAS performance-based standards that take not only the UAS but the operator and the whole operation into account. Within the aviation regulation world that is a relative new approach.

Common UAS types within this category range from quadcopter with the mass of several kilograms up to large scale UAS with 10 m wingspan and even more, since there is no upper limit yet. The three categories of UAS operation are officially introduced by the EASA in Commission Implementing Regulation (EU) 2019/947 of 24 May 2019 (The European Commission 2019). The concept of ALAADy researches a cargo drone concept that is especially designed for the specific category. In the following section the specific category is described further and whenever necessary the use case and UAS concepts developed in ALAADy are provided as to illustrate the specific categories inherent ideas.

3.1 UAS Operation in the Specific Category

As stated above, the specific category is meant to act as a transition between the open category and the certified category. Since the range of possible operations and their inherent risk varies in between those two categories, the specific category needs to adapt its level of rigor to all those levels of risk.

Figure 1 shows the stepwise approach for the level of rigor in the specific category compared to the categories open and certified. To determine the necessary level of rigor in the specific category, a risk assessment needs to be conducted for each operation. This is a crucial part of the process used to obtain operational approval by the competent authority. In Opinion No 1/2018 the EASA introduced the Specific Operations Risk Assessment (SORA) that is developed by JARUS as an acceptable means of compliance (AMC) for the necessary risk assessment. Since the SORA is the only known AMC until now and is especially named by the EASA, the entire process is described in this chapter.

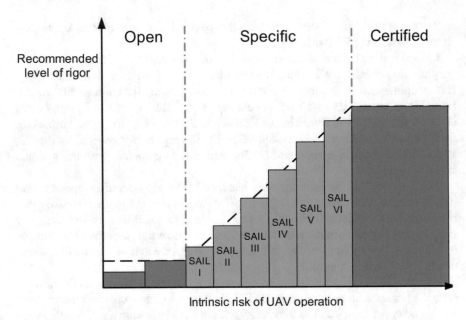

Fig. 1 Stepwise approach for the level of rigor in the specific category as described in (2019)

3.2 Related Work to Risk Assessment

Even though the SORA is the only AMC for risk assessment in the specific category, many other research approaches have been presented in the past. As early as 2006 Clothier and Walker (2006) took effort and determined UAS safety objectives. They use a historical analysis of civil aviation accidents and compared derived safety objectives to proposed draft regulations and similar studies. They also illustrate the impact of safety objectives to UAS design and operation with the use of a ground fatality expectation model and discuss the public acceptance of risk and highlight key factors that influence the definition of UAS safety objectives. Washington et al. (2017) published a review of UAS ground risk models where they analyzed and compared 33 existing UAS ground risk models. They discussed the relation of those ground risk models to the regulation efforts of the competent authorities and their influence on the development of UAS regulations.

Clothier published work on risk management and risk templates for civil UAS operations; this focuses on a UAS operation centric approach (Clothier 2015). In this work he introduces a risk assessment and management process based on bow-tie diagrams that shows many similarities to the SORA process that is discussed in later sections of this chapter. Belcastro et al. (2017) published their work on hazard identification and analysis for UAS operations. They collected UAS operation specific hazards through analysis of small UAS mishaps in the past and antic-ipate future hazards through analysis of small UAS use cases. Those results were used to develop a set of combined hazards for risk assessment. In 2018, la Cour-Harbo published a paper where he compares the SORA approach to high fidelity risk modeling and concludes that these existing different approaches are overall surprisingly well in agreement (la Cour-Harbo 2018). Previous work of the authors of this chapter compared the SORA approach to established civil aviation risk assessment methods (Nikodem, et al. 2018). It is shown that crucial parts of the SORA methodology resemble a root cause analyses that is pre-determined by the experts of JARUS.

3.3   Specific Operations Risk Assessment

The SORA process is a holistic approach that considers not only the UAS, but also the operator competencies, the remote crew and especially the properties of the operational volume. Earlier versions of the SORA guideline document show that the underlying risk model is based on bowtie diagrams (Joint Authorities for Rulemaking of Unmanned Systems 2017). In the current edition 2.0 of the SORA guidelines its origin bowtie diagram is not shown (Joint Authorities for Rulemaking on Unmanned Systems 2019). However, the principles are still apparent. To understand the core mechanism of the SORA, the idea behind a bow-tie argumentation is briefly explained.

The bowtie diagram is a risk assessment method based around *hazards* that could lead to *undesired events*. It is important to keep in mind that a hazard is a potential unsafe condition. To have a hazard evolve into an undesired event a *cause* is necessary. The cause can be a single event, a chain of events or multiple events at the same time. The undesired event then leads to undesired *consequences*. Causes are often named *threats* while the undesired consequences are often named *harms*. A simple example could be a human on board an aircraft. Since the aircraft is designed to fly under certain conditions only, being on board a flying aircraft is hazardous in general. An undesired event could be anything that worsens the conditions for the aircraft to fly that may ultimately lead to destruction of the aircraft and thus, dead or injury to people. A cause to this event could be an engine failure because of bird strike.

To summarize the bowtie diagram mechanics, regarding the hazard, multiple threats can lead to the centered undesired event which than can lead to multiple harms. Figure 2 shows a generic visualization to illustrate the name-giving form of the diagram.

The diagram can be understood as a sequence of events. Each of the causes can trigger the undesired event, which leads to at least one consequence. The path from threat to harm can be influenced by mitigations, which modify the likelihood that the undesired event occurs or the impact of harm and act as a kind of barrier. Depending on their position, these mitigations are either named *harm barriers* or *threat barriers* as shown in Fig. 3.

Now, with the fundamentals of the SORA risk model, the whole process can be described and is shown in Fig. 4. Initial input of the SORA process is a concept of

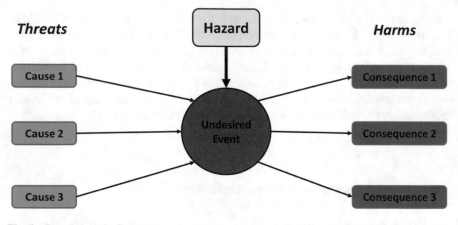

Fig. 2 Generic bowtie diagram

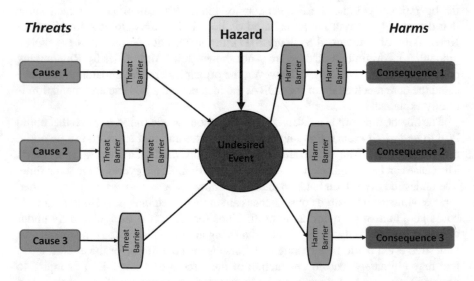

Fig. 3 Bowtie diagram with harm and threat barriers

operations document. The *Concept of Operations* document has to contain information about the operator or the operating company, the UAS, the remote crew and the location where the operation is conducted. A Location can be understood as the volume in air and over ground that is necessary to perform the operation. The core element of the SORA process is the assessment of risk to third parties for the respective operation conducted with a certain UAS. This assessment is based on the intrinsic risk to people on ground and the intrinsic risk to other manned aircraft. Based on the risk to people on ground and in air a specific assurance and integrity level (SAIL) is determined. There are six possible SAIL with increasing level of

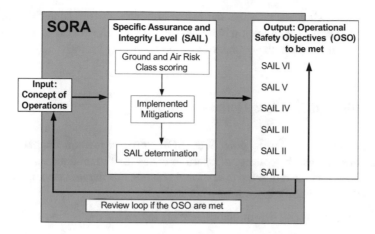

Fig. 4 Simplified SORA Process, modified from (2017)

rigor. Linked to the SAIL are so called Operational Safety Objectives (OSOs), which are safety requirements addressing the UAS, the remote crew, operational aspects and the operator. The twenty-four OSO are optional or have a low, medium or high level of rigor, depending on the necessary SAIL. The level of rigor of each OSO is expressed by a combination of the level of integrity and the level of assurance. The assurance is a requirement on how the effectivity of the OSO must be verified. In contrast, the integrity is a requirement on how the OSO has to be applied.

Within the SORA process the intrinsic risk to people on ground and in air, and therefore the SAIL, can be influenced by mitigations. The process distinguishes between mitigations for ground risk and mitigations for air risk. These mitigations along with the mechanisms of the process are further described by applying it to the ALAADy use case.

4   SORA of Cargo Drone Operations in Low Altitudes

The ALAADy project researches the question of cargo transportation of rather large UAS below common air traffic. The idea is to carry one metric ton of payload over a distance of about 600 km at roughly 200 km/h at an altitude of around 150 m. Within the project many different configurations were evaluated and narrowed down to a selection of three promising configurations. The configurations have a characteristic dimension of 12–16 m and a maximum takeoff weight of about 2500 kg. The configurations shall be operated within the EASA Specific Category and pose a low risk to people. The UAS itself as well as the operator will not be certified nor would the UAS be build according to common civil aviation standards such as the aforementioned DO-178C. To do so, the UAS operation itself must have a low risk to other people. Therefore, the UAS will be operated over environments with at least low population wherever possible and below common air traffic. In the following, this general operational concept will be put into the SORA process to analyze what SAIL are possible within the boundaries of the ALAADy project.

4.1 Ground Risk Class Determination

As shown in Fig. 4, the starting point of the SORA process is a concept of operations. Among other things, the concept of operations has to contain detailed information about the UAS and the requested operational volume. For the first step of the SORA, it is important to identify key information such as:

- the initial ground risk determination, the characteristic dimension of the UAS
- the typical kinetic energy
- whether the operation is conducted in visual line of sight (VLOS) or beyond (BVLOS)
- as well as a qualitative estimation of the population density of the overflown area is of importance.

Table 1 shows the scheme to determine the intrinsic ground risk class (GRC) of the operation. Taking the operational concept of the ALAADy project into consideration, possible GRC can be found in the outer right column, the over 8 m class. Further, the concept foresees the transportation of cargo over a long distance which requires a flight beyond visual line of sight (BVLOS). Regarding the area of operation, the operational scenario can be narrowed down to either a long distance flight over populated environment or sparsely populated environment, resulting in a GRC of 10 or 6.

Table 1 Intrinsic ground risk class (GRC) determination according to joint authorities for rulemaking on unmanned systems (2019)

Intrinsic UAS ground risk class				
UAS characteristic dimension	1 m	3 m	8 m	> 8 m
Typical kinetic energy	<700 J	<34 kJ	<1084 kJ	>1084 kJ
Operational scenarios				
VLOS/BVLOS over controlled ground area	1	2	3	4
VLOS in sparsely populated environment	2	3	4	5
BVLOS in sparsely populated environment	3	4	5	6
VLOS in populated environment	4	5	6	8
BVLOS in populated environment	5	6	8	10
VLOS over gathering of people	7			
BVLOS over gathering of people	8			

Since the ground risk has to be 7 or less to allow operation in the specific category, the BVLOS operation in populated environment is not possible with the original concept developed in ALAADy.

However, the SORA process allows reducing the GRC by applying mitigations that reduce the likelihood or the severity of harm from the UAS. Within the bowtie methodology, the mitigations resemble harm barriers. Within the process, mitigations reduce the GRC by a defined number of points depending on the assurance and integrity level of the mitigation. The specific assurance and integrity of each mitigation are divided into a low, medium and high levels. The necessary number of technical requirements scale with the level of the mitigations integrity. As for assurance, a low level of assurance means a self-declaration of the operator that the claimed level of integrity is achieved. A medium level of assurance needs supporting evidence given by the operator, such as simulation, testing or analyses. The high level of assurance is often covered by an appropriate third party, which confirms the achievement of a high level of integrity.

The SORA process provides three types of mitigations that can be implemented:

- M1 – Strategic mitigations for ground risk that intend to reduce the number of people at risk on the ground
- M2 – Mitigations that intend the effect of ground impact
- M3 – An Emergency Response Plan (ERP) is available

M1 enables a reduction of the number of people at risk by the use of a ground risk buffer. An operational risk buffer that uses at least a 1-to-1 rule, which means the horizontal distance to other people or critical infrastructure is equivalent to the altitude of the UAS, is a core requirement of the specific category. In medium and high integrity levels the formerly 1-to-1 buffer has to consider weather conditions, UAS performance, failures or malfunctions, etc. In general M1 aims on the substantiation that the overflown area inside the ground risk buffer contains an actually lower density of people compared to the expected average number of people at that area. A simple example would be a UAS infrastructure inspection conducted above an industrial facility. The facility would be regarded as populated environment. However, the operation might be carried out at Sunday nighttime when nobody except the inspection team is in the facility. The actual population density, therefore, is more comparable to an average sparsely populated environment or even a controlled area; therefore, M1 can be applied as mitigation. It has to be kept in mind, that M1 cannot reduce the GRC to a value lower than the rating for the controlled area in the particular vehicle class, since the controlled area is defined as area without non-active participants. Beside the reduction of the number of people at risk, another aim of the mitigation is the improvement of the buffer.

M2 shall reduce the impact dynamics in form of reduced impact area, reduction of kinetic energy before impact or reduced transfer energy. Examples to apply this mitigation would be a parachute system or a very soft and light-weight structure. However, special care must be taken when the operator wants to apply an M2 mitigation with high integrity. The crucial requirement is:

"impact dynamics and post impact hazards reduction to a level where it can be reasonably assumed that a fatality will not occur" (Joint Authorities for Rulemaking of Unmenned Systems 2019).

Even the use of a parachute system or similar is no guarantee that a fatality will not occur. Depending on the UAS remaining kinetic energy, it might still be able to fatally wound people.

M3 covers an Emergency Response Plan (ERP) that shall deal with emergency situations in the event of loss of control of the operation. The operation is out of control if it is in an unrecoverable state in which the outcome of the situation highly relies on coincidence, the situation cannot be handled by a contingency procedure or there is grave and imminent danger of fatalities. The ERP shall cover at least the conditions to alert the air traffic management and include a plan to limit the escalating effect of a crash of the UAS. The three mitigations and their possible impact on the GRC are shown in Table 2.

However, if and how these three mitigations can be applied depends on the operation and the UAS. In the following it will be discussed which mitigations can be applied to the concept developed in ALAADy. A full list of mitigations and their impact on the GRC in case of ALAADy is sown in Table 3.

The applicability of mitigation M1 is very limited. A crucial part of this mitigation is to show that the actual population density of the overflown area within the safety buffer is lower than the average population density of the claimed operational area type. In the case of BVLOS over sparsely populated environment, the operator has to show that the overflown area does not contain non-active participants, equal to a controlled area. This might be possible for a flight in a relatively small area, though it seems not likely for a flight from A to B over hundreds of kilometers and several hours as described in the concept of operations of the concept developed in ALAADy. Thus, in case of a BVLOS scenario over populated environment, it seems more likely that M1 is mainly applicable for small UAS or scenarios that are limited to a small area. Over villages and cities the application of M1 could mean that the operation is conducted when most people are inside buildings and the UAS is small enough to let shelter be relevant. However, a UAS of significant size and mass cannot rely on shelter. In SORA Annex B it is stated that shelter is only allowed for UAS with a take-off mass of less than 25 kg and is slower that 174 kt. In case of large UAS this

Table 2 Mitigations for final GRC determination according to Joint Authorities for Rulemaking on Unmanned Systems (2019)

Mitigation for ground risk	Level of integrity and assurance		
	Low/None	Medium	High
M1 – Strategic mitigations for ground risk	0: None/−1: Low	−2	−4
M2 – Effects of ground impact are reduced	0	−1	−2
M3 – An Emergency Response Plan (ERP) is in place, operator validated and effective	1	0	−1

mitigation can only be used in exceptional cases, such as flying over an industrial area outside operating hours. From this it follows:

Due to the mass of the UAS concepts developed in ALAADy and long-range flights the M1 mitigation is not applicable; no GRC reduction.

M2 mitigation is used to reduce the impact energy. At a high level of integrity, it would have to be shown that no fatality can be assumed in the case of direct involvement in a crash. Since the weight of each configuration developed in ALAADy alone is fatal, a high robustness cannot be assumed. However, for a medium level of integrity it would be sufficient to significantly reduce the impact dynamics, which would be achieved by ejecting a parachute or a controlled crash at minimum speed. Impact dynamics not only refer to the impact energy but also on the impact area, splash pattern and so on. Therefore, it depends on the certain case if additional measures to a parachute or a low speed impact have to be considered. Following this argumentation, we can state that:

As long as the required energy reduction for a medium level of integrity and assurance is not defined, the M2 mitigation seems applicable; GC reduction by −1

The M3 Mitigation aims at keeping the situation under control in case of loss of control by use of an ERP. The medium level of integrity is considered achievable, even though the ERP has to be developed to standards considered adequate by the competent authority and has to be validated through a representative tabletop exercise (Joint Authorities for Rulemaking of Unmenned Systems 2019). The high level of integrity requires the number of people exposed to risk to be significantly reduced by the ERP. It does not refuse to assume that no fatalities will occur. This mitigation is not necessarily linked to the UAS and depends more or less on the capabilities of the operator and the remote crew. Therefore, it can be stated that:

A highly qualified operator should be able to even apply an ERP of high integrity; GCR reduction 0 or −1.

Regarding the operational concept developed in ALAADy, for an operation conducted BVLOS over sparsely populated environment and with all possible mitigations applied, the GRC can be reduced by two at most. The final GRC with all mitigations applied is 4. The reduction by two also applies for BVLOS over populated environment. However, a GRC reduction by two leads to a final GRC of 8

Table 3 Final GRC after application of mitigations

	Possible GRC reduction
M1 – Strategic mitigations for ground risk	0
M2 – Effects of ground impact are reduced	−1
M3 – An Emergency Response Plan (ERP) is in place, operator validated and effective	0 or − 1
	Resulting GRC
Final GRC for BVLOS over sparsely populated environment	4 or 5
Final GRC for BVLOS over populated environment	8 or 9

which is still beyond the specific category. Therefore, within the specific category, operations developed in ALAADy can only be conducted over sparsely populated environment. That means the concept developed in ALAADy cannot be extended to operations over populated environments without certification of operator and UAS.

4.2 Air Risk Class Determination

In the next step, the air risk class (ARC) has to be determined. In general, the air risk class depends on the airspace encounter rate, which is a measure for the air traffic density. The air risk class is divided into in four different categories. The categories range from ARC-a for atypical airspace with no other air traffic up to ARC-d for high density airspace such as airport environment or class C or D controlled airspace. Possible ARC regarding the low altitude concept of ALAADy, flying at around 150 m above ground level (AGL), are shown in Table 4.

The concept of operation developed in ALAADy assumes for take-off and landing on small airfields or even agricultural fields and to operate in uncontrolled airspace over rural area. Depending on the specific scenario the initial ARC is b, c or d. Just like the GRC determination process, the ARC determination process is open for modification through mitigations, called strategic mitigations. ARC mitigations distinguish between operational restrictions and mitigation by structures and rules of air traffic management.

The strategic mitigation by operational restrictions is comparable to the M1 mitigation of the GRC process. The operator has to show that the actual airspace encounter rate or the air traffic density is lower than that of the claimed operational environment, which is an average estimation based on the airspace category. That can be done by the use of certain boundaries or airspace volumes or by the use of a certain operational time frame, where air traffic is lower, e.g. flights only at night.

The strategic mitigation by structures and rules rely on rules that are followed by all airspace participants. Strategic mitigation by structures and rules has to be understood as control of the airspace infrastructure by physical characteristics, procedures and techniques that reduce in-air conflicts or support conflict resolution. This mitigation

Table 4 ARC assignment to operational environment according to joint authorities for rulemaking on unmanned systems (2019)

Operational environment	Initial ARC
OPS in Airport/Heliport Environment in Class B, C or D airspace	ARC-d
OPS in Airport/Heliport Environment in Class E, F or G airspace	ARC-c
OPS < 150 m AGL in a Mode S Veil or Transponder Mandatory Zone (TMZ)	ARC-c
OPS < 150 m AGL in controlled airspace	ARC-c
OPS < 150 m AGL in uncontrolled airspace over Urban Area	ARC-c
OPS < 150 m AGL in uncontrolled airspace over Rural Area	ARC-b

is comparable to the common flight rules of manned aviation. In the future, the use of UTM/U-space services shall help to accomplish the goal of this mitigation.

These two types of strategic mitigation can be used to lower the intrinsic ARC. However, by application of strategic mitigations the ARC can only be reduced to an ARC-b. To be able to claim an atypical airspace with ARC-a, the operator needs to show that the airspace volume of the intended operation can meet the requirements of SORA atypical/segregated airspace. He also needs to comply with any other requirements of the competent authority. The requirements for SORA atypical/segregated airspace are located in Annex G of the SORA guidelines that has yet to be released.

Tactical mitigations are linked to the ARC tactical mitigations that are designed to cover the gap between the residual ARC and the overall airspace safety objective. The residual ARC can be understood as a simplified representation of the remaining collision risk with manned air traffic after all strategic mitigations are applied. Depending on the ARC the tactical mitigations have different level of performance required, starting with a low level of performance for ARC-b and a high level of performance for ARC-d. The tactical mitigations focus on different functions of UAS control in order to lower collision risk. Those functions are *detect*, *decide*, *command*, *execute* and a *feedback loop*.

4.3   Specific Assurance and Integrity Level Determination

The previous sections showed possible GRC and ARC scores and mitigations for the operational concept developed in ALAADy. Both scores combined resemble the Specific Assurance and Integrity Level (SAIL). When conducting the SORA process, it is important to know, that the higher score of both determines the SAIL, see Table 5.

According to the analysis of the previous sections, a SAIL III or IV resulting from a GRC of 4 and an ARC-b or ARC-c is the most likely outcome of the operational

Table 5 SAIL determination within ALAADy (blue) according to Joint Authorities for Rulemaking on Unmanned Systems, (2019)

Final GRC	Residual ARC			
	a	b	c	d
1 or 2	I	II	IV	VI
3	II	II	IV	VI
4	III	III	IV	VI
5	IV	IV	IV	VI
6	V	V	V	VI
7	VI	VI	VI	VI
> 7	Category C operation			

concept developed in ALAADy. However, depending on the mitigations and the specific circumstances of the operation a SAIL up to VI is possible. As shown in Fig. 4 the SAIL is linked to a number of Operational Safety Objectives (OSOs), which are generally discussed later in this paper.

4.4 SORA: Adjacent Area and Airspace Requirements

In addition to the OSOs the operator has to fulfill containment requirements. The objective of containment requirements is to reduce the risk for people in areas and airspaces adjacent to the operational area in case of loss of control of the operation. Those containment requirements can have two basic characteristics. The basic containment requirement is:

"No probable failure of the UAS or any external system supporting the operation shall lead to operation outside the operational volume." (Joint Authorities for Rulemaking on Unmanned Systems, 2019).

The operator can comply with this requirement by a design and installation appraisal that shall minimally include design and installation features, e.g. independence, separation and redundancy, as well as any relevant particular risk, e.g. hail, or snow, associated with the concept of operations.

If the adjacent areas or airspaces are gatherings of people, ARC-d airspace or the mitigation M1 has been applied in populated environments the containment requirement extends to:

"The probability of leaving the operational volume shall be less than 10^{-4} per flight hour. No single failure of the UAS or any external system supporting the operation shall lead to operation outside of the ground risk buffer. Software (SW) and Airborne Electronic Hardware (AEH) whose development error(s) could directly lead to operations outside the ground risk buffer shall be developed to an industry standard or methodology recognized as adequate by the competent authority." (Joint Authorities for Rulemaking on Unmanned Systems 2019).

As stated before, mitigation M1 cannot be applied over populated environment by the UAS concepts developed in ALAADy and due to the low altitude operation it is assumed unlikely to operate adjacent to ARC-d airspace. However, the definition of adjacent area/airspace needs to be handled with care. The explanation of when to consider an area/airspace as adjacent has not yet been precisely defined in the SORA guidelines. It seems to be linked to the endurance of the UAS and the judgement of the operator together with the competent authority. Taking the long routes and the endurance of the concepts discussed in ALAADy into account, it is assumed that the UAS concepts have to apply the advanced containment requirements.

4.5 Operational Safety Objectives (OSOs)

Operational Safety Objectives are measures to reduce the likelihood that an operation becomes out of control. In a bowtie diagram the OSOs would be represented as threat barriers. The SORA process proposes 24 different OSOs that are grouped in four different aspects as a starting point for safety within the specific category and can be extended on behalf of the local aviation authority. The sets can be identified to be the equivalent of threats or causes of a bowtie diagram:

- Technical issue with the UAS
- Deterioration of external systems supporting UAS operation
- Human Error
- Adverse operating conditions

Equivalent to the mitigations, OSOs have three levels of integrity and assurance: low, medium and high. The required level of assurance and integrity is determined by the SAIL. The number of OSOs to be fulfilled and their level of integrity and assurance is shown in Table 6.

The effort to accomplish integrity and assurance requirements raises with each level, however, the increase is not linear. A SAIL VI operation does not necessarily need twice the effort of a SAIL III operation. The exact scaling depends on the capabilities of the operator and the specific operation that is assessed. However, in general it is beneficial, in terms of effort, to aim for the lowest achievable SAIL. In most cases, the effort to improve one or two mitigations is worth the reduction in the required OSO levels.

Instead of presenting the entire set of OSOs, this section focusses on analyzing the increase in required integrity and assurance level of certain OSOs that are raised by a switch to a higher SAIL. That might be interesting for operators as well as manufacturers. Operators could consider whether or not it is beneficial to improve mitigation actions or restrict the area of operation to lower the SAIL or not. Manufacturers could consider to develop and manufacture their drones to fulfill a certain SAIL.

The first important step within the SAILs is the change from SAIL II to SAIL III. As shown in Table 6, the number of OSOs that have to be fulfilled on medium

Table 6 Operational safety objectives assigned to SAIL

SAIL	Number of operational safety objectives required			
	Optional	Low	Medium	High
I	9	15	–	–
II	6	14	4	–
III	–	6	14	4
IV	–	–	17	7
V	–	–	4	20
VI	–	–	–	24

assurance and integrity level increases by 10 and 4 OSOs have to be on high robustness implying a competent third party verification. While in SAIL II the several OSOs dealing with procedures for normal, abnormal and emergency procedures are the only ones that have to be implemented with medium robustness. In SAIL III, these OSOs need to have high robustness. Other OSOs that now have to be implemented on medium robustness cope with remote crew training, operator competencies and UAS maintenance.

From SAIL III to SAIL IV most of the former OSOs that have a low robustness assigned in SAIL III raise to a medium robustness. However, a more significant impact on effort is the change in OSO#1 from medium robustness to high robustness. OSO#1 addresses operator competencies and requires an UAS operator certificate that may be relatively expensive to obtain. The other OSOs on high robustness cope with external services supporting the operation, such as a communication service provider. They also deal with UAS design to withstand adverse weather conditions such as rain or adverse wind conditions as well as procedures to evaluate the environmental conditions before and during the operation. From SAIL IV onwards the UAS also has to be built according to design standards. However, it is not clear which design standards have to be used. In Annex E to the SORA (Joint Authorities for Rulemaking of Unmanned Systems 2019), it is stated that the UAS has to be designed according to design standards the competent authority considers adequate for a low level of integrity in case of SAIL IV. Nevertheless, in all operational scenarios where the reliability of the UAS is important, e.g. the transportation of expensive or time critical goods, design standards need to be used to verify robustness requirements out of own interest and regardless of the SAIL. Simply because operator as well as customer want the goods reliably delivered, or in more general, the operation successful conducted at all times. Therefore, the UAS operator certification is regarded as the most difficult hurdle.

The transition from SAIL IV to SAIL V seems to have a quite large gap. Most OSOs have to be fulfilled on high assurance and integrity. All those OSOs include verification by a third party. The exact effort regarding time and cost cannot be anticipated. We assume that at least the increased necessary communication with competent authority and competent third parties impact the effort to fulfill SAIL V requirements. OSO#5 of SAIL V is expected to have the largest impact on the necessary effort to verify a SAIL V operation. At this point, in the qualitative safety assessment SORA, with the exception of the containment requirements, quantitative failure rates are required for the first time. With OSO#5 the operator has to show that:

- Major failure conditions are not more frequent than remote
- Hazardous failure conditions are not more frequent than extremely remote
- Catastrophic failure conditions are not more frequent than extremely improbable
- Software (SW) and Airborne Electronic Hardware (AEH) whose development error(s) could directly lead to operations outside the ground risk buffer shall be developed to an industry standard or methodology recognized as adequate by the competent authority

To be able to verify these quantitative failure rates, the full application of safety and design standards, such as SAE ARP 4754A or DO-178C, seems necessary. In general, safety standards contain processes and methods that need to be adhered to in order to assume the required level of safety is achieved. Thus, not only the design of an item is influenced by the application of such a standard, but the whole developing and manufacturing company needs to be organized in a certain way to be able to apply these standards. So, there is a huge amount of effort necessary to use such standards to their full extend.

Looking at SAIL V and SAIL VI now, the difference between these two seems to be relatively minor. The remaining four OSOs now have to be fulfilled on high robustness. Even though the seemingly most comprehensive OSO#4, in which the UAS needs to be designed according to design standards, now needs to be fulfilled on high integrity and assurance, the difference to SAIL V seems to be relatively small. Design standards that are suitable for this requirement are expected to be SAE ARP 4754A, SAE ARP 4761, DO-178C and DO-254. However, these standards must already be applied on SAIL V. We assume that the increase in effort for a third party verification is relatively small, once the developing and manufacturing organization is set up to apply these standards. It must be kept in mind that even when a third party verification is not necessary on SAIL V, the operator respectively the organization must have supporting evidence of the correct installment of these standards. It also needs to be considered that many aspects of the UAS operator certification, certified maintenance personnel, certified manufacturer as well as certified remote crew and especially the application of civil aviation standards are part of the SAIL V and VI OSOs. This leads to an unclear distinction between SAIL V and VI requirements and the requirements that need to be fulfilled in the certified category.

Other entities also recognized the lack of explicit named standards throughout the SORA process. Therefore, the AW drones project was started in 2019 with well-known research institutes like the DLR or the Netherlands Aerospace Center (NLR) as well as industry partners such as UAS Manufacturer DJI or the Israel Aerospace Industries (IAI). This project aims on finding appropriate standards to fit the SORA requirements (AW-drones 2020).

As conclusion for ALAADy, SAIL IV should be especially avoided if the operator cannot obtain a UAS operation certification, unless the operator plans to achieve such a certificate and is aware of the additional effort. SAIL V and VI should be avoided whenever possible and if those SAIL cannot be avoided it should be analyzed if operation in the certified category is more beneficial. In the current version of the SORA the additional effort to achieve a full UAS certification compared to SAIL V and VI requirements seems to be manageable. The only exception might be a case where the high SAIL only results from a high GRC while the ARC is low. In this case a costly high performing and reliable Detect and Avoid (DAA) system is no necessary equipment, in contrast to certification requirements. The impact of the SAIL driven requirements on the system design and architecture is discussed in Chapter 11 of this book (Rothe and Nikodem 2020).

To not fall into SAIL V or SAIL VI, operations with large cargo UAS as discussed in ALAADy are limited to flights over sparsely populated environments and in low traffic airspaces. Care must be taken in crowded regions such as Central Europe, because of the long endurances the concept UAS shown in ALAADy are able to achieve, the containment requirements to protect people in areas nearby the operational area, have to be fulfilled on the highest level.

5   Future Work

5.1   Further Development of the SORA Process

The full SORA process was applied to the operational concept and the UAS concept designs developed in ALAADy. Application and analysis showed that the SORA process is far from being finished. One major issue is the lack of mathematical source data. The GRC lookup table is based on a simple scoring. While overall the scoring seems to have a logical background, mathematical proof on why the scores are the way they are is missing. Those scores should be substantiated by quantitative data. The same goes for the GRC mitigation scoring. JARUS refers to annex G of the SORA guidelines that shall include supporting data; this annex has not yet been released.

Extensive work with the SORA guidelines on several examples showed that especially OSO and mitigation requirements descriptions need further clarification to limit the space for interpretation on how to fulfill the requirement. A prominent example could be OSO#4 where the operator assures that the UAS has been built according to design standards recognized as adequate by the competent authority for low/medium/high integrity. At the moment it is unclear which standards should be applied. Other examples are the terms "sparsely populated" and "populated" environment which are not clearly defined. However, the SORA process is a relatively new developed approach that needs to grow with its use by operators and competent authorities.

5.2   Development of the Certified Category and Overlap to the Specific Category

The certified category of UAS operations is currently under development. It is expected that the certified category will rely on certification specifications similar to manned aviation including appropriate adaption of UAS specific equipment such as ground stations and DAA systems. The EASA will release a draft of a UAS certification specification in the foreseeable future. Comparable to manned aviation it is likely that the EASA will develop more than one certification specification for UAS for different types of UAS. In 2016, JARUS published recommendations on a certification basis of light UAS (Joint Authorities for Rulemaking on Unmanned Systems 2016). Having that in mind, it would be no surprise if the EASA creates new certification specifications for widespread demanded UAS types like light UAS, vertical take-off and landing (VTOL) UAS or heavy cargo UAS. Especially for light UAS, an overlap to the regulation of the specific category is possible. A draft of a certification specification for light UAS (Joint Authorities for Rulemaking on Unmanned Systems 2016) has been published by JARUS in 2016. In 2019, a draft certification specification has also been published by JARUS (Joint Authorities for Rulemaking on Unmanned Systems 2019). However, none of the documents is adopted by the EASA until now. The analysis of the requirements of SAIL V and VI of the SORA shows that the difference to the drafted certification specifications for UAS is potentially very small. Despite being relatively close to each other in terms of effort a crucial difference remains. Within the specific category and by successfully applying SORA the operator achieves a permission to conduct one specific type of operation. For any other operation the operator has to go through the SORA process again. In the certified category, once operator, design and manufacturing organization as well as the maintenance organization are approved, any suitable operation can be performed without applying for a new permit to operate. In this case a bit of additional effort to gain an UAS with a full type certificate might be beneficial for the operator or manufacturer. SAIL V and VI would become unnecessary, or just applicable for very specific use cases that way.

5.3   Other Risk Assessment Methods

The related work section of this chapter showed other risk assessment methods and approaches. Some of these approaches could be preferred over SORA. This could happen if they are more practical to use than the relatively detailed SORA process or if they are similar practical but substantiated with more quantitative data. Therefore, more effort should be taken to ease the use of SORA as well as of the other risk assessment methods and, in case of SORA, to substantiate it with quantitative data.

6   Conclusion

ALAADy (Automated Low Altitude Air Delivery) is a DLR project, that focused on a heavy lift cargo drone concept that shall be able to transport one ton of payload over long distances. One of the goals within the project was the application of EASA's new regulatory framework for UAS operation within the UAS category specific. The specific category requires a risk assessment of the intended operation in order to obtain an operational approval. The Specific Operations Risk Assessment (SORA) was applied to the operational concept and the UAS concepts developed in ALAADy, which is an accepted means to comply with the specific category's requirement on risk assessment.

Throughout this paper it was shown that the application of the bow-tie diagram-based SORA process favors not just one correct solution, but rather allows for different outcomes depending on operational and technical decisions made in the concept of operations. Especially the technical decisions as well as the possible outcomes of the risk assessment bear various consequences for the technical UAS design as well as consequences for UAS operators and manufacturers.

Furthermore, the application of SORA to the operational concept developed in ALAADy showed that the use of large cargo drones within the specific category is limited to flights over sparsely populated environments and outside of highly frequented airspaces such as controlled airspace C or D in Europe. Even with these constraints the use case scenarios of ALAADy can be placed in four out of six possible Specific Assurance and Integrity Level (SAIL). Evidence is shown that possible SAIL V and VI is unattractive in most cases, because the estimated difference of necessary effort to fulfill all relevant safety requirements seems to be relatively small compared to the certified category. It should be kept in mind that only an operational approval for the operation described in the operators' concept of operations is granted in the specific category. On the contrary, a UAS with type certificate can be operated without additional operational approval and under much more general conditions. Depending on the use case, it could be beneficial to achieve a type certificate for the UAS. However, with enough effort in risk mitigation strategies the operational concept developed in ALAADy is able to obtain a SAIL IV or even SAIL III rating, which should have a more promising balance between operational safety and effort to fulfill the related safety requirements. In SAIL III and IV the use case scenario is limited to operation over unpopulated or sparsely populated environments. Efforts must be taken to reduce impact energy and post impact hazards as well as to have emergency response strategies for an operation out of control situation in place that equal a medium level of robustness.

References

Belcastro CM, et al (2017) Hazards identification and analysis for unmanned aircraft system operations. In: Paper presented at the 17th AIAA Aviation Technology, Integration and Operations Conference, Dever, Colorado, 5–9 June 2017

Clothier RA (2015) Risk management and risk templates for civil UAS operations. RMIT University. ACADEMIA. https://www.academia.edu/22274007/Risk_Management_and_Risk_Templates_for_Civil_UAS_Operations

Clothier RA, Walker RA (2006) Determination and evaluation of UAV safety objectives. In: Proceedings 21st International Unmanned Air Vehicle Systems Conference, Bristol, p 18.1–18.16

AW Drones (2020). https://www.aw-drones.eu/, Accessed 15 Sept 2020

European Aviation Safety Agency (2017) NPA 2017–05 (B) Introduction of a regulatory framework for the operation of drones Unmanned aircraft system operations in the open and specific category. Avialable via EASA Document Library. https://www.easa.europa.eu/sites/default/files/dfu/NPA%202017-05%20(B).pdf, Accessed 15 Sept 2020

European Aviation Safety Agency (2019) Certification Specifications and Acceptable Means of Compliance for Large Aeroplanes CS-25 Amendment 23. Available via EASA Document Library. https://www.easa.europa.eu/document-library/certification-specifications/cs-25-amendment-23, Accessed 15 Sept 2020

Joint Authorities for Rulemaking of Unmanned Systems (2015a) AMC RPAS.1309 Issue 2 Safety Assessment of Remotely Piloted Aircraft Systems. JARUS. http://jarus-rpas.org/sites/jarus-rpas.org/files/jar_04_doc_1_amc_rpas_1309_issue_2_2.pdf, Accessed 15 Sept 2020

Joint Authorities for Rulemaking of Unmanned Systems (2015b) Scoping Paper to AMC RPAS.1309 Issue 2 Safety Assessment of Remotely Pioted Aircraft Systems. JARUS. http://jarus-rpas.org/sites/jarus-rpas.org/files/jar_04_doc_2_scoping_papers_to_amc_rpas_1309_issue_2_0.pdf, Accessed 15 Sept 2020

Joint Authorities for Rulemaking of Unmanned Systems (2017) JARUS guidelines on Specific Operations Risk Assessment (SORA) Ed 1.0. JARUS. http://jarus-rpas.org/sites/jarus-rpas.org/files/jar_doc_06_jarus_sora_v1.0.pdf, Accessed 15 Sept 2020

Joint Authorities for Rulemaking of Unmanned Systems (2019) JARUS guidelines on SORA Annex E Integrity and assurance levels for the Operation Safety Objectives (OSO) Ed 1.0. JARUS. http://jarus-rpas.org/sites/jarus-rpas.org/files/jar_doc_06_jarus_sora_annex_e_v1.0_.pdf, Accessed 15 Sept 2020

Joint Authorities for Rulemaking of Unmenned Systems (2019) JARUS guidelines on SORA Annex B Integrity and assurance levels for the mitigations used to reduce the intrinsic Ground Risk Class Ed 2.0. JARUS. http://jarus-rpas.org/sites/jarus-rpas.org/files/jar_doc_06_jarus_sora_v2.0.pdf, Accessed 15 Sept 2020

Joint Authorities for Rulemaking on Unmanned Systems (2016) JARUS CS-LUAS Recommendations for Certification Specification for Light Unmanned Aeroplane Systems Ed 0.3. JARUS. http://jarus-rpas.org/sites/jarus-rpas.org/files/jar_05_doc_cs-luas_v0_3.pdf, Accessed 15 Sept 2020

Joint Authorities for Rulemaking on Unmanned Systems (2019) JARUS CS-UAS Recommendations for Certification Specification for Unmannd Aircraft Systems Ed 1.0. JARUS. http://jarus-rpas.org/sites/jarus-rpas.org/files/jar_doc_16_cs_uas_edition1.0.pdf, Accessed 16 Sept 2020

Joint Authorities for Rulemaking on Unmanned Systems (2019) JARUS guidelines on Specific Operations Risk Assessment (SORA) Ed 2.0. JARUS. http://jarus-rpas.org/sites/jarus-rpas.org/files/jar_doc_06_jarus_sora_v2.0.pdf, Accessed 16 Sept 2020

La Cour-Harbo A (2018) The value of step-by-step risk assessment for unmanned aircraft. In: Paper presented at the 2018 Interantional Conference on Unmanned Aircraft Systems (ICUAS), Dallas, 12–15 June 2018

Nikodem F et al (2018) The new specific operations risk assessment approach for UAS regulation compared to common civil aviation risk assessment. In: Paper pesentend at Deutscher Luft- und Raumfahrtkongress (DLRK), Friedrichshafen, 4–6 Sep 2018

Single European Sky ATM Research (SESAR) Joint Undertaking (SJU) (2016) European Drones Outlook Study Unlocking the value for Europe. SESAR Joint Undertaking. https://www.sesarju.eu/sites/default/files/documents/reports/European_Drones_Outlook_Study_2016.pdf, Accessed 16 Sept 2020

Rothe D, Nikodem F (2020) Risk analysis based system architectures. In: Dauer JC (ed) Automated Low-Altitude Air Delivery - Towards Autonomous Cargo Transportation with Drones, p x-y. Springer, New York

The European Commission (2019) Commission Delegated Regulation (EU) 2019/897. EASA Document Library. https://www.easa.europa.eu/document-library/regulations/commission-delegated-regulation-eu-2019897, Accessed 16 Sept 2020

The European Commission (2019) Commission Implementing Regulation (EU) 2019/947 of 24 May 2019 on the rules and procedures for the operation of unmanned aircraft. EASA Document Library. https://www.easa.europa.eu/document-library/regulations/commission-implementing-regulation-eu-2019947, Accessed 16 Sept 2020

The European Parliament and the Council of the European Union (2002) Regulation (EC) No 1592/2002. EASA Document Library. https://www.easa.europa.eu/document-library/regulations/regulation-ec-no-15922002, Accessed 16 Sept 2020

The European Parliament and the Council (2018) Regulation (EU) 2018/1139. EASA Document Library. https://www.easa.europa.eu/document-library/regulations/regulation-eu-20181139, Accessed 16 Sept 2020

Washington A et al (2017) A review of unmanned aircraft system ground risk models. Prog Aerosp Sci 95:24–44

European Aviation Safety Agency (2015) Technical opinion introduction of a regulatory framework for the operation of unmanned aircraft. Available via EASA Document Library. https://www.easa.europa.eu/sites/default/files/dfu/Introduction%20of%20a%20regulatory%20framework%20for%20the%20operation%20of%20unmanned%20aircraft.pdf. Accessed 15 Sept 2020

Use-Cases for Heavy Lift Unmanned Cargo Aircraft

Henry Pak

Abstract There is a growing interest in large unmanned cargo aircraft (UCA) with payload capacity and range significantly larger than drones for last-mile parcel delivery. Several fields of application for such types of UCA were suggested. In this chapter, the usability of such a large UCA is discussed for three use cases, namely spare parts logistics, supply of people in hard-to-reach areas, and disaster relief. Performance parameters of the UCA are based on the large cargo drone concept developed by the DLR (German Aerospace Center) within the project Automated Low Altitude Air Delivery – ALAADy. We describe specific characteristics of each use case and how the use of an UCA could result in improved transport solutions. UCA can be a suitable transport solution when spare parts with high criticality and high value and with low and irregular demand have to be supplied. Furthermore, UCA can improve the transport of essential goods in hard-to-reach areas. For usage in disaster relief operations the UCA needs additional capabilities like to detect people in need and to supply them by airdrops.

Keywords Unmanned cargo aircraft · Logistics · Spare parts · Humanitarian logistics · Disaster relief · Humanitarian aid · ALAADy

1 Introduction

Drones and unmanned cargo aircraft (UCA) have attracted a lot of attention, since they may serve new potential markets. They are expected to be rapidly deployed in diverse sectors of our daily life for the transport of goods of various size and weight over a range of distances. According to the mission, UCA vary in size and performance. At the one end of the spectrum, small parcel and delivery drones, e.g. from Amazon (Amazon 2019), Flirtey (Garrett-Glaser 2019) or Alphabet Wing (McFarland 2019), can transport a few kilograms of light loads over distances of about thirty kilometers and thus may contribute to increasing the efficiency of transport

H. Pak (✉)
DLR (German Aerospace Center), Institute of Air Transport and Airport Research, Linder Höhe, 51147 Köln, Germany
e-mail: henry.pak@dlr.de

© Deutsches Zentrum für Luft- und Raumfahrt e. V. (DLR) 2022
J. C. Dauer (ed.), *Automated Low-Altitude Air Delivery*, Research Topics in Aerospace, https://doi.org/10.1007/978-3-030-83144-8_3

over the "last mile" (Clausen et al. 2016). So far, this type of drones has attracted most attention. At the other end of the spectrum, very large UCA are capable to perform long-distance flights with payload capacity of several tons and functioning in a comparable way to existing cargo airplanes (Schultz, et al. 2018). As such, they are expected to be not game-changing in the same way as are smaller drones and UCA designed for last mile, rural or spare parts deliveries (DHL 2014).

Meanwhile, there is also a growing interest in large UCA, which have significantly higher payload capacity and larger range compared to parcel and delivery drones, but are well below very large UCA. Examples of this are FlyOx (Singular Aircraft 2019), Rhaegal (Sabrewing 2019), or AT200 (Aerospace Technology 2019). Possible applications for such types of UCA are rural deliveries in areas that lack adequate infrastructure (e.g., in Africa) as well as supplier-to-plant emergency deliveries, which are typically performed by helicopter today (DHL Customer Solutions and Innovation 2014). Using UCA in disaster relief to deliver food, medical items and other relief goods to people in need is also considered (IATA 2019), (World Food Programme 2019).

In this chapter, we discuss the usability of this type of UCA. Based on DLR's ALAADy project, short for Automated Low Altitude Air Delivery, UCA with payload capacity of one metric ton and range of about 600 km are considered. This is comparable to the performance characteristics of light commercial vehicles (LCV) in Europe.[1]

In order to discuss relevant key factors, we focus on three fields of application, namely spare parts logistics, regular supply of people in hard-to-reach areas, and disaster relief. These use cases are characterized by the fact that the goods to be transported and their fast delivery are often of great importance in order to preserve life and health of people and to maintain production. In addition, poor transport infrastructure can make it difficult to transport such goods quickly and reliably. UCA are often seen as suitable solution for these applications because of their high speed and their independence from ground-based transport infrastructure, which in particular may be advantageous in situations with poor road conditions.

Furthermore, we assume that the UCA considered will be operated in the EASA Specific Category (see next section), which can entail flight routes avoiding populated areas and areas with critical and sensitive infrastructure as risk mitigating factor. This can lead to longer transport times and higher transport costs so that the transport by UCA may no longer offer an advantage over the transport by truck. Therefore we compare combined air-road transport with transport only by truck in terms of transport times and costs by using representative transport relations in Germany.

---

[1] Vehicle type definitions depend on national and professional definitions. In the EU the total weight of so-called light commercial vehicles may not exceed 3.5 metric tons.

2 Technical Specification of the UCA

Essential parameters of the UCA relevant for the transport task were specified in advance and serve as boundary conditions for the market and use-case investigations:

- Maximum payload: 1000 kg
- Cargo hold dimension: $1.3 \times 1.3 \times 3$ m^3
- Cargo hatch dimension: 1.3×1.3 m^2
- Range: 600 km
- Cruising speed: 200 km/h
- Maximum runway length: 400 m
- Operational concept according to the EASA Specific Category

The operational concept according to the Specific Category is an important aspect for integrating UCA into the existing aviation system. The European Aviation Safety Agency (EASA 2015) has developed a concept of operations for drones, which regulates operations in a manner proportionate to the risk of the specific operation. The safety risks considered must consider mid-air collision with manned aircraft, harm to people and damage to property in particular critical and sensitive infrastructure. Generally speaking, two approaches are possible to meet a level of safety acceptable to society (EASA 2015), (Dauer et al. 2016): Either an unmanned cargo aircraft is designed to meet thorough airworthiness requirements comparable to what is done for piloted aircraft. Or certain airworthiness risk factors are compensated for by operational risk mitigating factors, which can include specific limitations on the operations. Three categories of operations and their associated regulatory regime are introduced in order to define specific safety requirements tailored to the application (EASA 2015): Open, Specific and Certified. For UCA, the open category is of no relevance as the drone must be flown under direct visual line of sight (VLOS) and at an altitude not exceeding 150 m above the ground or water. The Specific Category covers operations posing more significant aviation risks which can be mitigated by additional operational limitations or higher capability of the personal and equipment involved. In the certified category the aviation risks are akin to those of normal manned aviation and operations, and the aircraft involved therein would be treated in the classic aviation manner.

In case of the Specific Category, flight routes underneath the altitude of the remaining air traffic and limited to unpopulated areas and to areas without critical and sensitive infrastructure are foreseen as risk mitigating factors (Dauer et al. 2016). Depending on the geographical situation, these measures can lead to longer transport times and higher transport costs. In extreme cases, it may even be impossible to carry out the transport if no flight route over unpopulated areas is available.

3 UCA and Spare Parts Logistics

Due to their high speed, UCA appear suitable for transporting urgently needed spare parts to maintain production. In this section we first give a brief overview of spare parts logistics. We then discuss the application of UCA taking spare parts logistics for agricultural machinery as an example.

3.1 Spare Parts Logistics Overview

A spare part is a component of a product which is not installed as a product part in the original equipment, but which is intended to restore or maintain the functionality of the product to its original extent by replacing the damaged, worn or missing product part in the course of repairs (Bloech and Ihde 1997). Spare parts logistics is a branch of logistics. The task of spare parts logistics is to make available spare parts and, if necessary, replacement devices that are coordinated in terms of time, space and quantity with defective primary products (Bloech and Ihde 1997).

There are several strategic options and related policies available in spare part logistics, which depend on the characteristics of spare parts (Huiskonen 2001). A classification scheme according to Huiskonen uses the following four features: (a) criticality of the spare part, (b) its specificity, (c) volume and predictability of spare part demand; (d) value of the spare part. These characteristics significantly influence the logistics concept and thereby the role and importance of the means of transport used.

The criticality of the spare part is related to the consequences that occur when a component (e.g. in a factory) fails and cannot be immediately replaced by a spare part. The more severe the consequences are, the higher the criticality of the spare part. A highly critical component must be replaced immediately in case of failure, whereas a medium critical component can be replaced after a certain time. Therefore, the more critical the spare parts and the more serious the (financial) consequences of failure are, the higher the requirements on the availability of spare parts.

Specificity characterizes whether the spare part is a standard part (e.g. a commercially available electric motor) or a component made specifically for the user (e.g. a special gear). In the case of standard parts, there is usually a strong demand for this spare part and a large number of suppliers. Suppliers are therefore willing to cooperate with customers in securing supplies. For user-specific spare parts, however, demand is low. The suppliers are generally reluctant to set up a spare parts logistics system and keep spare parts in stock. The customer is therefore usually forced to take care of stocking spare parts himself.

For spare parts, the quantity demanded can be very small. The smaller the quantity demanded, the less willing is the supplier to set up spare parts logistics and to keep spare parts in stock. Depending on whether it is a wear part with a predictable service life or randomly failing components, demand is more or less regular. If random failures dominate, a stocking strategy is preferred for the spare part. In the case of wear-related failures, a time-controlled maintenance and repair strategy may be the first choice.

The higher the value of an individual spare part, the lower is the willingness on the part of both the supplier and the customer to store the spare part, in order to avoid capital commitment. If stockpiling is necessary, centralized warehousing is preferred. For spare parts of low value, the cost problem then arises: the costs of storage and administration must not be disproportionately high in relation to the low value of the spare part.

The requirements for logistics solutions for spare parts are determined by the specific characteristics of the spare parts. Type of spare parts, the availability strategy, and the network structure have an impact on the relationship between supplier and customer in a supply chain and the system for controlling the inventory. Spare parts logistics solutions must be able to cope with sporadic, unpredictable quantities demanded. Requirements are also being driven by the need for ever smaller material and time buffers in regular production systems and supply chains.

Due to the specific properties of spare parts, various logistics solutions have evolved (Huiskonen 2001). In the case of user-specific parts with low demand, whose production requires lead time, it is the customer's responsibility to decide whether to stock spare parts or to accept the consequences of a temporary failure until the spare part arrives. For standard parts with high criticality, which are in demand more frequently and/or by several customers, a spare parts warehouse operated by the supplier or a third party (carrier, maintenance company) is a suitable logistics solution. Several customers can also join forces to operate such a spare parts warehouse. For standard parts with medium criticality, the costs of organizing replenishment (storage and transport) come to the fore, especially when the value of the spare parts is low. For spare parts with high value and low demand volume, the optimal logistics solution tends to be centralized warehousing close to the supplier. For spare parts with a low value, costs are reduced by combining several deliveries of spare parts into larger lots.

In the case of high criticality parts that are more or less standard, i.e. used by several customers, suppliers accept to hold stocks and offer special services, such as 24-h or faster deliveries (Huiskonen 2001). This time-guaranteed delivery is a strategy that can be implemented typically by using a specialized spare part service company. It is particularly attractive for high value parts with low and irregular demand pattern. For this type of spare parts, speed of transport means can play an important role to achieve the desired service level.

3.2 Spare Parts for Agricultural Machines

Spare parts for harvesters are a good example for high criticality parts. Harvesters are central elements of the agricultural harvesting system. At harvest time, time pressure is high as grain harvest is usually optimal at and limited to a few days a year. At peak times harvesters are used around the clock, seven days a week, to perform harvest. Bad weather conditions can increase the time pressure at harvest. Harvesters are heavily loaded machines which, depending on the weather and soil conditions, are subject to high wear that can lead to failure of the machine. The failure of a harvester during the grain harvest is associated with costs estimated at 500 Euros and more per hour lost (Lockenkötter 2013). The costs result from expenditures for tractors, trailers and personnel that were calculated but not used because of the failure of the harvester. Additional costs or losses may arise if the quality of the harvest suffers as a result of the harvester failure. The grain may need to be after-treated (e.g., dried) or in the worst case may no longer be usable.

Spare parts that are essential for the functioning of the harvesting machines can therefore be classified as high criticality parts. This is why customers want immediate availability of spare parts during the harvest. Hence, customers are willing to pay higher prices for spare parts and additional surcharges on transport costs and waive discounts, which agricultural machinery manufacturers grant, if customers allow longer delivery times. Customers are also willing to order more expensive original parts from manufacturers instead of cheaper replacements with longer delivery times. The price sensitivity for spare parts is therefore low during the harvest season.

Spare parts differ considerably in terms of size, weight, value and demand. The assortment ranges from small parts with low weight such as knife blades for cutting units or air filters up to complete mowing blades with up to 12 m in length or gear units weighing up to 2.5 tons (Fig. 1). Some spare parts are, or contain, hydraulic or pneumatic components as well as fluids, which means that in air traffic these spare

Fig. 1 Spare parts can cover a wide range of size and weight (Photos: © DLR 2020)

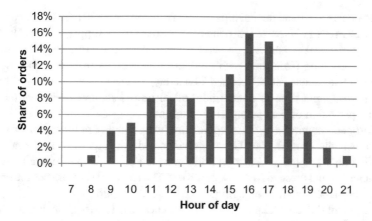

Fig. 2 Most spare parts for agricultural machines are ordered in the afternoon. Data from Lockenkötter (2013)

parts are classified as dangerous goods. The value of spare parts ranges from a few Euros up to 20 to 30 thousand Euros for engines and transmissions. Maintenance and wearing parts such as filters and knives are more frequently in demand (at least 100 times a year, usually more), while especially large parts such as feed rollers are requested less than 10 to 100 times a year (Lockenkötter 2013).

The demand for spare parts varies significantly both during the year and throughout the day. In Europe, 50 percent of spare parts sales occur in the months of July to September (Lockenkötter 2013). During the harvest season, most spare parts orders are placed in the late afternoon. Since harvesters are operated until late in the evening, spare parts are also ordered until late at night (Fig. 2).

3.3 Spare Parts Logistics at the CLAAS Group

How a logistics system for high criticality spare parts can look like is illustrated using the example of the CLAAS Group. CLAAS is one of the world's leading manufacturers of agricultural machinery based in Harsewinkel near Gütersloh, Germany. CLAAS generates around three quarters of its sales in Germany and Europe (CLAAS 2015).

For customers in Germany, CLAAS guarantees parts availability of over 90 percent within 2.5 h (CLAAS KGaA mbH 2016a). For customers in Europe, part availability of more than 90% is aimed for no later than the day after receipt of the order (Lockenkötter 2013). This availability should be reached at an order deadline, which is on average after 6 p.m. Orders received after the order deadline are delivered on the day after next.

In order to achieve the required spare parts availability, CLAAS has set up a three-stage storage system consisting of a central warehouse, several regional warehouses and warehousing at local dealerships on site (Lockenkötter 2013). The central warehouse of CLAAS in Hamm is responsible for spare parts supply worldwide. There, 98 percent of all parts are kept on stock – spare parts for harvesting machines and tractors, originating from their own production as well as from external suppliers. The central warehouse can be reached 24 h a day, all year round. In the central warehouse, a throughput time of a maximum of 30 min is achieved between receipt of order and provision of the spare part for shipping.

In European countries regional spare parts warehouses are located near Paris, in Saxham near the English East Coast, near Madrid, in Vercelli in Piedmont, in the Polish Buk near Poznan, in Austrian Spillern near Vienna, in the Romanian Afumati near Bucharest and close to Moscow and Kiev. The regional warehouses are operated by distribution companies belonging to the CLAAS Group.

Independent dealers make up the local warehouses. The local dealers are usually also the recipients of the spare parts deliveries. Only large spare parts are directly delivered to the customer.

In addition, six spare parts warehouses in Germany ensure that customers in Germany can be supplied with spare parts within 2.5 h.

Regional and local warehouses are supplied in various ways (Lockenkötter 2013). Regional warehouses are served directly from the central warehouse, while local warehouses are supplied either directly from the central warehouse or indirectly via a regional warehouse. CLAAS strives to use standard networks of forwarding companies. If available time and size of consignment permit, transport is carried out by truck as general cargo due to the lower costs. In case transport times by truck are unacceptably long, express services are used. Express deliveries from the central warehouse to the individual destination countries are carried out with those providers who are specialized in transports to the respective countries and can therefore realize minimum delivery times. There are different order deadlines for the acceptance of urgent spare parts orders for the individual destination regions. The majority of express transports are carried out with light trucks with costs around 1 Euro per kilometer[2] (CLAAS KGaA mbH 2016b).

At CLAAS it is uncommon to transport urgently needed spare parts by air (CLAAS KGaA mbH 2016b). The reasons for this are higher restrictions regarding weight and size of the spare parts, which is limited by the size of the cargo hatch and the cargo hold of the aircraft. Also, expenses are significantly higher. Additional work due to stricter safety regulations is necessary for instance for tamper-proof packaging by wrapping with foil. When shipping via large commercial airports, handling and waiting times at airports are usually long, so that the time advantage over road transport diminishes, at least for transports within Europe. Furthermore, there is no guarantee that air freight capacities are available when they are needed. However, an exception is the transport of spare parts to Great Britain and Ireland by air freight with "Innight" service by TNT Express via Maastricht airport. Order deadline here is 5:15 pm. Shipments leave the central warehouse in Hamm at around 6:00 pm and are delivered to the consignee after the night flight the next morning. Spare parts are

---

[2] Cost for return journey is included.

also delivered by air to Romania. Here the shipments have to leave Hamm at 3:30 pm to reach the consignee in Romania the next day. The delivery of the spare parts to the local dealers in Romania is made similar to the milk run concept along a defined route.

Occasionally high speed trains are also used for the delivery of small parts. However, helicopter services for spare part transportation are not considered due to high costs of 2,000 to 4,000 Euros per flight hour, according to (CLAAS KGaA mbH 2016b).

Spare parts deliveries at weekends are particularly problematic, as standard networks of forwarding companies are not available to the usual extent. Here, the supply of spare parts is maintained with expensive special solutions, e.g. by distributing the spare parts to a limited number of local dealers along a specified route.

3.4   UCA for Spare Parts Logistics in Europe

Companies like CLAAS aim to significantly reduce the time between order deadline and delivery to the customer, which would result in an improvement of customer service. At the same time, inventory investment and administrative costs should be kept to a minimum. With regard to these objectives, the use of an UCA can be a suitable solution in case of high criticality parts with high value and low and irregular demand. As UCA enable a reduction in transport times, fast deliveries of spare parts from central warehouses directly to the customer become possible thereby avoiding capital-binding storage of spare parts at regional warehouses.

A scenario for the transport of this kind of spare parts in Europe is a combined air-road transport and could be outlined as follows: A shipment prepared in the factory or central warehouse is transported by truck to a nearby airfield. There, the shipment is reloaded into an UCA and transported by air to an airfield close to the destination. Here, the shipment is reloaded onto a truck for the last part of the transport chain to the destination. While the shipment is delivered to the customer, the UCA flies back after a technical check and refueling. Usually, no goods are to be transported back to the factory or central warehouse of the manufacturer, so the return flight takes place without payload on board.

Aircraft requirements concerning the payload capacity depend on the weight and size of the spare parts. In the case of CLAAS, the average shipment weight is about 500 kg (CLAAS KGaA mbH 2016b). Hence, the intended maximum payload of one ton is sufficient for the majority of spare parts deliveries. However, large spare parts for cutting units of combine harvesters with a length of up to twelve meters require a larger cargo hold and hatch.

Furthermore, departure and arrival airfields have to be suitably equipped with regard to the infrastructure. This also refers to the devices for the loading and unloading of the cargo and in particular for the refueling or energy supply of the UCA. It can be assumed that adequately trained personnel at the departure or arrival airfield will be available in order to carry out non-automatable process steps during cargo and aircraft handling.

The required range of the UCA depends on the sales area. In the case of CLAAS, the range previously assumed is sufficient to serve destinations in central Europe. However, with an increased range of about 1,500 km, regional warehouses of CLAAS in Spain, Romania and the Ukraine could also be reached.

To illustrate the time advantage that can be achieved in case of using an UCA, transport times were determined for ten representative relations in Germany and compared with transport times if only light commercial vehicles (LCV) are used (Fig. 3 and Table 1). Direct distances from the origin to the destinations are between 150 and 528 km.

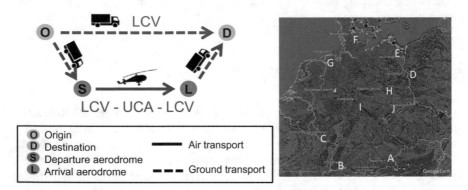

Fig. 3 Transport by UCA and LCV is compared with transport only by LCV for ten representative transport relations in Germany (Map: © GeoBasis DE/BKG, © Google, Image: Landsat/Copernicus, Data: SIO, NOAA, U.S. Navy, NGA, GEBCO)

Table 1 Distances between Hamm, Germany, and the ten destinations studied

	Destination	Direct distance [km]	Driving distance [km]
A	Raubling	528	674
B	Schopfheim	446	595
C	Schönenberg-Kübelberg	265	369
D	Jacobsdorf	445	500
E	Neubrandenburg	418	620
F	Bollingstedt	328	473
G	Wangerland	221	291
H	Raguhn-Jeßnitz	296	393
I	Waldkappel	150	180
J	Hartmannsdorf	347	439

Distances are based on Google Earth

In case of combined air-road transport, airfields that are close to the origin and the destinations and currently used mainly by aero clubs serve as take-off and landing sites. The road distance between the origin and the departure airfield is 12 km, while road distances between the arrival airfields and the destinations are between 10 and 30 km. The associated driving times were determined using a route planner. A detour factor of 1.1 was used to determine the exact flight path length from the linear distance between the departure and the arrival airfield. This value takes into account that flight routes have to avoid populated areas and areas with critical and sensitive infrastructure and is based on simulations of comparable flight routes (Schopferer and Donkels 2021). Flight times were then calculated based on a flight velocity of 180 km per hour. For the reloading of the freight between truck and UCA at the departure and arrival airfields, half an hour each was considered.

Driving times for the transport only by LCV are based on data from a German courier service (inTime Express Logistik GmbH 2019).

For all relations combined air-road transport is faster than transport only by LCV (Fig. 4). On the shorter distances the transport time is reduced by about 30%, on the longer distances up to 50%. The reduction of transport time is achieved by two means: firstly, the relatively high transport speed of the aircraft compared to the truck, and secondly, the use of small airfields nearby the origin, e.g. the central warehouse, and the destination for take-off and landing. This saves considerable time compared to the use of large airports, where freight handling in general is lengthy due to the size of the airport and the volume of traffic. It is also conceivable to carry out transports directly from the central warehouse with the help of a runway built on the company's premises, but this is innovative and cannot be implemented easily with the current air traffic regulations.

By using the cost model described by Liebhardt and Pertz (2021), costs for the combined air-road transport were estimated, too. Compared to Liebhardt and Pertz, the expected yearly utilization per aircraft was reduced to 1,000 h. This takes into account that the UCA is kept in standby in order to be able to react flexibly to the irregular demand for high criticality spare parts. As a consequence, the estimated costs may be higher compared to Liebhardt and Pertz.

Fig. 4 Combined air-road transport is faster compared to transport only by LCV

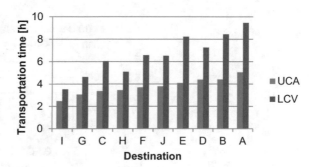

Costs for pre-carriage and onward carriage by truck as well as for the return flight of the UCA are included. The transport costs thus determined are shown in Fig. 5 and compared with prices based on data from a German courier service for transport only by truck. Costs for the combined air-road transport are four to five times as high as the costs for transport only by truck.

To adequately assess the additional costs of the combined air-road transport, production downtime costs of the client have to be considered. Figure 6 shows the resulting total transport and downtime costs in case the average downtime cost is 500 Euros per hour respectively 1,000 Euros per hour. In the first case, the total transport and downtime costs of combined air-road transport and transport only by truck

Fig. 5 Combined air-road transport is significantly more expensive than transport only by LCV

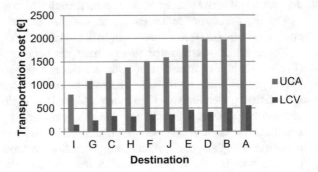

Fig. 6 Total transport and downtime cost for low (top) and high (bottom) hourly downtime costs

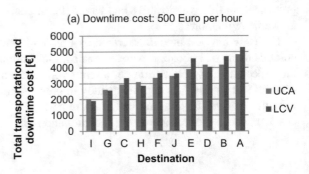

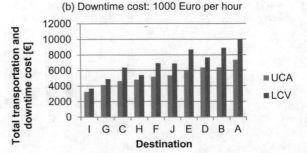

are approximately the same. However, in the latter case, the sum of transport and downtime costs is always lower for the combined air-road transport. Less production downtime due to faster transport justifies the use of the more expensive UCA.

4 UCA for the Transport of Essential Goods to Hard-To-Reach Areas

Transportation of goods by air is a feasible option for areas that are difficult to reach due to a lack of ground-based transport infrastructure. In this section, we discuss the use of UCA to transport medical and other relief supplies to hard-to-reach regions as part of humanitarian aid.

4.1 Remote and Physically Challenging Locations

Factors that render people hard to reach include difficult terrain, adverse climate, and lack of functioning transport links and infrastructure. These factors often overlap with man-made factors such as insecurity due to military operations, ongoing hostilities and violence. Fragile and vulnerable countries generally have poorer transportation infrastructure than others, with an average road length of 157 km per 100,000 people; which is less than four times the average road length (653 km) in countries that are not fragile or vulnerable (IFRC 2018). Mountainous regions and countries consisting of multiple remote island communities often have to deal with great distances, irregular transport and limited infrastructure.

Greenland is an example where the supply of the rural population is affected by climatic factors. Localities in Greenland are mainly located on the west coast. Although they can be supplied with coastal vessels in summer, dog sleds become important means of transport in winter, supplemented by cars that can sometimes drive over the frozen sea (Spaeth 2016). However, for fast freight transport helicopters (e.g., Bell 212) and aircraft (Dash-8-200) are used. People are also transported by air, but demand is low because air fares are relatively high.

Aircraft also play an important role in the supply of remote areas of Canada. The Canadian airline Air Tindi (2019), for instance, operates scheduled and on demand charter services for passengers and freight in the Northwest Territories. Small propeller aircraft are used, from Cessna C-208 Caravan to Dash 7 Combi. Many of the aircraft are equipped with floats or skids, so that they can take off and land on water or snowfields.

The task to provide efficient transport links in rural areas is aggravated in developing countries, which due to their economic weakness are often not able to build a sufficiently good transport infrastructure even with otherwise favorable topographical and climatic conditions. Transport of goods by air is uncommon due to the high costs.

4.2 UCA for Transportation of Humanitarian Relief Supplies in Africa

Poor transport links are also a problem for aid organizations providing humanitarian aid in developing countries mainly in Africa, Asia and South America. Humanitarian aid is often provided in form of multi-year projects, in which aid agency staff operate from locally based project offices and in health stations. For their work, the aid organizations have to provide a wide variety of goods, for example technical equipment such as water treatment plants and mobile power generators, as well as equipment for project offices, schools and health stations. Wherever possible, the required goods are procured locally, also for ethical reasons, in order to strengthen the local economy. This also applies in principle to the purchase of transport services, which are usually provided at low cost by truck or off-road pickup truck.

However, in developing countries where aid organizations are active the transport of goods by car or truck is usually very tedious. For example, it takes about 3½ hours with an all-terrain Toyota Landcruiser for the route from Goma to Masisi in the Democratic Republic of Congo (55 km as the crow flies, 78 km road distance), and a drive from Yambio in South Sudan to Nagero in the DR Congo (155 km as the crow flies, 288 km road distance) takes about nine hours (Johanniter-Auslandshilfe 2016). Bad road conditions are the reason for the long driving times. Roads outside larger cities are mostly uneven and thus only slowly passable sand tracks, which often become completely impassable in the rainy season (Fig. 7).

Fig. 7 Unpaved roads (left) and impassable roads after rain (right) hamper transport of goods. (Photos: © Medair/Lucy Bamforth)

Under these circumstances, fragile or perishable goods are likely to be damaged due to long transport durations or poor road conditions. In addition, high-value goods are often threatened to get lost due to a critical security situation, high levels of corruption and high crime rates. Hence, in the context of humanitarian relief, aid organizations have a strong interest in alternative transport options that provide fast and safe transport of goods (Action medeor e.V. 2016), (Johanniter-Auslandshilfe 2016). This applies in particular to the transport of medicines. During their transport and storage, certain maximum temperatures must not be exceeded. Due to the climatic conditions, therefore, fast transport in cooling boxes or even the use of a cooling aggregate during transport of drugs is often required. Hence, the use of an UCA can be a suitable transport solution to ensure safe transport thereby reducing risks of loss and damage. The prerequisite is, however, that the additional costs of UCA transports are kept within limits. Nevertheless, in regions with critical security situation there is a risk that the UCA will be shot down because the aircraft is flying at low altitude.

A scenario for the transport of medicines, medical products and other goods needed for humanitarian relief in Africa could be outlined as follows: Points of departure are large airports or ports, where relief goods are stored temporarily after transport via traditional logistics networks, e.g. by air freight. From there, health stations and project offices are supplied, which are usually dispersed in rural and hard-to-reach regions. The maximum distances to be covered are in the range of 600 to 900 km and the maximum weight of the cargo is below one ton (Action medeor e.V. 2016), (Johanniter-Auslandshilfe 2016).

Availability of infrastructure differs between departure and arrival airfields. At airports, which serve as departure airfields, a large part of the required infrastructure already exists. This also concerns devices for loading and unloading and, in particular, for refueling or energy supply for the UCA. Sufficiently trained personnel will usually be available at the airport.

Unlike the departure location, arrival airfields are likely to have minimal infrastructure. In particular, it cannot be assumed that refueling or provision of energy at the destination is possible. Runways will often be simple grass strips (Fig. 8). It can

Fig. 8 Mikumi Airstrip in Tansania. (Photo: Graham Laurence; https://commons. wiki-media.org/w/index. php?curid=26116253; licensed under the terms of the CC-BY-2.0: https://creati vecommons.org/licenses/by/ 2.0/deed.en)

also be assumed that personnel with only the most necessary knowledge regarding the operation of the UCA will be available at the destination.

The number of UCA needed by an aid organization can be roughly estimated. For example, action medeor e.V. (2016), a German medical aid organization, supplies about 2,000 healthcare facilities (hospitals, pharmacies and health centers) in Africa. Typically, each healthcare facility is supplied twice a year so 4,000 supply flights are necessary per year. If spreading them over 200 days and assuming that one aircraft can carry out two missions per flight day, about ten aircraft are necessary, which is a conservative estimate. The actual number of UCA needed may be larger for example due to an unfavorable geographical distribution of gateways and destinations. The need for transport services for projects of other aid organizations are not considered yet, which increases the number of UCA required. Furthermore, aid agencies are interested in further expanding the supply of those in need of assistance, which may result in additional demand for UCA.

5   UCA and Disaster Relief

In the event of a disaster, it is important to quickly provide the people affected with essential relief supplies. UCA can contribute to this, especially if the infrastructure in the disaster area is severely damaged or even destroyed. In this section, we discuss possible uses of UCA in disaster relief.

5.1   Disaster Relief Overview

Disasters can be distinguished in natural disasters (such as earthquakes, drought and the resulting epidemics, famine), technological disasters (such as nuclear disasters, oil spills) and social disasters caused by wars, armed conflicts, civil wars and other extraordinary social events (Bundesministerium für europäische und internationale Angelegenheiten 2007). As a result of such events, communities are no longer functioning and, in particular, unable to cope with the disasters on their own.

In recent years, the Center for Research on the Epidemiology of Disasters (CRED 2020) has registered between 332 and 523 natural disasters[3] per year (Fig. 9). These were mainly hydrological and meteorological disasters. The number of those affected, including the death toll, was between 84 and 660 million. The years 2002 and 2015 were exceptional, as China and India were hit particularly hard by floods and droughts in those years.

---

[3] In this statistic, natural disasters are recorded with at least 10 fatalities or with at least 100 affected people, or when the state of emergency is declared or when international assistance is requested. Because of these broad criteria, the numbers are high compared to the general perception.

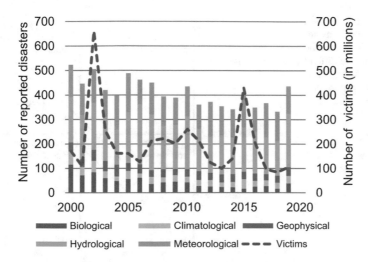

Fig. 9 Occurrence of natural disasters and victims. (Diagram generated on the basis of data extracted from EM-DAT, CRED / UCLouvain, Brussels, Belgium – http://www.emdat.be/ (D. Guha-Sapir))

With regard to humanitarian logistics, a distinction between natural disasters and man-made disasters is of no concern according to Apte (2009). Instead, it makes more sense to classify disasters according to the speed of onset of the crisis situation and to its spatial extent (Fig. 10). Depending on the combination of these characteristics, there are different challenges for humanitarian logistics.

A disaster can begin slowly and be limited to a confined area. An example of such a disaster is the famine in South Sudan, which has been going on for several years now. In this kind of situation, humanitarian logistics requirements are relatively low in terms of organizational preparation and delivery of relief supplies, as there is some lead time before the peak of the crisis. Therefore, the time pressure is less than in the case of a disaster with sudden-onset. Humanitarian logistics becomes more challenging as the disaster stretches across a wide area and locations to be served are dispersed, making the provision of relief supplies more difficult.

Sudden-onset of a disaster means additional challenges for humanitarian logistics. An example of a sudden-onset disaster with a localized area is Hurricane Katrina in 2005, while the tsunami in 2004 is an example of a disaster with large and scattered geographical areas affected. Preparation, disaster response, and ongoing humanitarian relief are more difficult in case of sudden-onset disasters. In particular, natural and technological disasters are usually sudden-onset disasters and are unpredictable in terms of time, location and extent. They pose significant challenges with regard to effectiveness and efficiency of transportation and distribution of critical supplies and services (Apte 2009).

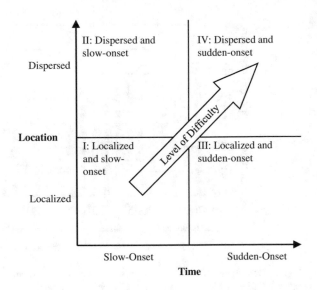

Fig. 10 Disasters with dispersed location and sudden-onset poses the highest level of difficulty of humanitarian logistics operations. Based on the graphics by Apte (2009)

The term disaster response means help that is provided in time immediately during or shortly after the (sudden) onset of the disaster. The first aim of such relief operations is to identify and locate survivors and to ensure a basic supply of water, food, shelter and medical care (Bundesministerium für europäische und internationale Angelegenheiten 2007). A major difficulty for disaster response is that there is limited information on actual needs, and simultaneously, time is the determining factor for the success of the relief effort (Apte 2009). Shortly after the onset of the disaster, the situation is usually chaotic, responsibilities are unclear, and resources are scarce. Unstable political circumstances (e.g. unclear power relationships, distrust of humanitarian workers) can complicate aid operations.

Based on the demand patterns, relief supplies can be divided into two groups (Balcik et al. 2008). Those needed immediately after the disaster event and usually only once are called Type 1 goods and comprise utilities such as tents, blankets, tarpaulins, canisters, etc. Those regularly needed over the entire period of the relief effort are items that are consumed regularly such as food and hygiene kits, also known as Type 2 goods. At the beginning of disaster relief, there is usually a wave of demand for relief supplies, which has to be tackled in the light of uncertain supplies and damaged infrastructures.

Transport of relief supplies is an integral part of disaster response. Specific requirements may result from the state of the infrastructure (degree of destruction), the available transport capacity in relation to the quantity of goods to be transported, geographic conditions (e.g. mountains during earthquakes in Nepal 2015, widely distributed islands during tsunami 2004) and the security situation (e.g. humanitarian crisis in South Sudan).

Particularly challenging is usually the final stage of the relief chain from the local distribution center to the people affected by the disaster, the so-called "last mile" (Apte 2009), (Balcik et al. 2010). The challenges lie in the planning of timing and routing supply vehicles, effective management of stockpiling and in efficient transport.

5.2 Transportation in Disaster Relief

Transport of relief supplies consists of the transport to distribution centers and the subsequent last mile distribution to the people in affected regions (Fig. 11). Transport of relief goods to the port of entry is usually not a problem and is carried out with the help of commercial transport providers (Deutsches Rotes Kreuz 2016).

For the onward transportation, capacities are usually rented locally or provided by local authorities or the local military. In case sufficient ground transportation capacities are not available, trucks of e.g. the World Food Program (WFP) are deployed to the disaster area.

In general, high effort and costs speak against the transport of relief goods by air. However, if the disaster area to be supplied is inaccessible by surface transport due to damaged infrastructure, flooding or a precarious security situation, transport of relief supplies on the last mile by air is considered as a last resort. Then, usually helicopters are used because of their capability to take-off and land vertically. Helicopters with high load capacity are preferred, such as Mil Mi-8 (Fig. 12) with about 4 tons payload capacity or Chinook CH-47 with a possible payload up to 12.7 tons.

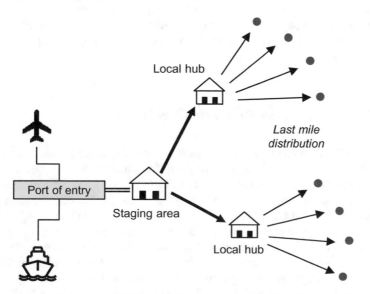

Fig. 11 Transport chain for aid deliveries. Based on the graphics by Balcik et al. (2008)

Fig. 12 The Mi-8 is widely
used to supply people in
need. (Photo: © WFP/Ala
Kheir)

In order to transport large quantities of goods in a short period of time, a large number of helicopters are needed. For example, after the 2015 earthquake in Nepal the World Food Program (WFP) deployed four Mil Mi-8 s and two Airbus Helicopters AS 350 (Logistics Cluster 2016). In addition, Nepal itself, India and China provided about 30 helicopters in total, while the USA deployed four V-22 Osprey aircraft that can take off and land vertically while carrying up to 9,000 kg. However, this was still found as insufficient (Spiegel 2015).

Airdrops - that is dropping off of basic relief supplies from airplanes - are indeed practiced, but only with great reluctance (Deutsches Rotes Kreuz 2016), (Diakonie Katastrophenhilfe 2016). The effort involved is not only high with regard to the preparation of relief supplies (suitable packaging), but also the discharge point must be prepared and the population must be informed. In this case, there is the additional danger that the relief goods will not reach the people in need, but will be taken over by unauthorized persons or criminals.

5.3 UCA for Disaster Response Operations

A major challenge with disaster relief lies in transporting large quantities of goods in a short time. Compared to the amount of goods to be transported, the load capacity of the envisaged UCA is relatively small. For example, the German Federal Office for Civil Protection and Disaster Assistance (Bundesamt für Bevölkerungsschutz und Katastrophenhilfe BBK 2015) recommends 28 l of water and about 20 kg of food as a 14-day basic supply per person, which means 3.5 kg per person per day. Hence, with one ton of relief goods, around 285 people can be supplied with drinking water and food for one day. Another example is the transport of tents. Tents for 5 persons are transported on pallets of 10 pieces each. The dimensions of the complete pallet are $2 \times 1 \times 1.36$ m with a weight of 850 kg (Deutsches Rotes Kreuz 2016). Hence, tents for 50 people can be transported on one flight with the envisaged UCA. These examples illustrate that generally an increased load capacity of the UCA for disaster

Fig. 13 During flood disasters landing areas for helicopters are not available (top), so that relief goods must be handed over from the hovering helicopter (bottom). (Photos: © Mercy Air Switzerland)

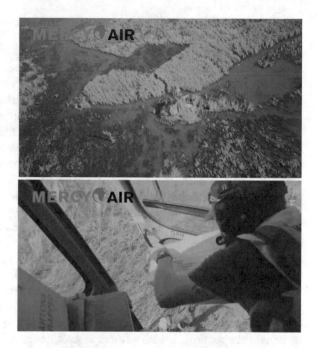

missions would be preferable. The transport capacity requirements are reduced if high-energy biscuits (HEB) are to be transported, as the daily per capita energy requirement can be covered by about 500 g of this kind of emergency compact food.

In case of flood disasters, not only road infrastructure is unusable, but also suitable landing sites are scarce. Pictures (Fig. 13) taken from a video by Mercy Air[4] during the response to Cyclon Idai in Mozambique in 2019 illustrate that there was even no sufficient landing area for helicopters. Relief goods were handed over to the people in need from the hovering helicopter. To handle such situations, a UCA would have to be capable of dropping packages with high accuracy so that the drop does not drift off out of reach.

6   Conclusions

UCA with payload capacity of about one metric ton and range of about 600 km represent a novel means of transport. In this chapter the usability of this type of unmanned aircraft is analyzed for three fields of application: spare parts logistics in Europe, supply of hard-to-reach areas in Africa, and disaster relief.

In spare parts logistics, the criticality, value, demand volume and regularity of demand of spare parts significantly influence the logistics concept and thereby the

---

[4] Mercy Air is an international aid organization specialized in air transport services. https://mercya ir.ch/.

role and importance of the means of transport used. Speed of transport means plays an important role in case of spare parts with high criticality and high value and with low and irregular demand as optimal logistic solutions tend to be centralized warehousing close to the supplier. In such cases, UCA can be a suitable transport solution in order to reduce transport times and thus production downtime. This is achieved by two means: high cruising speed of the UCA compared to trucks and the usage of airfields close to the origin and destinations and thus limiting time consuming pre-carriage and onward carriage by truck. Providing that downtime costs per hour are sufficiently high, the reduction of transport time decreases the total amount of production downtime costs to an extent which compensates higher transport costs when using UCA. Furthermore, example calculations for Germany show that transport times can be shortened by using UCA even for relatively short distances, even if a good road network is available and detours have to be flown to avoid densely populated areas. Increasing the range of the UCA will expand the area where suppliers of spare parts can offer a better service to customers.

Transport of goods in hard-to-reach areas is characterized by poor road conditions and on occasion also by a critical security situation. Here the objective is to deliver goods without damage and protected against loss and robbery to the consignee. The usage of UCA enables to avoid poor and insecure roads thus ensuring safe transport chains. This is particularly beneficial for the transport of relatively valuable, fragile or perishable goods. It is also advantageous that the UCA considered here can be operated from small landing sites with little infrastructure. To what extent additional costs of UCA transports are accepted by the user depends on the value of the goods. Therefore, drugs and other essential goods have a good chance of being among the first items to be transported by UCA. However, flying in low altitudes can bear the risk that the UCA will be shot down in regions with critical security situation. Here too, larger range of the UCA will increase the benefits by extending the operational area.

Disaster relief poses greatest challenges for the application of UCA. In disaster relief operations the objective is to transport large quantities of relief goods to people in affected areas while transport infrastructure is partly or completely unusable. In extreme cases like floods, even helicopters cannot land as appropriate landing areas do not exist. To be applicable in such situations, the UCA needs additional capabilities like to detect people in need and to supply them by performing high precision airdrops. Further work is required to cope with these tasks.

References

Action medeor e.V. (2016) Personal communication

Aerospace Technology (2019) AT200 Cargo Unmanned Aerial Vehicle. https://www.aerospace-tec hnology.com/projects/at200-cargo-unmanned-aerial-vehicle/, Accessed 23 Oct 2019

Air Tindi (2019) Air Tindi. https://www.airtindi.com/, Accessed 23 Oct 2019

Amazon (2019) A drone program taking flight. https://blog.aboutamazon.com/transportation/a-drone-program-taking-flight, Accessed 10 June 2019

Apte A (2009) Humanitarian logistics: a new field or research and action. foundations and trends® in technology. Inf Oper Manag 3(1):1–100

Balcik B et al (2010) Coordination in humanitarian relief chains: Practices, challenges and opportunities. Int J Prod Econ 126:22–34

Balcik B, Beamon BM, Smilowitz K (2008) Last mile distribution in humanitarian relief. J Intell Transp Syst 12(2):51–63

Bloech J, Ihde GB (eds) (1997) Vahlens großes Logistiklexikon. Beck Vahlen, München

Bundesamt für Bevölkerungsschutz und Katastrophenhilfe BBK (2015) Ratgeber für Notfallsorge und richtiges Handeln in Notsituationen

Bundesministerium für europäische und internationale Angelegenheiten (2007) Internationale humanitäre Hilfe: Leitlinie der Österreichischen Entwicklungs- und Ostzusammenarbeit. Wien

CLAAS KGaA mbH (2015) Annual Report 2015

CLAAS KGaA mbH (2016a) CLAAS Ersatzteilservice Deutschland. https://www.claas.de/service/service-ueberblick/ersatzteilservice-deutschland, Accessed 20 June 2016

CLAAS KGaA mbH (2016b) Personal communication

Clausen U, Stütz S, Bernsmann A, Heinrichmeyer H (2016) ZF-Zukunftsstudie 2016: Die letzte Meile. Available via ZF Friedrichshafen AG. https://www.zf-zukunftsstudie.de/site/zukunftsstudie/de/home/homepage.html, Accessed 27 June 2017

Dauer JC, Lorenz S, Dittrich JS (2016) Automated Low Altitude Air Delivery

Deutsches Rotes Kreuz (2016) Personal communication

DHL Customer Solutions & Innovation (2014) Unmanned Aerial Vehicles in Logistics

Diakonie Katastrophenhilfe (2016) Personal communication

EASA (2015) Concept of Operations for Drones: A risk based approach to regulation of unmanned aircraft

CRED (2020) EM-DAT: The International Disaster Database. Centre for Research on the Epidemiology of Disasters - CRED / UCLouvain, Brussels, Belgium, (D. Guha-Sapir). https://www.emdat.be, Accessed 23 Sept 2020

Garrett-Glaser B (2019) Flirtey wants to enable ten minute commercial drone deliveries. https://www.aviationtoday.com/2019/09/10/flirtey-wants-ten-minute-point-point-drone-delivery-businesses-across-america/, Accessed 17 Sept 2019

Huiskonen J (2001) Maintenance spare parts logistics: Special characteristics and strategic choices. Int J Prod Econ 71:125–133

IATA (2019) Cargo Drones. https://www.iata.org/en/programs/cargo/cargo-drones, Accessed 17 Oct 2019

IFRC (2018) World Disasters Report 2018. International Federation of Red Cross and Red Crescent Societies. https://media.ifrc.org/ifrc/world-disaster-report-2018/, Accessed 30 Jan 2020

inTime Express Logistik GmbH (2019) inTime Preisrechner. https://www.intime.de/services/preisrechner/, Accessed 19 Sept 2019

Johanniter-Auslandshilfe (2016) Personal communication

Liebhardt B, Pertz J (2021) Automated cargo delivery in low altitudes: business cases and operating models. In: Dauer JC (ed) Automated Low-Altitude Air Delivery - Towards Autonomous Cargo Transportation with Drones. Springer, New York

Lockenkötter A (2013) Erntezeit, aber der Mähdrescher läuft nicht. https://logistik-heute.de/sites/default/files/public/data-statische-seiten/transport_logistic_2013_vortrag_andr_lockenk_t_17033_0.pdf, Accessed 25 Apr 2016

Logistics Cluster (2016) Nepal Lessons Learned Report. https://reliefweb.int/report/nepal/nepal-lessons-learned-report-january-2016, Accessed 24 Nov 2016

McFarland M (2019) Alphabet's Wing to make Walgreens' drone deliveries in small Virginia town. https://edition.cnn.com/2019/09/19/tech/alphabet-wing-drone-delivery/index.html, Accessed 25 Sep 2019

Sabrewing (2019). https://www.sabrewingaircraft.com/cargo-uav/, Accessed 23 Oct 2019

Schopferer S, Donkels A (2021) Trajectory risk modelling and planning for unmanned cargo aircraft. In: Dauer JC (ed) Automated Low-Altitude Air Delivery - Towards Autonomous Cargo Transportation with Drones. Springer, New York

Schultz M, Temme A, Kügler D (2018) Unbemannter Frachttransport im Luftverkehrssystem. Internationales Verkehrswesen 70(1):40–42

Singular Aircraft (2019) FLYOX I. http://singularaircraft.com/flyox-i-3/. Accessed 16 Oct 2019

Spaeth A (2016) Hubschrauber oder Hundeschlitten. AEROReport 02/16. MTU Aero Engines AG. https://aeroreport.de/content/6-mediathek/7-ausgabe-02-2016-deutsch/02-2016-deutsch.pdf

Spiegel (2015) Wir haben nicht genügend Helikopter. http://www.spiegel.de/panorama/erdbeben-nepal-braucht-fuer-hilfe-mehr-hubschrauber-a-1031895.html, Accessed 24 Nov 2016

World Food Programme (2019) UAVs for Cargo Delivery. https://innovation.wfp.org/project/uavs-cargo-delivery, Accessed 16 Oct 2019

Automated Cargo Delivery in Low Altitudes: Business Cases and Operating Models

Bernd Liebhardt and Jan Pertz

Abstract The project Automated Low Altitude Air Delivery (ALAADy) focuses on a transportation drone supposed to transport one metric ton of goods over a distance of 600 km with a cruise speed of 200 km/h. The aircraft is planned to operate under a newly established category of certification (EASA's Specific category) that is intended to make certification less constraining and operations more affordable and equally safe compared to other means of air transportation. The present chapter introduces two possible applications of this aircraft, namely a commercial flight network for the delivery of agricultural spare parts in Europe, and a catastrophe relief mission for overflooded areas. In order to illustrate the parameters of the respective operations, both concepts are assessed using cost models as well as flight schedules and mission trajectories, respectively. This chapter's purpose is to give impressions on the operational environments, particularly regarding economics and business-case establishment, that longer-range transportation drones have to comply with.

Keywords UAS · Drone · Cargo delivery · UCA · Autonomy · Aircraft design · Concept study · Use case · ALAADy · Agricultural spare parts · Humanitarian relief · Cost model · Mission trajectories · Package drop

1 Introduction

Thanks to technical innovations including (but not exclusively) systems miniaturization, advanced lithium-ion accumulators, and computerized flight stabilization, flight drones have become commonplace in the past years. However, their application is usually confined to surveillance missions where the only useful load is cameras; air transportation is rarely attempted. The main reasons for this are the lacking energy

B. Liebhardt (✉) · J. Pertz
German Aerospace Center (DLR), Air Transportation Systems, Blohmstrasse 20, 21079 Hamburg, Germany
e-mail: Bernd.Liebhardt@dlr.de

J. Pertz
e-mail: Jan.Pertz@dlr.de

© Deutsches Zentrum für Luft- und Raumfahrt e. V. (DLR) 2022
J. C. Dauer (ed.), *Automated Low-Altitude Air Delivery*, Research Topics in Aerospace,
https://doi.org/10.1007/978-3-030-83144-8_4

density of electric accumulators, a yet-to-be-established safety record, and unclear business models.

The ALAADy project was established to address the mentioned shortcomings: Instead of electric energy, traditional carbon-based fuels and the according engines are used for the initial implementation to achieve reasonable flight ranges. High safety shall be achieved using both technology (so-called threat and harm barriers, respectively) and operational means like routing over sparsely populated areas. And finally, operational models were designed and assessed to find promising business cases for air transportation.

This chapter focuses on the latter. We introduce operational models for spare parts delivery and for disaster relief, accompanied by cost modelling and by flight schedules and mission trajectories, respectively.

2   Agricultural Spare Parts Delivery

In the following, we describe an operational model for agricultural spare parts delivery by a cargo drone (Pak 2021). This includes operational planning, a model for cost of operation, and parameter studies on business cases. For the sake of simplicity, we concentrate on the ALAADy gyrocopter design (Hasan and Sachs 2021).

2.1   Operating Cost Model

Operating cost is an essential characteristic for viability and profitability of transportation systems. An early cost assessment is vital regarding a reliable estimate of market potential, the identification of realistic operational scenarios, and the cost-oriented technical specification of the air vehicle. Coincidentally, the challenge of civil freight transport with payloads of up to one metric ton in an unmanned aircraft over distances of several hundred kilometers presents itself as unique and novel, whereas the setup of operating cost interacts with a multitude of external and internal factors. This subpart attempts to cover and quantify the major cost items of the gyrocopter.

The aircraft is planned to operate under EASA's "Specific" certification category, and thus, its safety is to be assessed using the SORA (Specific Operational Risk Assessment) framework. The "Specific" category being located between the "Open" category for small drones and the "Certified" category for traditional certification, the requirements for the aircraft's safety are somewhat relaxed, which is intended to make aircraft design and operation markedly less expensive than what is usual in air transportation. As the aircraft is expected, allowed, and designed to safely terminate its flight once in roughly 10,000 flight hours, it will need fewer redundancies in flight control, it will use cheaper (not necessarily air transport-rated) systems, and it will require less maintenance.

The probably most common cost metric in aeronautics is direct operating cost (DOC), which contains all cost items that can be directly attributed to the operation of the air vehicle. For the purpose at hand, a DOC model was developed bottom-up, with the following cost items being covered and assessed separately:

- Depreciation
- Interest
- Insurance
- Operator wage
- Fuel
- Maintenance
- Airport fees
- Control station

As a reference, the model takes flight mission, flight distance, or flight time.

(Imputed) Depreciation is determined by four parameters: depreciating method, depreciation basis, depreciation period, and residual value. As a depreciation method, a simple linear approach is chosen (whereas the method influences the cost of interest). The depreciation basis is the acquisition cost of the air vehicle which, in turn, has to be derived from the cost of development and production. (The latter are discussed in a separate paragraph below.) Common figures for commercial air transportation indicate depreciation periods of 10–20 years and residual values of 10–30%. Cost of depreciation is calculated as follows:

$$Depr_{calc} = \frac{DeprBasis - ResidValue}{DeprPeriod} \quad \left[\frac{€}{year}\right]$$

$$Cost_{depr} = \frac{Depr_{calc}}{YearlyUtilization} \quad \left[\frac{€}{flighthour}\right]$$

(Imputed) Interest takes into account the average capital commitment, the latter being linearly reduced from the cost of acquisition to the residual value over the service life. The inputs are cost of acquisition, residual value, and interest rate:

$$Int_{calc} = \left(\frac{DeprBasis - ResidValue}{2} + ResidValue\right) \cdot IntRate \quad \left[\frac{€}{year}\right]$$

$$Cost_{int} = \frac{Int_{calc}}{YearlyUtilization} \quad \left[\frac{€}{flighthour}\right]$$

Operator wage refers to the pay for the persons steering or monitoring the aircraft. It accounts for the number of full time equivalents (FTE) and the yearly wage of one FTE. Obviously, both parameters are influenced by a multitude of factors, including yearly aircraft utilization, degree of autonomy, legal regulations, training and qualification, local wages level, and potential synergy effects from parallel operation. The cost of wages is calculated as follows:

$$Wages_{calc} = n_{FTE} \cdot Wage_{FTE} \quad \left[\frac{\text{€}}{year}\right]$$

$$Cost_{wages} = \frac{Wages_{calc}}{YearlyUtilization} \quad \left[\frac{\text{€}}{flighthour}\right]$$

Maintenance in aeronautics discerns between so-called line maintenance during operating hours and base maintenance for which operations are suspended. Assessing the cost of maintenance entails high complexity and uncertainty, particularly for a novel concept like the one at hand. Our approach does a separate estimation for the main assemblies and components, respectively, which are engines, structure, systems, and rotor.

The engines, be they piston or turbine driven, require both concurrent maintenance as well as complete overhaul after a predetermined number of operational hours, called time between overhaul (TBO). The cost of overhaul can be attributed to flight hours:

$$Cost_{OH,h} = \frac{Cost_{OH}}{TBO} \quad \left[\frac{\text{€}}{flighthour}\right]$$

Table 1 lists the cost of overhaul for some engines in appropriate performance classes.

The cost for concurrent maintenance and inspection of the engines is more difficult to estimate due to the lack of data, which is equally true regarding structure and

Table 1 Aircraft engines and overhaul data

Maker		Cessna	Piper	Quest	Textron	Diamond
Make		Caravan	M 500	Kodiak	Cessna 172	DA-40
Engine		PT6A-114A	PT6A-42A	PT6A-34	Lycoming IO-360	AE300
Type		TP	TP	TP	Piston	Piston
Fuel		Jet A	Jet A	Jet A	Avgas	Diesel
Power	kW	500[a]	370[b]	519[c]	133[d]	124[e]
Cost$_{OH}$	€	182,900[f]	162,600[f]	191,000[f]	25,600[f]	16,700[f]
TBO	h	3,600	3,600	4,000	2,000	1,800
Cost$_{OH,h}$	€/h	50.8	45.2	47.8	12.8	9.9

[a](Textron Aviation 2020)
[b](Piper Aviation 2020)
[c](Quest Aircraft 2020)
[d](Lycoming 2005
[e](Austro Engine 2020)
[f](Aviation Week Network 2016)

Table 2 Cost and life of Bell 206 rotor assembly (*Source* private inquiry)

Part	Life	Number	Cost/piece	Cost/h
[-]	[flight h]	[-]	[USD]	[USD/bl-h]
Trunnion	2,400	1	3,910	1.6
Strap Fitting	2,400	2	2,134	1.8
Strap Pin	1,200	2	333	0.6
Grip	4,800	2	9,540	4.0
Tension–Torsion Strap	1,200	2	4,827	8.0
Latch Bolt	1,200	2	843	1.4
Main Rotor Blade	4,000	2	55,420	27.7
Main Rotor Mast	5,000	1	17,284	3.5
Cyclic Tube	4,800	2	2,460	1.0
Total				**49.6**

systems. The cost items have to refer to the actually implemented systems and structures individually. For these three items, educated guesses are made, considering that the other items are usually more expensive.

Regarding the rotor, parts of it have to be replaced after a predefined number of flight hours. In the case at hand, a relatively simple rotor assembly of a Bell 206 Ranger helicopter (MTOW of 2,000 kg, similar to the gyrocopter), whereof all cost is known (see Table 2), is taken as a reference.

Finally, the model inputs for maintenance are the hourly cost of: engine overhaul, concurrent engine maintenance and inspection, structural maintenance, systems maintenance, and rotor maintenance.

Insurance must take into account the purportedly higher failure probability of ALAADy aircraft compared to traditional ones. Whereas a similar safety level shall be attained by operational means (particularly flight routing over sparsely populated areas), material damage has to be expected for the aircraft and its payload nevertheless. Thus, the required inputs are failure probability and estimated damage. Both parameters can be reduced with the help of technology, e.g. sophisticated controls and decelerating parachutes, respectively. The calculation formulae for the cost of insurance are as follows:

$$Cost_{ins} = \frac{EstimDamage}{FailureRate} \quad \left[\frac{€}{flighthour}\right]$$

$$Ins_{calc} = Cost_{ins} \cdot YearlyUtilization \quad \left[\frac{€}{year}\right]$$

Fuel cost has higher impact on the economy of air transportation systems compared to ground transportation due to significantly higher fuel consumption per mass transported. Fuel price and fuel consumption are the basis for the calculation

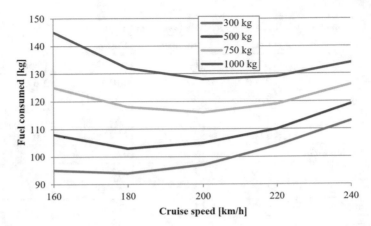

Fig. 1 Diesel consumed by the gyrocopter on a 600 km mission for different payloads

of fuel cost. The former tends to fluctuate significantly together with the oil price, and it depends on the location and on the consumer's bargaining power. The latter is a function of flight speed and payload: the faster the flight and the heavier the payload, the more fuel is consumed. Figure 1 displays the fuel consumed over the cruise speed for payloads between 300 and 1000 kg on a 600 km mission.

The required inputs for fuel cost are fuel price, fuel consumption (specific to speed and payload), and flight distance. The latter can be expressed through straight-line distance and a detour factor since the assessed flights are usually rerouted. Fuel cost is calculated as follows:

$$Cost_{fuel,h} = FuelPrice \cdot SpecifFuelCons[\tfrac{1}{h}] \quad \left[\frac{€}{flighthour}\right]$$

$$Cost_{fuel,km} = FuelPrice \cdot SpecifFuelCons[\tfrac{1}{km}] \quad \left[\frac{€}{flightkm}\right]$$

Airport fees are directly subject to the performed mission and should ideally be retrieved for the origin and destination airfields, respectively. Usually depending on MTOW, they can vary drastically between locations.

A *remote control* is needed for command and control (C2) of the aircraft. The control station can be implemented as a small mobile device up to a complete control room, whereas for the purpose at hand, something in between should be envisioned. Regarding the datalink, both radio and satellite systems are possible. However, satellites should only be used when radio is not available because they are at least one and possibly even three orders of magnitude more expensive. Optimally, mobile telephony networks could be used as probably the least expensive option, whereas these could be unavailable if operations take place in sparsely populated areas. Moreover, the needed data rates depend on the aircraft's degree of autonomy. Lastly, the required inputs are the mean required data rate as well as cost rates and usage distribution, respectively, of the implemented datalinks, plus an estimate for the cost of the control station.

According to the International Telecommunication Union (ITU), expected data rates for Command & Control, ATC, and Sense & Avoid are 12,167 bits/second, 2 × 4,955 bits/s, and 9,120 bits/s, respectively, which is about 31,000 bits/s in sum. Video and weather radar for low altitudes requires about 270,000 bits/s (ITU 2010).

Acquisition cost of aircraft factor into the cost of operations, namely into the accounting of depreciation, interest, and insurance. Acquisition cost, in turn, result from the cost of *development and production*. For estimating the latter, we use the modified Development and Production Costs for Aircraft (DAPCA-IV) model (Hess 1987) that refers to development and production cost of general aviation aircraft and that has been updated not long ago (Gudmundsson 2013). In our case, parts of the DAPCA methodology are taken to estimate the cost of the airframe, the engines, and the propeller. All cost regarding the aircraft's autonomy (ground station, datalink, and other systems) are not covered and have to be assessed separately. Table 3 displays the influence of inputs to the model (rows) on cost items (lines).

In the following, a parameter study is shown that compares the acquisition cost for production runs of 100 and 500 units, respectively (see Tables 4 and 5). It can be seen that the price per aircraft drops significantly for the longer production run, namely from about 590,000 Euros to 385,000 Euros (a decrease of 35%).

2.2 Operational Use-Case

As a use-case for the operation of the ALAADy aircraft, the express delivery of spare parts is assessed. A certain company that produces agricultural machines has its service center in Hamm, Germany. Its regional subsidiaries in all of Europe depend on the rapid delivery of spare parts, particularly in the harvesting season. The latter's locations are in Madrid (Spain), Bury St. Edmunds (England), Paris (France), Vercelli (Italy), Vienna (Austria), Poznan (Poland), Bucharest (Romania), Kiev (Ukraine), and Moscow (Russia). All subsidiaries are to be supplied from the service center with at least one daily delivery.

It is assumed that each aircraft has one destination that it carries its payload to, whereupon it returns to the service center unladen. Even though it would basically be possible to unload some payload at intermediate stops, it has to be supposed that the whole aircraft capacity is needed for the destination; this is taken as a basic assumption for flight plan development.

In the following, preliminary flight routes are developed to estimate the required flight times. ALAADy is based on the premise that despite a higher risk of failure compared to certified manned transportation systems, the risk of damage in the air and on the ground shall be kept low. This is attempted by using the mitigations and operational safety objectives, respectively (see Chapter 2 of this book). For one, in this case, flight shall only happen at night in order to reduce the probability of interfering with ground traffic in the event of failure. Moreover, the flight routes

Table 3 Inputs (rows) and outputs (lines) of development and production cost. (x): direct correlation; (o): indirect correlation

	Aircraft produced	Airframe mass	Max. speed	Certification type	Composites share	Engine number	Engine power	Engine type	Propeller diameter	No. of prototypes	Development wages	Tooling wages	Production wages	Quantity factor	Inflation	Currency exch. rate
Development	o	o	o	o	o						x				x	x
Tooling	o	o	o		o							x			x	x
Dev. – man h	x	x	x	x	x											
Tool. – man h	x	x	x		x											
Dev. overhead		x	x	x	x					x					x	x
Flight testing		x	x	x						x					x	x
Production	o	o	o	o	o								x		x	
Prod. – man h	x	x	x	x	x											
Quality control	o	o	o	o/x	o/x									o		
Materials	x	x	x	x		x									x	
Engines	o					x	x	x						x	x	
Propellers	o						x		x					x	x	
Rotor														x		
Systems														x		
Control station														x		

Table 4 Breakdown of aircraft acquisition cost; production run of 100 units

	Program	Base price	Quantity factor	Cost/Aircr.	Share
1.Airframe				547,606	93%
1.1. Certification	8,309,835			83,098	14%
1.1.1. Development	3,587,278			35,873	6%
1.1.2. Devel. overhead	140,424			1,404	0%
1.1.3. Flight tests	798,715			7,987	1%
1.1.4. Tool manufact.	3,783,417			37,834	6%
1.2. Production	18,927,400			189,274	32%
1.3. Quality control	1,230,281			12,303	2%
1.4. Materials	1,532,386			15,324	3%
1.5. Engines		173,972	0.68	117,544	20%
1.6. Propeller		24,079	0.68	16,269	3%
1.7. Rotor		160,000	0.71	113,794	19%
2. Control systems				42,673	7%
2.1. Ground station		10,000	0.71	7,112	1%
2.2. Systems		50,000	0.71	35,561	6%
Total cost per aircraft				590,279	100%

Table 5 Breakdown of aircraft acquisition cost; production run of 500 units

	Program	Base pr.	Quantity factor	Cost/AC	Share
1.Airframe				347.185	90%
1.1. Certification	11,358,188			22,716	6%
1.1.1. Development	4,815,890			9,632	3%
1.1.2. Devel. overhead	140,424			281	0%
1.1.3. Flight tests	798,715			1,597	0%
1.1.4. Tool manufact.	5,603,159			11,206	3%
1.2. Production	43,989,724			87,979	23%
1.3. Quality control	2,859,332			5,719	1%
1.4. Materials	5,482,170			10,964	3%
1.5. Engines		173,972	0.60	104,346	27%
1.6. Propeller		24,079	0.60	14,442	4%
1.7. Rotor		160,000	0.63	101,017	26%
2. Control systems				37,881	10%
2.1. Ground station		10,000	0.63	6,314	2%
2.2. Systems		50,000	0.63	31,568	8%
Total cost per aircraft				385,066	100%

are designed such that settlements as well as critical infrastructure are circumvented. Beside power stations, critical infrastructure includes overhead power lines, expressways, and railroads. As those three are connected in networks that cross all of Europe and that cannot possibly be circumvented, they shall at least be crossed orthogonally to minimize the duration of exposure. Originally, 500 m were aimed for as a minimum distance to settlements. However, it was quickly found that this target cannot be met for Germany and for most of Europe, because the density of settlements is usually higher. Thus, for the time being, the distance to settlements shall only be maximized. Lastly, the aircraft shall preferably operate on small airfields to avoid compatibility problems with regular air traffic and high landing fees.

Origin Airport. About 10 km west of the service center and about a 15 min car drive, there is the airfield Hamm-Lippewiesen (ICAO code EDLH) that has a 631 m asphalt runway (see Fig. 2). Due to its location in the meadows of the river Lippe, it can be approached from the East as well as from the West. However, the free space in the West is narrower, and there is a large coal power station on the way where numerous power lines originate. Therefore, Eastern trajectories are to be preferred.

As an alternative for an additional acceleration and simplification of the transport chain, the construction of a dedicated airfield next to the service center could be undertaken, enabled by several free areas nearby. Most of the subsidiaries have access to such areas, which could eventually eliminate the need for last-mile ground transport both at origin and at destination. However, this possibility is postponed for the purpose of limiting initial expenses.

Bury St. Edmunds, England. The beeline from Hamm goes through North Rhine-Westphalia over the Netherlands and the North Sea to Eastern England. All land areas are quite densely populated, although agricultural fields can be found everywhere with single farms in between. There are virtually no farms located farther than a few hundred meters from others. Accordingly, it proves difficult finding straight flight paths with adequate distance buffers.

Fig. 2 Surroundings of EDLH airfield [compiled with Google Earth]

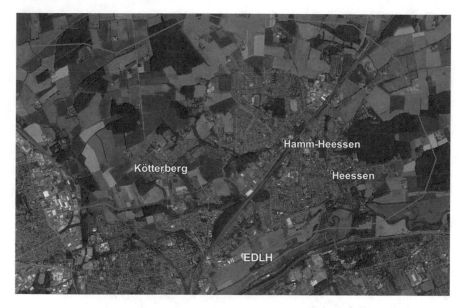

Fig. 3 Circumvention of Hamm-Heessen [compiled with Google Earth]

Eventually, the flight is commenced from EDLH to the East, making a long turn to the North whilst circumventing Hamm-Heessen, until the Western direction is attained (see Fig. 3).

Flying through North Rhine-Westphalia, towns and cities are avoided, farms are circumvented, and, wherever possible, silvan areas are used for safe transition. The Rhine is crossed between Rees and Xanten. South of Nimwegen, just behind the Dutch border, the route swivels in over the river Maas, which, compared to the Rhine, has only sparse ship traffic and can thus be regarded as quite uncritical for overflight. The route further follows the Maas till its confluence with the Rhine's main arm and further to the North Sea estuary at Haringvlietdamm. The North Sea is crossed in a straight line before landfall is made south of Aldeburgh. From there on, only small villages and farms are circumvented before eventually, Rougham Airfield (ICAO code GB-0146) in the East part of Bury St. Edmunds is reached.

The total flight distance is 522 km, whereas the beeline is 488 km, resulting in a detour of 7% (see Fig. 4 for a total route overview). The concluding car drive to the subsidiary takes 10 min.

Vercelli, Italy. The city is located in North-Western Italy between Milan and Turin. The beeline leads south through Western Germany, Switzerland, and the Alps. From the start of the Rhenish Massif about 25 km south of Hamm –and very opposed to the Bury St. Edmunds route–, there are hardly any single farms for the rest of Germany, but nearly exclusively compact towns and cities which are fairly easy to circumnavigate.

Fig. 4 Overview of the Hamm to Bury St. Edmunds route [compiled with Google Earth]

After the Rhenish Massif, the route leads through the Rhine Valley with the Rhine being crossed west of Karlsruhe, and then through the Black Forest. In case of carrying significant payloads, Donaueschingen airfield (EDTD) can be used for refueling after 430 km of flight.

Until Switzerland, routing remains simple. It gets more problematic from Zurich on: the Limmat, the densely populated effluent of Lake Zurich, is crossed between Neuendorf and Killwangen. After that, numerous settlements have to be circumvented for eventually reaching Lake Zug between Cham and Zug. The lake is crossed and left at Küssnacht for continuing on Lake Lucerne which is used over its total length till Altdorf. From there, the Gotthard route is taken to eventually traverse the Gotthard Pass at 2,100 m above sea level, following steep climbs at Göschenen and Höspental.

Thereafter, the flight continues through the Laventina valley to the Southeast, following the Ticino river. Before Bellinzona, either the Ticino or the mountainside has to be taken till Lake Maggiore due to dense population. The lake is used for about 50 km for overflight and left just before Arona. After strenuous circumventing of widely ramified settlements in the Alpine foothills, the Po Plain is reached that is characterized by large croplands and that is easy to cross. Finally, Vercelli's airfield (LILI) is reached in the South of the city.

Figure 5 displays an overview of the route. The total flight distance is 784 km (422 + 362); the beeline is 713 km, which makes a detour of 10%. The drive to the subsidiary takes 5 min.

Paris, France, is located south-west of Hamm. On the beeline, there is the very densely populated Rhine-Ruhr metropolitan region (see Fig. 6) that must be circumnavigated laboriously. Later, Belgium is sprinkled with countless small towns that often have an elongated extension. The overflown region in France however exhibits large agricultural fields and compact settlements.

Another difficulty arises through the subsidiary being located in a south-western suburb of Paris, requiring a large circumvention. The shores of the river Seine that

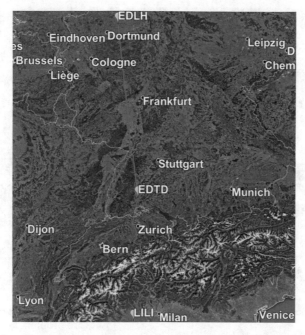

Fig. 5 Overview of the Hamm to Vercelli route [compiled with Google Earth]

Fig. 6 Beeline from Hamm to Paris through Rhine-Ruhr [compiled with Google Earth]

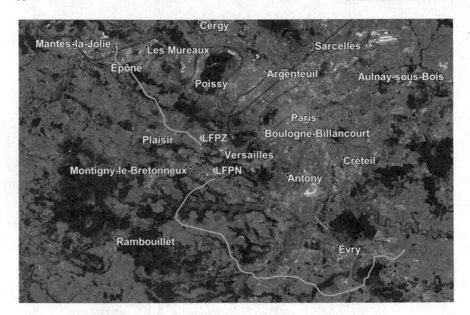

Fig. 7 Flight routes around Paris crossing the river Seine [compiled with Google Earth]

divides Paris from the South-East to the North-West are built up very densely and allow crossing only in considerable distance. A western circumnavigation of Paris appears shorter than an eastern one (see Fig. 7). Using the former, the route ends at Saint-Cyr-l'Ècole airport (XZB/LFPZ) west of Versailles, wherefrom the drive to the subsidiary takes 10 min.

Due to the complexities and the lack of fixed parameters in route design, and due to the methodology now being highlighted sufficiently, we refrain from designing whole routes from here on. In the Paris case, we assume a detour of 20% on the direct line of 520 km, resulting in a flight distance of 624 km. This is doable non-stop in case the transported mass is significantly lighter than the design payload of one metric ton. If not, Belgium offers at least three appropriate airfields for refueling, namely Liernu (EBMB), Namur (EBNM), and Maillen (EBML).

Madrid, Spain. The beeline crosses the Rhine-Ruhr area again, yet a bit more to the South, so that an overflight of the Rhenish Massif can be conducted. Initial possibilities for crossing the densely built-up Rhine shores are found below Remagen. After that, there are no elongated settlements any more till Madrid, which should make the routing fairly simple. The Pyrenean mountains between France and Spain pose a minor obstacle; however, they can be crossed in the eastern part where they descend toward the Atlantic Ocean below altitudes of 1,000 m.

The great circle distance between the targeted airfield (Aerodromo Loring, ES-0162) and EDLH is 1,506 km which calls for two regular refueling stops. Amongst many others, these could be the aerodromes of Auxerre-Branches (LFLA, kilometer 528) and Nogaro (LFCN, kilometer 1058). For exceptional cases of one stop, two airfields near Guéret (LFCE, kilometer 748, or LFBK, kilometer 727) can be used. Conservatively estimating the detours on the single segments as 15/10/15%, the total distance is 1,711 km. Eventually, the drive to the subsidiary takes 30 min.

Poznan, Poland, appears not to have any airfields beside its main airport that is surrounded by housing and its military airport Krzesiny. Adjacent to the company subsidiary in the West of town however, there are numerous elongated fields that could come into use as runways. The great circle distance is 606 km. Should an intermediate stop be required, Magdeburg City airport (EDBM, kilometer 266) is located directly on the straight line between Hamm and Poznan.

The western-Polish landscape being characterized by large fields and compact settlements, defining safe flight routes appears relatively unproblematic. Therefore, a detour of 5% is assumed, resulting in a total distance of 636 km.

Kiev, Ukraine, is located on a similar direction to Poznan. In Poland's South-East, settlements as well as fields are scattered significantly, with a predomination of elongated and multidirectionally oriented ribbon-built villages that have elongated fields directly attached to the courts. Crossing Lublin, for instance, where appropriate airfields for refueling stops can be found on both the eastern and the western sides of town, is hindered by a nearly 40 km long, North–South-oriented street continuously seamed with houses. In Ukraine, on the other hand, settlements appear thinner and more compact.

In any case, routing proves complicated. Greater detours might have to be accepted to bridge the direct distance of 1,579 km to the Kiev-West airfield (UKKJ). The first of two stops can be performed in western Poland's Zielona Góra (EPZP, kilometer 551) or Lubin (EPLU). For the second stop, Radzyń Podlaski (PL-0057, kilometer 1,023) or Żłobek Mały (PL-0137) seem appropriate. Assuming detours of 10/20/10%, the total flight distance is 1,784 km. The closing drive to the subsidiary takes 30 min.

Moscow, Russia, has a great circle distance of about 2,000 km to Hamm which requires at least three intermediate stops. At the same time, the number of airfields in the relevant areas is much smaller compared to central and western Europe.

One possible route uses Szczecin's airfield (EPSD) in Poland as a stop at kilometer 499, then the Jurbarkas airfield (EYJB) in Lithuania at kilometer 1063, then the Russian airport of Welikije Luki (ULOL) at kilometer 1,575, before eventually, Ostafievo airport (UUMO, km 2016) closes the route (see Fig. 8, purple line).

The probably better, second route takes Heringsdorf (EDAH, kilometer 492), Taurage in Lithuania (LT-0006, km 1,030), then again Welikije Luki (ULOL, km 1,575) and finally Ostafievo (UUMO, km 2017) (see Fig. 8, light-blue line). This option enables most of the second segment being flown over the Baltic Sea. Both routes can easily avoid Belarussian territory, if politically opportune.

Going east, settlements become more and more scattered and compact, so that routing appears easy. Approaching Moscow, however, might prove complicated due to the increasing density of settlements and despite widespread woodlands.

Assuming detours of 10% on each segment, the total distance is about 2,219 km. Driving to the subsidiary from Ostafievo takes about 30 min.

Vienna, Austria, has the subsidiary located about 20 km north of the city center, wherefrom driving to Stockerau airfield (LOAU) takes 15 min. The great circle

Fig. 8 Flight routes from Hamm to Moscow [compiled with Google Earth]

distance to Hamm is 701 km; Helmbrechts airfield (EDQO, kilometer 320) in Bavaria appears to be an appropriate location for refueling. All areas to be crossed, in Germany as well as in the Czech Republic and Austria, predominantly feature large fields and compact settlements. Thus, for routing being unproblematic, a detour of 7% is assumed, rendering the total distance 750 km.

Bucharest, Romania, is located about 1,600 km from Hamm, requiring two stops and an evenly segmented route. Coincidentally, it has to be considered that the Carpathian highlands in Romania need crossing, which entails increased fuel burn. Additionally, there is a 70 km long, transverse Danube segment around Budapest that can hardly be crossed due to its tight built-up.

The chosen route leads over Vlašim (Czech Republic, LKVL) and Hajdúszo-boszló (Hungary, LHHO) to Gagu Airfield (no ICAO code, yet frequented) north of Bucharest. The great circle distances are $544 + 542 + 492 = 1578$ km. Considering the crossing of the Carpathian Mountains and the relatively favorable settlements dispersion, detours of 10/10/15% are assumed, resulting in a total distance of 1,760 km. The concluding drive takes 20 min.

Routes Summary. Table 6 provides an overview on the designed routes. It also contains the total mission durations, assuming a constant speed of 200 km/h, 10 min for one process of taxiing out and in, and 10 min for refueling.

To sum up, flight routing, at least in Europe, exhibits diverse challenges and various degrees of difficulty that very much depend on the regional settlement structure and on the distribution of airfields, respectively. However, it appears that reasonable flight routes can eventually be found in most cases, one way or the other.

Flight Plan. A flight plan was drafted with the goal of flying to every destination daily. As parameters, we suppose

- 10 min for taxiing (out and in; included in flight times)
- 10 min for refueling
- night flying only between the local times of 22 and 6 o'clock
- takeoffs in minimum intervals of 5 min

The resulting flight plan is detailed in Table 7. It can be seen that Bury St. Edmunds

Table 6 Overview of flight routes and flight times (color shades in Distance row indicate mission feasibility w.r.t. range: red for infeasible, green for feasible)

Origin		Destination	Gr. cir. d. [km]	Detour	Distance [km]	Duration [hh:mm]	Transfer [hh:mm]	Time shift
Hamm	-	Bury St. E.	488	7%	522	02:46	00:10	-01:00
Hamm	-	Vercelli	713	10.0%	784	04:25	00:05	00:00
Hamm	-	*Donauesch.*	416	1%	422	02:16		*00:00*
Donauesch.	-	*Vercelli*	297	22%	362	01:58		*00:00*
Hamm	-	Paris	520	20.0%	624	03:37	00:10	00:00
Hamm	-	*Namur*	252	20%	302	01:40		*00:00*
Namur	-	*Paris*	270	20%	324	01:47		*00:00*
Hamm	-	Madrid	1506	13.6%	1711	09:23	00:30	-01:00
Hamm	-	*Auxerre*	528	15%	607	03:12		*00:00*
Auxerre	-	*Nogaro*	530	10%	583	03:04		*00:00*
Nogaro	-	*Madrid*	453	15%	521	02:46		*-01:00*
Hamm	-	Poznan	606	5%	636	03:40	00:05	00:00
Hamm	-	*Magdeburg*	266	5%	279	01:33		*00:00*
Magdeburg	-	*Poznan*	340	5%	357	01:57		*00:00*
Hamm	-	Kiev	1579	13.0%	1784	09:45	00:30	01:00
Hamm	-	*Zielona Góra*	551	10%	606	03:11		*00:00*
Zielona Góra	-	*Radzyń Podl.*	471	20%	565	02:59		*00:00*
Radzyń Podl.	-	*Kiev*	557	10%	613	03:13		*01:00*
Hamm	-	Moscow	2017	10.0%	2219	12:15	00:30	02:00
Hamm	-	*Heringsdorf*	492	10%	541	02:52		*00:00*
Heringsdorf	-	*Taurage*	539	10%	593	03:07		*01:00*
Taurage	-	*Welikije Luki*	545	10%	600	03:09		*01:00*
Welikije Luki	-	*Moscow*	441	10%	485	02:35		*00:00*
Hamm	-	Vienna	701	7.0%	750	04:15	00:15	00:00
Hamm	-	*Helmbrechts*	320	7%	342	01:52		*00:00*
Helmbrechts	-	*Vienna*	381	7%	408	02:12		*00:00*
Hamm	-	Bucharest	1578	11.6%	1760	09:38	00:20	01:00
Hamm	-	*Vlašim*	544	10%	598	03:09		*00:00*
Vlašim	-	*Hajdúsz.*	542	10%	596	03:08		*00:00*
Hajdúsz.	-	*Bucharest*	492	15%	566	02:59		*01:00*

can be served by just one aircraft that can return on the same night without violating the curfew. The same is true for Paris and Poznan. For Vercelli and Vienna however, the return flight has to be performed on the following night, whilst another aircraft starts from the origin at Hamm. Therefore, two aircraft are needed for serving these destinations.

Madrid, Kiev, and Bucharest require three aircraft each, whereas the aircraft only reaches its destination in the middle of the second night, swiftly leaving for the first

Table 7 Draft flight plan. (STD: scheduled time of departure; STA: scheduled time of arrival)

AC #	Origin	STD	Destination 1	STA	STD	Destination 2	STA
1	Hamm	22:35	Bury St. Edm.	00:21	00:31	Hamm	04:18
2	Hamm	22:15	Donauesch.	00:31	00:41	Vercelli	02:40
3	Vercelli	22:00	Donauesch.	23:58	00:08	Hamm	02:25
4	Hamm	22:05	Paris	01:42	01:52	Hamm	05:30
5	Hamm	22:25	Auxerre	01:37	01:47	Nogaro	04:52
6	Nogaro	22:00	Madrid	00:46	00:56	Nogaro	03:42
7	Nogaro	22:05	Auxerre	01:09	01:19	Hamm	04:32
8	Hamm	22:00	Poznan	01:40	01:50	Hamm	05:31
9	Hamm	22:30	Zielona Góra	01:41	01:51	Radzyń P.	04:51
10	Radzyń P.	22:00	Kiev	02:13	02:23	Radzyń P.	04:37
11	Radzyń P.	22:05	Zielona Góra	01:04	01:14	Hamm	04:26
12	Hamm	22:10	Heringsdorf	01:02	01:12	Taurage	05:20
13	Taurage	22:00	Welikije Luki	02:09	02:19	Moskau	04:55
14	Moskau	22:00	Welikije Luki	00:35	00:45	Taurage	02:55
15	Taurage	22:05	Heringsdorf	00:12	00:22	Hamm	03:15
16	Hamm	22:40	Helmbrechts	00:32	00:42	Wien	02:55
17	Wien	22:00	Helmbrechts	00:12	00:22	Hamm	02:15
18	Hamm	22:20	Vlašim	01:29	01:39	Hajdúsz.	04:48
19	Hajdúsz.	22:00	Bucharest	01:59	02:09	Hajdúsz.	04:09
20	Hajdúsz.	22:05	Vlašim	01:13	01:23	Hamm	04:33

of three return stages. Thus, customers will only receive their orders on the second day. The latter is also true for Moscow, whereas this destination even requires four aircraft.

In total, 20 aircraft would be needed for the described, somewhat simplistic operating model, excluding spare aircraft. Route-specific operations are arbitrarily scalable up and down with varying demand (as long as airfield capacities are adhered to).

Discussion. In the following, some noteworthy issues that showed up in operations drafting are discussed briefly.

Winds. In the planned flight altitude of 150 m (about 500 feet) above the ground, significant wind speeds are expected. Assessing atmospheric data by ECMWF[1] for the relevant area between Madrid and Moscow (4° West to 38° East in longitude, 40° to 56° North in latitude), it was found that average wind speeds of 10 m per second (36 km/h) are not seldom and that sometimes even 20 m/s (72 km/h)

---

[1] European Center for Medium-Range Weather Forecasts, http://www.ecmf.int

are recorded. The mean wind speeds are around 6 m/s (22 km/h). Considering the gyrocopter's cruise speed of 200 km/h, this means that ground speeds and flight ranges will often be significantly impacted. Therefore, with the aim of robust flight planning, additional refueling stops should be considered for cases of strong headwinds.

Variable cruise speed. The gyrocopter can choose cruise speeds between 160 km/h and 240 km/h for optimizing operations. Lower speeds enable lower fuel burn and probably also lower noise emissions. This could come into play when there is enough time to reach the destination before the start of the curfew (for instance, the flight to Vercelli, see Table 7). Higher speeds can be used to offset headwinds and to satisfy curfews, respectively. However, this entails a reduction of flight range.

Technical stops. Airfields used for technical stops (esp. refueling) require service personnel who ensure the runway's usability, do the refueling, perform visual checks on the airframe, and approve for flight continuation. Optimally, these tasks can be performed by one trained person.

"Light" return flight. The flight back to the origin is usually done unladen. This allows for longer ranges and for reducing the number of refueling stops.

Nighttime disturbance. A motorized aircraft usually emits considerable noise. Resulting from the restriction to flying at night, it is all the more important to keep a distance from settlements. The acceptance of the public and required distances, respectively, need to be assessed in future work.

2.3 Business Case Drafting

A business case basically balances the cost of an undertaking with its revenues and a profit margin on top. For the case at hand, the cost of the described operations can be calculated with the cost model to eventually find the required revenues to close the business case.

It has to be mentioned that the described use-case would only be needed in harvest-time which only lasts a few weeks. Therefore, it is assumed that the aircraft are wet-leased[2] from a lessor who has other applications for it for the rest of the year.

Thus, we make the following assumptions:

- 22 aircraft (including two spares) have been leased that stem from a production run of 500 aircraft, costing the lessor 385,000 Euros each.
- The depreciation period is 10 years after which the residual value of an aircraft is 20% of its original value, i.e. €77,000 Euros.
- Interest rate is 2%, compliant with today's low interest economic environment.

---

[2] Wet-leasing means that the aircraft comes with all personnel and equipment needed for direct operation

- The mean daily aircraft utilization according to the flight plan being 5:16 h, taking into account 2 spare aircraft, and assuming 14 days out of operations for maintenance per aircraft, the expected yearly utilization per aircraft is 1,850 flight hours.
- Thanks to automation, *one* pilot, with a yearly wage of €60,000, controls *two* aircraft on average.
- For a lack of data, maintenance cost of engine concurrent inspection, structures, and systems are best-guessed to be €10 per flight hour each.
- The cost of engine overhaul and rotor replacement is €40 and €50 per flight hour, respectively.
- The cost of 1 L of Diesel is €0.80 (tax free).
- The fuel consumption is ca. 40 kg per hour.
- Airport fees are €40 per mission.
- Assuming the need for video and weather radar during 10% of the flight time, the data rate required on average is 58 kbit/s, translating to 25.5 MB per hour. We further assume that mobile communications provide 1 MB for €0.1, that satellite communications cost €1/MB, and that satellite connection will be needed for about 20% of flight time due to unavailability of mobile communication networks.

As a result, the cost for a transport mission of *500 km'* distance is displayed in Fig. 9. It can be seen that maintenance is by far the greatest cost item, representing 55% of the total mission cost of €581. Fuel adds another 14%; all other items' respective shares are below 8%.

As the lessor is a for-profit company, a margin has to be added to the cost to arrive at the price for the client. Assuming 10% margin, the one-way mission price would be €639. As the aircraft returns empty, fuel consumption is slightly lowered,

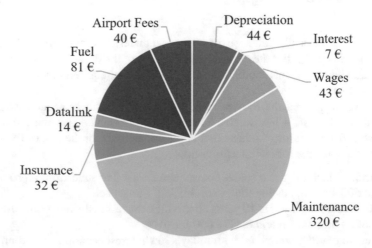

Fig. 9 Cost distribution of a one-way 500 km mission

whereas all other cost items stay the same. Consequently, the transportation price would be round about €*1250*, or €*1.25/km*.

The obvious competition to this application would be road transportation by utility vans (e.g. Mercedes Sprinter). When inviting offers from trucking companies, prices are quoted in the range of €250 to €400, hence between a fifth and a third of the price calculated for the air transport by our drone. Assuming an average speed of 100 km/h on the road, the 500 km mission would take 5 h compared to 2.5 h for the aircraft. Therefore, the air transportation's speed advantage and its reduced risk of delays in road traffic would have to justify a cost multiplication. Additionally, road transportation can be done all day, whereas at least in the use-case at hand, flights would only happen at night.

From a business perspective, there are other applications of air transportation on fixed routes that appear to be more promising. One could imagine, for instance, flight networks between the distribution centers of commerce companies like Amazon, where the flights would be dedicated to the bundled relocation of premium goods for express delivery. Another application could be the on-demand transportation of spare parts between industrial factories to quickly solve costly disruptions of fabrication processes. All this would not necessarily involve the original aircraft designs, as these could be scaled up or down, depending on the applications' requirements.

3 Humanitarian Aid

This section is about a business and yield model specially designed for the humanitarian aid. In detail, the models underneath are drafted for a relief in a catastrophic area.

3.1 Initial Situation in Mozambique After Cyclone Idai

Mozambique is located at the south east coast of Africa. In March 2019, the cyclone Idai made landfall near Beira, the capital of Sofala province. The tropical thunderstorm's wind speeds surpassed 200 km/h and extensive areas were flooded. Due to massive rainfall, rivers flowed over and the population lost their homes, access to food and fresh water and was exposed to further wind and rain without any shelter. The people were separated and split in several local spots as seen in Fig. 10.

Due to the wide scattering of the people affected, delivering supplies to the population proved challenging. Therefore, the future usage of unmanned aircraft like the gyrocopter configuration of ALAADy seems promising, since a land-based supply is not possible.

The United Nations Office Humanitarian Response Depots (UNHRD) in VAE, Ghana and Italy are starting points for supplying goods. Chartered aircraft transported the freight as well as the transport helicopters of the type Mil Mi-8MTV1 to Maputo,

Fig. 10 Aerial photography of the flooded area around Beira, gatherings of people located in several spots (Reid 2019)

the capital of Mozambique (Volga Dnepr 2015). Next, the World Food Programme (WFP) used their Cessna Grand Caravan and their CASA C295H aircraft to further transport the goods to Beira. The last mile delivery got done by the Mi-8 helicopters (World Food Programme 2019). Figure 11 represents the overall logistic chain.

The last mile delivery is the most suitable chain element for the usage of a gyro-copter configuration. With its range of 600 km, it can serve several people within one mission, that will be described in the following.

Fig. 11 Scheme of the supply chain for humanitarian aid

In total, Idai caused huge flooding. Looking at the area around Beira, around 750 square kilometers involving possible survivors were detected. For this consideration, standard unit areas of 1 × 1 km are defined. All units border other units whereas the average distance to Beira is 75 km. One flight serves one unit by supporting the people in need based on package drops. For each unit, 25 people are expected according to the population density in Sofala. Each flight to one unit is described as a mission.

3.2 Mission Specification

A mission is separated into several elements. These elements can be grouped into Pre-, In- and Post-Flight phases. All the elements get tagged with time periods which are necessary for the operation. The Pre-Flight describes all elements which have to be done before the aircraft takes off.

1. Check

   The start preparation is necessary to check the gyrocopter's function and technical state to guarantee high reliability and safety of the operation. The time for checks, including maintenance, flight planning and estimating the weather conditions, is assumed to be 10 min in average. Based on many uncertainties, the time for pre-flight checks cannot be investigated more in detail.

2. Loading

   After checking the systems, the payload is placed in the fuselage of the aircraft. The time for loading depends on the size of the payload, the ground handling facilities and the worker on the apron. It is roughly estimated to be 15 min by virtue of the same reason as before.

After finishing both steps, the aircraft is ready to leave the parking position. After the breaking blocks are removed, the In-Flight sequence is described as block time.

1. Taxi

   Considering the dimension of the airport in Beira, the time for taxiing to the runway is about 2 min considering a roll speed of 10 m/s.

2. Start

   To reach the cruise altitude of 500 ft, the aircraft has to climb. Its acceleration ranges from 1.2 to 1.4 m/s^2. Here, the aircraft needs 30 s to reach the altitude of 500 ft and a speed of 200 km/h.

3. Cruise

   During cruise, the aircraft travels to the target unit. On average, a unit is located 75 km out of town. Flying this distance with 200 km/h takes 22,5 min.

4. Unit

   A unit is the area where the aircraft operates and provides the freight to the targeted people. To increase the safety in the air, the operating aircraft in the unit reduces its flying altitude to avoid collisions with an intruder. Furthermore, the unit gets blocked for other aerial vehicles by zoning the airspace with a geofence. No other aircraft is allowed in the unit except the gyrocopter itself. The gyrocopter scans the area using appropriate equipment to detect the people in need. A flying altitude of 100 m is recommended to detect a person behind common obstacles (Israel 2018). With its scan width of about 100 m, the gyrocopter has to fly 10 stripes to scan the whole unit. If a person is detected, the scan is interrupted and the gyrocopter descends to a lower altitude to reduce the impact energy of the dropped package (see the explanation below where the package drop is explained in detail). After each drop, the aircraft increases its altitude to continue the scanning. In total, the freight is limited to 25 drops since there is not enough space inside the aircraft. The total load of the payload is 375 kg, while each package weights 15 kg. After the dropping, the aircraft returns to the airport by climbing to the cruise altitude again. The whole operation in the unit is calculated to take 17 min including climb and descend phases and scanning the unit.

5.   Cruise
     analogous to step 3
6.   Descent and Landing
     analogous to step 2
7.   Taxi
     analogous to step 1

Since the impact energy of dropped packages is high, the risk of hitting or even damaging infrastructure or injuring a person need to be evaluate. There are two considered concepts for the package drop. Both concepts are calculated with a mass of 15 kg per package.

(i)   The aircraft descends to a lower flying altitude to minimize the impact as much as possible. Additionally, the gyrocopter reduces its speed as much as it can to keep its altitude. That minimum speed is 20 m/s. Dropping a package at an altitude of 45 m results in an impact speed of 36 m/s and 9,75 kJ of impact energy, respectively. During the fall time of 3 s, wind effects can become an issue, with the consequence of missing the targeted area. The flight trajectory of the package in this case is shown in Fig. 12.

(ii) Instead of flying horizontally, the gyrocopter decelerates to zero horizontal speed. This flying state entails a continuous vertical sink rate of 11,5 m/s. While the gyrocopter descends, the package is dropped at the same altitude of 45 m. In this case, the package needs only 2 s to hit the ground. Wind effects are less noticeable. Since there is no horizontal speed, aiming of the drops is improved as well. On the other hand, impact energy and impact speed are increased to 13 kJ and 41,5 m/s, respectively. Hence, the ground impact can cause more damage or injuries to the people. Additionally, the altitude of 45 m is not high enough for such a maneuver. It takes more altitude and time to flare out the aircraft and intercept the sinking. Because of that, the first case is preferred.

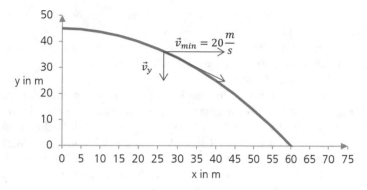

Fig. 12 Trajectory of a dropped package

Parking at the airports ends the In-Flight phase. At the end of each mission, the Post-Flight phase is performed to prepare the aircraft for the next mission.

1. Refuel
   This procedure takes about 2 min.
2. Maintenance
   Post-flight checks are performed to ensure the helicopter's function. Since there are no operating experiences with the gyrocopter, this is expected to take 15 min after each mission.

The altitudes of each mission phase are shown in Fig. 13. It gives an overview

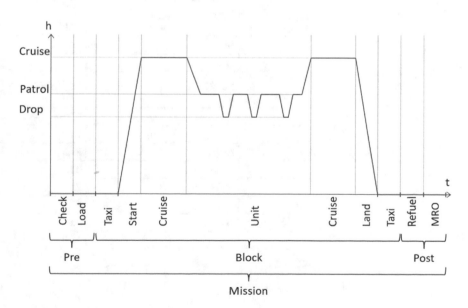

Fig. 13 Overview of the mission phases and their altitudes

of all described time elements of the three flight phases pre-flight, block time and post-flight. In sum the three phases are a mission.

Finally, one mission's duration is 109 min, including 67 min of block time, the latter being equivalent to the flying time of the aircraft. All mission elements are listed with its duration in Table 8.

The relief mission is assumed to take 10 days. After this period, the streets are traversable again, so the supply can get realized by land-based vehicles. Furthermore, there is a 10 h operating frame per day only as the operation has to be done in daylight. On the one hand, the people in need should be aware of incoming air vehicles, to pick up the dropped packages out of the water. Otherwise, the boxes could sink or float away. On the other hand, people should use the night time to sleep and recover. The loud engine noise disturbs the people on ground.

Regarding those limitations, one aircraft would need about 20 weeks to serve all units. To handle the supply in 10 days, there has to be a fleet of several aircraft (Table 9).

By using 14 aerial vehicles, all 750 units can be served with humanitarian goods within 10 days. More vehicles would reduce the relief time, so that some elements could be served twice, but the operation scale at the airport in Beira would increase significantly.

Table 8 Mission segmentation

Mission element			Min
t_{Pre}			25,0
	t_{Check}		10,0
	t_{Load}		15,0
t_{Block}			67,0
	t_{Taxi}		2,0
	t_{Start}		0,5
	t_{Cruise}		22,5
	t_{Unit}		17,0
		t_{Patrol}	7,0
		t_{Drop}	10,0
	t_{Cruise}		22,5
	t_{Land}		0,5
	t_{Taxi}		2,0
t_{Post}			17,0
	t_{Refuel}		2,0
	t_{MRO}		15,0
$t_{Mission}$			109,0

Table 9 Aircraft number versus overall task duration

n_{UCA}	t^*_{Relief} in min	t^*_{Relief} in h	t_{Relief} in d
1	81.750	1.362,50	136,25
5	16.350	272,50	27,25
10	8.175	136,25	13,63
13,625	6.000	100,00	10,00
20	4.088	68,13	6,81
50	1.635	27,25	2,73

3.3 Benchmarking the System

Performing 750 missions with 375 kg of packages on each flight, the relief mission supplies 281.25 metric tons of humanitarian goods. Distributed on vehicles, one aircraft delivers 20 t in 10 days. Every aircraft can do 5.5 missions a day with respect to the 10 h a day limitation. To evaluate the usage of the gyrocopter, it is compared with the transport helicopter Mil Mi-8MTV1, used by the WFP, shown in Table 10 (World Food Programme 2019).

The Mi-8 cargo volume does not consist of freight only, but also of passengers for search and rescue tasks. Both systems deliver almost the same amount of cargo. The big difference in both systems is the metric *Drops per Mission*. While the gyrocopter delivers at a wide area to supply the people in need, the helicopter can be used for one heavy payload. Spotted a located landing site, the helicopter can touch down here. Its payload can consist of bulky and heavy constructions, which cannot be provided by package drops. Equipment for shelter, tents, tools and other unmanageable goods can be unloaded safely. Furthermore, the helicopter can focus on passenger transport for Search and Rescue missions. During this, the gyrocopter provides individual spots with blankets, food, water and medicine if necessary. Its advantage of package drops gets significant when no landing is possible, even for STOL. That limited space that is not flooded is primary used by the people in need. To sum up, a combination of

Table 10 Overall task metrics

Metrics	Mil Mi-8 MTV1	Gyrocopter
Amount of aircraft $n_{Vehikel}$	5	14
Block time t_{Block} [h]	0.53	1.12
Missions per day and vehicle	5.26	5.5
Drops per Mission	1	25
Cargo per Mission [kg]	1055	375
Cargo per day and vehicle [kg/d]	5549	2062
Cargo per relief (10 days) [t]	277.47	281.25

both systems offers opportunities to optimize the cargo delivery for humanitarian help. In a next step, the operating costs of the gyrocopter have to be determined to evaluate its economic behavior.

3.4 Operating Cost Model

This cost model is similar to the operating cost model in Sect. 2.1. The acquisition costs for the gyrocopter are assumed to be 526,928 € (assuming a 100-aircraft production run). The operating costs have to be calculated separately for this use case. Therefore, different cost objects have to be identified.

1. General mission parameter
   The mission details are recorded in the general mission parameters. It consists of the average air speed for the whole mission, which is 146 km/h, and the average estimated distance of 163 km. Additionally, two minutes of taxiing are added to the block time of 1.18 h. By the annual use of the aircraft, the metric block hours per year are determined. Further details are explained later on.
2. Fixed direct operative costs

   a. Depreciation
      As already mentioned at the beginning of this section, the acquisition cost of 526,928 € are depreciated over ten years.
   b. Interest
      The calculated interest rate is set to 1.6 %, which is a reference to March 2017, same as in Sect. 2.1.
   c. Operator wage
      Since it is yet difficult to set the operator wage, it is set to 60,000 € per year for a monitoring operator.

3. Variable DOC (common)

   a. Maintenance
      There is no reason why the maintenance cost should be changed compared to the already assumed values of Sect. 2.1. The whole aircraft is maintained for 120 € per block hour.
   b. Insurance
      In the case of a contingency and a flight termination, the aircraft is lost. The amount of damage is equivalent to the acquisition costs. In addition, the mean time between failures is 10000 h.

4. Variable DOC (single)

   a. Fuel
      For humanitarian missions, a combustion engine is needed. The needed diesel cost 0.93 €/l in October in Mozambique (GlobalPetrolPrices.com 2020).

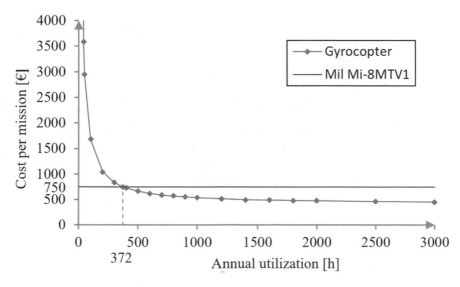

Fig. 14 Mission operating cost over annual utilization

    b.   Airport fees

           For humanitarian operations, the start and landing fees at the airport in
           Beira are assumed to be remitted.

    c.   Communication

           There is no way to communicate by terrestrial infrastructure, since it
           is destroyed by the cyclone. This is why the communication has to be
           performed by satellites, which is more expensive. According to a major
           satellite communications provider, the price for 1 MB of data is about 3 €.

After all inputs are identified and inserted to the cost model, the overall operating
cost for one transport mission are calculated as a function of the annual usage. This
relation is shown in Fig. 14.

Increasing the annual utilization of the aircraft has a significant impact on the
overall costs. Especially, the fixed direct operative costs have a major impact on the
costs for rarely used aircraft. With increasing numbers of missions a year, the trend
of the costs approaches an asymptote. The value of the mathematical asymptote is
the sum of the variable DOCs that accrue independent of the annual utilization.

Secondary, the red marker is set as a threshold, over which the costs for each
delivered ton of humanitarian goods is more expensive than the delivery by the Mil
Mi-8. Due to the flight schedule of the Mi-8 for an equivalent mission in spring
2019 its costs equal 750 €. To be more efficient economically, the gyrocopter has to
perform at least 333 missions a year, which can be expressed in 372 block hours a
year. Every cost element is visualized in the pie chart of Fig. 15. The constellation
is based on an annual usage of 372 block hours per year.

The biggest fraction is the loan for the operators. All fixed costs (depreciation,
interest and operator wage) will lose significance by increasing the annual utilization

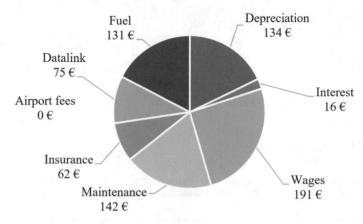

Fig. 15 Composition of the cost elements

of the gyrocopter. In the same way, the variable DOC will mainly shape the operating costs.

3.5 Summary

Finally, the usage and the capability of the gyrocopter are recapped. This type of vehicle allows a cargo distribution on a wide area since a fleet of several aircraft is deployed. Based on the configuration of the aircraft, autonomous operations are possible. The cruise and scan phase during a mission can be executed without human interaction. When packed with several packages including blankets, fresh water and some sort of high energy nutriments, the aircraft can serve numerous seperate groups of people in need. Due to the dimensions of the cargo hold, it is not recommended to load the aircraft with bulky equipment like big shelter tents. Instead of that, the Mil Mi-8 helicopter is more suitable for transportation of large and unwieldy cargo or even of rescued people.

Additionally, the aircraft's STOL ability takes advantage of the limited space at Beira Airport. After rainfall, the airfield cannot be used completely in general. On the other hand, there are no runways and airfields on which the aircraft can land and the cargo can be unloaded during a mission. The mission takes place in flooded areas where water covers streets and fields. At this point, there has to be an alternative option to deliver the packages. As already described, parcel dropping is a solution that is already used for freight delivery by the Mil Mi-8. Its disadvantage is the ground impact of the falling package. Regarding the risk of hitting or injuring a nearby person, the right spot to drop the package must be identified. In case of landing in water, the package will float away due to the current or sink to the ground where it can't be retrieved. Furthermore, no research is currently being carried out into a mechanism that enables the package drop on the aircraft itself. A new subsystem increases the complexity of the aircraft and thus the aircraft costs. Depending on its configuration, the time for maintenance, checks and loading has to be adjusted. Moreover, there is a difference how the people in need perceive the Mil Mi-8 and the gyrocopter, the latter being a largely unknown aircraft. So far, there is no research on the reaction and acceptance of the people in need confronted by an unmanned aircraft that drops a package.

The aircraft itself costs around 527,000 € for a production volume of 100. Depending on the utilization per year, the costs per hour vary. A workload of 373 h per year corresponds to six operations in the manner described above and means a total cost for a mission of 750 €. In this case, every ton of delivered freight costs 2000 €. That is equivalent to the delivery cost of the Mil Mi-8.

4 Conclusion

Two operating models for a transportation drone carrying one metric ton of cargo over 600 km of distance with a cruise speed of 200 km/h were assessed. Regarding the first, a flight network was designed for the transportation of agricultural spare parts from Germany to numerous European destinations. Flight routes were designed for each destination, taking into account topography, airports for refueling stops, and population density, assuming that cargo drones will preferably fly over sparsely populated areas for safety and disturbance reasons. Flight routing sometimes proved challenging due to a high density or an elongated arrangement of settlements, respectively. Yet, reasonable routes could be found for all destinations, requiring 20% of detour at most. A flight plan was drafted for serving all destinations daily. It showed that many destinations take more than one day to reach due to a safety requirement for night flights exclusively, which could impair the speed advantage over road transportation. Eventually, a model was presented for the aircraft's total operating cost. It showed that unit prices of up to 590,000 Euros can be expected and that the total cost would be a multiple of road transportation. In cases of urgent transportation needs, that cost could be outweighed by the time advantage.

A second use case was elaborated for humanitarian help. To provide supplies to people in need, a simultaneous usage of a conventional Mil Mi-8 helicopter and the gyrocopter enables efficient operations. While the Mil Mi-8 focuses on bulky payloads and on rescuing people, the gyrocopter fleet supplies large areas with package drops of small first aid kits. So far, there is no research on a suitable hardware configuration for a parcel release construction. To analyze the impact of that described system precisely, a more detailed elaboration of that component is necessary. Additionally, in terms of regulation and certification there are issues concerning the dropping. The regulation framework SORA doesn't allow that kind of operations above a gathering of people. However, for a successful provision of parcel delivery, the gyrocopter has to operate close to them.

References

Austro Engine (2020) AE300/AE330 Key Benefits. Wien. https://austroengine.at/uploads/pdf/mod_products1/AE300FactSheet.pdf. Accessed 25 Sept 2020

Aviation Week Network (2016) Operations Planning Guide. Business & Commercial Aviation, pp. 22–41ah

GlobalPetrolPrices.com (2020) Diesel prices. (G. P. Prices, Ed.) GlobalPetrolPrices.com. https://www.globalpetrolprices.com/diesel_prices/. Accessed 25 Sept 2020

Gudmundsson S (2013) General aviation aircraft design: applied methods and procedures. (Butterworth-Heinemann, Ed.) Butterworth-Heinemann

Hasan YJ, Sachs F (2021) Performance-based preliminary design and selection of aircraft configurations for unmanned cargo operations. In: C. Dauer J (ed) Automated low-altitude air delivery - towards autonomous cargo transportation with drones. Springer, Heidelberg

Hess RW (1987) Aircraft Airframe Cost Estimating Relationship: All Mission Types. The RAND Corporation, Santa Monica. www.rand.org/content/dam/rand/pubs/notes/2005/N2283.1.pdf. Accessed 25 Sept 2020

Israel M (2018) Suchstrategien, unser Verfahren. (F. Wildretter, Ed). https://fliegender-wildretter.de/unser-verfahren/. Accessed 25 Sept 2020

ITU (2010) Characteristics of unmanned aircraft systems and spectrum requirements to support their safe operation in non-segregated airspace (report ITU-R M.2171). International Telecommunication Union, Geneva

Lycoming (2005) O-360, HO-360, IO-360, AIO-360, HIO-360 & TIO-360 Operator's Manual. Document Part No. 60297-12 & 60297-12-5. Williamsport, PA, USA. https://www.lycoming.com/content/operator%27s-manual-O-360-HO-360-IO-360-AIO-360-HIO-360-TIO-360-60297-12. Accessed 25 Sept 2020

Pak H (2021) Use-cases for heavy lift unmanned cargo aircraft. In: Dauer JC (ed) Automated low-altitude air delivery - towards autonomous cargo transportation with drones. Springer, Heidelberg

Piper Aviation (2020) M500 Aircraft. Piper M500. https://www.piper.com/model/m500/. Accessed 20 Nov 2019

Quest Aircraft (2020) The Kodiak. (Daher, Ed). https://kodiak.aero/kodiak/. Accessed 25 Sept 2020

Reid K (2019) 2019 Cyclone Idai: Facts, FAQs, and how to help. (W. Vision, Ed.). https://www.worldvision.org/disaster-relief-news-stories/2019-cyclone-idai-facts. Accessed 25 Sept 2020

Textron Aviation (2020) Cessna Caravan. Cessna txtav. https://cessna.txtav.com/en/turboprop/caravan. Accessed 20 Nov 2019

Volga Dnepr (2015) We deliver humanitarian & peacekeeping missions. Volga Dnepr Airline, Stansted. https://www.volga-dnepr.com/files/brochure/WedeliverHumanitarian&Peaceeepingmissions.pdf. Accessed 20 Nov 2019

World Food Programme (2019) WFP - United Nations Humanitarian Air Service. Aircraft Utilization Summary. (N. A. Officer, Ed.) Johannesburg, South Africa

Part II: Unmanned Aircraft Configuration

Performance-Based Preliminary Design and Selection of Aircraft Configurations for Unmanned Cargo Operations

Yasim Julian Hasan and Falk Sachs

Abstract This chapter deals with the selection process of three different aircraft configurations that are appropriate for automated cargo operations at low altitudes and above sparsely populated areas. The selection process starts with a conceptual assessment of various aircraft types with respect to their suitability as an unmanned cargo aircraft (UCA) and subject to a set of specific top-level aircraft requirements (TLAR's). Based on the TLAR's and further characteristics like fuel efficiency, cargo space accessibility, technical simplicity or size, five fixed-wing aircraft and two rotorcraft are chosen for further investigations. A subsequent preliminary design study dimensions the fixed-wing aircraft for different wing spans based on flight performance requirements according to the TLAR's. A possible mean to enhance take-off and landing performances is the integration of additional small propellers and to make use of the propeller slipstream effect. Therefore, this chapter furthermore investigates the benefits from the integration of additional electrically driven propellers. The second part of the study focusses on the comparison of a helicopter and a gyrocopter, presenting performance key parameters and discussing their respective advantages and disadvantages. This chapter concludes with the selection of a twin boom aircraft of 16 m wing span, a box wing of 12 m span and a gyrocopter with a rotor radius of 7 m and small supplementary wings for the use as UCA.

Keywords Conceptual air vehicle assessment · Preliminary aircraft design · Flight performances · Electric propulsion · Unmanned aircraft system · Cargo aircraft · Conventional configuration · Canard · Twin boom · Biplane · Box wing · Helicopter · Gyrocopter

Y. J. Hasan (✉) · F. Sachs
DLR, German Aerospace Center, Institute of Flight Systems, Flight Dynamics and Simulation Department, Lilienthalplatz 7, 38108 Braunschweig, Germany
e-mail: yasim.hasan@dlr.de

F. Sachs
e-mail: falk.sachs@dlr.de

© Deutsches Zentrum für Luft- und Raumfahrt e. V. (DLR) 2022
J. C. Dauer (ed.), *Automated Low-Altitude Air Delivery*, Research Topics in Aerospace,
https://doi.org/10.1007/978-3-030-83144-8_5

1   Introduction

Within the project ALAADy (Automated Low-Altitude Air Delivery), the DLR (German Aerospace Center) deals with the development and investigation of an automated low-altitude air delivery concept using unmanned cargo aircraft (UCA). A major particularity of this concept is the limitation of the aircraft's operation envelope to altitudes below general air traffic and to air spaces only above loosely inhabited areas. This way and by realizing inherent safety characteristics and passive termination possibilities for the aircraft, the safety of other humans and objects both airborne and on ground are ensured.

One of the most fundamental aspects of the automated low-altitude air delivery concept is the choice and design of appropriate aircraft configurations. The autonomous transportation task poses a multitude of requirements to the unmanned cargo aircraft that are spread over a wide range of disciplines. Out of the variety of different air vehicle types, many vehicles can have advantages with respect to the transportation task. Therefore, no air vehicle type was to be excluded a priori. This chapter describes the general approach and the analyses performed to choose aircraft for the role as UCA within ALAADy. This aircraft selection is mainly based on flight performance analyses. In addition, other characteristics facilitating the transportation task like cargo space accessibility, technical simplicity, size or the need of supplementary infrastructure are also considered on a qualitative level.

However, ALAADy deals with a feasibility study of the complete automated air transportation system including the aircraft itself, ground segments, path planning and more. Therefore, the suitability of the aircraft with respect to the transportation task might not be defined by the aircraft alone. Instead, it depends on the interplay between the different project focuses to an equivalent extent. Hence, not only one but three different promising configurations are chosen to assess each aircraft's specific advantages and disadvantages from the complete system's perspective.

In the context of ALAADy, some TLAR's are formulated. These requirements are especially motivated by commercial viability of the complete air transportation system. They target on a rather small aircraft that can be fully operated with the least possible amount of infrastructure, e.g. for the transport between premises of different company sites. Hence, a high operational flexibility, short take-off and landing capabilities and a good accessibility of the aircraft's cargo space are necessary. Driven by these considerations, the aircraft requirements in ALAADy read:

- Payload of one metric ton
- Cargo space large enough to store two Euro-pallets
- Range of 600 km
- Cruise speed of 200 km/h
- Maximum take-off and landing distances of 400 m

In addition to these indispensable requirements, the aircraft selection should be made, respecting further aspects that support the general goal of commercial viability.

These comprise, among others, fuel consumption, complexity and compactness of the UCA, inherent safety properties and cargo space accessibility.

This chapter is subdivided into three parts. The first part examines a broad range of different air vehicle types, evaluates their suitability for use as the ALAADy UCA and makes a preselection about whether the air vehicle type should be included in the flight performance study or not. The second part deals with the design study of a certain set of fixed-wing aircraft and the third part analyses two promising rotorcraft configurations. Finally, three configurations are proposed for more detailed analyses within the further project.

This chapter is mainly based on the works of Hasan (2017) and on Hasan et al. (2018).

2 Preselection of Air Vehicle Types

As described before, the selection of an aircraft for the use as ALAADy UCA shall not be reduced a priori to certain aircraft types. However, in order to reduce the amount of workload, not all types are to be considered within the preliminary design process. Instead, this section makes a preselection of vehicles that are analyzed afterwards in more detail. For this purpose, the different aircraft types are assessed based on a set of criteria on a qualitative level that are in accordance with the project goal. The criteria are as follows:

- Short take-off and landing distances
- High aerodynamic efficiency in cruise flight
- Low structural weight
- Operational use without infrastructure extension
- Low pollutant and noise emissions
- Sufficient flight dynamic stability and low gust sensitivity
- Low complexity, easy accessibility and maintenance
- Inclusion of cargo area and easy stowing possibilities without additional equipment
- Compactness
- Inherent safety characteristics and possibility to realize flight termination

A multitude of different air vehicle types are analyzed during the preselection phase. This section describes the considerations made for those types that seem the most promising.

Fig. 1 Sketch of a
conventional tube-and-wing
aircraft configuration (Hasan
et al. 2018)

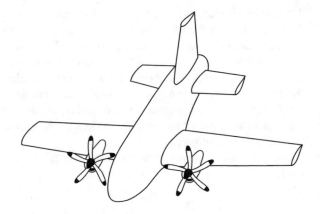

2.1 Tube-and-Wing Aircraft

Tube-and-wing aircraft, as the name suggests, consist mainly of tube-shaped fuse-lages containing the payload and wings, generating lift dynamically. The arrange-ment and number of fuselages and wings can differ among the various configurations. Figure 1 shows a sketch of a conventional tube-and-wing aircraft configuration.

The use of these aircraft for the automated transportation task of ALAADy would bring several benefits. Tube-and-wing aircraft are well-proven in practice for decades. Hence, there is a lot of experience available on how to design such aircraft with respect to adequate manoeuvrability and stability. Depending on the particular config-urations, the complexity and the general production costs can be low. In addition, aircraft generating lift dynamically have comparatively high aerodynamic efficiency compared to those vehicles with static lift like e.g. airships, especially at higher velocities. On the other hand, these aircraft have a rather poor take-off and landing performance, requiring longer runways. Furthermore, tube-and-wing aircraft offer only limited possibilities to seal-off the engines from the ground, leading to rela-tively high noise emissions. However, due to their success in aviation and their high aerodynamic efficiency, some different tube-and-wing configurations are included within this design study.

2.2 Blended Wing Body Aircraft

A blended wing body (BWB) aircraft, as shown in Fig. 2, is a fixed-wing aircraft type that intends to further increase aerodynamic efficiency compared to tube-and-wing aircraft by the use of a flat airfoil-shaped fuselage. This way, the fuselage itself generates lift at low angles of attack and the wing size can be reduced. Consequently, parasite drag can be reduced. BWB aircraft usually have no tail plane. Longitudinal stability can e.g. be realized by an S-shaped airfoil profile or high wing sweep, while

Fig. 2 Sketch of a blended wing body aircraft configuration (Hasan 2017)

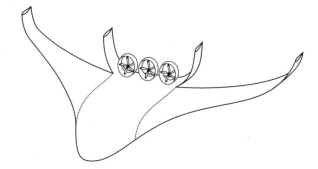

the wing tips generate a downward force. This, in turn has a degrading effect on aerodynamic efficiency again. However, their efficiency is still superior to comparative tube-and-wing aircraft.

In addition, the large platform allows an engine installation on top, which would reduce noise emissions. The more uniform lift distribution across the fuselage section and the resulting reduction of the wing root bending moment allows for smaller wing structural weight. In addition, the known issue of cabin pressurization for BWB (Mukhopadhyay 2005), leading in turn to a weight penalty, does not apply to the ALAADy UCA since it is supposed to operate at very low altitudes. In this case, the structural weight of the BWB could thus be lower than the weight of a tube-and-wing UCA. Nevertheless, the BWB aircraft has some major drawbacks regarding the take-off and landing performance. Due to the absence of the horizontal tail plane, additional pitching moments due to flap and slat deflections can hardly be compensated (Liebeck 2004). As a consequence, from the strong limitation of the use of a high-lift system, it is very likely that the BWB aircraft cannot meet the runway requirements of ALAADy. This configuration is thus excluded from further investigations.

2.3 Rotorcraft

Rotorcraft are air vehicles that generate lift using one or more rotorblades. Those rotorblades can either be powered by a motor or brought to rotation by the incoming airflow. Figure 3 exemplarily shows the sketch a helicopter, as one prominent representative of the class of rotorcraft.

With regard to the requirements formulated in ALAADy, the predominant advantage of rotorcraft is their take-off and landing performance. While rotorcraft with motor-powered rotors have vertical flight capabilities, those vehicles with rotors that are unpowered need a runway. However, these runways are rather short compared to those needed for fixed-wing aircraft. In addition, the autorotation abilities of the rotor bring benefits in terms of safety management. In contrast, the cruise performances of rotorcraft are inferior to those of fixed-wing aircraft. The relatively high rotor drag

Fig. 3 Sketch of a
helicopter configuration

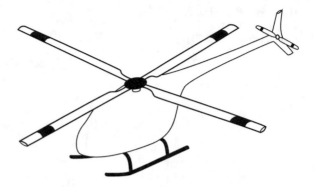

Fig. 4 Sketch of a tiltrotor
aircraft configuration (Hasan
2017)

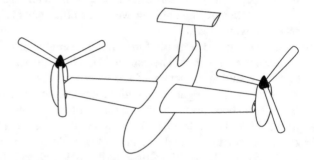

leads to higher fuel consumption and the maximum speed is limited by local subsonic effects at the rotor tips. Nevertheless, different rotorcraft types are investigated and included within the study since their vertical flight and the autorotation capabilities would be of a major advantage for the automated transportation task, especially with respect to the risks on ground.

2.4 Tiltrotor Aircraft

The tiltrotor aircraft, as shown exemplarily by Fig. 4, is a fixed-wing aircraft that is equipped with tilting rotors on the wing tips. For take-off and landing, the rotors and their thrust vectors are oriented upwards, generating lift and allowing vertical flight. During cruise flight, the rotors are tilted in a horizontal direction and generate thrust. In this case, lift is generated by the wings.

It combines the advantages of fixed-wing and rotary wing aircraft, enabling vertical take-off and landing while maintaining a high aerodynamic efficiency. However, the tilting mechanism of the rotors requires a high mechanical complexity, which is also associated with high maintenance costs and mass. In addition, the integration of the heavy rotors at the wing tip leads to a high wing root bending moment

Fig. 5 Sketch of an airship
(Hasan 2017)

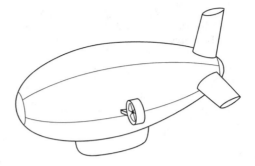

and augments flutter characteristics. Therefore, a comparatively high wing structural mass is required (Acree 2006). The high complexity as well as the expectedly high structural weight of the tiltrotor aircraft is not in accordance with the ALAADy transportation concept. For this reason, the tiltrotor aircraft excluded from the further study.

2.5 Airships

Airships air vehicles of the type lighter-than-air. Their voluminous bodies are filled with a lifting gas that generates aerostatic lift, while propellers are used to generate thrust. Figure 5 shows an exemplary sketch of an airship. Airships have some characteristics that would be advantageous for the use as a UCA for the transportation task. These characteristics comprise a good capability to provide a cargo area and its accessibility for loading and unloading.

On the other hand, airships have a very large wetted surface and thus suffer from high parasite drag at moderate airspeeds. Typical airships have vertical flight capabilities that would, in principle, be of benefit for ALAADy. However, the operation of an airship would require some kind of device to strap the airship after landing. Such a device would require an extension of the local infrastructure, which is undesirable. The airship is excluded from further investigations for the previously mentioned reason but more due to the ALAADy cruise speed requirement. Such a relatively high cruise speed cannot be achieved by an airship while respecting aspects of fuel consumption and commercial viability.

2.6 Others

A large number of other air vehicle types like textile-based vehicles, e.g. paragliders, as shown in Fig. 6, inflating vehicles or lifting body aircraft were assessed. However, none of these vehicles proved suitable for ALAADy.

Fig. 6 Sketch of a
paraglider with payload and
propulsion nacelle

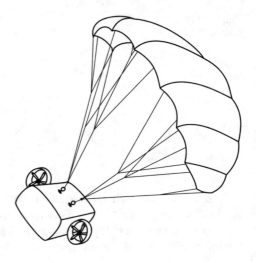

3 Performance Based Preliminary Design Study of Fixed-Wing Aircraft

The preselection process suggests that both fixed-wing aircraft of tube-and-wing type and rotorcraft have the potential to meet the TLAR's of ALAADy. For both types, the requirements that are most restrictive are different ones. In addition, vehicle modelling, performance computation and the design approaches differ clearly. Therefore, this work deals with two separate studies, being one for the fixed-wing aircraft and one for the rotorcraft.

This section deals with the first study, the performance based preliminary design study of fixed-wing aircraft. It starts with a description with those fixed-wing aircraft configurations that are included in the study. Subsequently, it demonstrates the flight performance computation process and compares the aircraft with each other for different wing spans.

A possible mean to enhance take-off and landing performances is the integration of additional small propellers and to make use of the propeller slipstream effect. Therefore, in a second step, the study is repeated with four additional electrically driven propellers installed for each configuration.

3.1 Considered Aircraft Configurations

Altogether, this fixed-wing aircraft study examines five different aircraft configurations. These are a conventional configuration, a canard configuration, a twin boom configuration, a biplane configuration and a box wing configuration. The following sections describe these aircraft and explain their respective advantages for ALAADy.

3.1.1   Conventional Configuration

The conventional aircraft configuration with forward wings and a conventional aft tail, as shown in Fig. 1 is included. It is the predominant aircraft configuration in general aviation and there is extensive experience available with this aircraft type.

3.1.2   Canard Configuration

The second tube-and-wing type aircraft is the canard configuration. Figure 7 shows a sketch of this aircraft. The canard configuration has the potential of having a higher aerodynamic efficiency thanks to the arrangement of the horizontal stabilizer in a forward position. This forward stabilizer is called the canard. While a typical aft horizontal tail plane generates a downward force during most flight phases in order to realize longitudinal trim, the canard generates lift. This way, the main wing needs to generate a lower amount of lift to compensate the aircraft weight compared to the conventional aft tail. In sum, the resulting trim drag can be lower.

However, this wing-tail plane arrangement also has drawbacks. Since the main wing operates in the wake of the canard, its efficiency is reduced and mutually induced drag is caused due to the interaction of both lift-generating surfaces. In addition, to achieve adequate stall characteristics, the canard needs to stall first. One way to achieve this characteristic is to design the canard with a relatively high span such that it has a higher lift curve slope than the main wing. As consequence from the canard stalling first, the main wing does not reach its maximum lift potential and the maximum lift coefficient of the aircraft is somewhat lower than for an aft tail plane configuration.

In this study, the following canard specific properties are respected:

- Mutually induced drag between canard and wing
- Reduced maximum lift coefficient
- Higher canard span compared to the aft tail planes of the other configurations

Fig. 7 Sketch of a canard configuration as considered in the study (Hasan et al. 2018)

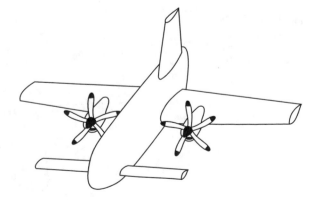

More detail on the canard modelling within this study can be found in the works of Hasan (2017) and of Hasan et al. (2018).

3.1.3   Twin Boom Configuration

The third tube-and-wing aircraft regarded is the twin boom aircraft. Such a configuration is shown in Fig. 8. This configuration is characterized by a short fuselage and a V-tail plane, which is mounted on two booms which replace the conventional rear part of the fuselage. The twin boom aircraft has a particular advantage for unmanned aircraft. Generally speaking, the fuselage of unmanned aircraft needs a lower capacity than for manned flight. However, for aircraft without booms, a minimum fuselage length is required to provide a sufficient tail lever arm regardless of the required fuselage capacity. In case of the twin-boom aircraft, this lever arm can be adjusted independently by the boom length and the fuselage size is consequently mainly driven by the need for payload space.

Modelling the twin boom aircraft, the following characteristics are respected:

- Shorter fuselage
- Weight and drag for booms
- No nacelle drag due to pusher propeller configuration behind the fuselage

The induced drag, masses and parasite drag of the V-tail plane are approximated by consideration of equivalent horizontal and vertical tail planes with the respective projected surfaces.

3.1.4   Biplane Configuration

The biplane configuration, as shown in Fig. 9, is characterized by two main wings instead of one, which are arranged in a stacked way. In the early days of aviation biplane aircraft were the most predominant air vehicle. This was essentially due to

Fig. 8 Sketch of a twin boom configuration as considered in the study (Hasan et al. 2018)

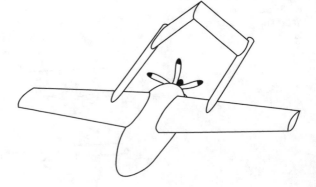

Fig. 9 Sketch of a biplane configuration as considered in this study (Hasan et al. 2018)

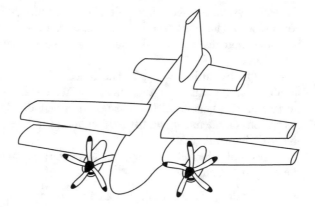

the fact that the aircraft structure technology at that time did not allow the use of wings with sufficiently high spans to produce enough lift to carry the given weight while keeping induced drag low.

The concept of biplane theory was studied by Prandtl (1924). The two wings of the biplane share the total lift, such that, ideally, each wing generates half of the lift. If both wings are completely isolated, this results in an induced drag which is one quarter at each wing and, in sum, half for the total wing system compared to an equivalent monoplane. In practice, the two wings cannot be isolated, such that there is a mutual interference between the two wings. This interference is the stronger the smaller the vertical distance between the wings gets and the larger the wing span is. As long as a certain vertical distance can be provided, the biplane thus has lower induced drag compared to an equivalent monoplane with the same wing span.

On the other hand, parasite drag increases due to the smaller chord lengths and the double amount of wing-fuselage intersections. In addition, the wing concept requires a higher weight. Nevertheless, such a nonplanar wing concept is of interest, because it allows the design of an aircraft with low wing span, while maintaining a reasonably good aerodynamic efficiency.

Modelling the biplane in this study, the following aspects are taken into account:

- Induced drag of the complete biplane wing system
- Additional wing-fuselage interference drag
- Parasite drag and weight for each wing part

The wings are arranged as low wing and high wing, such that vertical distance between them is always as large as possible.

3.1.5 Box Wing Configuration

The last tube-and-wing aircraft considered in this study is the box wing aircraft configuration. It expands the biplane concept by the addition of vertical fins that connect low and high wing. Figure 10 shows a sketch of this aircraft type. Closing the wing system has the potential to further reduce induced drag as described by Kroo (2005). However, compared to the biplane, parasite drag of the box wing is increased more due to the additional edges of the vertical fins and the supplementary wetted area.

If the high and low wings are designed with a sweep, such that the resulting aircraft neutral point lies sufficiently aft of the center of gravity, no supplementary horizontal tail plane is needed to establish longitudinal trim. In this case, the same restrictions on the maximum lift coefficient as for the canard configuration apply to the box wing configuration.

In this study, the following specific characteristics are respected for modelling of the box wing configuration:

- Induced drag of the complete box wing system
- Additional interference drag for the wing-fuselage junctions and the wing-vertical fin junctions
- Parasite drag and weight for each wing part and the vertical fins
- Reduced maximum lift coefficient

The specific design of the box wing permits an installation of the aft wing on top of the vertical tail plane, such that the average vertical distance between both wings is maximized. For structural reasons and to reduce the wetted surface of the vertical fins, the wings are designed with positive dihedral of the lower and negative dihedral of the upper wing.

3.2   Performance Calculation and Preliminary Aircraft Design

The aim of this section is to describe the design process of appropriate aircraft configurations for different wing spans, such that all ALAADy requirements on flight performances are met. The fuselage dimensions are preset for all aircraft a

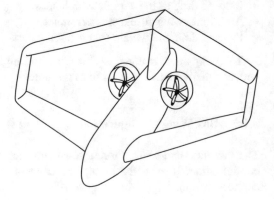

Fig. 10 Sketch of a box wing configuration as considered in this study (Hasan et al. 2018)

priori on the basis of static stability aspects and required cargo space. The fuselage dimensions are set the following way:

- The fuselage length for the twin boom configurations is set to 5 m, which is solely driven by the need for cargo space. In order to account for stability, the tail plane lever arms can be adjusted by means of the booms. The fuselage length for all other aircraft configurations is set to 7 m. It is initially assumed that this length provides a sufficiently large lever arm of the tail plane.
- The fuselage width is set to 1.8 m.
- The fuselage height is also set to 1.8 m.

The following study determines aircraft parameters and flight performance characteristics of all five aircraft using a broad range of preliminary design equations. For this purpose, mass estimations according to Raymer (1989) were used and drag is considered using both approximations of Torenbeek (1982) and experimental values from Hoerner (1965). Effects of high-lift devices were considered based on equations of Roskam (1987) and propeller aerodynamics were modeled using approximations of Solies (1993). Box wing-related modeling was based on Schiktanz and Scholz (2011). However, this chapter rather focusses on the qualitative and methodological aspects of the study and therefore does not provide equations or modelling details. For more information on this study the reader is referred to the works of Hasan (2017) and of Hasan et al. (2018).

Strategy

Figure 11 shows the general strategy pursued to design configurations for the different fixed-wing aircraft types. The process is iterative and requires as input one of the five aircraft types described in the Sects. 3.1.1–3.1.5 and a wing span. For this input, the process designs and sizes the aircraft such that the ALAADy TLAR's described before are met and provides the aircraft as an output.

The complete design process as depicted in Fig. 11 was implemented in a tool. The calculations start with a review of the landing performance. Assuming that the aircraft used up most of its fuel, it calculates a first estimate of the maximum allowed wing loading during landing using flaps, such that the aircraft's landing distance is well below the 400 m required in ALAADy. Calculating the landing distance, it is assumed that the aircraft has a constant glide angle of 7.5° during landing. This relatively high glide angle is chosen to reduce the landing distance and to facilitate a landing in an area, which is predominantly covered with buildings, e.g. premises. Together with an initial fuel mass, an average wing loading for cruise flight operation can be calculated from the wing loading during landing.

Subsequently, the tool calculates mass and aerodynamic parameters. First, the wing loading yields a wing area and, with the required cruise speed of 200 km/h, an average lift coefficient. The following estimation of the drag coefficient includes, depending on the respective configuration, components of the following list:

- Induced drag
- Wing-canard mutually induced drag

- Fuselage form and skin-friction drag
- Wing form and skin-friction drag
- Engine form and skin-friction parasite drag
- Tail plane form, skin-friction and interference drag
- Boom form and skin-friction drag
- Wing-fuselage interference drag
- Wing-vertical fins interference drag
- Wing-nacelle interference drag
- Fuselage-tail plane interference drag

Having estimated drag, the tool calculates the required engine power and the associated engine mass. The total mass calculation takes into account all of the following masses, if applicable for the different aircraft types:

- Payload of 1 t according to the requirements
- Equipment as actuators, instruments, electronics, etc.
- Fuel mass and 20% of reserve fuel mass
- Fuselage mass
- Engine mass
- Wing mass and masses of vertical fins and junctions
- Tail plane mass
- Boom mass
- Gear mass
- Mass of 50 kg that is reserved for the integration of a termination system

In the next step, the tool regards cruise performance and dimensions the fuel mass such that the aircraft reaches the demanded range of 600 km and maintains a reserve of 20%. This change in fuel mass induces mass and wing parameter changes, which in turn require an update of cruise flight calculations. These steps thus form an inner loop.

Afterwards, the take-off performance is checked. If the aircraft's take-off distance exceeds the allowed 400 m, additional power is provided. As a consequence, the aerodynamics and mass estimation step is redone until the calculation converges.

Finally, the updated wing loading is compared with the initial wing loading and the process is repeated until there are no more significant changes along with the iteration steps.

Results

Figure 12 shows the resulting configurations for the different fixed-wing aircraft, calculated for wing spans from 6 to 24 m. It only shows the resulting configurations for which a solution could be found. This signifies that, for example, no configuration with a wing span of 6 m could be found that fulfills the ALAADy flight performance TLAR's.

The first part of Fig. 12 shows the drag coefficient in cruise flight. As expected, it decreases along with the growing wing span for all configurations due to the lower induced drag. At low wing spans, induced drag reduction capabilities of the nonplanar wing systems of biplane and box wing are observable. Their drag coefficients are characterized by a flatter curve and especially the box wing has the lowest drag coefficient overall.

The second part of Fig. 12 shows wing loading, which is predominantly restricted by the maximum landing distance. It increases slightly along with the wing span since the lift increment due to flap deflection increases with a higher lift curve slope. The lower maximum lift coefficient for the canard and the box wing are responsible for their respective overall lower wing loading curves.

The third part of Fig. 12 shows the resulting wing area distributions. It is, per definition, correlated with wing loading and aircraft mass. Both the low wing loading at low wing spans and the higher masses at larger wing spans lead to an augmentation of the wing areas, such that the wing area is minimal around medium wing spans.

The fourth part shows the maximum take-off mass. It generally increases along with the wing span due to the higher wing masses. At low wing spans, the comparatively poor high lift characteristics lead to high induced drag. As a consequence, the fulfillment of the take-off distance requirement makes the supply with additional power necessary, which also leads to a mass increase. Masses are larger for biplane and box wing, which is mainly due to the comparably higher wing masses of the nonplanar systems. This difference in masses increases along with the wing span, where the wing mass becomes more predominant. It should be noted that the twin boom aircraft's mass is lowest for all wing spans.

The last part of Fig. 12 shows the fuel masses, including reserves, required for the range of 600 km. At high wing spans, the twin boom aircraft shows the least fuel needed, which is likely to be caused by the lower overall mass and the good drag characteristics at high wing spans. The nonplanar wing system aircraft are at a disadvantage at high wing spans. However, at very low wing spans, the box wing configuration shows some fuel saving potential in relation to the other aircraft.

As shown in these results, all aircraft configurations have a relatively low wing loading for aircraft of the given size, which is driven by the stringent requirement on landing distance. During cruise flight, however, such a low wing loading is rather disadvantageous.

Integration of Supplementary Electrically Driven Propellers
Hence, in order to improve the cruise performances, wing loading needs to be increased while achieving the required landing distance. This can be achieved by an augmentation of the maximum lift coefficient beyond the use of the high-lift system.

For this purpose, this work investigates the suitability of the distributed electric propulsion (DEP) concept as introduced by Stoll et al. (2014a) and Stoll et al. (2014b) for the UCA. This concept addresses the span wise integration of small electrically driven propellers along the wing leading edge, such that the wing sections behind the propellers operate in the propeller slipstream effects. When all propellers are active, very high lift coefficients can be reached.

Due to the high battery weight and capacity issues, a purely electric UCA that is capable to fulfill all TLAR's is not feasible for ALAADy. Therefore, in this study, four electrically driven propellers are added to the configuration, which are only active during take-off and landing, while during cruise flight, the UCA still operates using fossil fuel. Figure 13 shows this concept using the example of the twin boom aircraft.

The increase in the lift coefficient and the corresponding drag coefficient depends on the amount of provided propulsion power. Therefore, the aircraft design process described in Fig. 11 is expanded by a step that dimensions the electric power in such a way that the constant glide angle of 7.5° is reached during landing. More information on the underlying calculations can be found in Hasan (2017) and Hasan et al. (2018).

Figure 14 shows the extended configuration definition approach that includes the electric propulsion. It starts with the afore-mentioned step that dimensions the power supply for the additional electric engines. In the next step, the tool approximates the masses for the electric engine system on the basis of this resulting amount of power supply. The minimum battery mass is calculated by respecting the overall used electric energy, which is estimated by considering the duration of use of the electric engines. Assuming that take-off and landing durations are similar, the duration of use of the electric engines is approximately twice the take-off duration time. Furthermore, the drag increment due to the electric engine nacelles is calculated.

Subsequently, the tool performs the same steps as previously described in Fig. 12. However, different from before, the additional power in case of non-compliance with the take-off distance requirement is now provided by the electric wing-mounted engines rather than by the fossil fuel driven ones.

Results

Figure 15 shows the configuration results with electric engines for the electric propulsion power, wing loading, wing area, the maximum take-off mass and fuel mass. As

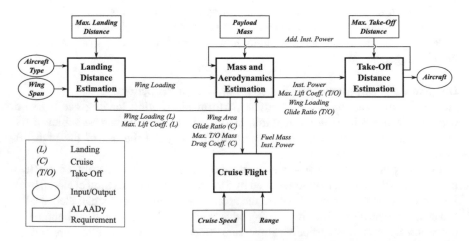

Fig. 11 General approach for the definition of configurations with different wing spans under consideration of performance calculations and the ALAADy requirements, based on Hasan et al. (2018)

Fig. 12 Resulting aircraft configurations for cruise drag coefficient, wing loading, wing area, maximum take-off mass and fuel mass for different wing spans (Hasan et al. 2018)

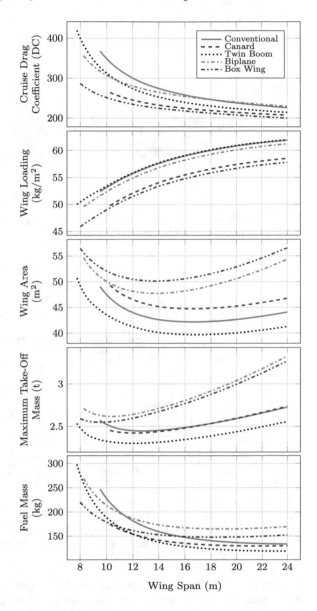

can be seen immediately, there are fewer solutions at low wing spans that meet the ALAADy requirements. This is due to the fact that at low wing spans, a large amount of additional electric power (see first part of Fig. 15) is required. The high electric propeller power leads to a high maximum lift coefficient, which in turn causes a very high induced drag, particularly at low wing spans.

The second part shows wing loading, which, unlike in the previous results, decrease along with the wing span. This is due to the fact that the benefit from

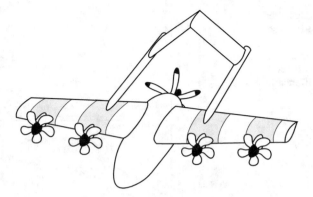

Fig. 13 Twin boom aircraft with four additional electrically driven propellers and grey sections representing the wing surfaces affected by the propeller slipstream, based on (Hasan et al. 2018)

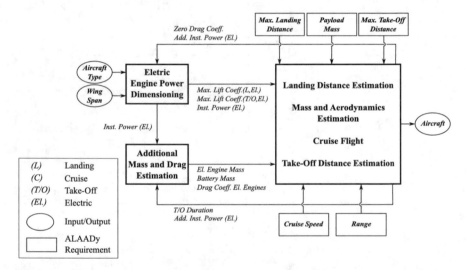

Fig. 14 Configuration definition approach extended by a dimensioning of electric engines and the consideration of their associated mass and drag contribution, based on Hasan (2017)

the propeller slipstream effect depends on the amount of provided electric power, which drops for higher wing spans. Nevertheless, wing loading is generally higher than without electric engines. The third part of Fig. 15 shows the wing area. As desired, the curves are generally lower than without electric engines. The fourth part shows the maximum take-off mass. In the respective wing span ranges, the curves are basically similar to the case without electric engines.

The resulting aircraft, however, generally perform better with respect to fuel consumption, as shown in the lowest part of Fig. 15. The twin boom aircraft has the lowest fuel consumption at high and moderate wing spans. In comparison, the

Fig. 15 Resulting aircraft configurations with additional electric propulsion for electric power, wing loading, wing area, maximum take-off mass and fuel mass for different wing spans (Hasan et al. 2018)

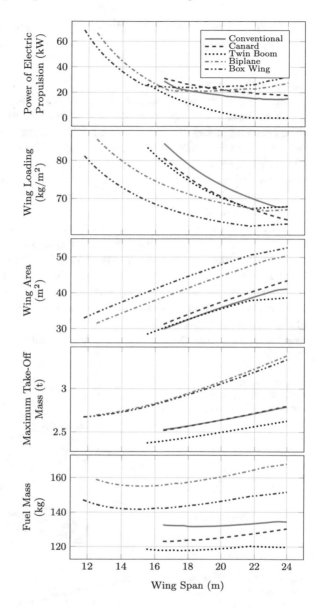

box wing has a higher need for fuel mass, but it can be operated with a clearly lower wing span.

4   Performance Based Preliminary Design Study of Rotorcraft

The preselection process described at the beginning of this chapter pointed out, that amongst all possible flying vehicles those that would be chosen for ALAADy would either be fixed-wing aircraft or rotorcraft or compounds. Since rotorcraft produce lift by rotating blades they are able to take-off and land without forward movement. As a result, their according take-off and landing distances can be reduced to zero. This is true for helicopters (Fig. 16), but vehicles employing auto-rotating rotors can also have impressing slow flight capabilities, which enables them to perform starts and landings within significantly shorter distances compared to fixed-wing aircraft of the same weight and size.

Since the top level aircraft requirements demand good take-off and landing capabilities, the helicopter and the gyrocopter (Fig. 17) seem to be suitable candidates for the transportation task in ALAADy and are therefore studied in more detail. Tilt rotor and convertiplanes could, in principle, also be in the scope, but their technical complexity and the additional weight of the necessary mechanics are some of the reasons to exclude them from further investigations. Multicopters do also provide the ability to take-off vertically, but concerning the rotor size several effects are working aversively. To ensure good hover performance and the ability to perform a safety

Fig. 16 Sketch of a helicopter configuration as considered in this study (Hasan et al. 2018)

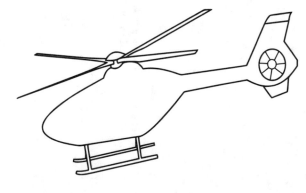

Fig. 17 Sketch of a possible gyrocopter configuration with tractor propellers (Hasan et al. 2018)

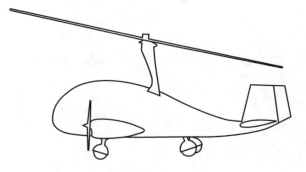

landing in autorotation large rotor blades should be applied, but the larger the rotor the more relevant becomes the parasitic drag caused by the framework that carries the rotors. The packing fraction is always lower than the one of a single rotor rotorcraft of the same size which is unfavorable in terms of power efficiency from a momentum theory point of view (Strickert 2016).

Considering the rotorcraft study, it can thus be assumed beforehand that the two vehicles helicopter and gyrocopter are capable to meet the take-off and landing requirements. The comparison thus focusses on the vehicles' respective performances and their technical complexity.

So, simple models of a helicopter and a gyrocopter are set up and used for performance calculations. The lift is exclusively provided by the rotor. For both rotorcraft mass estimations are carried out. In the case of overall mass, the gyrocopter benefits from its low technical complexity. The complexity of the rotor head is reduced. Furthermore, the weight for the complex gearbox and the tail rotor can be saved. A minor portion of weight is added by the noncomplex pre-rotation mechanism. Overall the takeoff weight of the helicopter is 5% higher than the compared gyrocopter's take-off weight.

In Fig. 18 the helicopter reference case is compared to two gyrocopter designs. The gyrocopter is powered by two propellers with a radius of 1 m. The first gyrocopter design is equipped with a rotor of the same size as the helicopter rotor. The fixed rotor blade pitch angle is set to a value that is necessary to realize a comparable rotor

Fig. 18 Rotor RPM, glide ratio and cruise power of a helicopter and a gyrocopter with same rotor radius (5.1 m) and same rotor solidity (0.05) and a second gyrocopter with higher rotor radius (7.0 m) but lower rotor solidity (0.03) (Hasan et al. 2018)

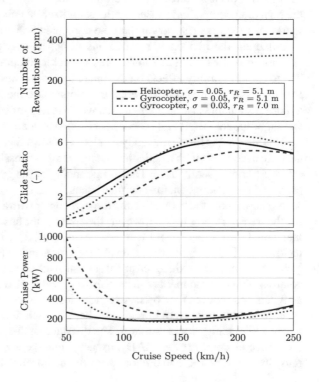

rotational rate. As a consequence, next to the rotor solidity and the rotor radius also the rotational rate of the rotor is the same for both vehicles at low airspeeds. Just the Gyrocopter's rotor is auto rotating the rotational rate grows slightly with increasing airspeed in contrast to the helicopter case.

Figure 18 shows clearly that a gyrocopter with a comparable rotor size shows reduced performance compared to the helicopter in this example, especially at low airspeeds. At higher airspeeds, especially above the cruise speed of 200 km/h as formulated in the ALAADy requirements, the performance disadvantage is nearly disappearing. However, the rotor concept demands for enormous power in the start and acceleration phase at low flight speeds. Operation at those speeds wouldn't be possible or a massive overpowering has to be taken into account. The related high costs and mass penalties are not acceptable with respect to the concept in ALAADy. Furthermore, the gyrocopters often highlighted advantage of low operation capabilities seems not to be given in this case.

Figure 18 also displays the results for a second gyrocopter concept. For this gyrocopter the rotor radius is increased from 5 to 7 m such that the disk loading is reduced to the half. The rotor solidity is reduced to 0.03 and the blade pitch angle set to 1.5°. The solidity represents the ratio of the rotor blade surface to the overall rotor area. The results show that, thanks to the lower inclination of the rotor, overall drag can be reduced significantly.

The gyrocopter also benefits from the slightly lower weight compared to the helicopter. This is an effect that couldn't be found with the smaller rotor because of the high drag component caused by the rotor itself. At airspeeds above 125 km/h the gyrocopter requires less power than the compared helicopter.

Enlarging the helicopter rotor radius is not as easy as in the gyrocopter case because to realize a larger rotor radius the rotor rotational rate has to be reduced due to compressibility effects on the blade tips in forward flight. A reduction of the rotational rate of the rotor increases the moment on the rotor shaft, which requires a more sustainable and heavier gear as well as a bigger tail rotor. That increases the weight of the rotorcraft and reduces performance again.

This simple performance analysis already shows that an appropriate gyrocopter design can compete with a helicopter design if all the possible advantages the concept offers are consequently used. The compounding of the gyrocopter with a fixed wing promised even more benefits at cruise speed.

As shown, both the helicopter and the gyrocopter with a larger rotor appear suitable for the transportation task. However, the gyrocopter has two additional advantages that would be of great benefit with respect to ALAADy. These advantages comprise its low technical complexity and its ability to constantly fly in autorotation. The low technical complexity reduces costs for several gearboxes, the tail rotor or special flight controls. Those components can be removed, which also facilitates the maintenance activities and reduces the costs. A reduced technical complexity is also desirable in order to be able to operate in regions where no maintenance might be possible and where a high reliability is required. The constant autorotation flight is advantageous for the overall concept of ALAADy. In case of emergency, a termination of the vehicle must always be possible in order to prevent uncontrolled behavior of the aircraft and to avoid harm to other air traffic participants as well as people on the ground. The termination shall be working passively. For such a maneuver the low complexity of the auto-rotating rotor can also be seen as an advantage. The helicopter's rotor needs to be controlled to change it's working state from driven by the engine to an autorotation scenario for an emergency landing. The time slot to perform such a change is usually very tight. Such a system is not necessary for the gyrocopter case which facilitates the design of the emergency maneuver and the corresponding control mechanism.

5 Air Vehicle Selection

Based on the preceding analyses, three air vehicle configurations are to be chosen for further investigations in ALAADy. Within the first study, the aircraft equipped with additional electric engines generally show better performances. The electric version of the twin boom aircraft has the highest fuel saving potential. Therefore, and for reasons of compactness, the electric twin boom aircraft with 16 m wing span is selected.

As a second aircraft, the electrified box wing is chosen with a wing span of 12 m. Indeed, it requires more fuel than the conventional and the canard aircraft, but it can be built with a considerably smaller wing span and still has a reasonable aerodynamic efficiency. The compactness of the aircraft facilitates ground handling activities as well as it reduces the space to park and store the aircraft. Different wing spans were not explicitly required for a selection of aircraft, but regarding different aircraft with different spans gives the possibility to change the focus on issues like e.g. ground handling.

Regarding the rotorcraft, the gyrocopter is chosen. This choice is made both due to its simplicity, its autorotation capabilities and the associated advantages concerning flight termination and safety. The second study shows that flight performances of the gyrocopter can be enhanced using a larger rotor radius. In addition, small supplementary wings are integrated into the gyrocopter to further increase its cruise performance.

Summarized, the three configurations, proposed are

- the twin boom aircraft (Fig. 8) with a wing span of 16 m and additional electric engines,
- the box wing aircraft (Fig. 10) with a wing span of 12 m and additional electric engines
- and the gyrocopter (Fig. 17) with a rotor diameter of 14 m and supplementary wings with a span of 8 m.

It should be noted that these aircraft form the best compromises with respect to the particular requirements in ALAADy. Nevertheless, all configurations investigated in the previous studies proved to be useable for the task.

6 Conclusion

This chapter focuses on one of the fundamental steps within the project ALAADy. It is the choice and initial design of three appropriate air vehicle configurations, respecting a broad range of fields and with a special focus on flight performances. Since the automated transportation task in the project is rather unique in terms of its requirements and the dimensions of the underlying vehicles, it is not useful to narrow down the pool of aircraft types a priori. Instead, a first preselection process is performed, that evaluates a great variety of different air vehicle types based on a multitude of qualitative considerations. As a result, tube-and-wing aircraft as well as rotorcraft are proposed for the subsequent design studies.

Next, a preliminary design study is performed for the tube-and-wing aircraft, respecting conventional, canard, twin boom, biplane and box wing aircraft. Each aircraft is designed for different wing spans and such that all ALAADy requirements are met. The results show that the requirement on landing distance is the most stringent and leads to relatively low wing loading that tends to degrade cruise flight performance. As a consequence, an extension of the preliminary design study equips the aircraft with additional electric engines that aim to improve high-lift performances by an augmentation of the maximum lift coefficient. The resulting configurations have a higher wing loading and a superior aerodynamic efficiency. Finally, a twin boom aircraft with a moderate wing span and a box wing with low wing span, both equipped with additional electric engines, are chosen for the automated transportation task.

The second study compares a helicopter and a gyrocopter, presenting key parameters and discussing their respective advantages for the given task. The gyrocopter, extended by a small wing, is selected for the further project.

The flight performance investigations presented here make use of preliminary design approximations and assumptions. Nevertheless, the aircraft specific differences are modelled thoroughly. Accordingly, the main findings of this work are rather the qualitative differences and trends between the aircraft configurations than the quantitative results.

Finally, it must be emphasized that the aircraft selection is strongly linked to the requirements of ALAADy imposed on the UCA, being the need for a relatively high payload of one metric ton, the range of 600 km, cruise speed of 200 km/h and maximum take-off and landing distances of 400 m. It does not represent validity for UCA in general.

References

Acree CW, Jr (2006) Impact of aerodynamics and structures technology on heavy lift tiltrotors. Technical report, NASA Ames Research Center, Mottfett Field, California, United States

Hasan YJ (2017) Konzeptuntersuchungen für ein zukünftiges unbemanntes Lufttransportsystem (in German), DLR-IB-FT-BS-2017-34 (Engl. Title: Conceptual Investigations for a Future Unmanned Air Transportation System). DLR, Institute of Flight Systems

Hasan YJ, Sachs F, Dauer JC (2018) Preliminary design study for a future unmanned cargo aircraft configuration. CEAS Aeronautical Journal

Hoerner S (1965) Fluid-dynamic drag: practical information on aerodynamic drag and hydrodynamic resistance. Hoerner Fluid Dynamics, Bakersfield, California, United States

Kroo I (2005) Nonplanar wing concepts for increased aircraft efficiency. In: Lecture Series on Innovative Configurations and Advanced Concepts for Future Civil Aircraft, Von Karman Institute for Fluid Dynamics

Liebeck R (2004) Design of the blended wing body subsonic transport. J Aircraft 10–25. https://doi.org/10.2514/1.9084

Mukhopadhyay V (2005) Blended Wing Body (BWB) fuselage structural design for weight reduction. https://doi.org/10.2514/6.2005-2349

Prandtl L (1924) Induced drag of multiplanes. NACA TN 182

Raymer DP (1989) Aircraft design: a conceptual approach. Educ Series. American Institute of Aeronautics and Astronautics, Washington D.C., US

Roskam J (1987) Airplane design part VI: preliminary calculations of aerodynamics, thrust and power characteristics. Roskam Aviation and Engineering Corporation, Ottawa, Kansas, US

Schiktanz D, Scholz D (2011) Box wing fundamentals - an aircraft design perspective. In: Deutscher Luft- und Raumfahrtkongress Bremen, 27–29 September 2011, pp 601–615

Solies UP (1993) Numerical methods for estimation of propeller efficiencies. J Aircr 31(4):996–998

Stoll AM, Bevirt J, Moore MD, Fredericks WJ, Borer NK (2014a) Drag reduction through distributed electric propulsion. AIAA Aviation Technology, Integration and Operations Conference, Atlanta, Georgia

Stoll AM, Bevirt J, Pei PP, Stilson EV (2014b) Conceptual design of the Joby S2 electric VTOL PAV. AIAA Aviation Technology, Integration and Operations Conference, Atlanta, Georgia, 16–20 June 2014

Strickert G (2016) Faktencheck Multikopter: Ähnlichkeiten und Unterschiede zu etablierten VTOL-Konfigurationen. Deutscher Luft- und Raumfahrtkongress, Braunschweig, 13–15 September 2016

Torenbeek E (1982) Synthesis of Subsonic Airplane Design. Springer, Netherlands

Configurational Aspects and Vehicle Specific Investigations for Future Unmanned Cargo Aircraft

Falk Sachs

Abstract This chapter focuses on the work done to examine in more detail how three configurations, namely double tail boom fixed wing aircraft, box wing aircraft and gyrocopter as proposed in a pre-study (Hasan YJ, Sachs F (2021) Performance-based preliminary design and selection of aircraft configurations for un-manned cargo operations. In: Dauer JC (ed) Automated low-altitude air delivery - towards autonomous cargo transportation with drones. Springer, Heidelberg) are suited for the task described in the project Automated Low Altitude Air Delivery (ALAADy). It is investigated which specific properties and features of each configuration are rather contributing or contravening for the ALAADy purpose. These investigations comprise, above all, the conduction of wind tunnel tests and a flight experiment. In order to identify and investigate inherent safety features 6-degree-of-freedom flight dynamic simulations of the three configurations are set up. To provide sufficiently precise simulation models many details of the configurations and the according power train variations are created and examined. For the gyrocopter wind tunnel tests of a scaled gyrocopter demonstrator are conducted as well.

Keywords Gyrocopter · Box wing · Tail boom · Flight simulation · Flight termination · Parachute · Autorotation · Flight test · UAS · Transport drone

1 Introduction

In the ALAADy project a vast collection of usual and even rather unusual aerial vehicles was considered to determine a suitable air vehicle. Without any preconceptions many types of aircraft were examined on a qualitative basis. In a second step, the more promising aircraft's flight performance key figures were roughly calculated to assess their suitability for the ALAADy purpose. Finally, three configurations turned out to be the most promising for the intended task. The results of this pre-study, as

F. Sachs (✉)
German Aerospace Center (DLR), Institute of Flight Systems,
Lilienthalplatz 7, 38108 Braunschweig, Germany
e-mail: falk.sachs@dlr.de

© Deutsches Zentrum für Luft- und Raumfahrt e. V. (DLR) 2022   133
J. C. Dauer (ed.), *Automated Low-Altitude Air Delivery*, Research Topics in Aerospace,
https://doi.org/10.1007/978-3-030-83144-8_6

described in (Hasan et al. 2018), lead to the conclusion that some types of both fixed wing and rotary wing aircraft, were suited for the given task.

This chapter deals with detailed analyses of the three air vehicles, being a twin-boom aircraft, a boxwing aircraft and a gyrocopter, that were proposed for the ALAADy transportation task in the pre-study described in (Hasan and Sachs 2021). Based on the foregoing results and the preliminary geometry definitions, nonlinear flight dynamics simulations are performed, the geometries are refined and the suitability of the aircraft for the transportation task is reevaluated.

Each aircraft is modelled to take its specialties into account. For each aircraft an own powertrain was specially adapted even if the general design was fitting for all of them. The calculation of structural loads, aircraft mass and aircraft performance are interfering with each other and demand to use sizing loops. Especially, since technical solutions for some parts of the aircraft like the emergency landing system or the powertrain were under development nearly till the end of the project.

Practical tests are performed, such as wind tunnel tests as well as flight tests.

2   The Configurational Properties of the ALAADy Vehicles

2.1   Fixed Wing Designs

Two fixed wing designs were chosen for further investigation (Fig. 1 and Fig. 2). The first results of the design study showed, that the tail boom configuration (Fig. 1) seemed to be a simple as well as the most fuel-efficient design at higher wing span.

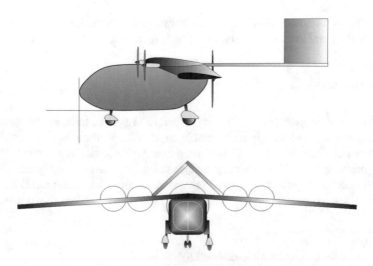

Fig. 1 Front and side view of the double tail boom configuration

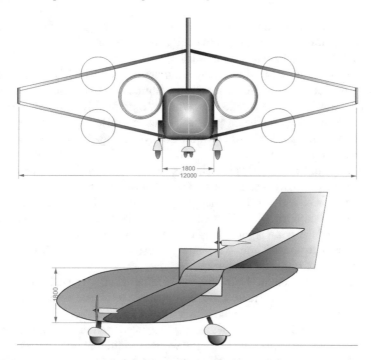

Fig. 2 Front and side view of the box wing configuration

The high wing span is a prerequisite for this configuration, in order to minimize induced drag and thus overall drag of the aircraft (Fig. 1).

In the beginning of the project it was not clear if the size of the ALAADy aircraft could be limited by ground handling issues, which at the end of the project turned out to not to be the restricting one. Nevertheless, it was decided to design a second vehicle with a lower wing span to be prepared for external inputs concerning the vehicle size.

An alternative approach is the use of so-called nonplanar wing systems. While usually suffering from higher parasite drag, these concepts have the potential of lower induced drag, particularly at lower wing spans.

To keep a restricted lateral size and to guarantee a comparably small induced drag at the same time, the wing has to be separated as it is done in a multiplane configuration and each wing has to be designed with a high aspect ratio. Furthermore, the design must consider requirements regarding landing distances, which demand a minimum size of wing area in order to realize slow approaches. There have been multiple designs of nonplanar configurations in the preliminary investigation, but especially from an aerodynamic point of view, the box wing seemed to be the first choice for a plane with a limited wing span (Fig. 2). The comparably low induced drag helps to reduce the fuel consumption. Even if the boxwing design comes with disadvantages in terms of structural weight it turned out to be the most efficient design at low wing span.

The two fixed wing designs have in common that the freight volume and payload are identical. Both configurations were designed accordingly. Furthermore, both of them are able to be loaded via a front door.

The gears are attached to the fuselage and are not retractable due to simplicity reasons. A retractable gear would result in additional weight and would increase complexity which has to be avoided in favour of a preferably light, but also robust and simple design. The increased parasitic drag in cruise flight is accepted. Due to the limited flight speed of 200 km/h the so caused negative impact to the flight performance is still acceptable.

The landing field constraint as presented in (Hasan and Sachs 2021) as well as the requirement to operate as cost efficient as possible led to the decision to implement a concept similar to which was already proposed by (Stoll et al. 2014): The wing leading edge is equipped with electrical engines. Their purpose is to increase the lift coefficient of the wing significantly during takeoff and landing and to provide additional thrust in takeoff run. They are not used in the cruise flight and their propeller blades are retracted in this case.

Due to the electrical engines mounted on the wings leading edge, the flow velocity over the wings surface is increased. So, it is possible to extend the wings operating range to lower flight speeds without risking to stall the airplane.

By increasing the flow over the wings surface, the dynamic pressure is increased resulting in a higher lift compared to a wing without additional flow causes by engine slipstream blow at the same angle of attack. Consequently, the angle of attack can be reduced by the electrical engines effect at a given flight speed. So, the highest possible lift coefficient of the wing is reached at a lower airspeed with additional engines working. In that way the phenomenon of flow separation can be shifted to lower flight speeds.

A STOL-plane (Short Takeoff and Landing) would normally be flown with a comparably low wing loading. So, the lift coefficient at high air speed would be relatively low and the best ratio of lift coefficient to drag coefficient can't be met.

Using additional electrical engines enables to perform short takeoff and landings at a higher wing loading. In cruise flight at sufficient airspeed the electrical engines can be turned off. An additional slipstream of the electric driven propellers is not necessary at high air speeds, since the incoming airstream causes sufficient dynamic pressure on the wing leading edge. Thus, the wing surface can be tailored for the optimum wing loading to achieve the best gliding ratio in cruise. So, it is possible to gain a better overall efficiency due to an adoption of the wing to the longest segment of the flight. Of course, the electric engines propellers effect on lift is not unlimited especially during the landing phase. So, the optimum wing loading for cruise cannot absolutely be met. But by reducing the wing size it becomes considerably better suited for cruise.

The ALAADy requirements literally invited to take profit of hybridization concepts for the power train. Especially the demand to operate on a short field (400 m) and to operate at a rather moderate airspeed in cruise flight gives way to choose efficient engines for cruise, which can be boosted during the start phase or during climbs. So, all the ALAADy-vehicles power train concepts are hybrid ones.

The box-wing airplane has two ducted electrical engines mounted at the rear fuselage. The electrical energy is provided either by a combustion engine with a generator or by fuel cells that are located in the fuselage. Additional electrical engines are mounted at the wing's leading edge.

The concept of the twin-boom also includes the electrical engines with retractable propellers on the wings leading edge. The difference is that the tail boom design enables to install a pusher propeller directly behind the engine. Thus, the combustion engine directly drives the main propeller, which is only during takeoff and in steep climb phases supported by the electrical engines.

In normal operation the batteries are reloaded to be able to support climb phases multiple times by the use of the electrical engines.

Detailed information concerning the design of the powertrain is given in (Aptsiauri 2021).

In the conceptual design phase some configurational details had to be changed in order to gain favorable aerodynamic characteristics. For this purpose, the airplane characteristics for short period, phugoid, dutch roll, spiral and roll motion were investigated and if necessary the configuration changed.

The biggest impact was on the stabilizer surface or wing shape after the initial design. Analyses by AVL-Athena Vortex Lattice (Drela and Youngren 2017) showed that the initial size of the vertical stabilizer was underestimated for all the configurations in the first design stage. The newly sized surfaces proved good results in simulation.

2.2   The Gyrocopter Design

The design of the gyrocopter in ALAADy is based on the payload bay size and the payload mass requirement as well as it is done for the fixed wing designs. So, the ALAADy gyrocopter possesses a very spacious fuselage for gyrocopter compared to the rotor size, which can be seen in Fig. 3.

The gyrocopter was chosen for investigation because of its excellent short takeoff and landing capabilities, but also because a safe flight termination with reduced

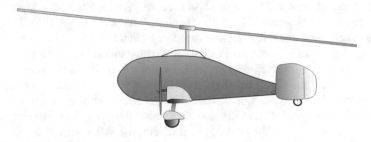

Fig. 3 Side view of the ALAADy gyrocopter

impact energy of the fuselage can be realized via the rotor in near vertical autorotation. Since the rotor is already autorotating during flight, a change of the rotor working state for emergency maneuvers at a possible engine shut down is not necessary. This autorotational capability facilitates the design of an emergency mechanism enormously. Furthermore, there is no need for another additional safety system such as a parachute that would result in a weight penalty.

It is obvious, that the flight performances of a rotorcraft in cruise are below those of fixed-wing aircraft. Nevertheless, the gyrocopter can be built relatively light, thanks to its lift producing rotor and the rotors favorable characteristic to dissipate and reduce the aerodynamic loads on the airframe (de la Cierva and Rose 1931). The gyrocopter's resting performance disadvantage was regarded as acceptable compared to the advantages the configuration has for the ALAADy concept.

In order to counteract these performance shortcomings of the gyrocopter, supplementary fixed wings can be installed. In addition to its rotor the ALAADy gyrocopter has a fixed wing delivering up to 30% of the overall lift in cruise at a moderate angle of attack, thus at a good wing gliding ratio. Producing lift with the wing helps to lower the rotor disc loading to reduce the rotor related drag. These two effects multiply and help to reduce the overall drag.

The dynamic pressure on the fixed wings profile is produced by the incoming airstream and the wing mounted engines influence. Their propellers slipstream over the wing's surface cause a significant portion of the lift.

The wing furthermore accommodates the fuel tanks and the batteries and additionally it carries the landing gear.

The rotor of the gyrocopter has a relatively high radius compared to a helicopter of the same maximum takeoff weight. The motivation to apply a low disk loading is that the necessary tilt of the rotor disc to the aft in flight can be reduced and the rotor drag due to the inclined rotor force vector can be decreased. The rotational rate of the rotor can also be lower than the one of a helicopter in order to reduce profile losses. In contrast to helicopter applications the rotational rotor rate has no impact on the moment in the rotor shaft, which is a limiting factor for a helicopter. However, it has to be ensured that the advance ratio of the rotor remains within an acceptable frame to keep effects like rotor flapping controllable. Especially, if only a simple central teeter hinge is installed and no further technical means to reduce the flapping exist.

The gear is designed as a tail dragger. Such a setup is advantageous especially for the takeoff maneuvers. After pre-rotation modern gyrocopters with tricycle undercarriage have to incline and balance the fuselage in takeoff run on the main gear to maintain a sufficient angle of attack of the rotor to increase the rotor rotational rate. This maneuver has to be performed with a certain precaution to avoid the risk of a tail strike. The takeoff with a tail dragger undercarriage is easier to perform. The rotor is pre-rotated until the aimed rotational speed is reached. Next, the rotor head pitch angle is adjusted to flight position. In the following acceleration phase, the fuselage's inclination is also transferred to the rotor. This way, the gyrocopter accelerates until lift-off without modifying the controls in pitch control. After lift-off the rotor angle of attack is automatically reduced to the correct value for flight.

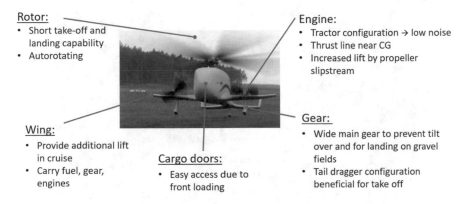

Rotor:
- Short take-off and landing capability
- Autorotating

Engine:
- Tractor configuration → low noise
- Thrust line near CG
- Increased lift by propeller slipstream

Wing:
- Provide additional lift in cruise
- Carry fuel, gear, engines

Cargo doors:
- Easy access due to front loading

Gear:
- Wide main gear to prevent tilt over and for landing on gravel fields
- Tail dragger configuration beneficial for take off

Fig. 4 Configurational elements of the ALAADy gyrocopter

Having the engines mounted on the wing with propellers in tractor configuration has a positive effect to reduce noise in comparison to the use of pusher propellers. In (Yin et al. 2012) a study showed the significance of this configurational choice. The noise emission can be reduced by several decibels. A pusher propeller interacts with the disturbed flow of the fuselage which causes most of the noise. Scaled flight testing showed qualitatively that the ALAADy gyrocopter configuration is significantly quieter with tractor propellers than with pusher propellers (Sachs et al. 2016). Figure 4 shows the ALAADy gyrocopter. Rationales for the different configurational choices are highlighted as bullets.

3 Emergency Systems and Strategies

As the drones investigated in ALAADy are operated over sparsely populated areas only the possibility to perform safe flight terminations is an adequate means for risk mitigation. To be able to ensure a minimum ground risk even in such a flight termination scenario several systems and strategies to realize minimized impact energy were discussed. They have in common that they are designed to work passively without any sensor feedback signals when they are engaged. Furthermore, they shall work reliably at any situation of flight. For the fixed-wing configurations three different termination strategies were investigated aiming at a reliable reduction of impact energy with predictable trajectories to evaluate existing safety buffers (Schopferer and Donkels 2021; Schirmer and Torens 2021).

The most unusual way to perform a termination was the trial to bring an aircraft into a flat spin to reduce energy and to impact with reduced speed. But it turned out, that the modifications of the aircraft to reliably realize such stable spin maneuvers would also reduce lateral and longitudinal stability, which is not desirable. Furthermore, the altitude of less than 500 ft is quite limited to achieve a stable flat spin and to evolve a sufficient angular momentum. The success of the maneuver is strongly depending

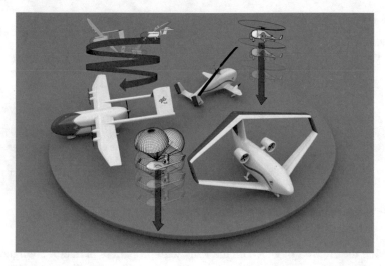

Fig. 5 Ensemble of ALAADy-configurations and termination ideas

Fig. 6 Parachute flight
termination

on the attitude and the speed of the airplane at the moment of termination. Reliability
over a vast range of speeds and attitudes couldn't be achieved and also, a sufficient
reduction of impact energy wasn't reached. So, this termination idea was abandoned
(Fig. 5).

Quite similar was the investigation, whether a termination can be succeeded by
another rather procedural approach, namely performing a steady sideslip maneuver.
Experiments in simulations regarding this maneuver were presented in (Russel 2018).
A transition into a stable sideslip could be demonstrated at any flight speed, but not
at all attitudes of flight. Being in a turn to the adverse direction may lead to an unintended behavior of the airplane when setting the controls in the predefined way. Such

behavior would not be tolerable for an automated flight termination system, which is supposed to terminate each flight in any possible flight condition in a predictable and reliable way. So, the choice of flight termination for the fixed wing aircraft was to use a parachute system adapted to the weight and the cruise speed of the airplane (Fig. 6). It could be demonstrated in simulation that the kinetic energy could be decreased significantly. Even in such low altitude flight corridors in which the ALAADy vehicles operate the parachute could be used efficiently. Nevertheless, a parachute system adds weight to the overall aircraft system. This weight is in the range of up to 100 kg, but not the parachute system alone is a weight influencing system. Also, the structural stresses, which have to be absorbed in the case of parachute deployment, have a significant influence on the structural sizing of the airplane. The structural weight added due to the parachute deployment related forces exceeds the weight added by the parachute itself. Details concerning the added weight can be found in (Hecken et al. 2021).

In (Russel 2018) the most useful design and the effectivity of the parachute system applied for the fixed wing aircraft were investigated. Figure 7 shows simulation results for the box wing aircraft starting from the moment of activation of the parachute system in flight until touch down. The left graph illustrates the velocity of the aircraft displayed over the altitude. It can be found, that at the end of the trajectory the remaining speed of the aircraft and the parachute is identical for all initial flight speeds tested. The right graph shows the potential to reduce the kinetic energy of the vehicle by means of a parachute. In the case of flight at the given cruise speed of 200 km/h the impact energy is only 5% of the original kinetic energy. The amount of potential energy is also absorbed by the parachute but not displayed for this comparison. The oscillatory change in speed and kinetic energy is due to the pendulum movement of the fuselage under the parachute. If the system is applied more than 100 m above ground the movement is sufficiently damped to small amplitude.

The choice of the attachment point plays a significant role. The most favorable position to attach the parachute to the fuselage is slightly over the center of gravity to produce a gentle pitch up moment during the braking phase in order to keep the

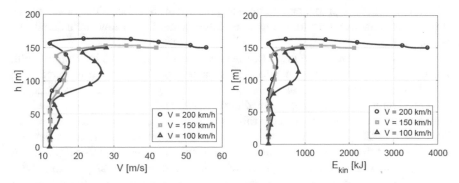

Fig. 7 Speed and energy reduction after parachute release, results from a nine degree-of-freedom non-linear simulation starting at 200, 150 and 100 km/h

altitude as long as possible and to decelerate the aircraft as much as possible already in a near flight altitude before descending.

The fourth emergency flight termination strategy is specifically related to the gyrocopter configuration. The auto-rotating rotor is used to descend in a very steep manner. Ideally, a gyrocopter is capable to descend vertically in a maneuver called vertical autorotation. The rotor acts like a parachute. The kinetic energy of the impact is determined by the vertical descent rate, which depends on the given vehicle's weight and the installed rotor system. After a deceleration phase the gyrocopter sinks in a vertical flight.

Entering a proper vertical autorotation within an altitude range of not more than 500 ft requires a well predefined procedure to modify the pitch controls in a way to reduce speed while assuring not to overshoot the maximum altitude and to enter in a vertical descent after having decelerated. With the controls driving in a predefined way to a predefined position, this procedure has to be adapted to a narrow speed range and payload conditions. If the speed is lower there is a risk to end up in a backward flight of the gyrocopter and to come to an unforeseen situation since the stabilizing effect of the stabilizer would be inverted.

Furthermore, the flight attitude and the position of the controls in the moment of flight termination play a significant role for the dynamic behavior of the vehicle after termination. To enable a wide range of in-flight situations from which an automated termination procedure would be possible, the pure vertical autorotation had to be put aside. Instead, the primary goal was to establish a near vertical autorotation with reduced forward speed, just enough to create a certain damping by the stabilizer surfaces for the lateral and longitudinal motion. The rotor roll control is moved to a position such that the gyrocopter performs a left hand turn and the rotor pitch control is moved rearward position suited for very slow flight (<50 km/h). The rudder deflection to the left-hand side supports the turn and leads to a stable state without any feedback control. Preliminary works on these maneuvers were demonstrated in simulation as explained in (Schopferer et al. 2021) and (Schirmer and Torens 2021).

For feasibility studies in 2017 DLR conducted experiments with a gyrocopter equipped with actuators and a reduced flight control device to be able to automatically set the controls to a predefined value and hold them. So, the initial phase of the termination maneuvers up to a slow and stable descent with low radius could be demonstrated.

Figure 8 shows the procedure. The maneuver was initiated in a horizontal flight. The thrust lever was reduced to zero. At the same time the rudder was deflected to the left and the rotor head roll angle ξ_{RH} was slightly tilted to the left while the rotor head pitch angle η_{RH} was reduced in this case. The gyrocopter decelerated and started a circular flight with a stable roll angle $\phi = 27°$ and a pitch angle $\theta = 33°$. Turn rate $r = 20°/s$ and pitch rate $q = 10°/s$ also showed only minor variations. The displayed flight speed V contains a significant sink rate of 8 m/s (29 km/h). A stable descent within a radius of 45 m was demonstrated. By increasing the rotor head pitch angle η_{RH} during the descent phase the vertical speed can be reduced further. Such a more progressive setting is chosen for the emergency procedure applied for an unmanned DLR gyrocopter demonstrator presented in (Lorenz et al. 2021).

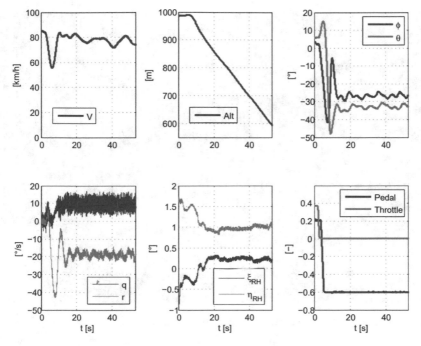

Fig. 8 Flight test results for the initialization and demonstration of an emergency termination maneuver including multiple slow circles in descent with preset controls

4 Model Demonstrations

To demonstrate cargo gyrocopters under real environmental conditions two test platforms were designed and built up. The ALAADy Demonstrator is derived from a known ultra-light gyrocopter configuration and equipped with additional actuators and computers to demonstrate a full- automated flight with a representative payload. This 450 kg heavy gyrocopter-demonstrator is presented in more detail in (Lorenz et al. 2021).

In order to investigate the flight dynamics of a cargo gyrocopter with a very large fuselage and additional fixed wings, a scaled demonstrator called Air Cargo Gyro 2 (ACG2) was designed and tested in wind tunnel and flight experiments. All configurational aspects were applied as long as scaling effects could be considered (Fig. 9).

The model is equipped with two electrical engines driving counterrotation propellers. Each of them has an allowed maximum electrical power consumption of 7 kW.

Using the advantage of electrification, the rotor pre-rotation is realized via a separate electrical engine. Thanks to the accumulator pods on the wing and the small size of the further electrical components, the voluminous fuselage can be used entirely for the payload as shown in Fig. 10.

Fig. 9 ACG 2 on ILA (engl.: International Aerospace Exhibition) 2018 in Berlin

Fig. 10 Front view into the payload bay of the ACG 2

The predecessor of the ACG2, the ACG1, was used to demonstrate its capabilities and to investigate the proposed configuration, which is very similar to the ACG2. Major results of those investigations are published in (Sachs et al. 2016). The tail dragger design turned out to be beneficial for takeoff and the two-tractor propeller configuration demonstrated in a field test proved that the noise level could be reduced significantly in contrast to a single push propeller at the same thrust.

For the ALAADy-project the experiences made with the experimental gyrocopter configuration turned out to be beneficial. Former designs had to be adopted to realize

Table 1 Technical data of the model demonstrator ACG2

Technical data - ACG2	
Rotor radius	2.1 m
Propeller	Diameter 0.63 m, fixed pitch, counter-rotating
Motor	2 × Hacker Q80-6L V2 (7000 W each)
Overall weight	38.6 kg
Fuselage length	2.30 m
Wingspan	1.86 m
Wing area	0.423 m^2

a spacious freight area and an enhanced aerodynamic design of the vehicle with increased power for payload transport (Table 1).

4.1 Wind Tunnel Experiments

The Wind tunnel experiments were conducted in order to understand the impact of the fuselage and the fixed wing on the aerodynamic characteristics of the ACG2. The wind tunnel experiments were supplemented by theoretical calculations using the Athena Vortice Lattice (AVL) program.

The experiments were carried out in the DNW-LLF wind tunnel facility in Brunswick. This tunnel is equipped with a 3.5 MW motor within the boss of the ventilator. It achieves a maximum wind speed of 80 m/s in the open test section and 90 m/s if the test section is closed/slotted. The rectangular test section has an inlet size of 3.25 m width and 2.80 m height. It is specially equipped and designed to serve as an aero-acoustic wind tunnel test facility (German Dutch Wind Tunnels DNW, Bergmann, Andreas 2012) (Fig. 11).

The wind tunnel experiments pointed out that design modifications were needed in order to achieve stable flight dynamics. Figure 12 shows the initial and the final design of the ACG2. The following modifications were realized (Fig. 13):

Dihedral of the Wing Tips
This change to the initial configuration was needed in order to improve the rolling moment due to sideslip. A low wing under the dominating fuselage has the advantage that the wing beam doesn't necessarily cross the payload bay. This way, the wing beam doesn't have to be separated into two parts and thus it is reduced in weight at comparable stiffness. The low wing causes a positive roll moment due to sideslip causing a tendency to roll into the direction of sideslip, which destabilizes the lateral motion.

The dihedral of the wing tips was adjusted to 30° in order to reduce the positive roll moment due to sideslip by 50%.

Fig. 11 ACG2 in the wind tunnel DNW-LLF in Braunschweig

Fig. 12 ACG2 design and modifications after wind tunnel experiments (left hand: initial, right hand: final)

Fig. 13 Dihedral of 30° at the outer wing tips

Increased Area of the Vertical Stabilizer

In order to increase the directional stability a third fin was installed. Furthermore, the outer vertical stabilizer surfaces were enlarged by 20%. The middle fin was equipped with an additional rudder augmenting the yaw control capabilities. Especially in the landing phase with low speed the idling propellers additionally reduce the dynamic pressure vertical stabilizer. Increased yaw stability also helped to reduce sideslip angles in maneuvers and thus helped to reduce the tendency to roll into the direction of the turn, by reducing the angle of side slip (Fig. 14).

Rotor Hub Fairing

A rotor hub fairing was tested in several measurements to check if a drag reduction at this part of the rotor is possible. It turned out that the proposed design didn't contribute to the reduction of rotor head drag, so it was removed from the vehicle in order to save weight and to reduce technical complexity (Fig. 15).

Sense of Propeller Rotation

The propulsive force causes a pitch moment because there is an offset between the propeller thrust axis and the center of gravity. In the case of the ACG2, the thrust

Fig. 14 Vertical fins of the ACG2 including extensions

Fig. 15 Rotor Hub Fairing
Design

axis is slightly below the center of gravity. So, a thrust increase causes a pitch up moment. In the first design approach the right propeller was turning counterclockwise and the left propeller vice versa. This way, both propellers induced a downwind on the horizontal stabilizer that also resulted in a positive pitching moment as thrust is increased.

In order to balance those effects and to minimize the thrust-pitch coupling as much as possible, the sense of propeller rotation was changed.

Incidence of the Horizontal Stabilizer

The horizontal stabilizer was adjusted to minus 0.5°. Experiments showed that the absolute pitching moment of the fuselage including stabilizer and wing and necessary propeller thrust at an angle of attack of 0° corresponds to the absolute pitching moment produced by the rotor at cruising airspeed (approximately 70 to 100 km/h). The rotor force was determined assuming a wing lift corresponding to a fuselage angle of attack of 0°.

Furthermore, the wind tunnel experiments were used to explore the new configuration in detail as much as possible. For this purpose, the aerodynamic forces and moments caused by each single element like the wing, the gears, the propellers etc. were investigated. To do so, multiple experiments were conducted.

First, the coefficients of forces and moments at angle of attack or angle of sideslip series of measurements were conducted once. Second, one possible component of the plane was removed. The difference between the two measurements reflects the aerodynamic influence of the according component including aerodynamic interactions like interference drag.

Figure 16 shows this approach in a graphic way for the coefficient's determination of a complete wing section including gears and engine nacelles.

Figure 17 displays some of the variations tested in the wind tunnel. Successively, the different parts of the gyrocopter ACG2 were disassembled and measured each by each.

In order to highlight one specific aspect investigated during the campaign, this chapter will show some representative results concerning the influence of the idling propeller on lift, drag and pitching moment. This investigation became inevitable since the propeller thrust axis of the ACG2 is lower than the center of gravity. Thrust variations in combination with a vertical lever arm of 45 mm are causing pitching moments. In the wind tunnel the drag of the propellers in wind milling as well as for propellers in brake position was investigated in order to predefine the expectable pitch down moment while the thrust level is reduced below idle or the respective pitch up moment at various positive thrust levels. Especially in the landing or takeoff phase the change in pitching moment caused by a thrust change has to be reduced as much as possible to prevent the interference with the control inputs by the pilot or the flight controller. At least for a flight control design the thrust-pitch-coupling has to be anticipated for effective feed forward control.

Figure 18 shows the three setups for the different measurements conducted. To create a reference, the first measurement was performed without propellers at all. Only the spinner and a foam adapter were mounted on the engine.

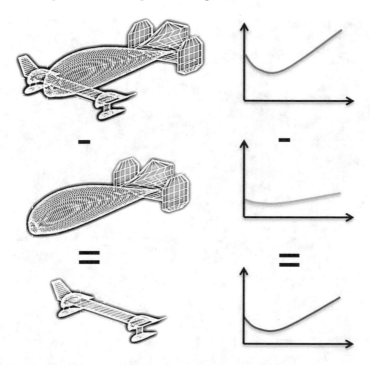

Fig. 16 Determination of aerodynamic influences of single components

In a second trial the propellers were fixed in a predefined 45° position by tape and the third test was carried out with propellers wind milling.

In all these tests, the free stream airspeed in the tunnel was set to 20 m/s. The reference area to calculate the respective coefficients is the ACG2 wing area and the reference chord is the wing chord of the ACG2 wing. Since this wing is quite small in comparison to fixed wing airplanes, the coefficients may appear unusually high at the first glance.

The C_D-plot in Fig. 19 shows, that at usual angles of attack during forward flight (−1 to 5°) the wind milling propellers are nearly doubling the drag of the ACG2 fuselage excluding the rotor. This increase of C_D can be cut by 50% if the propellers are able to be braked at a predefined position.

Furthermore, the idling propellers show a positive influence on the overall stall behavior. It is found, that stall occurs at higher angles of attack and that it is smoother than in both other cases, see the C_L-plot.

For the pitching moment curve the influence of the wind milling propellers is quite undesired. The clear negative slope of C_m over α vanishes if the propellers are in wind milling state. So, the static stability for the fuselage is reduced. The reason for this is the reduction in dynamic pressure behind the propeller discs. The resulting lift force of the stabilizer is reduced to 80% since the airspeed behind the idling propeller is reduced to nearly 90% of airspeed. Below roughly −5° and above

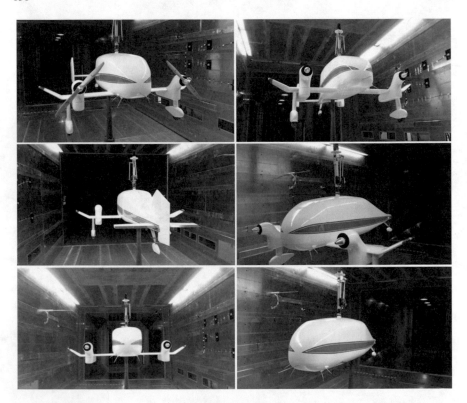

Fig. 17 Different measurement setups to identify the single component's aerodynamic characteristics of the ACG2

Fig. 18 Three setups for propeller influence measurement (left hand: without propeller, middle: propeller fixed, right hand: propeller wind milling)

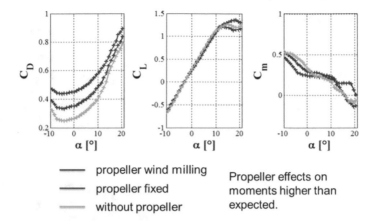

Fig. 19 Drag, lift and pitch moment coefficients due to the propellers

17° the horizontal tail leaves the shadow of the propeller disk. The increases of the C_m-slope at the respective angles of attack reflect this effect.

4.2 Flight Tests

The ACG2 demonstrator was tested in flight in order to prove its controllability and suitability for the planned tasks as cargo UAS (Unmanned Aircraft System). After the wind tunnel testing the aerodynamic and mass characteristics were known well. The rotor was the only component that couldn't be tested in advance. Analytical equations were used to predict the behavior of the auto rotating rotor (Duda and Seewald 2016). Especially the characteristics of the compound gyrocopter were investigated in detail. For those purposes the ACG2 was equipped with measurement devices.

Fig. 20 Measurement unit

Fig. 21 Nose boom

The main measurement unit of the ACG2 was installed inside its fuselage near the center of gravity. It consists of an inertial measurement unit (IMU), several pressure sensors (dynamic and total pressure), a magnetometer, a GPS-receiver and a microcontroller which processes and records the data. This measurement box was extended with a nose boom that sensed angle of attack and angle of sideslip with a five-hole-probe (Figs. 20 and 21).

Furthermore, engine parameters, rotor parameters, all control commands, actual voltage and current, accumulator status, GPS-data, the pressure altitude and the flight speed were recorded.

Takeoff Phase
Before takeoff the rotor of the ACG2 was pre-rotated up to 400 rpm. Then the pre-rotation was turned off and the rotor head pitch was inclined to flight position. The main piloting task was to increase thrust and control the yaw axis. Due to the positive inclination of the fuselage, the rotor angle of attack was superior to the necessary rotor flight angle of attack. So, the rotor was accelerated as soon as the gyrocopter accelerated on ground. After taking off the rotor angle of attack was reduced to the flight angle of attack and the ACG2 climbed according to the thrust setting. In flight the rotational speed of the rotor increased to ca. 450 rpm according to the flight speed and the trim elevator position.

The Handling Qualities of the ACG2 were rated satisfactory. The pilot reported that almost no control inputs for takeoff and climb phase were required. At a mass of 37 kg with a head wind of 7 km/h and a pre-rotation of 400 rpm a takeoff roll distance of about 7 m was determined, see Fig. 22.

Fig. 22 Only 7 m take off roll distance of the ACG2

Fig. 23 Landing of the ACG2

Landing Phase

One of the main advantages of the gyrocopter is the low approach speed that enables a short landing roll distance after touch down. The minimum airspeed at touch down recorded was 25 km/h. After touch down, the rotor was used to decelerate the ACG2 by inclining the rotor backwards and directing the rotor thrust vector rearwards. This mechanism is one reason for the short landing roll distance. Figure 23 gives an impression on how short a final approach might be. The demonstrated landing roll distance was the same as the take-off roll distance. (Note: the gear does not have any wheel brakes).

During landing roll the tail dragger configuration in combination with a wide gear turned out to be more challenging than predicted. In contrast to an airplane, the gyrocopters rotor lift doesn't depend on forward flight speed. This is a desired feature to fly slowly, but it might also be disadvantageous after touch down. During the touch down phase the gyrocopter was flying slowly and the stabilizers showed only minor effectivity. In contrast to the starting phase there was no significant propeller blow on the stabilizer surfaces. If one wheel of the wide gear got stuck an unintended yaw movement followed. The destabilizing effect of the center of gravity behind the gear can only be counteracted by rudder inputs which might not be sufficiently effective due to the prior described low dynamic pressure on the stabilizers and rudders.

Cruise Flight

The analysis of flight data of cruise flight performance was focused on the coupling of the rotor and the fixed wing. The ACG2 is equipped with a trimmable elevator on the horizontal stabilizer to modify the pitching moment of the fuselage independently from the rotor. The idea was to change the fuselage's angle of attack in flight to manipulate wing and rotor loading and to trace the effects on rotor dynamics and flight dynamics.

Unloading the rotor reduces rotor drag, since the rotor angle of attack can be reduced. At the same time, the rotor rotational rate declines and the rotor force declines, too. The reduced rotational rate of the rotor in relation to the forward speed increases the advance ratio of the rotor. This increase of the advance ratio causes an increase of the rotor's flapping angle. The pilot has to compensate such an increased flapping angle by means of pitch control.

Flight experiments showed that the pitching moment coefficient of the fuselage has to be chosen carefully especially when the rotor is the only active mean of control for pitch and roll (Fig. 24).

This pitching moment coefficient that has been modified by the trim elevator in the experiments has a strong impact on the angle of attack. In flight tests the trim elevator input was increased multiple times to unload the rotor consecutively and to trace the effects of the shift of lift from rotor to the wing. Thanks to the low risk for any flight personal, which is an immanent advantage of scaled model flight testing, it was possible to extend the testing to the edge of the flight envelope.

Assuming a positive pitch coefficient of the fuselage, the respective pitch down moment portion is produced by the rotor. With increasing flight speed the lift of the fixed wing increases and the rotor is getting unloaded about the same amount. Due to the reduced rotor force the pilot has to tilt the rotor forward to produce sufficient pitch down moment.

Fig. 24 ACG2 fly by during flight test

At a very high airspeed the risk for an uncontrollable pitch up rises because the rotor may be unloaded. In this case the elevator is most effective for pitch control.

The level of rotor rotational speed dropped from 450 rpm down to 330 rpm and below at high airspeed. The more the trim of the elevator was set to negative values (pitch up) the lower dropped the rotor rotational speed.

On the one hand a significant rotor unloading and thus rotor drag reduction could be demonstrated. On other hand, the low rotor force led to high required control inputs, that made a high-speed flight quite challenging for the pilot.

In general, the flight tests pointed out some major results for future applications.

1) In the flight tests very short start and landing capabilities could be demonstrated
2) The concept of a conventional undercarriage proved to be beneficial for the starting phase of the gyrocopter, but it also revealed some weak points for the landing phase, especially at very low landing speeds.
3) The handling qualities for start, landing and cruise flight turned out to be good
4) At high speed the interactions of fuselage and rotor may lead to a dangerous rotor unloading that reduces the pilot's authority that has to be prevented
5) Wind tunnel results and flight test observations are coherent and lead to clear design recommendations for future configurations

With the ACG2 the interaction of rotor and wing could be demonstrated as it was intended. The trimmable elevator helped to modify the wing angle of attack quite effectively. Nevertheless, the lift distribution in fast flight was poor, due to the wing position and the additional propeller blast over the wing surface. This created a high pitch up moment of the fuselage in combination with a high wing lift force and low rotor force at an unfavorable trim elevator position. That made the vehicles control difficult in high-speed flight.

In contrast to the experiment setup, for normal operation, the trim elevator has to be used to constantly ensure a sufficient rotor loading. Furthermore, in a future design a more balanced lift distribution by a repositioning of wing and rotor has to be considered. The functional integration of the gear and the wing as it was originally planned has to be re-assessed.

5 Conclusion

The three presented vehicles proposed for the ALAADy concept differ significantly in appearance, size and configurational aspects but all might be used as transport UAS. For the fixed wing vehicles, it turned out, that additional electrical driven propeller at the wing leading edge significantly increase lift and thus help to reduce the landing and start distance. Those propellers may be retracted in cruise but their additional thrust may also be used in the start phase and during climbs to increase the climb ratio. The gyrocopter is compounded with an additional wing, to increase the flight performance and to carry the motors as well as a wide gear to land on unpaved runways.

For all presented aircraft systems hybrid and fully electric propulsion systems were investigated.

Performing a reliable and quick emergency landing after in case of flight termination is a key feature for all ALAADy vehicles. For the fixed wing concepts the application of a parachute system in simulation shows promising results to reduce the aircrafts kinetic and potential energy in case of a safety emergency landing. The gyrocopter instead uses the rotor perform a slow descent by flying circles with a minimum radius. This concept had been simulated but was also demonstrated in flight test with a manned gyrocopter.

To demonstrate one of the specific configurations in flight test a scale model of the gyrocopter configuration was build, investigated in wind tunnel and flight tested. The model demonstrated good handling qualities for cruise and short take-off and landing capabilities. It showed also, that an unloading of the autorotating rotor at high speed flights is possible due to a strong pitch up tendency of the fuselage and the wing at some trim conditions. Such behavior must be prevented. The flight test results underlined all necessary information to redesign the cargo gyrocopter configuration by shifting the wing position to use the benefits of the hybrid lift configuration and ensure a safe flight.

References

Aptsiauri G (2021) Concepts of electric and hybrid propulsion and operation risk motivated integrity monitoring. In: Dauer JC (ed) Automated low-altitude air delivery - towards autonomous cargo transportation with drones. Springer, Heidelberg

Carter J, Lewis JR (2015) Beyond clean sky: cartercopter slowed rotor/compound exceeds efficiency and emissions goals. In: 41st European Rotorcraft Forum, Munich

de la Cierva J, Rose DF (1931) Wings of tomorrow. Brewer, Warren & Putnam

Drela M, Youngren H (2017) Massachusetts Institute of Technology. 12 February 2017. http://web.mit.edu/drela/Public/web/avl/avl_doc.txt

Duda H, Seewald J (2016) Flugphysik der Tragschrauber. Springer Verlag GmbH Deutschland, Berlin

German Dutch Wind Tunnels DNW, Bergmann, Andreas (2012) The aeroacoustic wind tunnel DNW-NWB. In: 18th AIAA/CEAS Aeroacoustics Conference (33rd AIAA Aeroacoustics Conference). AIAA, Colorado Springs, CO, USA

Hasan YJ, Sachs F, Dauer JC (2018) Preliminary design study for a future unmanned cargo aircraft configuration. CEAS Aeronaut J, December 2018, pp 571–586

Hasan YJ, Sachs F (2021) Performance-based preliminary design and selection of aircraft configurations for un-manned cargo operations. In: Dauer JC (ed) Automated low-altitude air delivery - towards autonomous cargo transportation with drones. Springer, Heidelberg

Hecken T, Cumnuantip S, Klimmek T (2021) Structural design of heavy-lift unmanned cargo drones in low altitudes. In: Dauer JC (ed) Automated low-altitude air delivery - towards autonomous cargo transportation with drones. Springer, Heidelberg

Lorenz S et al (2021) Design and flight testing of a gyrocopter drone technology demonstrator. In: Dauer JC (ed.) Automated low-altitude air delivery - towards autonomous cargo transportation with drones. Springer, Heidelberg

Russel J (2018) Master thesis, Flugmechanische Untersuchungen zu Flugabbruchsystemen von unbemannten Frachtflugzeugen. RWTH Aachen University, June 2018

Sachs F, Duda H, Seewald J (2016) Flugerprobung zukünftiger Tragschrauber anhand skalierter Versuchsträger. Deutscher Luft- und Raumfahrtkongress. DGLR, Braunschweig

Schirmer S, Torens C (2021) Safe operation monitoring for specific category unmanned aircraft. In: Dauer JC (ed) Automated low-altitude air delivery - towards autonomous cargo transportation with drones. Springer, Heidelberg

Schopferer S, Donkels A, Schirmer S, Dauer JC (2021) Multi-disciplinary scenario simulation for low-altitude unmanned air delivery. In: Dauer JC (ed) Automated low-altitude air delivery - towards autonomous cargo transportation with drones. Springer, Heidelberg

Schopferer S, Donkels A (2021) Trajectory risk modelling and planning for unmanned cargo aircraft. In: Dauer JC (ed) Automated low-altitude air delivery - towards autonomous cargo transportation with drones. Springer, Heidelberg

Stoll A, Bevirt J, Moore MD, Fredericks WJ, Borer NK (2014) Drag reduction through distributed electric propulsion. In: Aviation technology, integration, and operations conference, Atlanta, Georgia

Yin J, Stürmer A, Aversano M (2012) Aerodynamic and aeroacoustic analysis of installed pusher-propeller aircraft configurations. J Aircr 49(5):1423–1432

Structural Design of Heavy-Lift Unmanned Cargo Drones in Low Altitudes

Tobias Hecken, Sunpeth Cumnuantip, and Thomas Klimmek

Abstract This chapter presents the conceptual loads analysis and the structural design of the unmanned cargo aircraft concepts within the Automated Low Altitude Air Delivery (ALAADy) project of the German Aerospace Center, DLR. Three concepts of a gyroplane, a fixed high-wing with a twin boom v-tail aircraft and a box wing aircraft are concerned in the project. The main focus is on the estimation of the structural weight of each unconventional aircraft configuration. However, the application of empirical formulations based on conventional aircraft configurations is not suitable for this task. Instead, the task is performed with a parametric design process which is designed to be used for the structural design of an unconventional aircraft. In this process, the structural models are set up as finite element models in a parametric manner. The loads analysis and the aeroelastic structural dimensioning are then performed using these finite element models. This process is developed by the DLR and it is called the MONA process. The major result of the process is the structural weight of each unmanned aircraft (UA) concept. Finally, the structural weights of all aircraft are compared. The weight comparison contributes to the global evaluation of the aircraft concepts within the ALAADY project. Based on the considered aircraft requirements, the considered load cases and the considered system masses, the gyroplane has the minimum structure weight of 2,720 kg.

Keywords Preliminary design · Parametric aeroelastic design process · Structural model · Transportation drone

T. Hecken (✉) · S. Cumnuantip (✉) · T. Klimmek (✉)
German Aerospace Center (DLR) Institute of Aeroelasticity, Bunsenstrasse 10, 37073 Goettingen, Germany
e-mail: tobias.hecken@dlr.de

S. Cumnuantip
e-mail: sunpeth.cumnuantip@dlr.de

T. Klimmek
e-mail: thomas.klimmek@dlr.de

© Deutsches Zentrum für Luft- und Raumfahrt e. V. (DLR) 2022
J. C. Dauer (ed.), *Automated Low-Altitude Air Delivery*, Research Topics in Aerospace,
https://doi.org/10.1007/978-3-030-83144-8_7

1 Introduction

An unmanned cargo aircraft (UCA) is a transport vehicle with special guidance and control equipment and without crew and life systems on board. The prospect to reduce the operation costs thanks to the lack of a pilot, is one core motivation for the development of a UCA. The Automated Low Altitude Air Delivery (ALAADy) project of the German Aerospace Center (DLR) is about the conceptual design and development of an automated and unmanned air transport system. Special emphasis is put on the safety aspects for the realization of such a system, operational boundary conditions, system architecture and necessary algorithms for the implementation of such an air transport system.

A conceptual loads analysis and structural design are carried out for the following three configurations:

- a gyroplane, shown in Fig. 1(a)
- a fixed high-wing with a twin boom v-tail aircraft, shown in Fig. 1(b)
- a box wing aircraft, shown in Fig. 1(c)

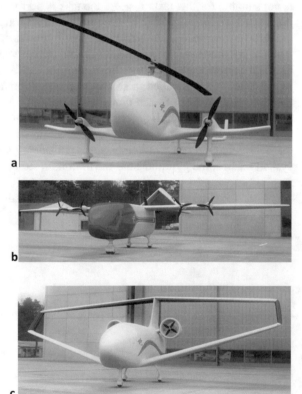

Fig. 1 Artistic depiction of the three chosen aircraft configurations, (Dauer et al. 2016), **a** The Gyroplane, **b** The Fixed-High Wing Configuration, **c** The Box Wing

All of these concepts are designed to have a hybrid electrical powertrain which is described in detail by Aptsiauri (2021). The main advantage of the gyroplane is the autorotation of the main rotor system in case of an engine failure. However, due to the large power requirements on the blades, the flight performance is limited. On the other hand, the twin boom configuration having a high aspect ratio wing is expected to show a better flight performance. The third configuration, the box wing concept, is particularly advantageous when the operation conditions require the UCA to be more compact, since its nonplanar wing system allows for low induced drag even at relatively low wing spans. However, both concepts require an additional safety system like a parachute system to guarantee its safe landing according to the minimum landing impact energy requirement which has been described by Nikodem et al. (2021) in case of an operation termination which is a mandatory criterion in the design of the vehicle in the ALAADy project. The hybrid electrical powertrain and the additional safety components can lead to complications regarding the structural integration and finally to a structural weight penalty to the aircraft.

In this chapter, thus, the structural concepts of the above configurations are designed and analyzed. The main focus of the work is on the estimation of the structural weight of each configuration. The final comparison of the structural mass result is an important criterion for the final vehicle configuration evaluation within the ALAADy project. However, the unmanned aircraft (UA) concepts are unconventional aircraft configurations, especially the gyroplane and the box wing configuration. As a result, the implementation of a conventional empirical formulation for the weight estimation of these aircrafts is not appropriate. Instead, the design and analysis of the structure is performed with the Parametric Finite Element Model Generation and Design Process, also called MONA process, of the Institute of Aeroelasticity of the DLR, see Klimmek (2014).

This chapter begins with the description of the MONA process in the first section. Here, an overview of the MONA process is shown in Fig. 2. The second section of the paper is concerned with the general description of the UA configurations including the mission requirements. This section is also about the generation of the aircraft's finite element (FE) models using the MONA process. The third section, the conceptual loads analysis and structural sizing of the FE models from the second section is explained. The considered load cases are in accordance with the certification specifications part 23, CS-23, for small aeroplanes and the certification specifications part 27, CS-27, for small rotorcrafts. Subsequently, the requirement of a safe termination with low impact energy and in uninhabited area in case of the flight termination, the parachute load case is considered as well. For the dimensioning of the set-up FE models of the UA configuration, analytical methods are used as well as gradient-based structural optimization methods using the design optimization solution 200 (SOL200) of the program NASA Structural Analysis System (NASTRAN) of the MacNeal-Schwendler Corporation (MSC Software), so-called MSC NASTRAN. The definition of the structural optimization task is laid out in the third section as well as its integration in the iterative aero-elastic design process MONA.

The conceptual structural mass results are presented in the last section. The leverage of the parachute loads case, being a mission requirement specific load case,

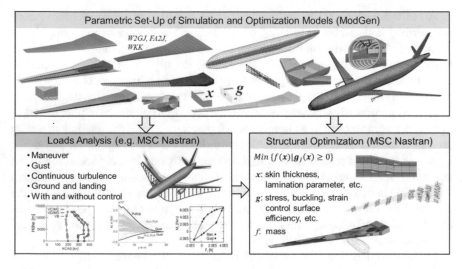

Fig. 2 The aeroelastic structural design process, so-called MONA, Klimmek (2014)

to the aircraft structural mass is discussed in particular. Finally, the structural weight results of the three configurations are compared.

2 Parametric Finite Element Model Generation and Design Process, MONA

At DLR the parametric aeroelastic design process named MONA has been developed in recent years in order to carry out a structural and aeroelastic assessment for a given aircraft configuration. The MONA process is described in detail in this section.

The basic aeroelastic structural design process can be broken down into two steps:

1. Loads analysis for the flexible structure
2. Structural design

For the latter the use of structural optimization methods is apparent, as far as such methods exhibit a certain maturity and aeroelastic parameters, like aileron efficiency, divergence, or even flutter can in principle be taken into account. Presupposed that a reasonable starting point for the structural design is available (e.g. thickness of skins, spars and ribs), such a process can be repeated until convergence is achieved with respect to the structural weight and the loads.

In order to ensure consistency of the simulation and optimization models for the loads analysis and the structural optimization a specific parameterization concept has been developed over the last almost 20 years at DLR. Therein, various aircraft components (wing, fuselage, engine/pylon) as well as various design concepts for the load carrying structure can be considered.

The computer program ModGen, standing for Model Generator, has been developed at DLR to guarantee consistency of the simulation models. It sets up the needed simulation and optimization models for the aeroelastic structural design process. Starting with ModGen the aeroelastic structural design process, named MONA, is pursued. It can be divided into three steps as shown in Fig. 2. The name MONA comes from the basic computer programs that are used. These are the DLR's in-house computer program ModGen for the model set-up and the already mentioned MSC NASTRAN software for the loads analysis and the structural optimization.

The parametric model set-up with ModGen starts with parametrically defined geometry models of the wing-like components and the fuselage. This comprises the outer geometry and the elements of the load carrying structure (e.g. spars and ribs). Thereby B-splines are used, because of their typical characteristic to define smooth curves and surfaces. Differential geometrical design methods are used for the successive geometry model set-up process. The generated geometrical entities are the basis for the meshing towards the FE model. For the analyses and optimization of the generated FE models the MSC NASTRAN software is used. Here, common analyses are linear static, eigenvalue analysis, static aeroelasticity, aeroelastic flutter, dynamic aeroelasticity, design sensitivity and optimization. The basic MONA process is furthermore described in Klimmek (2014).

As the MONA process was already applied to classical transport aircraft configurations like the XRF1-DLR, the DLR-D150, or the structural model for the "NASA Common Research Model" (Klimmek 2014), the mentioned UA configurations require various adaptations of MONA. This comprises for example the set-up of a box wing structure and the structural representation of the fuselage. The latter differs considerably from classical tube-like fuselage structures. The UA rectangular fuselage is designed for a cargo transport mission without the requirement of cabin pressurization. The UA's finite element model generation is explained in more detail in the next section.

3   Unmanned Aircraft Configurations and Finite Element Models

3.1   Aircraft Requirements

The global top-level aircraft requirements (TLARs) of the UA configurations are summarizes in the following. The maximum take-off mass (MTOM) was specified to be around 2500 kg with a payload of 1000 kg. The maximum range of 600 km, a cruise speed of 200 km/h as well as the design cruise altitude of around 200 m are specified. Some of these requirements are the result of the initial configuration study which has been described by Hasan and Sachs (2021) and Sachs (2021). In addition, so-called light soft requirements were specified, which define that the aircraft should have low complexity, be compact and have a simple construction design.

Certain special specifications are also required for the UA configurations within the ALAADy project. The configurations are required to be able to take-off and land on an unprepared runway such as a lawn near to the area of destination. In addition, the configurations are required to have a take-off field length (TOFL) and a landing field length (LFL) of 400 m. The most important special requirement of the UA configurations is the requirement of the safe termination of the aircraft. If the aircraft unexpectedly leaves the pre-defined operating corridor, then safe termination is the last resort. In the event of safe termination, the aircraft shall perform an emergency landing to ensure its safe and planned touchdown. Due to its low altitude and that it flies over sparsely inhabited areas the safe termination can be performed in consideration of the specific operations risk assessment (SORA) according to Murzilli (2019). See Dauer and Dittrich (2021) and Nikodem et al. (2021) for more information on operational risk-based concepts for such UCA. These special requirements have an impact on the loads and the structural mass of the UA configurations and finally on the final selection of a possible aircraft configuration for the ALAADy project. The UA configurations are developed based on the above described TLARs and SORA. Special focus of the development is on the flight termination criterion.

Several aerial vehicle configurations e.g. airships, para gliders and balloons etc. have been considered during the ALAADy's project initial configuration study to be the UA configuration. This study has been described in detail by Hasan and Sachs (2021). However, these configurations have been excluded due to their inability to meet all of the above-mentioned termination requirements.

The result of the initial configuration study was the proposition of three aircraft configurations which seem promising with respect to meet all TLARs and the termination criterion. These three configurations which have been shortly mentioned in the introduction section are:

1. The gyroplane
2. The fixed-high wing with a twin boom v-tail
3. The box wing

The main advantage of the gyroplane is its autorotation capability of the rotor system which allows the safe termination of the aircraft. The fixed-high wing aircraft offers a possibility of a higher flight efficiency than the gyroplane. It can also perform a safe emergency landing using a parachute system. However, the loads from the parachute may have a negative effect on the aircraft structure mass. The box wing aircraft offers a more compact configuration, since its nonplanar wing system allows for low induced drag even at relatively low wing spans. However, in order to perform a safe emergency landing according to the minimum landing impact energy requirement described by Nikodem et al. (2021), it requires the integration of a parachute system which finally might lead to an additional weight to the aircraft structure.

Table 1 summarizes the results of the initial conceptual design and analysis process of the UA configurations (Hasan and Sachs 2021).

Table 1 Results of the initial conceptual design and analysis process of the UA configurations, Hasan and Sachs (2021)

	Gyroplane	Twin boom	Box wing
MTOM	2449 kg	2343 kg	2591 kg
TOFL	220 m	278 m	233 m
LFL	100 m	395 m	370 m
Design range	600 km	600 km	600 km
Design speed	200 km/h	200 km/h	200 km/h
Minimum speed	60 km/h	75 km/h	72 km/h
Glide ratio		13.2	11.6

3.2 The Unmanned Aircraft Configurations

3.2.1 The Gyroplane

The gyroplane consists of a fuselage, a straight wing, a rotor which is attached through a beam construction to the fuselage and a backward swept horizontal tail plan, HTP. Figure 3 shows the planform and the dimension of the gyroplane.

The fuselage is divided into a front, a main and a tail section. The cargo and the safety system are accommodated in the fuselage. The cargo area has the height and the width of 1.3 m and a length of 3 m so that the aircraft have enough space to store two europallets. System components such as the safety system and fuel tanks are located within the fuselage. The cross section of the front fuselage section enlarges in longitudinal direction along the fuselage center line while the tail section tapers down in longitudinal direction along the fuselage centerline. The overall fuselage length is 7.4 m. The wing has a surface area of 8 m^2. The rotor has a surface area of 22 m^2. According to the initial conceptual design result (Hasan and Sachs 2021), the wing contributes up to 30% of the total lift force during the cruise flight. This has been taken into account for the forces and moments calculation of the static aeroelastic trim analysis. The gyroplane is equipped with two engines each with a power of 175 kW. These engines are mounted on the wing and have a rotor radius of 1 m. Table 2 presents the basic geometrical parameters of the gyroplane's wing, rotor as well as horizontal and vertical tail plan, HTP and VTP.

Fig. 3 Side view of the gyroplane, Sachs (2021)

Table 2 Basic geometrical parameters of the gyroplane

	Wing	Rotor (m)	HTP	VTP (m)
Span	8 m	7	3.4 m	1.2
Area	8 m^2	2.66	2.72 m^2	1.56
MAC	1 m	0.38	0.80 m	1.15

3.2.2 The Twin Boom

The twin boom configuration consists of a fuselage, a backward swept wing and an inverted v-tail with an angle of 45° in y–z plane which is attached to the wing through a twin circular tail boom. Figure 4 illustrates the planform and the dimension of the twin boom configuration. The fuselage is divided into a front, a main and a rear section. The cargo is accommodated in the fuselage. System components such as the safety system and fuel tanks are located within the fuselage. The cross section of the front fuselage section enlarges in longitudinal direction along the fuselage centerline while the tail section tapers downs in longitudinal direction along the fuselage centerline. The overall fuselage length is 5 m. A symmetrical airfoil is used for the v-tail. The aircraft is equipped with a 135 kW engine and four additional 20 kW electric motors. The main engine is mounted at the rear end of the fuselage while the electrical motors are distributed in spanwise direction at 1.6 and 2.8 m. The propeller radius is 0.5 m for the electrical motors and 1 m for the main engine. This powertrain arrangement leads to an increase in the lift coefficient during take-off because the wing sections behind the propellers operate in the propeller slipstream effects. It also increases the thrust and acceleration during take-off. These finally lead to the short TOFL capability of this aircraft concept. The wing propellers are folded during cruise flight to reduce the cruise drag and therefore increase the cruise

Fig. 4 Front and side view of the twin boom, Sachs (2021)

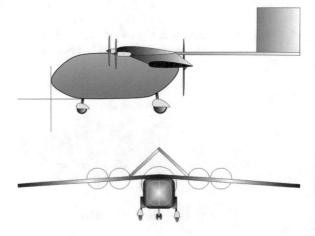

Table 3 Basic geometrical parameters of the twin boom

	Wing	V-Tail
Span	16 m	2.975 m
Area	28.7 m^2	5.06 m^2
MAC	1.79 m	1.7 m
AR	8.92	1.75
Dihedral angle	−4°	−45°
Taper ratio	0.8	1
Sweep	10.5°	0°

efficiency. One parachute is located at the v-tail for the application in case of the flight termination. Table 3 shows the basic geometrical parameter of the twin boom's wing, inverted V-tail and tail boom.

3.2.3 The Box Wing

The box wing configuration consists of a fuselage, a backward swept lower wing, a forward swept upper wing and a vertical tail plane. Figure 5 illustrates the planform and the dimension of the box wing configuration. The lower and upper wings are connected at their tip through a vertical element. The VTP is attached to the

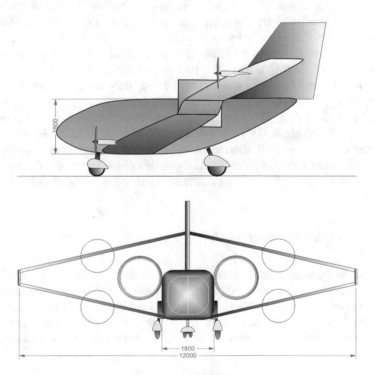

Fig. 5 Front and side view of the box wing, Sachs (2021)

Table 4 Basic geometrical parameters of the box wing

	Lower wing	Upper wing	VTP
Span	12 m	12 m	2.5 m
Area	16.74 m^2	16.74 m^2	5.625 m^2
MAC	1.395 m	1.385 m	2.25 m
AR	8.6	8.6	2.22
Dihedral angle	13°	−13°	
Taper ratio	0.8	0.8	0.5
Sweep	23.19°	−21.22°	41.77°

upper wing. A symmetrical airfoil is used for the VTP and the vertical element which connect the upper and the lower wing. A symmetrical airfoil is used for the vertical element in order to avoid the generation of aerodynamic forces by the vertical element. This force can lead to an additional aerodynamic load to the wing tip which finally may has a negative effect on the wing structural mass. Table 4 presents the basic geometrical parameter of the box wing's lower wing, upper wing and VTP.

The fuselage is divided into a front, a main and a rear section. The cargo is accommodated in the fuselage. System components such as the safety system and fuel tanks are located within the fuselage. The cross section of the fuselage front enlarges in longitudinal direction along the fuselage centerline, while the cross section of the fuselage tail tapers down in longitudinal direction along the fuselage centerline. The overall fuselage length is 7.5 m. This box wing configuration has the advantage of a low induced drag at relatively low wing span for a compact aircraft. The smaller wing span compared to that of the twin boom configuration is motivated by an improved ground handling of the aircraft. The aircraft is equipped with two engines and four additional 50 kW electric motors. The two main engines provide a total power of 231 kW to two propellers with a radius of 0.71 m. They are attached to the side of the main fuselage. The electrical motors are mounted on the upper and lower wing in spanwise direction at 3 m. This arrangement leads to an increase in the lift coefficient during take-off and climb. This powertrain concept leads to an increase in the lift coefficient during take-off. It also increases the thrust and acceleration during take-off. These finally enable the short TOFL capability of the box wing. The propellers are folded during cruise flight. The parachute is located at the fuselage rear section.

3.3 The Finite Element Models

The topology of the current finite element models of all configurations are the conventional frame, stringer and skin configurations. These FE models are generated by the DLR in-house tool ModGen which has been described in Sect. 2. Although the parameterization concept of ModGen allows for the generation of finite element models regardless of the solver, MSC Nastran is the main finite element computer program used in MONA. Be-sides the FE models also the aerodynamic and mass models are important for the loads analysis and therefore for structural dimensioning of the air-craft. The aerodynamic and mass models as part of the loads analysis with MSC Nastran SOL144 (MSC Software Corporation 2010), is also set up with ModGen. Within MSC Nastran the Vortex Lattice Method (VLM) and the Doublet Lattice Method (DLM) are used, the latter for unsteady aerodynamic analysis. A slender body aerodynamic model is used for the fuselage. The aerodynamic methods are based on the potential theory and are fast methods for the estimation of motion-induced aerodynamic forces for sub-sonic flow. These are panel methods in which the lift surfaces of the air-craft, e.g. wing, are assumed to be flat panels and are subdivided into small, trapezoidal lift elements. The mass model represents the mass distribution of the structural masses, system masses for example engine or landing gear, payload as well as all remaining masses. The mass items are attached to the structure at the nodes of a so-called loads reference axis (LRA). The LRA is derived from the detailed FE model and is defined approximately along the elastic axis of the wing and the central axis of the fuselage. The LRA also transfers the aerodynamic loads to the load carrying structure, like the wing box.

Concerning the connection of aircraft components in the FE models, the aircraft wing, the HTP and the VTP are connected to the fuselage with the rigid body elements of form 2, RBE2 elements, of the MSC Nastran, MSC Software Corporation (2010). The rigid body elements of form 3, RBE3 elements, connect the LRA to the nearby aircraft structure. The aerodynamic loads and moments which are applied to the LRA can be transfer to aircraft structure via these RBE3 elements.

Aluminum 2024 is utilized as material for the aircraft structure because, in addition to being a lightweight material commonly used in aviation, it is less expensive and easier to manufacture than modern sandwich structures. For the structural sizing a safety factor of 1.5 is applied. The minimum thickness of the shell elements for the skin, rib and spar are limited to 2 mm while the minimum thickness of the twin boom shell elements are limited to 3 mm. These minimum thickness limits are used to constrain the manufacturing costs of plate and tube elements due to overly complex manufacturing processes which cannot meet the low-cost criterion. Furthermore, the optimization of the twin boom elements does not take buckling into account, so a corresponding minimum thickness had to be assumed.

3.3.1 The Gyroplane Finite Element Model

The FE model of the gyroplane consists of a fuselage, a wing, a tailplane and a rotor attachment as shown in Fig. 7. All components except the rotor attachment are generated by the in-house tool ModGen. The rotor attachment is modelled manually as a rectangular beam and shell elements. The beam element transfers the forces and moments of the rotor to the fuselage structure. It is aligned along the rotor shaft axis.

The fuselage is modelled in three parts being a front, a main and a tail part. The front part is modelled with less stringers than the other two parts due to the assumption that the loads acting on this structure area are much smaller. All of the fuselage frames

Fig. 6 Beam cross-sections
"I," "T" and "Z"

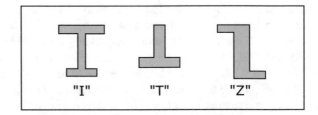

Fig. 7 The finite element
model of the gyroplane

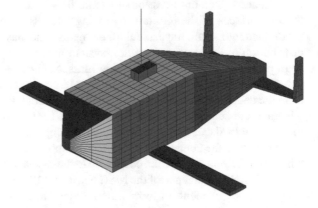

are modelled as a I Beam element which is a common applied element for the fuselage structure of a transport aircraft. In case of the wing and the HTP, the stringer and the spar caps are modelled with Z and T beam elements respectively. These are also common for the wing of a transport aircraft. The cross-sections of the beam elements are shown in Fig. 6. The cross-section dimension definitions are used according to MSC Software Corporation (2010).

The wing and the HTP are attached to the fuselage with RBE2 elements. The VTP is attached to the HTP. The rotor is rigidly connected to the fuselage.

The mass of each system component is modelled as a concentrated mass element connection of form 2, CONM2 element. Each CONM2 element including its corresponded moment of inertia with an offset of the center of gravity position is rigidly connected to the LRA. Table 5 summarizes the system component masses which are considered in the finite element model of the gyroplane configuration.

As mentioned previously, the rotor of the gyroplane generates approximately 70% of the total lift force during cruise flight. Therefore, corresponding aerodynamic panels are defined for VLM/DLM to account for the lift due to the rotor.

3.3.2   The Twin Boom Finite Element Model

The FE model of the twin boom consists of a fuselage, a wing, an inverted v-tail and a twin boom. Figure 8 shows the FE model of the twin boom. The twin boom was

Table 5 Masses of considered system components for the gyroplane

	Mass (kg)
Safety system	50
Engine	2×65
Rotor	100
Rotor equipment	140
Freight	2×500
Main landing gear	2×50
Rear landing gear	50
Fuel	125.4
Battery	30

Fig. 8 The finite element model of the twin boom

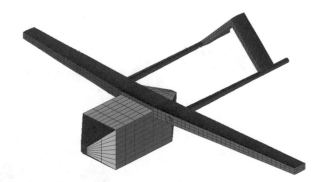

modelled manually with shell elements. The diameter of the twin boom varies from the v-tail toward the wing. In order to be able to withstand the bending moments due to the v-tail aerodynamic forces, the twin boom moment of inertia is increased by the increasing of the boom radius from the v-tail toward the wing.

The twin boom fuselage, wing and HTP finite element model are modelled with the structural topology and elements as same as the gyroplane finite element model described previously.

The wing is attached to the fuselage via RBE2 elements. The inverted v-tail is rigidly attached to the twin boom while the twin boom is then attached to the wing.

The mass of each system component is rigidly connected to the LRA with the same method used for the gyroplane described above. Table 6 summarizes the system component masses which are considered in the finite element model of the twin boom configuration.

3.3.3 The Box Wing Finite Element Model

The FE model of the box wing consists of a fuselage, a lower wing, an upper wing and a VTP. Figure 9 shows the FE model of the box wing. The box wing fuselage,

Table 6 Masses of considered system components for the twin boom

	Mass (kg)
Safety system	50
Payload	2×500
Main engine	128.8
Electrical engine	4×11
Main landing gear	2×45
Nose landing gear	37.7
Tank	45.3
Generator	18.7
Fuel	139.3
Battery	79.7
Parachute	100

Fig. 9 The finite element model of the box wing

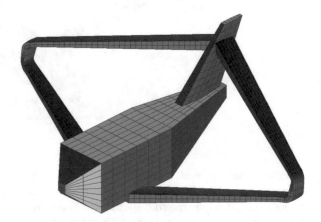

wing and VTP finite element model are modelled with the structural topology and elements as same as the gyroplane finite element model described previously. The lower wing and the VTP are attached to the fuselage via RBE2 elements. The upper wing is directly attached to the VTP. The lower and upper wings are rigidly connected to each other. The mass of each system component is rigidly connected to the LRA with the same method used for the gyroplane described above. Table 7 summarizes the system component masses which are considered in the finite element model of the box wing configuration.

Table 7 Masses of considered system components for the box wing

	Mass (kg)
Safety system	50
Payload	2 × 500
Main engine	2 × 120
Electrical engine	4 × 12.5
Main landing gear	2 × 45
Nose landing gear	37.7
Tank	53.7
Generator	20.2
Fuel	171.3
Battery	10
Parachute	50

4 Conceptual Loads and Structural Analysis

4.1 Load Cases and Conceptual Loads Analysis

The twin boom and the box wing configuration are fixed wing aircraft configurations with an estimated maximum take-off mass of 2343 and 2591 kg. Therefore, they are subjected to the Certification Specification Part 23, CS-23, of the European Aviation Safety Agency (EASA) for normal, utility, aerobatic and commuter category aeroplanes (EASA 2015) because the MTOM is lower than the MTOM limit of 5670 and 8618 kg mentioned in the CS-23.

Concerning the gyroplane configuration, the current EASA regulation (EC) 216/2008 annex I (EASA 2008) categorizes the gyroplane with a MTOM below 560 kg as an ultralight aircraft. However, the MTOM of the gyroplane configuration is 2449 kg. Thus, the gyroplane cannot be categorized as an ultralight aircraft. Currently, according to the author knowledge, there is no EASA certification specification which can be applied to the gyroplane. In addition, EASA answered a question from 2019 (EASA 2019), whether there is a regulation for gyrocopters, with the following lines:

2019-05-01: As of today, EASA has not issued any Certification Specification for gyroplanes. EASA recommends to check first the Maximum Take-off Weight of the gyroplane. If it is below 560 kg, national regulation shall be used: indeed, in accordance with the Regulation (EC) 216/2008, Annex I, gyroplanes with a maximum Take-off Weight below 560 kg belong to the ultralight class.

So far there is no gyroplane type certified by EASA. In case of an application for a type certificate of such a product, the EASA would address the certification basis through e.g. Special Conditions in accordance with Part 21. According to the author's opinion, it is suitable to refer the gyroplane configuration which has a MTOM lower than 3175 kg to the Certification Specification 27 of the EASA for small rotorcraft (EASA 2018).

Based on the above reasons, the gust and maneuver loads are calculated according to CS-23 amendment 4 (EASA 2015) for the twin boom and the box wing. In the case of the gyroplane, these loads are calculated according to the CS-27 amendment 6 (EASA 2018).

4.1.1 The Gust and Maneuver Load Cases

For the fixed wing configurations of the twin boom and the box wing, the positive limit maneuvering load factor is calculated as follows:

$$n_z = 2.1 \frac{24000}{MTOM + 10000} \tag{1}$$

Note that the unit for MTOM in the aforementioned equation is pounds [lbs]. The negative limit maneuvering load factor is 0.4 times the positive maneuvering load factor.

In case of the gyroplane a limit maneuvering load factor ranging from a positive limit of 3.5 to a negative limit of -1.0 is possible. Note that any positive limit maneuvering load factor not less than 2.0 and any negative limit maneuvering load factor of not less than -0.5 is appropriate. However, in this work the conservative values of the positive limit maneuvering load factor of 3.5 and the negative limit maneuvering load factor of -1.0 according to the CS-27 is used.

Concerning the gust loads determination, the EASA certification specifications require that gust load factors have to be calculated according to Pratt's formula as follows (Pratt 1953):

$$n = 1 \pm \frac{k_g \rho_0 u_g v_{EAS} c_{L\alpha}}{(W/S)} \tag{2}$$

Here, k_g is the gust alleviation factor, ρ_0 is the density of the air at sea-level, u_g is the derived gust velocities, v_{EAS} is the airplane equivalent airspeed, W/S is the wing loading and $c_{L\alpha}$ is the lift coefficient slope. The gust alleviation factor is defined as:

$$k_g = \frac{0.88\mu_g}{5.3 + \mu_g} \tag{3}$$

where μ_g is the airplane mass ratio and is calculated as follows:

$$\mu_g = \frac{2(W/S)}{\rho g \bar{c} c_{l\alpha}} \tag{4}$$

Here, ρ is the density of the air at the altitude considered, g is the acceleration due to gravity and $\bar{c}$ is the mean geometric chord.

Based on the Pratt's formula, an equivalent gust load factor can be determined for different mass configurations (the center of gravity position is not considered) and flight conditions as well as altitudes. In case of the CS-23 requirement, the positive (up) and negative (down) gust of 50 fps at a speed of v_c, 25 fps at a speed of v_d and 66 fps at a speed of v_b must be considered. In case of the CS-27 requirement the rotorcraft must be designed to withstand the loads resulting from a vertical gust of 30 fps at each critical airspeed. The unit fps is defined as feets per second whereby 1 fps is equivalent to 0.3048 m/s.

Although the UA of the ALAADy project operate at low altitudes above ground, several altitudes are taken into account for the load factor calculation for the maneuver and gust loads calculation. The consideration of further altitudes than sea-level is motivated by a possible mission scenario which is considered in the ALAADy project which requires the UA to be able to fly over, for example, the Alps or other mountainous landscapes. In Table 8 all considered altitudes are listed.

In Table 9 the maximum positive and minimum negative maneuver and gust load factors are listed for all configurations.

With the above mentioned load factors, a quasi-stationary trim analysis is carried out. The loads of each configuration are estimated by static aeroelastic trim analysis using MSC Nastran Solution 144 for the main structure. The basic equation is defined as

$$(K_{aa} - q Q_{aa})u_a + M_{aa}\ddot{u}_a = q Q_{ax} + P_a \tag{5}$$

Table 8 The considered altitude levels

Sea-Level	1000 m	2000 m	3000 m	4000 m	5000 m	6096 m

Table 9 The load factor of all configurations at sea-level

	Gyroplane	Twin boom	Box wing
Pos. 50 fps gust		3.102	2.691
Neg. 50 fps gust		−1.102	−0.691
Pos. 25 fps gust		2.314	2.057
Neg. 25 fps gust		−0.314	−0.057
Pos. 66 fps gust		2.344	1.981
Neg. 66 fps gust		−0.344	0.019
Pos. 30 fps gust	1.386		
Neg. 30 fps gust	0.614		
Pos. maneuver	3.5	3.683	3.627
Neg. maneuver	−1	−1.473	−1.451

Where K_{aa} is the structural stiffness matrix, Q_{aa} the dynamic influence coefficient matrix, M_{aa} the structural mass matrix, Q_{ax} a matrix which provides the forces at the structural grid points due to deflection of aerodynamic extra points (e.g. aileron), P_a is the vector of applied loads and q is the dynamic pressure. The following parameters are taken into account as known and unknown in the trim analysis:

- Angle of attack α
- Pitch rate $\dot{\Theta}$
- Pitch acceleration $\ddot{\Theta}$
- Load factor n
- Elevator deflection η_{ELV}
- Mach number

According to the load case, the angle of attack and pitch acceleration are unknown variables in case of a gust load. The load factor, mach number, elevator deflection and pitch rate are the known parameters. For a maneuver load, the elevator deflection is also unknown, while the pitch acceleration is additionally known.

In case of the maneuver loads during the static aeroelastic trim analysis the control surfaces are fixed to their neutral position, except for the elevator surfaces. For the gust load cases the elevator surfaces are also fixed to their neutral position. Furthermore, the pitch acceleration is only permitted for gust loads and not for maneuver loads, so it is a free parameter.

The resulting forces and moments acting on the finite element nodes are the aerodynamic and inertia loads with $P_a^{aero} = q Q_{aa} u_a$ and $P_a^{inertia} = M_{aa} \ddot{u}_a$ respectively. While the first iteration considers the structure as rigid, the second iteration takes the structural flexibility into account. In total 64 and 57 load cases are considered for the twin and box wing, and gyroplane, respectively.

4.1.2 The Parachute and Landing Load Cases

As the result of the safe emergency landing according to the minimum landing impact energy requirement which is described in Nikodem et al. (2021), the parachute load case must be taken into account for the twin boom and box wing aircraft. The aerodynamic force generated by the parachute during its opening phase is the result of the initial conceptual design and analysis process of the UA configuration performed by Sachs (2021). It is assumed that the termination of the aircraft and thus the release of the parachute occurs during a 1 g cruise flight.

Concerning the landing load case, based on the short review which is described in Cumnuantip (2014), during the aircraft conceptual design phase it is commonly assumed that 85% of the total landing load is applied at the main landing gear and 15% of the total landing gear load is applied to the rear or nose landing gear. In case of the gyroplane, 15% of the total landing load is applied to the rear landing gear and 85% of the landing load is applied to the main landing gear at the wings. The total landing load is assumed to be 1.5 g of MTOM. In case of the fixed wing configurations, 85% of the landing load is applied to the main landing gear and 15% of the landing load to the nose landing gear.

4.2 Conceptual Structural Analysis

The structural sizing is performed by the application of a structural optimization algorithm with MSC Nastran SOL200 (MSC Software Corporation 2010), where a mathematical optimization task is defined and executed with gradient-based mathematical optimization algorithms. The structure is sized component-wise.

A gradient-based sizing optimization is applied in this work. The optimization problem definition is expressed as follows:

$$Min\{f(x)|g(x) \leq 0; x_l \leq x \leq x_u\} \tag{6}$$

Here, f is the objective function, x is the vector of n design variables, g is the vector of m_g inequality constraints and R^n is the n-dimensional Euclidean space. The so-called design model consists of the objective function f, the design variables x and the constraints g. The mass of the aircraft structure is specified as the objective function. In case of wing structures, the thicknesses of the skin fields bordered by two consecutive ribs and two spars are defined as the design variables. In addition, the spar web thickness between two ribs and the web thickness of a rib are also defined as design variables. The fuselage skin field thickness bordered by two frames and two stringers is also specified as one design variable. The twin boom has 7 design fields distributed along the length of the boom.

The constraints are allowable maximum Von-Mises stress value for each shell element and minimum buckling safety factor for each element. In the case of the buckling safety the allowable buckling stress is calculated. According to Bruhn (1973), the estimation of the allowable buckling stress is based on a simplified analytical formula where a buckling field is assumed to be a rectangular skin field and is defined as follows:

$$\sigma_{buck} = \frac{k_c \pi^2 E}{12(1 - v^2)} \left(\frac{t}{b}\right)^2 \tag{7}$$

Here, E is the Young's modulus, v is the Poisson ratio, t is the actual thickness of the buckling field and b is the shortest edge of the buckling field. The parameter k_c depends on the boundary and load condition. The parameter b of the buckling fields is also estimated due to the parameterized structure geometry. In the case of the skin the compression buckling is considered. On the other hand, the shear stress buckling is considered for the spars and ribs. A typical buckling field at the wing box is the upper or lower skin between two consecutive ribs and to successive stringer.

At the beginning of the sizing process, a preliminary cross-section sizing based on cutting loads and the use of analytical methods within the in-house software MONA is performed prior to the optimization. Herewith, the size of the cross-section of the beam elements such as stringer, spar and fuselage frame are estimated. The beam elements are thus dimensioned for the corresponding loads and relieve the skin elements.

The sizing process is an iterative process where the different load cases and the static aeroelastic trim analysis are considered. The forces and moments resulting from the trim analysis also depend on the mass model (distributed masses and center of gravity position) of the aircraft. Because a change of the overall aircraft mass at the end of the optimization also changes the loads and moments from the trim analysis, the sizing of the structure is performed iteratively until convergence is achieved. While the loads calculation is done with the complete FE model, the structural optimization is a component-wise sequential process. After the sequential optimization of the major components is finished, a new global sizing loop begins and each component is sized again until it converges. This iteration process is illustrated in Fig. 10.

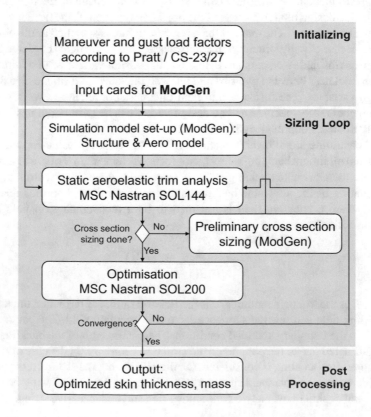

Fig. 10 The structural sizing and optimization process

5   Structural Optimization Results of the Configurations

In Fig. 11 the skin thickness distribution results of the gyroplane, twin boom and box wing configuration without the consideration of the parachute load case are shown. It should be noted that the design variable minimum limit is set to be 0.002 m for the skin thickness and 0.003 m for the twin boom cylinder thickness to constrain the manufacturing costs.

In case of the gyroplane structural sizing result, the fuselage skin panel thickness is mostly 0.002 m. However, certain reinforcements at the rear fuselage section are noticed. No reinforcement at the upper area of the main fuselage section is noticed, the skin panels with minimum thickness design limits are able to withstand the rotor system forces which are applied to this structural area. The skin panels at the wing root area are also reinforced. This is due to the bending moment from the wing aerodynamic force. A relatively thin cross-sectional area of the symmetrical airfoil is used at the wing root. In order to compensate for the effect from the bending moments, the skin panels at the wing root are thickened.

In case of the twin boom structural sizing result, the skin panels are significantly thicker at the area of the fuselage, wing and twin boom intersection. The major cause of this reinforcement is the bending moment generated by the v-tail lift force which

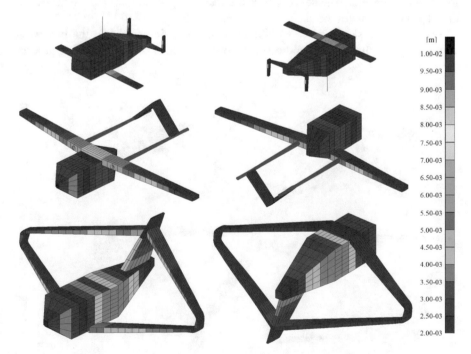

Fig. 11 The skin thickness distribution result of the gyroplane, the twin boom and the box wing configuration without the consideration of the parachute load case in the sizing process

Table 10 The converged masses of the gyroplane, the twin boom and the box wing configuration without the consideration of the parachute load case in the sizing process

	Initial mass (kg)	Converged mass (kg)	Deviation (%)
Gyroplane	2449	2720	11.05
Twin boom	2343	2813	20.05
Box wing	2591	3218	24.21

is transferred to this structural area via the twin boom rod. It can also be observed that the skin panels at the lower part of the fuselage rear section show a larger thickness. This is due to inertia forces from the attached system components and engine masses at this area.

Concerning the box wing structural sizing result, the skin panel thicknesses at the rear fuselage section are also increased. This reinforcement is due to the effect of the inertia forces from the attached system components and engine masses at this area. The aerodynamic forces from the VTP and the aerodynamic forces from wings which are attached to the rear fuselage also contribute to this reinforcement.

In Table 10 the converged masses of the sizing process of each aircraft and the deviation towards the initial masses are listed. According to the result, without the consideration of the parachute loads in the sizing process, the structural mass of the gyroplane is the least. The deviations of the fixed wing aircraft optimized masses to the TLARs mass are of the same order of magnitude.

In Fig. 12 the skin thickness distribution of the twin boom and the box wing configuration with the consideration of the parachute load case in the sizing process are shown.

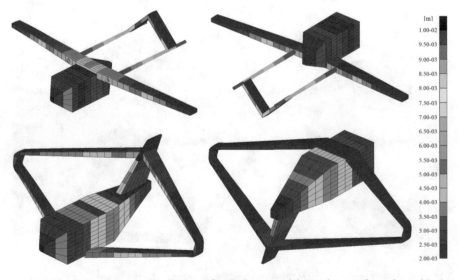

Fig. 12 The skin thickness distribution of twin boom and box wing configuration with the consideration of the parachute load case in the sizing process

Table 11 The converged masses of the gyroplane, the twin boom and the box wing configuration with the consideration of the parachute load case in the sizing process		Initial mass (kg)	Converged mass (kg)	Deviation (%)
	Gyroplane	2449	2720	11.05
	Twin boom	2343	2860	22.06
	Box wing	2591	3229	24.61

In case of the twin boom, the parachute load case has a remarkable impact on the structural skin panel thickness optimization result. This load case causes a reinforcement of the structure of the twin boom compared to the twin boom structure results from the sizing process without the consideration of the parachute load. The tension forces from the parachute also lead to causes a reinforcement of the structural panel at the attachment area of the twin boom to the wing. In addition, the parachute forces also lead to an increased thickness of the panels of the v-tail compared to the results from the sizing process without the consideration of the parachute forces.

In case of the box wing, the parachute load case has a minor impact on the structural mass but slightly on the mass distribution because the force of the parachute is attached to the rear fuselage section. In contrast to the twin boom, the rear fuselage area is already optimized for higher forces and moments from other load cases. Therefore, the parachute load case has a minor effect on the structural mass.

In Table 11 the converged masses of the sizing process of each aircraft and the deviation towards the initial values are shown. The consideration of the parachute loads leads to the increase of the structural mass of the twin boom and a slight increase of the structural mass of the box wing compared to the results shown in Table 10. The gyroplane does not require a parachute system due to the autorotation of the rotor system that can guarantee the safe emergency landing. These factors finally lead to the result that the structural mass of the gyroplane is the minimum among the three configurations.

The deviation of the optimized structural masses in Table 11 with respect to the initial masses is smallest for the gyroplane. It should be noted that the initial masses are calculated by empirical methods. Such empirical methods do not consider the special parachute load cases. As a result, the structural masses of the twin boom deviate from the initial masses. However, in case of the box wing, almost no effect is noticed since the parachute force is attached to a structure where already higher forces and moments are considered.

6 Summary

This chapter has shown the conceptual loads analysis and the conceptual structural design of three unmanned cargo aircraft concepts within the ALAADy project of DLR. Three aircraft concepts, namely a gyroplane, a fixed high-wing with a twin boom v-tail aircraft and a box wing aircraft are investigated in this work. The parametric finite element model generation and design process, MONA, of the DLR is used for the conceptual structural design of these three unconventional aircraft configurations.

Based on the current specific mission requirements and the current applied load cases, the gyroplane has the minimum structural mass of 2720 kg. The optimized twin boom and box wing aircraft have already a higher mass than the gyroplane even without the parachute load case. The special requirement of a safe termination has only a slightly disadvantageous effect impact on the twin boom and box wing configuration. The parachute forces result in additional structural reinforcements at the tail boom and v-tail. This finally ends up in a higher structural mass of the twin boom and box wing aircraft configuration compared to the structural mass of the gyroplane.

References

Aptsiauri G (2021) Concepts of full-electric and hybrid-electric propulsion and operation risk motivated integrity monitoring for future unmanned cargo aircraft. In: Dauer JC (ed) Automated Low-Altitude Air Delivery - Towards Autonomous Cargo Transportation with Drones. Springer, New York, Berlin, Heidelberg.

Bruhn EF (1973) Analysis and Design of Flight Vehicle Structures. Tri-State Offset Co, Cincinnati

Cumnuantip S (2014) Landing Gear Positioning and Structural Mass Optimization for a Large Blended Wing Body Aircraft, Dissertation, Technical University of Munich

Dauer JC, Lorenz S, Dittrich JS (2016) Automated low altitude air delivery. In: Deutscher Luft- und Raumfahrtkongress, Braunschweig, Germany, 13–15 September 2016

Dauer JC, Dittrich JS (2021) Automated cargo delivery in low altitudes: concepts and research questions of an operational-risk-based approach. In: Dauer JC (ed) Automated Low-Altitude Air Delivery - Towards Autonomous Cargo Transportation with Drones. Springer, New York, Berlin, Heidelberg.

EASA (2008) Regulation (EC) No 216/2008, Annex I, of the European Parliament and of the Council of 20 February 2008 on common rules in the field of civil aviation and establishing a European Aviation Safety Agency. https://www.easa.europa.eu/document-library/regulations/regulation-ec-no-2162008, Accessed 10 May 2021

EASA (2015) CS-23 Amendment 4: Certification Specifications for Normal, Utility, Aerobatic, and Commuter Category Aero-planes. https://www.easa.europa.eu/certification-specifications/cs-23-normal-utility-aerobatic-and-commuter-aeroplanes. Accessed 10 May 2021

EASA (2018) CS-27 Amendment 6: Certification Specifications and Acceptable Means of Compliance for Small Rotorcraft. https://www.easa.europa.eu/certification-specifications/cs-27-small-rotorcraft. Accessed 10 May 2021

EASA (2019) Frequently Asked Questions No. 46187. https://www.easa.europa.eu/faq/46187. Accessed 10 May 2021

Hasan YJ, Sachs F (2021) performance based preliminary design and selection of aircraft configurations. In: Dauer JC (ed) Automated Low-Altitude Air Delivery - Towards Autonomous Cargo Transportation with Drones. Springer, New York, Berlin, Heidelberg.

Klimmek T (2014) Parametric set-up of a structural model for FERMAT configuration for aeroelastic and loads analysis. J. Aeroelasticity Struct Dyn 3(2):31–49

MSC Software Corporation (2010) MD/MSC Nastran 2010 Quick Reference Guide. https://simcom panion.mscsoftware.com/infocenter/index?page=content&id=DOC9467&actp=RSS. Accessed 10 May 2021

Nikodem F, Rothe D, Dittrich J (2021) operations risk based concept for specific cargo drone operation in low altitudes. In: Dauer JC (ed) Automated Low-Altitude Air Delivery - To-wards Autonomous Cargo Transportation with Drones, p x–y. Springer, New York, Berlin, Heidelberg.

Pratt KG (1953) A Revised Formula for the Calculation of Gust Loads. National Advisory Committee for Aeronautics (NACA) Technical Notes: 2964

Sachs F (2021) Configurational aspects and vehicle specific investigations for future unmanned cargo aircraft. In: Dauer JC (ed) Automated Low-Altitude Air Delivery - Towards Autonomous Cargo Transportation with Drones. Springer, New York, Berlin, Heidelberg.

Concepts of Full-Electric and Hybrid-Electric Propulsion and Operation Risk Motivated Integrity Monitoring for Future Unmanned Cargo Aircraft

Gubaz Aptsiauri

Abstract Besides lower environmental impact, full and hybrid-electric propulsion systems offer an opportunity to set-up a highly redundant powertrain. As safety is one of the important topics for the design of unmanned aircraft systems (UAS), concepts of full-electric and hybrid-electric propulsion have been investigated in terms of meeting: mass, volume, power, energy and safety constraints. In the first step, the powertrain system has been sized according to preliminary design data described in the project Automated Low Altitude Air Delivery (ALAADy) using generic data of state-of-the-art system components. In the next steps the powertrain system has been analyzed and modelled with simulation models to investigate the system performance in more detail. Optimal powertrain system architectures concerning system overall mass and redundancy have been set up especially for the vehicles designed for the ALAADy project. The differences between full-electric and hybrid-electric powertrain systems is discussed as well.

Keywords Hybrid-electric · Fuel cell · Battery · Inherent safety · Simulation · UAS

1 Introduction

In the ALAADy project different types of aerial vehicles were investigated to choose a suitable air vehicle. This chapter focuses on the work which has been done to investigate the powertrain of three vehicles: double tail boom fixed wing aircraft, box wing aircraft and gyrocopter as proposed in a pre-study (Hasan and Sachs 2021). The ALAADy-vehicle powertrain has been set-up as a hybrid powertrain with the main power source: fuel cell stacks (FC-stack) or an internal combustion engine (ICE) and the boost power source for high power demands during start and emergency conditions provided by the batteries. As various hybrid powertrain systems can provide a

G. Aptsiauri (✉)
German Aerospace Center (DLR), Institute of Engineering Thermodynamics,
Pfaffenwaldring 38–40, 70569 Stuttgart, Germany
e-mail: gubaz.aptsiauri@dlr.de

© Deutsches Zentrum für Luft- und Raumfahrt e. V. (DLR) 2022   185
J. C. Dauer (ed.), *Automated Low-Altitude Air Delivery*, Research Topics in Aerospace,
https://doi.org/10.1007/978-3-030-83144-8_8

different level of redundancy and have several advantages and disadvantages, determination of the optimum system architecture for the ALAADy vehicles is the main topic of the presented work.

To assess the powertrain concept containing: batteries, FC-stacks/ICE, electric motor and power electronics, a literature research with respect of state-of-the-art (SOA) components has been done as the first step. According to the research results, estimations regarding the powertrain mass, volume and costs have been made and preliminary powertrain configurations have been set-up for all three ALAADy-vehicles. In the next step, simulation models of powertrain components and the entire powertrain have been developed to be able to simulate the entire flight mission of the ALAADy vehicle and to assess vehicle performance in more detail. Based on a preliminary powertrain system architecture and simulation models a powertrain system with inherent safety capability has been set-up and an online monitoring concept of this system has been developed for safe operation monitoring according to Schirmer and Torens 2021. In this context inherent safety means that a failure of one of the major components of the powertrain system does not lead to an immediate loss of control over the vehicle and does not necessarily cause an immediate emergency landing even.

2 Hybrid Powertrain Configuration

2.1 State of the Art System Components

In the first step, characteristic data for the SOA system components has been gathered to determine: gravimetric and volumetric energy densities and power densities of system components. This allowed us to determine the possibility of hybrid powertrain integration into the ALAADy-vehicles and to decide which hybridization concepts are feasible. In the second step the powertrain system configuration has been developed according to the preliminary design data of the vehicles (power demand, energy demand). Market analysis and studies to determine the costs of the powertrain components have also been done to provide information to other working groups of the project to investigate the business case of the ALAADy concept.

2.1.1 Energy Storage and Conversion Components

Next to diesel and gasoline, two energy sources have been considered for the powertrain systems: secondary batteries and hydrogen which is used to supply fuel cells. Lithium-ion-batteries with several chemistries (Nickel Manganese Cobalt, Nickel Cobalt Aluminum, Lithium Iron Phosphate, Lithium Cobalt Oxide, etc.) have been investigated in more detail for the project. Other technologies as Nickel–metal hydride or Nickel–Cadmium batteries have very low gravimetric energy densities and post lithium-ion battery technologies e.g. lithium-air batteries are not commercially available yet, therefore, those are not considered for further investigations.

Batteries

Batteries or rather battery systems contain several battery cells which are connected in series or parallel in battery modules. Battery modules are assembled to battery systems which contains also all the necessary sensors, electronics and in some cases cooling components to monitor and operate the batteries. Major characteristics of the battery system are volumetric and gravimetric energy densities. SOA commercially available battery cells reach densities of 263 Wh/kg and 676 Wh/l (Panasonic 2021; Moore 2016). Specially optimized battery cells can reach energy densities >300 Wh/kg. Because of the additional mass and the volume of the battery system components (structure, electronics, cooling) the energy densities of commercial battery systems do not exceed 150 Wh/kg and 230 Wh/l. Future estimations for the battery systems are 350 Wh/kg and 750 Wh/l (Meesus 2018). Based on gathered data an energy density of 200 Wh/kg is considered as a conservative assumption for a UAS powertrain battery system and is used to calculate the battery system weight in calculations shown below.

Fuel Cells and Hydrogen Tank

SOA applications of fuel cells for mass optimized systems use gaseous (with a pressure of 350 or 700 bar) or liquid-hydrogen tanks with low pressure ($\leq$10 bar). Other tank systems e.g. metal hydride tanks will not be considered in the project because of their high weight or their low technology readiness level.

Pressurized hydrogen tanks with 700 bar have higher volumetric energy density than the tanks with 350 bar but the costs for such tanks because of the higher manufacturing effort are higher (R. Kochhan). SOA pressurized hydrogen tank systems reach gravimetric energy densities of $1,4 \pm 0,04$ kWh/kg and volumetric energy densities of $0,81 \pm 0,01$ kWh/l. In near future densities of 1,8 kWh/kg and 1,3 kWh/l on tank system level are feasible (New 2011).

Liquid hydrogen tanks have significantly higher gravimetric and volumetric energy densities. Liquid hydrogen Tanks with 15wt. %H_2 (percentage of H_2 weight referred to the overall tank system weight including H_2) have already been produced for automotive applications and as a spin-off of the space applications liquid tank systems with up to 28wt. % H_2 are feasible (Crespi 2017). For comparison: pressurized hydrogen tanks are in a range of 4–6 wt. % H_2.

For powertrain weight estimations described in Sect. 2.2, only liquid hydrogen storage systems with 15wt. %H_2 and 28wt. % H_2 are considered due to highest Energy density.

SOA fuel cell stacks can reach output power densities of up to 3.5 kW/kg at an efficiency above 50%. Besides the fuel cell stack the whole fuel cell system contains many balance of plant (BoP) components (e.g. compressor for air supply, coolant pump for cooling system etc.). These components are mainly defining the volumetric and gravimetric energy densities of the fuel cell systems. In this project the fuel cell system has been investigated in more detail to calculate the actual system weight and volume. The approach and calculation results will be discussed in more detail below.

Electric Motor

Because of high efficiency and high power density a permanent magnet synchronous motor is expected to be most suitable for an ALAADy vehicle. Siemens has tested an electric motor with: 260 kW continuous power, 50 kg weight and 95% efficiency in 2015 successfully. This results in a power density of 5 kW/kg. This motor has been tested in an aircraft as the main engine in 2016. Commissioned by NASA a demonstrator with 6,6 kW/kg with 95–97% efficiency and with up to 100 kW power has been designed and tested.

According to NASA roadmap power densities of: 16,5 kW/kg till 2025 and 19,7 kW/kg till 2030 are feasible. With Kryo-technology even higher power densities, up to 41,1 kW/kg in 2035, of the electric motors are expected (Rosario 2014).

2.1.2 Cost Estimations

The costs of the energy storage and conversion systems such as fuel cells, batteries and electric motors are topic of numerous scientific works. Actual estimations as well as predictions for the development of those costs vary significantly. Hence, it is not possible to make precise and reliable statements in this field. Nevertheless, data from several studies has been gathered to be able to roughly estimate the costs for the powertrain components. Table 1 shows an overview of cost estimations according to

Table 1 Cost estimations for powertrain components (Bubeck 2016)

Description	Unit	2015	2030	2050
Combustion engine (petrol)	€/kW	54	58	58
Combustion engine (diesel)	€/kW	74	86	86
EREV[a] ICE (petrol)	€/kW	50	55	55
EREV ICE (diesel)	€/kW	71	82	82
Electric motor	€/kW	20	14	7
PEM fuel cell	€/kW	284	65	48
Fuel tank	€/kWh	1.9	1.9	1.9
Hydrogen tank	€/kWh	22	11	11
BEV[b]-battery	€/kWh	476	260	200
EREV battery	€/kWh	595	325	250
HEV[c]-battery	€/kWh	952	520	400
Controller, converter, Inverter	€/kW	27	18	18

[a]Extended-range electric vehicles
[b]Range-extended battery-electric vehicle
[c]Hybrid-electric-vehicles

(Bubeck 2016). In Table 1 the prices are referred either to the power of the components in kW or to the energy of the components in kWh. It can be observed that prices of fuel cells and hydrogen storage tanks are currently significantly higher compared to the prices of ICE and fuel tanks. However it is assumed that fuel cells will have lower price by 2050 compared to ICE and Hydrogen tanks will be available for two times lower price by 2030.

Cost for the SOA battery packs for automotive applications are 190–200 $/kWh and 100 $/kWh are predicted for 2030 (Kochhan 2014).

Fuel cells have very high costs of approx. 280 $/kW due to very low production figure. Assuming a production of 500.000 pieces/year 55 $/kW are possible (Spendelow 2013). The US Department of Energy has defined the targets for the fuel cells with 30 $/kW at 60% efficiency with 5000 h lifetime (Dinh 2014).

Costs for present pressurized Hydrogen tanks are estimated with 1000 €/kgH$_2$ where a SOA tank system with 5 kg Hydrogen capacity is the baseline (Kochhan 2014). For 2030 300 €/kgH$_2$ are predicted.

Present costs for the automotive powertrain electric motors are 15 €/kW. Power electronics cost is almost in the same range (Kochhan 2014).

Summarizing the above, it can be said, that fuel cells and hydrogen tank systems will be significantly more expensive compared to combustion engines and fuel tanks in the near future (by 2030). There is a big potential of reducing fuel cell costs below the ICE costs but hydrogen tank systems will still be more expensive than fuel tank systems even in distant future.

2.2 Powertrain System Design Based on Available System Components and Vehicle Preliminary Design Criteria

According to the data which has been gathered for the SOA components and the preliminary design of the vehicles first concepts of the system architecture have been set up. With this, it was possible to determine if the overall mass and power of the hybrid-electric powertrain system can be integrated into an ALAADy vehicle. Two main concepts have been considered:

1. Fuel cell system with battery system, with the fuel cell system supplying cruise power and the battery acting as a boost for Take-off and climb phases.
2. Internal combustion engine (ICE) or gas turbine (GT) with a battery system acting as a boost for Take-off and climb phases.

Required components for both concepts have been identified and the performance data of the components has been gathered.

2.2.1 Components

Fuel Cell System
For further calculations data of two different fuel cell stacks has been chosen:

- Stack of Toyota with 2 kW/kg and 3,1 kW/l power density (Toyota 2016)
- Stack of Elring Klinger with 3,3 kW/kg and 3,4 kW/l (ElringKlinger AG 2017)

To determine the hydrogen consumption and the efficiency of the fuel cell system at several power settings tests in the laboratory with a 30 kW fuel cell stack have been done and analyzed. Furthermore, available data of the 4-Seater fuel cell-battery-hybrid aircraft HY4 regarding the weights and power consumptions of the balance of plant (BoP) components has been analyzed and the overall system mass and maximum power output has been determined. As DLR was involved in the development of the HY4 powertrain a very detailed data of the powertrain system was available including the data of the smallest parts of the system e.g. relays and fuses.

150 kW systems with all the necessary BoP components have been set up and the masses and volumes have been calculated. Table 2 shows the summarized results of the fuel cell system data. It has to be considered that the calculation of the weights of the BoP components have been done with conservative assumptions which lead to a higher system weight which can be optimized (reduced) for a real application if the focal point will be set on that issue.

Hydrogen Tank System
As already mentioned above, hydrogen storage systems with gaseous tanks have significantly higher mass than the liquid hydrogen storage systems. The first preliminary calculations have shown, that gaseous tank systems are not suitable for ALAADy vehicles. For dimensioning of the powertrain system two options of the tank systems have been considered: Liquid H_2 Tank System with 15 wt.% H_2 (system already designed and tested for automotive applications) and a tank system with 28 wt.% H_2 which is expected to be available for aircraft applications until 2025 (Crespi 2017).

Battery System
SOA battery cell data has been gathered and a system with all the necessary BoP components with the same approach as for the fuel cells has been designed. Cooling system, safety electronics, cabling and housing weights have been estimated. These calculations have shown that battery system gravimetric energy densities of 200

Table 2 Fuel cell system weight and power density

Stack power density [kW/kg]	Stack nominal power [kW]	Stack weight [kg]	System weight [kg]	System power [kW]	System power density [kW/kg]
2	150	75	182,5	133,5	0,731
3,3	150	45,5	153	133,5	0,872

Table 3 Turboprop weight and fuel consumptions

Turbine	Max. power [kW]	Weight [kg]	Fuel consumption during cruise [kg/kW/h]
RR500	298	102	0,411
PBS TS 100	160	61,3	0,548

Table 4 ICE weight and fuel consumption

ICE	Max. power [kW]	Weight [kg]	Fuel consumption at cruise power [kg/h]
Graflight- V8	261	240	35
TIO-540-AE2A	261	269	53
TSIO-360-A	156,6	128,8	32,4

Wh/kg and 1 kW/kg are feasible with SOA components. This result also corresponds to the data shown in Sect. 2.1.1.

Electric Motor and Generator

According to the generic data research, 5 kW/kg for the electric motor including power electronics has been assumed as the power density for further calculations. A power density of 7,9 kW/kg (Anghel 2015) has been assumed to be feasible for the generator which will be used to convert mechanical energy from an ICE or turboprop in electric energy to supply the electric motors.

Gas Turbine and Internal Combustion Engine

Data for the SOA turboprop engines, which can provide enough power for cruise has been gathered based on manufacturer's datasheets. Two different gas turbines have been considered for further calculations:

- RR500 for box wing and gyrocopter configurations
- PBS TS 100 for the tail boom configuration

Table 3 shows turboprop engine data. Next to turboprop engine data, also data for the ICEs has been gathered to be able to decide which engine will be most suitable for an ALAADy vehicle for the desired flight mission with 600 km range in terms of overall mass of the powerplant including engine and fuel. Both gasoline and diesel ICEs which are certified for aerospace applications have been considered. Data of the ICEs is given in Table 4.

2.2.2 Powertrain System Preliminary Design

For all ALAADy vehicle configurations hybrid powertrain system weight and volumes have been estimated both for full electric (Fuel Cell + Battery) and hybrid electric powertrains (ICE/GT + Battery). For Hybrid electric powertrain the option with the lowest overall mass (engine eight + fuel weight) for the desired 600 km mission resulted in 2 different ICE engine types for ALAADy vehicles:

Fig. 1 Powertrain mass overview of the Gyrocopter

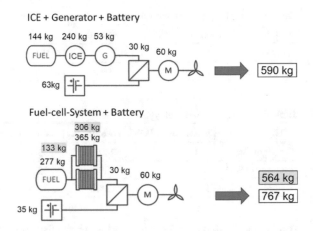

- Diesel engine Graflight-V8 for Box wing and Gyrocopter vehicles
- Gasoline engine Continental TSIO-360-A for the double tail boom vehicle

Although the gas turbines have a significantly higher gravimetric power density in comparison to the ICEs, the higher fuel consumption of the small gas turbines (<350 kW max. Power) lead to a higher overall mass of the powertrain system including the fuel and tank system. Figure 1 shows the powertrain mass estimations for the 600 km Mission with the ALAADy gyrocopter described in (Hasan and Sachs 2021). Besides ICE + Battery hybrid also two different masses for the full electric (Fuel Cell + Battery) have been calculated. One calculation has been made with available liquid hydrogen SOA tank with 15 wt. % H_2 and a fuel cell system with a fuel cell stack with 2 kW/kg power density. The second calculation has been made with a tank system which is feasible to be available in the near future with 28 wt. % H_2 (Crespi 2017) and a SOA fuel cell stack with 3,3 kW/kg power density. The weights of the components with lighter fuel cell stack and hydrogen tank are highlighted in yellow in the figure below. Figure 1 shows that due to lower battery mass and the absence of the generator the full electric powertrain of the gyrocopter can be lighter if a liquid hydrogen tank system with 28 wt. % H_2 would be integrated in the vehicle.

Similar calculations have been done for the other two ALAADy vehicles: Box wing and double tail boom. Also, these calculations show that a full electric powertrain of the ALAADy vehicle is slightly lighter than the hybrid electric powertrain if the liquid hydrogen tank with a high energy density mentioned above would be integrated. Figures 2 and 3 show the powertrain component weights. As the double tail boom configuration has the lowest power demand, the overall weight of the powertrain is more than 100 kg less than for the other configurations resulting in more payload for the mission. But as already mentioned in (Hasan and Sachs 2021) the gyrocopter has several advantages for flight operation regarding safety which make this vehicle a preferred vehicle for ALAADy missions.

Due to a lower power demand of the double tail boom vehicle only one fuel cell stack with a power of 150 kW has been chosen as the best solution in terms of

Fig. 2 Powertrain mass overview of the double tail boom

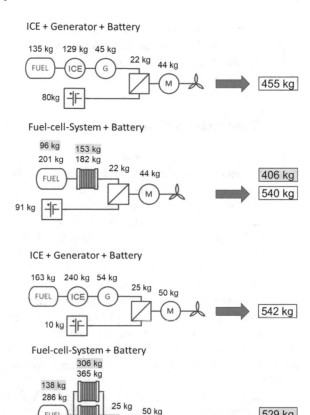

Fig. 3 Powertrain mass overview of the box wing

powertrain mass. Therefore, gyrocopter and box wing in these shown full electric configurations have a higher level of redundancy in powertrain system due to less power loss in case of a fuel cell stack failure.

2.3 Powertrain System Detailed Design with High Level of Safety

A possibility to achieve a high level of redundancy and safety with hybrid or full electric powertrain was the main topic of the work described in this chapter. Calculations in Sect. 2.2 show that there is potential of having lower weight of the powertrain system when using full electric powertrain instead of a hybrid one. To be able to investigate the powertrain performance in more detail a detailed design of the full electric powertrain system has been performed.

In the first step a stationary model of a full electric powertrain has been set up in EES (Engineering Equation Solver) to determine the needed size and amount of fuel cell stacks and batteries. In the second step the optimum powertrain architecture in terms of redundancy and system weight was determined.

2.3.1 Powertrain Configuration

A stationary powertrain system simulation model has been set up in EES to analyze the performance of the different system configurations. With these calculations it is possible to match the fuel cell system with the battery and electric motor system in terms of voltage and current specifications. So, this is the next step in powertrain design after setting up a rough powertrain system with generic data described in Sect. 2.2.2. Figure 4 shows the user interface of the model. Battery and fuel cell degradation and the resulting voltage difference between BoL (Begin of Life) and EoL (End of life) has also been implemented in this model to be able to simulate realistic operational scenarios. With this model it was possible to calculate the system configuration characteristics given in Table 5. Overall system weight has been calculated including the e-motor system but excluding the tank system mass. In these calculations an EMRAX 348 e-motor data has been used. This e-motor is used in different full electric aircrafts e.g. "Alpha Electro" of Pipistrel Aircraft (Pipistrel Aircraft 2021) and represents a state-of-the-art engine for airborne applications in serial production.

In Table 5 different configurations of the fuel cell stacks are shown. "2s2p-340" means in this case that we have two fuel cell stacks connected in series and this double stack is connected in parallel with another double stack so that we have four fuel cell stacks in the whole system. 340 is the amount of the cells per stack. More than four stacks in the system are not considered because with higher amount of stacks

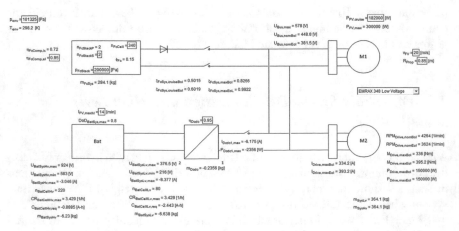

Fig. 4 User interface of the powertrain model

Table 5 Powertrain system configurations with detailed design

Parameter	2s2p-340	3s1p-355	3s1p-450	2s1p-500	1s2p-500
Parallell FC stacks	2	1	1	1	2
Serial FC stacks	2	3	3	2	1
FC cells per stack	340	335	450	500	500
Battery weight [kg]	−6.656	215.9	3.126	216.2	216.2
FC system weight [kg]	284.1	236.7	270	220.5	219.6
Overall system weight [kg]	364.1	540.2	350	524.4	523.5
FC max voltage [V]	578	854.3	1148	850	425
FC nominal voltage BOL[a] [V]	448.8	663.3	891	660	330
FC nominal voltage EOL[b] [V]	381.5	563.8	757.4	561	280.5
Drive max current BOL [A]	334.2	226.1	168.4	227.3	454.5
Drive current EOL [A]	393.2	266	198.1	267.4	534.8
Drive RPM BOL [1/min]	4264	2852	3831	2838	3135
Drive RPM EOL [1/min]	3624	2424	3257	2412	2665

[a]Begin of life—when the component has no degradation
[b]End of life—when the component has to be replaced due to ageing degradation

the weight of the BoP components increases leading to a higher overall system mass. Furthermore, also the probability of a failure of one of the stacks increases with a higher number of stacks. Battery and fuel cell system masses are calculated including all of the BoP components of the systems.

Negative mass values of the battery system mean in this case that there is more installed power of the fuel cell than needed for the mission and no battery is needed in this case. So, according to the detailed powertrain design it can be seen that in terms of overall mass of the system, a "fuel-cell-only" configuration without battery is the better option. The needed maximum power for the mission (take-off power) can be provided either by 2s2p-340 or by 3s1p-450 configurations which have almost the same system weight. The configuration with four stacks is slightly heavier than the configuration with three stacks. However, the four stack configuration has two parallel branches, which means that in case of a failure of one string the voltage will not drop and it will be possible to fly with the half of the installed power which is sufficient to hold the altitude with reduced air speed. This safety margin gives an opportunity to continue the flight without a need of immediate emergency landing procedure in case of a single-point-of failure (one of the fuel cells is inoperative). Therefore, this option is preferable for the ALAADy vehicle.

Although the calculations above show no need of the battery as an energy source for e-motors used for propulsion, a small battery (capacity < 1 kWh) will still be needed to power up the fuel cell system and avionics before fuel cell system starts providing power.

2.3.2 Fuel Cell Simulation Model

Calculations shown above are based on a fuel cell model which has been set up for this purpose. This model is explained in more detail below to give you an overview regarding the approach and needed parameters which have been validated based on fuel cell test data.

The electrical behavior of the fuel cell stack has been modeled as a quasistationary model of a single cell. The media distribution (H_2, oxygen, coolant) and the effects of inhomogeneous distribution over the cell area or cell length have not been modeled in this case. The voltage of a cell is given by superposition of the ideal cell voltage (Nernst equation), activation overpotential (Butler-Volmer equation), ohmic losses, diffusion losses and concentration overpotential. From these terms a model can be generated where the single terms can be verified by the test data. For reversible cell voltage following applies (Barbir 2005):

$$E = 1.229 + 0.85 * 10^{-3}(T_{fc} - 298.15)$$
$$+ \frac{R}{2F}T_{fc}\left(\ln\left(\frac{p_{H2}}{p_0}\right) + 0.5\ln\left(\frac{p_{O2}}{p_0}\right)\right) \tag{1}$$
$$p_0 = 1.01325 \textbf{ bar}$$

Where E is the cell voltage, T_{fc} is the fuel cell temperature, R is the specific gas constant, F is the Faraday constant, p_{H2} is the anode partial pressure and p_{O2} is the cathode partial pressure.

The activation overpotential is given by:

$$v_{act} = a\ln\left(\frac{j + j_{perm}}{j_0}\right) \tag{2}$$

Where: $a = \frac{RT}{\alpha F}$ and α is the transfer coefficient, j is the electrode current density, j_0 is the exchange current density and j_{perm} is the current density caused by hydrogen diffusion through the membrane. j_{perm} depends on hydrogen permeability through the membrane:

$$j_{perm} = 2FK_{H2}\frac{p_{O2}}{t_m} \tag{3}$$

Where: K_{H2} is the permeability of hydrogen and t_m is the membrane thickness. K_{H2} can be determined as the product of diffusivity D_{H2} and solubility S_{H2}:

$$K_{H2} = S_{H2}D_{H2} \tag{4}$$

Solubility can be assumed as constant with $S_{H2} = 2.2 \cdot 10^{-4} \frac{mol}{m^3 Pa}$. For the diffusivity a temperature dependent correlation can be assumed where T is the temperature:

$$D_{H2} = 0.41 \cdot 10^{-6} \exp\left(-\frac{2602}{T/K}\right) \tag{5}$$

Permeability can also be calculated for a reference case (25 °C) which is a preferred approach in the praxis where model validation is conducted based on empirical test data.

$$K_{H2} = K_{H2ref} \exp\left(\frac{2602}{298.15} - \frac{2602}{T_{fc}}\right) \tag{6}$$

With this approach K_{H2ref} can be determined using experimental data for parameter fitting. The exchange current density j_0 can be determined by known j_{0ref} (exchange current density at reference conditions):

$$j_0 = j_{0ref}\left(\frac{p_{O2}}{p_0}\right)^\gamma \exp\left(-\frac{E_c}{R_m T_{fc}}\left(1 - \frac{T_{fc}}{298.15}\right)\right)$$

$$\gamma = 0.5 \tag{7}$$

$$E_c = 66\frac{kJ}{mol}$$

Where R_m is the universal gas constant.

The reference exchange current density j_{0ref} can also be determined by parameter fitting. The ohmic losses result from the humidity-dependent ionic conductivity of the PEM (Polymer Electrolite Membrane). The humidity (RH) of the membrane is assumed to be in equilibrium with the media humidity of the anode or cathode gas on its surfaces. The humidity inside the membrane is taken as the mean value of the humidity on the surfaces. The electrical resistance at the contact surfaces is considered as an additional constant contribution **r** to the ohmic resistance. Also parameter **r** will be determined by parameter fitting.

$$v_{ohm} = j\left(\frac{t_m}{\sigma} + r\right) \tag{8}$$

$$\sigma = 100(0.005139\lambda - 0.00326)$$

$$\lambda = \begin{cases} 0.04 + 17.81RH - 39.85RH^2 + 36RH^3 & for\,RH < 1 \\ 14 + 1.4(RH - 1) & else \end{cases} \tag{9}$$

The concentration over potential can be determined by a constant coefficient **c**, limiting current density j_L, and $j + j_{perm}$ (described above).

$$v_{conc} = c\exp\left(\frac{j + j_{perm}}{j_L}\right) \tag{10}$$

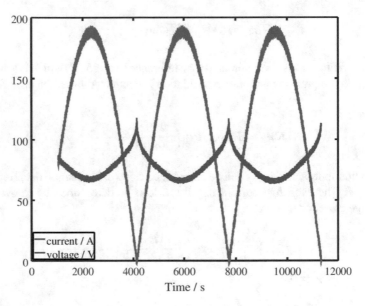

Fig. 5 Measurement protocol of the stack varying stack voltage and current

Coefficient **c** and j_L will be determined by parameter fitting. The cell voltage will then result in:

$$U = E - v_{act} - v_{ohm} - v_{conc} \tag{11}$$

As a first step the model has been implemented in octave.[1] Membrane thickness will be assumed to be 25 μm. The free parameters j_{0ref}, K_{H2ref}, c, j_L and r have been determined from measurement data of a fuel cell stack with 120 cells and 195 cm² active cell area. The fuel cell stack has been operated at atmospheric pressure (no overpressure on anode and cathode sides) and no humidifier was used as the mentioned stack is designed to be operated without. The test consists of three consecutive sinusoidal current curves shown in Fig. 5. With this data, the electrical polarization curve of the fuel cell has been determined and the free parameters j_{0ref}, K_{H2ref}, c, j_L and r have been determined using the least-square-fitting which is a standard approach to approximate the solution of overdetermined systems. Current and voltages shown below are the stack output current and stack output voltage.

Figure 6 shows the calculated polarization curve after parameter fitting and the measured data from the test. The "noise" of the calculated data is caused by different operating parameters (temperature, pressure) at the same current values. The numerical values of the fitted parameters are shown in Table 6.

---

[1] Octave is a freely available mathematical scripting language that is largely compatible with (The MathWorks).

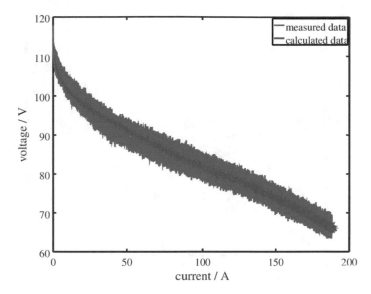

Fig. 6 Polarization curve of the stack with measured and calculated data

Table 6 Fuel cell parameters

Parameter	Par.	SI-Unit	Value
Exchange current density at reference conditions	j_{0ref}	A/m^2	2.1645e$-$004
Hydrogen permeability at reference conditions	K_{H2ref}	mol/m/s/Pa	7.1020e$-$014
Coefficient c	c	V	3.6648e$-$006
Limiting current density j_L	j_L	A/m^2	1187.4
Contact resistance	r	Ohm/m^2	1.9108e$-$005

With these parameters, it is possible to simulate the variation of the polarization curve with other temperatures and pressures.

The curve at different temperatures and at 1 bar absolute cathode pressure is shown in Fig. 7. It has been assumed that the relative humidity is approx. 80% at each temperature shown below.

Figure 8 shows the simulation results for different cathode pressures (Supply Air Pressure). In this simulation model, anode pressure (Hydrogen pressure) is 200 mbar higher than the cathode pressure and the stack temperature is 65 °C.

As mentioned above, the first parameter set has been generated with fuel cell stack data which has been operated at atmospheric pressure. Stacks which are designed to be operated at higher pressures are the high performance stacks with high power densities (>2 kW/kg). However, it has been decided to validate the simulation model and the abovementioned assumptions with a "lower performance" stack, as detailed test data for such stacks is available. As the results show very good correlation

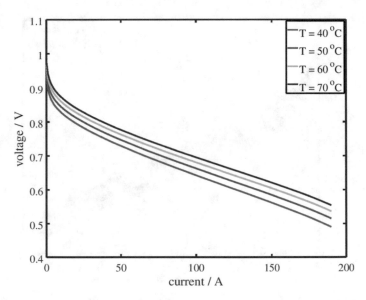

Fig. 7 Simulated polarization curves at different stack temperatures at 1 bar pressure

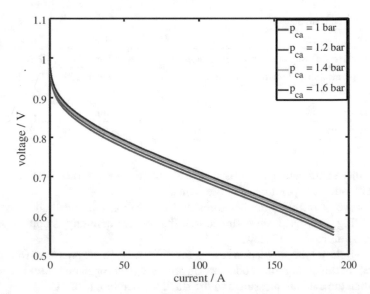

Fig. 8 Simulated polarization curves at different cathode pressures and 65 °C stack temperature

between simulated and measured values, polarization curves have been simulated for a pressurized stack also.

The parameter identification has been performed with a 300 cell-stack with an active area of 300 cm^2. This stack has been designed for pressurized operation and

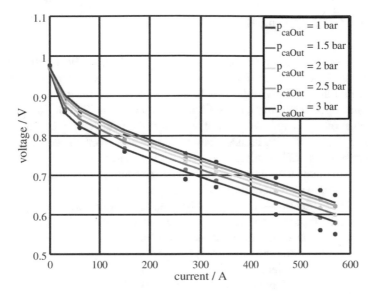

Fig. 9 Polarization curves of the stack with pressurization (dots: measurement data, lines: simulation data)

is operated with active humidifier. For this stack polarization curves are available at different pressures. Measurement data (points) and simulated data (lines) are shown on Fig. 9. It can be observed, that the measured and simulated polarization curves match very well together. Parameters for the SOA pressurized stack are given in Table 7.

In the next step the simulation model of the fuel cell has been implemented in the simulation framework to be able to simulate flight missions and perform a design loop of the powertrain system in more detail. Besides fuel cell model, also electric drive and power electronics models have been implemented in the framework and merged to a complete powertrain system model which will calculate the behavior of the powertrain system according to the input parameters such as: ambient temperature, ambient pressure, power demand and propeller speed. Integration of powertrain system model into the overall aircraft simulation framework is a topic of future work.

Table 7 Fitted parameters of a pressurized fuel cell stack

Parameter	Par.	SI-unit	Value
Exchange current density at reference conditions	j_{0ref}	A/m^2	1.7352e−004
Hydrogen permeability at reference conditions	K_{H2ref}	mol/m/s-/Pa	1.1936e−013
Coefficient c	c	V	4.8557e−006
Limiting current density j_L	j_L	A/m^2	2793.5
Contact resistance	r	Ohm/m^2	7.1953e−006

3 Powertrain Online Health Monitoring

Besides a redundant design of the powertrain, also an online health monitoring of the system is needed to detect failures in the system and react accordingly. The state operation monitor of the vehicle should supervise the overall state of the aircraft systems and operational limitations (Schirmer and Torens 2021). For example as mentioned above if we use two parallel branches of the fuel cell stacks, it is possible to continue the flight with reduced power. To decide when to reduce power and when to turn off the failed strings it is necessary to detect the failures with reliable methods. To maintain the safety but also make the system as simple as possible, a set-up with a minimum number of sensors for full electric and hybrid electric powertrain systems has been identified. These investigations have been based on measurements performed in the laboratory and analyzing the behavior of the sub-systems (battery, fuel cell, DC/DC converter, electric motor, inverter).

To illustrate the differences in monitoring effort for different powertrain architectures a list of parameters, which have to be monitored has been set up for three different architectures. Table 8 shows the parameters of the powertrain system which have to be monitored to detect a failure in the system and be able to react accordingly. It should be noted that Table 8 does not show the parameters of the E-Motor system which drives the vehicle because these are the same for all Three configurations listed below. It can be observed that fuel cell-battery-hybrid and ICE-battery-hybrid configurations would need the same number of parameters to be monitored while an architecture with only fuel cell system will need much less parameters.

Although this observation leads to a conclusion, that the fuel cell powertrain will need a smaller number of sensors, at first glance, a detailed look at the powertrain architecture leads to an opposite conclusion. As mentioned above a 2S2P configuration of the fuel cell stacks would be the most suitable for the ALAADy-vehicle in terms of redundancy and inherent safety. This means that there are four fuel cell stacks in the system and the same type of parameters at all of the four fuel cell stacks have to be monitored. Considering this point, on the overall system level more sensors are needed for a fuel cell powertrain compared to a ICE-battery-hybrid powertrain, which would lead to a higher sensor system complexity and costs.

Table 8 Powertrain system parameters which have to be monitored to identify a component failure

ICE-battery-hybrid	Fuel cell	Fuel cell-battery-hybrid
ICE revolutions per minute (RPM)	Fuel cell temperature	Fuel cell temperature
ICE temperature	Fuel cell voltage	Fuel cell voltage
Generator RPM	Fuel cell current	Fuel cell current
Generator temperature	Fuel cell lowest cell voltage	Fuel cell lowest cell voltage
Inverter temperature	Fuel cell cooling water flow	Fuel cell cooling water flow
Inverter output current	H_2 concentration in FC compartment	H_2 concentration in FC compartment
Inverter output voltage	H_2 concentration in tank compartment	H_2 concentration in tank compartment
DC/DC temperature		DC/DC temperature
DC/DC output current		DC/DC output current
DC/DC output voltage		DC/DC output voltage
Gen-set cooling water flow		Gen-set cooling water flow
Battery highest cell voltage		Battery highest cell voltage
Battery lowest cell voltage		Battery lowest cell voltage
Battery highest cell temperature		Battery highest cell temperature
Battery lowest cell temperature		Battery lowest cell temperature
Battery current		Battery current
Battery cooling water flow		Battery cooling water flow

4 Conclusion

Concepts of full electric and hybrid electric powertrain systems for ALAADy-vehicles have been analyzed. Different options of the powertrain system have been presented and discussed in terms of overall system mass and system safety. Simulation models of full electric powertrain with fuel cell or fuel cell-battery hybrid have been set up based on generic data for the weight of the system and test results for the electrical properties of the system. Also cost estimations of system components have been presented to have a basis for decision-making when ALAADy-vehicle will be designed in more detail for a specific mission.

It can be observed that fuel cell powertrain option might be the best solution in terms of redundancy leading to a high level of safety and overall system mass. However, such a powertrain has also disadvantages in terms of system cost and complexity which has to be taken into account when it comes to a detailed vehicle design for a certain operational scenario. Also, hydrogen infrastructure which is needed for fuel cell powertrain is an important issue which can make an ICE-battery-hybrid powertrain preferable for a certain design mission.

As overall conclusion it can be said, that full electric powertrain is a possible solution for a UAS with regards of system weight and has advantages in safety and weight compared to an ICE-battery-hybrid but if it comes to a simple and cost-efficient design of the vehicle ICE-battery-hybrid is a better solution.

References

Anghel C (2015) Hybrid electric propulsion techologies 1 MW high efficiency generator. https://fdocuments.in: https://fdocuments.in/document/honeywell-cristian-anghel.html

Crespi P (2017) Liquid hydrogen :a clean energy for future aircrafts. In: E2Flight symposium, pp 8–12, Stuttgart

Dinh JK (2014) Fuel cell technology status cost & price status. In: DOE annual merit review meeting, p 2

ElringKlinger AG (2017) Highly integrated PEM fuel cell stack module for aerospace applications, Stuttgart, 5 October 2017

Barbir F (2005) PEM fuel cells: theory and practice. Elsevier Acaddemic Press, Amsterdam

Hasan YJ, Sachs F (2021) Configuration aspects and vehicle specific investigations for future unmanned cargo aircraft. In: Dauer J (ed) Automated low-altitude air delivery - towards autonomous cargo transportation with drones. Springer, Heidelberg

Spendelow JM (2013) Fuel cell system cost. DOE Fuel Cell Technol Off Rec 13012:1–8

Meeus M (2018) Overview of battery cell technologies. In: European battery cell R&I workshop, p. 17, Brussels

Moore M (2016) Distributed electric propulsion (DEP) aircraft near-term electric propulsion evolution strategy. In: 5th symposium on collaboration in aircraft design, Naples

New D (2011) DOE hydrogen and fuel cells program record. Hydrogen Energy Gov 2:1–4

Panasonic (2021) https://www.nkon.nl/de/panasonic-ncr18650b-made-in-japan.html

Pipistrel Aircraft (2021). https://www.pipistrel-aircraft.com/aircraft/electric-flight/velis-electro-easa-tc

Propfe B (2016) Marktpotentiale elektrifizierter Fahrzeugkonzepte unter Berücksichtigung von technischen, politischen und ökonomischen Randbedingungen, Stuttgart

Kochhan SFR (2014) An overview of costs for vehicle components, fuels and greenhouse gas emissions

Rosario RD (2014) A future with hybrid electric propulsion systems: a NASA perspective, pp 1–21

Bubeck S, Tomaschek J, Fahl U (2016) Perspectives of electric mobility: total cost of ownership of electric vehicles in Germany. Transp Policy 50:63–77

Schirmer S, Torens C (2021) Safe operation monitoring for specific category unmanned aircraft. In: Dauer J (ed) Automated low-altitude air delivery - towards autonomous cargo transportation with drones. Springer, Heidelberg

The MathWorks (n.d.) MATLAB, Boston

Toyota (2016) Mirai Product Information

Cargo Handling, Transport and Logistics Processes in the Context of Drone Operation

Peter A. Meincke

Abstract The steadily growing share of air freight transport in the entire logistics industry is mainly due to the three major advantages of speed, safety and reliability. In order to meet increasing demands, more and more automated transport and delivery processes are used. As part of the DLR (German Aerospace Center) research project Automated Low Altitude Delivery (ALAADy), a fully automated Unmanned Cargo Aircraft (UCA) with a payload of one ton is being developed in cooperation with eight DLR institutes. As a general area of application, the UCA is responsible for the so-called "penultimate mile" in the air freight logistics chain. In order to achieve an optimal integration of a UCA into the freight supply chain, this research focuses on ground handling and in particular on the loading and unloading processes. The theoretical and practical concepts of the integration of UCA were examined within this research under the premise that no infrastructure exists at the destination in order to obtain the most automated process possible for future logistics. The research shows that the interaction between a UCA and an automated robot container system can solve both problems of the "last mile" and the "penultimate mile" within the logistics chain.

Keywords UCA · ALAADy · UAS · Air cargo supply chain · Logistics · Cargo handling · Unmanned cargo aircraft · Unit load device · Ground handling · Last mile

1 Introduction

Automated cargo delivery is one of the civil applications for Unmanned Aircraft Systems (UAS) that is often considered to play a significant role for aviation in the future. A project of the DLR (German Aerospace Center) called Automated Low Altitude Delivery (ALAADy) investigates the application of a very low level flight unmanned aircraft for an innovative approach of cargo delivery (Dauer et al. 2016).

P. A. Meincke (✉)
German Aerospace Center (DLR) Institute of Transportation Systems, Lilienthalplatz 7,
38108 Braunschweig, Germany
e-mail: peter.meincke@dlr.de

© Deutsches Zentrum für Luft- und Raumfahrt e. V. (DLR) 2022   205
J. C. Dauer (ed.), *Automated Low-Altitude Air Delivery*, Research Topics in Aerospace,
https://doi.org/10.1007/978-3-030-83144-8_9

Fig. 1 Gyrocopter version—DLR project ALAADy—automated low altitude delivery

Operating at altitudes below general air traffic, payloads of around one to two tons are transported. The following figure (Fig. 1) shows a gyrocopter, one of the three selected aircraft configurations of the project, to which the further investigations in this research refer.

This project investigates new safety concepts and the economical validity. As part of the project ALAADy, the DLR Institute of Air Transport and Airport Research is developing concepts for unloading Unmanned Cargo Aircraft (UCA) and analyzing how these concepts would optimize the classic air cargo supply chain.

The aim of this research is to determine new concepts on how to handle cargo of a UCA on a destination with no cargo infrastructure at all and to determine how to achieve positive impact on the air cargo supply chain. Previous studies (Feng et al. 2015) from 1996 to 2019 indicate that there is no literature about the cargo ground handling of UCAs or even generally about aircraft turnarounds at locations without (cargo) infrastructure.

Air freight handling in combination with the required documents causes a significant bottleneck factor in air freight transport and thus reduces productivity. The average consignment transportation time of six days in the traditional air freight transport chain has remained practically unchanged in 20 years (IATA 2008). Air freight goods spend around 80% of their time between sender and recipient on the ground (Vahrenkamp 2007). The air freight clients therefore demand a rethinking of traditional processes and structures and the development of innovative model solutions for air freight transport (Heerkens 2017). Commercial shippers can only prevail in a difficult global competitive environment through continuous improvement in production, the creation of innovative products and optimization in the logistics field. The demands on air freight service providers are therefore increasing too. The aim is to offer transport solutions with high reliability and reaction times in order to be able to reduce delivery times, inventories and costs (Lennane 2017). However, the efficiency of the air freight transport chain is decided on the ground, for example in air freight handling (Amaruchkul et al. 2011).

Very often, the quality of the available freight infrastructure (even non-existence) has a strong influence on this efficiency. Some destinations are therefore generally not suitable for conventional freight flights. For example, there are destinations without any infrastructure for cargo handling, such as a landing zone in the mountains, on an island or a green field, on construction sites, at major cultural events or in disaster areas. Then the use of a UCA (Collins 2016) with autonomous freight unloading technology opens up a new "last mile"—or rather a "penultimate mile" (Warwick 2017). For such niches or exceptional situations, UCA are a suitable solution (ZF 2016).

This chapter is structured as follows: After introducing the motivation and background of this research in this section, related basics of ground and air freight handling, the air cargo transport system and air cargo supply chain are presented in Sect. 2. Section 3 explains in detail the degree of automation of the processes involved in ground handling of a UCA, discussing the effects of fully and semi-automated handling in comparison to manual handling. Furthermore, an overview of the automation processes of departure and arrival is given. Section 4 describes the cargo Ground Handling with or without Infrastructure on destination. It gives details on how the cargo unloading of an UCA under the premise of maximum autonomy from the conventional airport infrastructure ("no cargo infrastructure at destination") can be divided into different concepts. Furthermore, it shows the integration of a UCA into the classic air freight supply chain. Section 5 discusses further problems in the design of the gyrocopter for smooth unloading of the load. These include the optimal positioning of the loading doors, the stabilization of the UCA during loading and the possible container design for the application. Finally, Sect. 6 concludes the chapter with a review of the results and highlights future opportunities for the air freight supply chain.

2 Basics of Ground and Air Freight Handling

Before the described topic can be analyzed, some basics of ground and airfreight handling and the logistics chain have to be explained.

2.1 Ground and Air Freight Handling in Civil Aviation

Ground handling is the preparation of an aircraft for the next flight in air traffic (IATA 2013). If the handling of a dedicated cargo aircraft is considered, some core processes are left out or replaced, while others become the focus of attention, for example the safe and efficient loading. The individual process steps for cargo aircraft only include:

- Securing the incoming machine (with chocks and pylons)

- Ground power supply/refueling
- Unloading or loading the cargo
- Aircraft tow or pushback
- De-icing (if necessary)

Nevertheless, the ground time of a cargo plane is often higher than that of passenger aircraft. On the one hand this is due to a different operating model and, on the other hand, to the fact that the loading process occupies a large part of the total ground handling, which is particularly due to the various activities and the cargo to be loaded (Kazda and Caves 2011).

2.1.1 Basic Components of the Air Cargo Transport System

The physical components of the air freight transport system include all technical and transport infrastructures. Among these are facilities for the organization of transport processes, such as airport infrastructure, logistics and cargo centers, as well as air traffic control, information and communication systems. In addition, the goods to be transported and the unit load devices (ULD) required, such as containers and pallets, are also material components of the air freight transport system. However, the means of transport do not only include aircraft, but also trucks and industrial trucks (e.g. fork lifters) for the transport of air cargo on the airport grounds. In this research, the relevant processes for the turnaround of a UAS (Valavanis and Vachtsevanos 2015) based an aircraft are to be considered (Beier et al. 2016). A turnaround is the process by which the aircraft is prepared directly for a new flight after landing (Gomez and Scholz 2009). According to Kazda and Caves (2011) there are three conceivable approaches to ground handling:

- The aircraft's own equipment is used.
- Technical equipment of the airport is used.
- Airport fixed distribution networks with a minimum of mobile facilities are used ("vehicle free apron").

In the present work, only the first approach (i.e., using aircraft's own equipment) is considered in order to create concepts for the loading and unloading of a UCA with the requirements of the ALAADy concept. In addition to the chosen approach of using only on-board systems for the turnaround (Schmidt 2017), the following should apply: "There is no (cargo) infrastructure at destination". In other words, there is no prior technical equipment or infrastructure for unloading at the destination. On the one hand, this approach is based on covering the widest range possible of use cases with the UCA design together with the used handling technique, and on the other hand, it encourages a maximum autonomous unloading of the UCA. Ultimately, a UCA with a self-sufficient handling technology should bring a maximum of efficiency into the air cargo supply chain.

2.1.2   Airfields, Landing Areas and Landing Zones

In this context, it must be explained how unmanned aircraft landing sites have to be designed, what legal requirements exist and what other infrastructure requirements must be met in order for goods to be transported logistically. In the further course of the investigation, reference will be made to the extreme case of "greenfield" at destination for freight delivery, i.e. a landing site without any infrastructure. This means that the UCA must be able to land on unpaved ground or that there is no safety area or cargo infrastructure for loading and unloading.

At this point this will be examined using the example of German air traffic law. According to this law, the landing site is not a landing site approved according to LuftVG § 6 (BMJV 2019a). In other words it is an outlanding which needs approval according to LuftVG § 25 (BMJV 2019b). All underlying four use cases in the ALAADy project (humanitarian logistics, internal transport of components between production plants and assembly sites, spare parts logistics, supply in hard-to-reach areas) have a landing position "in the green" in common, which is not a landing site approved according to LuftVG § 6 (BMJV 2019a). This would also apply to the use case "internal transport of components between production plants and assembly sites". For the other use cases (disaster or humanitarian aid), in which it is not possible to plan where to land the UCA in advance and/or in which an application for an external landing permit would take too long, there is an exception in the LuftVG § 30. Which says: "…deviations from the regulations on behavior in the airspace are only permitted insofar as this is absolutely necessary for the fulfilment of sovereign tasks…" (BMJV 2019c). According to LuftVG § 1a, this law applies to all aircraft registered in Germany, even if they are stationed abroad (BMJV 2019d). Thus, in order to fulfil the sovereign task, no approved landing sites, no external landing permits and no strictly regulated integration of the drone into the airspace are required in operation—once the specific requirement has been determined.

At this point, the terms "landing zone" and "greenfield landing" shall also be explained. Coming from the military terminology, the landing zone is an area where aircraft can land. The prerequisites for a landing zone depend on the approach speed, approach direction, landing weight and size of the aircraft, a stable and level ground, obstacle clearance and visibility. Landing zones differ from landing areas in the absence of compacted or paved (concreted or asphalted) surfaces, structural facilities (hangar, tower) and stationary supply facilities (lighting, refueling) (NATO STANANG 1980).

2.1.3   Unit Load Device Containers

To make loading easier, facilities for sorting the cargo, distributing it to suitable unit load device (ULD), transporting it via the apron and loading it into the aircraft must normally be available at the airfield. In doing so, it must be possible to react flexibly to the various aircraft types and types of cargo. Loose cargo is often loaded by hand conventionally, whereas ULDs require equipment such as lifts.

It should be noted at this point that the nature of air cargo has a significant impact on its transport and handling (Kazda and Caves 2011). Thus, air freight affine goods can be divided into the following main groups:

- Dangerous goods
- Valuable cargo
- Perishable goods
- Animal transports
- Mail
- General cargo
- Express cargo

Since the ULD has a considerable influence on the basic loading technology, it also represents an important factor in all concepts for the handling of freight.

Table 1 shows the general benefits and disadvantages of containers in an overview.

Table 1 General benefits and disadvantages of containers

General Benefits of Containers	
Loading and unloading faster and with less staff possible	Time saving aspect
Standardized container units can be integrated into existing warehouse technology and automated storage systems	Time saving aspect
Less staff around the aircraft	Safety aspect
High handling speed makes the use economical	Economic aspect
Container can be loaded before, this saves time during a turnaround	Time saving aspect
Weight and Balance could be done easier	Safety aspect
Better securing of cargo possible (in the container and in the aircraft)	Safety aspect
Cargo space can be optimally used	Economic aspect
Goods can be made storable by containers	Economic aspect
Use of special containers (e.g. cooling containers)	Economic aspect
Protection of the goods against external influences (e.g. weather or theft)	Security aspect
Containers can be waterproof and floatable	Safety aspect
General disadvantages of Containers	
Filling, securing, marking and emptying the containers is associated with additional effort	Time consuming aspect

2.1.4 Aviation Security

There are many areas at airports which are not accessible to the public and therefore require appropriate physical barriers to prevent unauthorized access to critical areas such as the airside of the airport. Fencing and additional sensors can make intrusion attempts more difficult and detectable. Uncontrolled access from the terminal area to the airside must also be prevented. Identity checks are an essential part of airport security. Security gates are used to identify dangerous objects in the luggage or on the bodies of passengers and employees before they can enter critical areas. This means that strict separation of persons before and after the security check is necessary (Kazda and Caves 2011). With regard to an unmanned aircraft, however, the measures for the inspection of persons are of secondary importance.

Air freight is also subject to a variety of controls before it can be loaded onto an aircraft. The methods used and the number of consignments checked vary, depending on the airport and type of cargo. X-ray equipment, explosives detectors, specially trained dogs and vacuum chambers are used, among other things, to activate any triggering mechanisms of explosives by simulating cabin pressure during cruise flight (Mensen 2001). Due to the high volume of cargo, it is not possible to check every piece of cargo individually on site. For this reason, the responsibility for the security of air cargo lies with the shipper, who must take the appropriate control measures and prevent unauthorized access until loading. When a company can prove that it meets the required standards, it is referred to as a "known consignor" (EU regulation 1998).

In order to meet the security standards applicable to air transport, the use of personnel, technology and time is essential for cargo transport. This results in a high number of personnel for implementation. In order to reduce this factor to a minimum, uniform standards apply, which are laid down in the implementing regulation (EU) 2015/1998 (EU Regulation 1998). Among other things, it describes the tasks and requirements of a "known consignor". After a successful certification procedure, as well as an on-site inspection of the company, this known consignor can dispatch air freight with the status "secured" and without further control by an air transport company. This legal framework makes it possible to separate the process areas of a known consignor from those of an air transport company.

Within the framework of the EU Regulation and the German Aviation Security Act, air freight companies, forwarding agents, etc. can apply for the status of "regulated agent" in order to simplify the handling of freight at the airport. Through this approval, a Regulated Agent may classify cargo handed over to him by a shipper as safe or unsafe and make unsafe cargo safe through appropriate control measures.

Freight that is "secured" to an airline is not subject to any further security checks. The agent ensures, among other things, that if access to the operating rooms or to the cargo warehouse by a person from outside the company is necessary, that this person is constantly accompanied and supervised by a person from the company who has been inspected for this task, and that no prohibited items are placed in or on the cargo (EU regulation 1998).

2.1.5 Cyber Security

One security aspect that has gained relevance over the past decades is information security, also often referred to as cyber security. In almost all technological and economic areas, the networking of systems among each other has increased strongly. More and more data is exchanged almost in real time. In aviation, new systems such as Automatic Dependant Surveillance (ADS-B) and more accurate satellite navigation have opened up new operational possibilities by reducing lateral distances or reliably, monitoring aircraft position and system functions (Sampigethaya et al. 2011). In order to make air traffic increasingly efficient, the scope, importance and complexity of the software of future aircraft will increase exponentially (Sampigethaya et al. 2008). However, this development will also be accompanied by an increase in the vulnerability of increasingly digital systems (De Cerchio and Riley 2011). The definition of uniform security standards by the responsible aviation security authorities is made more difficult by the complexity of the systems concerned and the resulting large number of potential points of attack, which is why, compared to other safety–critical areas of air traffic, there are only a few legal bases. Apart from the possibility of operational disruptions due to attacks on airport networks, fatal consequences are conceivable through software compromise, which is directly linked to the safe operation of an aircraft. On the one hand, the high number of parties involved increases the general vulnerability of the aviation infrastructure in the form of airports, satellites, navigation and communication systems (De Cerchio and Riley 2011). On the other hand, integrated aircraft systems themselves could be the target of attacks from a distance or from mobile devices in the cabin (Sampigethaya et al. 2011). A secure system architecture is therefore indispensable, whether a UAS operates autonomously or remotely. Successful attacks on the software of the UAS would make it possible to control it or at least to seriously disrupt its operation. With increasing automation and complexity of the units involved, the challenge of ensuring the integrity of communication, navigation and control of aircraft also increases.

2.1.6 Air Cargo Supply Chain

In traditional air freight traffic, which is characterized by the cooperation of freight forwarders, transshipment companies and airlines, there is a need for a shift towards reduced-complexity logistical structures with increasing integration of processes.

The variety of processes that are required between take-off and landing of a cargo aircraft is shown by the logistic supply chain of air cargo handling in Fig. 2.

Figure 2 shows that before cargo is loaded onto an aircraft, and after it is unloaded, there are chains of processes with critical elements, such as pre-sorting and loading of vehicles at depots, that can hinder timely loading. The individual processes are dependent on a variety of factors, such as the supply chain distances and available resources in the form of vehicles and labor. Supply chain bottlenecks can lead to significant delays. Overall, for successful ground handling, there are many steps that

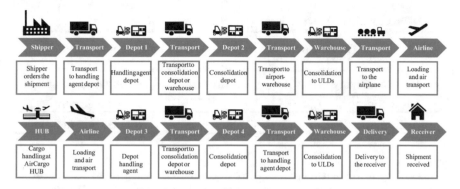

Fig. 2 Logistic supply chain of air cargo handling—Inbound (top) and outbound (bottom down)

must be considered that form a network of mutually dependent processes (Gomez and Scholz 2009).

In traditional air freight transport chains, the airlines are responsible for transporting the goods from the departure airport to the destination airport (airport-to-airport). The freight airlines operate internationally and offer their airport-to-airport freight services via cargo-only aircraft and passenger aircraft belly capacities on scheduled and charter flights. The freight normally comprises standard cargo and a small amount of express products. In continental air freight transport, consignments are often transported between the airports by truck as a replacement for planes due to the lower cost structures. This is also known as Trucking or Road Feeder Service (Nsakanda et al. 2004).

On the air freight market, forwarders sell freight capacities for the airlines, so the airlines normally do not directly market their services. In the air cargo chain, the forwarders also perform the transfer to the airport, possibly consolidate the consignments, and perform the final transfer from the destination airport and delivery to the recipient. They also perform additional forwarding and logistical services such as customs clearance. This extends the air freight service into a door-to-door service (Merkert et al. 2017).

The traditional air freight service has typically shown little standardization (Boonekamp and Burghouwt 2017). There are various different interfaces and parties in the transport chain and these significantly hinder a transparent and comprehensive flow of information and the goods (Sampigethaya et al. 2011).

For example, 60% of the air freight traveling to intercontinental destinations arrive at the Frankfurt hub as belly freight in passenger aircraft from European destinations (Meincke and Tkotz 2015). The remaining 40% are delivered to Frankfurt by truck from all over Germany or from Northern and Eastern Europe. Within Europe, air freight is only flown to hubs as a supply of belly freight in passenger aircraft, but not as air freight between two destinations. Rather, deliveries within Europe are almost exclusively carried out by truck instead of by air. This is due to the fact that truck transports can compete with air freight in terms of costs and transit times (Boonekamp and Burghouwt 2017).

Table 2 Time shares in the door-to-door supply chain (Kraus 2001)

Time Share in the Door-to-door Supply Chain

Before arriving at the airport	At the airport	After leaving the airport
26 percent	17 percent	57 percent

According to Bisignani (IATA 2005), door-to-door transit times for general air freight have remained constant at an average of six days since 1972. Despite all the technical advances, transit times have only been reduced from 6.5 days to 6 days in recent years. To understand the 6-day transit time, one has to consider the structure of the supply chain shown in Fig. 2 from the sender to the recipient: Many handovers and changes of responsibilities are needed during transit, until the cargo finally arrives at the airport security area. Cargo handling at a mega-hub, such as Frankfurt Airport, alone takes an average of 24 h (Frye 2011).

In addition to air routing, there are further transshipment activities at the airports and ground-based delivery services by air freight forwarders to and from the respective local airports. These delivery routes have their starting point or destination in the transshipment terminals of the air freight forwarders, which may be located in smaller cities. Transshipment in the terminals requires additional time shares. From these transshipment terminals, local groupage and distribution transports constitute the contact to the consignors or consignees. According to Vahrenkamp and Kotzab (2012), onward carriage at the airports is slow, so that there is still potential for optimization in the supply chain. The high proportion of time required for on-carriage is also confirmed in the study by Kraus (2001). Table 2 shows that 57% of the total delivery time is spent on onward carriage.

3   Level of Automation of the Processes in the Ground Handling of a UCA

An important part of the processes in ground handling of a UCA is the level of automation in terms of operation, technical effort and personnel and initial procurement costs.

In the following, the processes are divided into three levels of automation: manual, semi-automated and fully automated. Since there are several combinations here, scenarios or plausible combinations have been selected for the respective degree of automation based on practical empirical values (Beier et al. 2016). It is assumed that there is a UCA port both at the origin and at the destination, whereby the necessary infrastructure for the respective degree of automation is assumed. The important sub-process for loading and unloading a UCA will be discussed individually below.

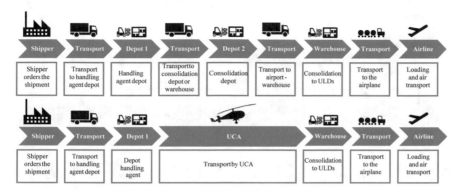

Fig. 3 Manual ground handling process

The focus will be on the extreme example of there being no freight infrastructure at the destination point (see Sect. 4.2).

3.1 Manual Ground Handling Process

In general, only one person is available in the manual process on site (Fig. 3). If several persons are on site, some of the process steps can run in parallel in the manual case (Beier et al. 2016). However, this must then be considered separately.

3.2 Semi-automated Ground Handling Process

Figure 4 shows the semi-automated process steps during take-off and landing. The process steps are marked in different colors to illustrate the degree of automation. Green represents the fully automated step, orange the partially automated step and red the manual step. In the semi-automated process, automated steps run largely simultaneously. However, additional resources in the form of people are required for manual steps.

3.3 Fully Automated Ground Handling Process

Figure 5 shows the analyzed process steps in fully automated operation. The coloration indicates that this process is completely self-sufficient and that several process steps can run in parallel.

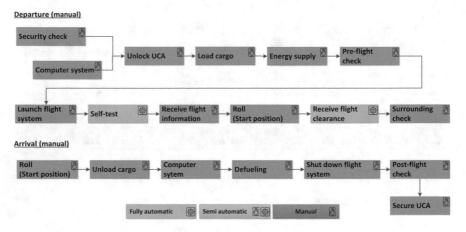

Fig. 4 Semi-automated ground handling process

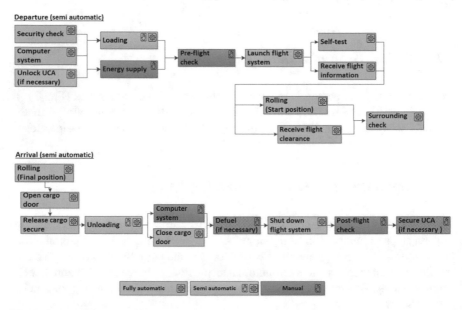

Fig. 5 Fully automated ground handling process

3.4 Comparison of the Automation Processes of Departure and Arrival

Now the chronological sequence of processes before departure, which are shown in Tables 3, 4 and 5 and then the processes after arrival, which are shown in the Tables 6, 7 and 8, will be compared in more detail.

Table 3 Process steps—fully automatic departure

Departure (fully automatic)	1	2	3	4	5	6	7	8	9	(⚙ automatic / ✋ manual)
Security check	⚙									
Computer system	⚙									
Unlock UCA (if necassary)	⚙									
Open cargo door		⚙								
Cargo loading system			⚙							
Secure cargo				⚙						
Close cargo door				⚙						
Energy supply			⚙							
Pre-flight check					⚙					
Launch flight system						⚙				
Self-test							⚙			
Receive flight information							⚙			
Roll (start position)								⚙		
Receive flight clearance								⚙		
Surrounding Check									⚙	

Table 4 Process steps—semi automatic departure

Departure (semi automatic)	1	2	3	4	5	6	7	8	9	(⚙ automatic / ✋ manual)
Security check	⚙									
Computer system	⚙									
Unlock UCA (if necessary)	⚙									
Open cargo door		⚙								
Cargo loading			⚙							
Secure cargo				⚙						
Close cargo door				⚙						
Energy supply				✋						
Pre-flight check					✋					
Launch flight system						⚙				
Self-test							⚙			
Receive flight information							⚙			
Rolling (start position)								⚙		
Receive flight clearance								⚙		
Surrounding Check									⚙	

Table 5 Process steps—departure manual

Departure (manual)	1	2	3	4	5	6	7	8	9	(⚙ automatic / ✋ manual)
Security check	✋									
Computer system	✋									
Unlock UCA (if necessary)		✋								
Open cargo door			✋							
Load cargo				✋						
Secure cargo				✋						
Close cargo door					✋					
Energy supply						✋				
Pre-flight check						✋				
Launch flight system							✋			
Self-test							✋			
Receive flight information								✋		
Rolling (start position)								✋		
Receive flight clearance									⚙	
Surrounding Check									✋	

Table 6 Process steps—fully automatic arrival

Arrival (fully automatic)	⚙ automatic	✋ manual
Roll (final position)	⚙⚙⚙ (start)	
Open cargo door	⚙⚙⚙	
Release cargo securing	⚙⚙⚙	
Cargo Unloading-System	⚙⚙⚙	
Close cargo door	⚙⚙⚙	
Computer system	⚙⚙⚙	
Defueling (if necessary)	⚙⚙⚙	
Shut down flight system	⚙⚙⚙	
Post-flight check	⚙⚙⚙	
Secure UCA (if necessary)	⚙⚙⚙	

Table 7 Process steps—semi automatic arrival

Arrival (semi automatic)	⚙ automatic	✋ manual
Roll (final position)	⚙⚙⚙	
Open cargo door	⚙⚙⚙	
Release cargo securing	⚙⚙⚙	
Unload cargo		✋✋✋
Close cargo door	⚙⚙⚙	
Computer system		✋✋✋
Defueling (if necessary)		✋✋✋
Shut down flight system	⚙⚙⚙	
Post-flight check		✋✋✋
Secure UAV (if necessary)		✋✋✋

Table 8 Process steps—manual arrival

Arrival (manual)	⚙ automatic	✋ manual
Roll (position)		✋✋✋
Open cargo door		✋✋✋
Release cargo securing		✋✋✋
Unload cargo		✋✋✋
Close cargo door		✋✋✋
Computer system		✋✋✋
Refueling		✋✋✋
Shut down flight system		✋✋✋
Post-flight check		✋✋✋
Secure UCA (if necessary)		✋✋✋

The process steps security check, computer system and, if necessary, unlock UCA, run in parallel in fully and semi-automatic operation (Tables 3 and 4). When the cargo passes through the security check, the freight is removed from the computer system. Meanwhile, the securing or anchoring points on the UCA are released automatically and the UCA is thus unlocked. This cannot be done manually. The release from the computer system can partly take place parallel to the security check. However, the UCA can only be unlocked afterwards. The remaining work steps in the manual process sequence must also be carried out one after the other if only one person is on site (Tables 5 and 8). Subsequently, only the semi-automatic and fully automatic processes are compared.

After the UCA has been unlocked, the "open cargo door" process step takes place. This process step is the same for both levels of automation. However, the difference at the next point is clearly visible. In the fully automatic process sequence, the step "load cargo" occurs automatically and parallel to the automatic energy supply (Table 3). In the semi-automatic process, these process steps must be carried out manually and therefore cannot be carried out in parallel (Table 4). After the step "load cargo", the UCA automatically secures the cargo and locks the cargo door. In the semi-automatic process, the person on site must release the UCA for the next steps after loading. The following process step is the pre-flight check. This step must not run parallel to other steps, since the positive result of the check is required to release the subsequent sub-processes. In contrast to the points "roll" (start position) and "receive flight clearance", the steps "self-test" and "receive flight information" can again be processed in parallel. The process step "receive flight clearance" begins when the UCA is in the position in which it receives the last take-off clearance and then rolls onto the runway. There it would still be possible to abort the take-off.

The manual process steps for the arrival (Table 8) are the same as for the departure if only one person is on site (Tables 3 and 5). Each point must be completed one after the other, as one person cannot process two things simultaneously. In the first part of the process chain of the partially and fully automatic sequence their processes are identical. Thus, the UCA automatically unlocks the cargo while opening the cargo door. In contrast to the departure, the step "Unload cargo" does not take place in parallel with refuelling or defuelling. This would only be the case if the aircraft has a kerosene engine. Subsequently, the UCA closes the cargo door automatically (Table 6), in the semi-automatic process only when the personnel on site give their approval (Table 7). At the same time the freight is entered into the computer system. The UCA then switches to the semi-automatic procedure if the above-mentioned conditions are met. In the fully automatic process, this step takes place while the UCA closes the cargo door and the freight is entered into the IT system (Tables 3 and 4). The flight system then shuts down and the post-flight check is carried out. In the semi-automatic process, this does not take place in parallel for safety reasons, so that the personnel can carry out the inspection properly. As a final step, the UCA will be secured, if it is parked for a longer period or if the weather conditions require it.

3.5   Comparison of the Different Levels of Automation

In the following, the process sequences with their process steps can be seen in a Table 9 for take-off and landing with freight at different degrees of automation.

Table 9 Classification of automation level

	Manual	Semi automatic	Fully automatic
Personnel costs (operation)	High	Low	None
Personnel costs (maintenance)	Expected	Expected	Expected
Initial acquisition costs	Very low	Low	Very high
Operation	Complex	Simple	Very simple
Technical effort	Low	Low	High

In fully automatic operation, the acquisition costs are probably very high, since full automation is combined with high technical requirements. Nevertheless, personnel costs for operation are close to zero. In manual operation, the conditions are exactly reversed because considerable personnel costs must be expected, while the acquisition costs for facilities and systems are probably very low in comparison. The advantage in turn is that the personnel can be deployed flexibly for operation and can react immediately to unforeseen circumstances. In a fully automated process, on the other hand, the process chain would come to a standstill if a system were to fail. The semi-automatic process sequence offers a combination of these extremes. This means that the process steps that are most expensive in fully automatic operation are replaced by manual steps. In all degrees of automation, however, personnel costs for maintenance must also be taken into account. Even in the fully automated process, maintenance personnel are required. The amount of personnel required on site and how they need to be trained depends on the infrastructure on the one hand and on the maintenance work which needs to be carried out on the other hand.

A fully automatic sequence is probably the most economical if it can be assumed that these systems are used continuously and uniformly. In the case of challenging sub-processes, manual input is usually preferred in order not to endanger a trouble-free overall process. A fully automated process would therefore probably be unsuitable. In order to ensure the quality of the unloading process, staff should generally be on site to ensure efficiency.

Accordingly, the semi-automatic process is well suited. Each use case, however, must be carefully examined and the special features identified. Only then a well-founded statement can be made as to which degree of automation is the most suitable.

4 Cargo Ground Handling with or Without Infrastructure on Destination

After analyzing the process steps of UCA ground handling—from manual to semi-automatic to fully automatic—it is now necessary to focus on the loading and unloading processes.

For the project, two questions must be examined at this point. Firstly, how the integration of a UCA could affect the classic air cargo chain. On the other hand, which freight unloading techniques could be used to run the use cases identified in the ALAADy project. In order to be able to handle all use cases, an extreme case is assumed, that no freight infrastructure is available at the destination.

4.1 Integration of an UCA in Air Cargo Supply Chain

The logistic supply chain of air cargo handling and the multitude of processes required between take-off and landing of a cargo aircraft has already been illustrated in Fig. 2.

Figure 6 shows the origin part of classic process chain of air freight without UCA (top) and with UCA (bottom). It can be seen that some processes of the classical air freight process chain can be substituted by integrating the UCA from the ALAADy. Depending on the variety of process steps in the transport chain and the parties

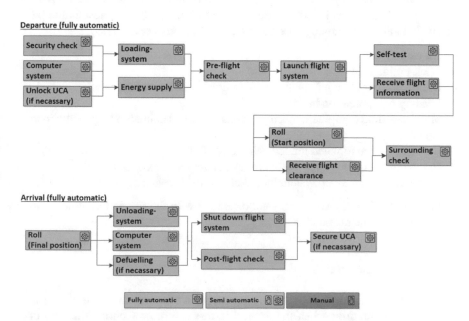

Fig. 6 Air cargo supply chain—comparison of the part to the airport (above) with integration of the UCA ALAADy (bottom)

involved, considerable time savings can be achieved. Furthermore, the introduction of a UCA reduces both the number of parties involved within the process chain and the number of process steps. The air cargo supply chain is thus significantly simplified.

The use cases examined in the ALAADy project (e.g. humanitarian logistics, transport of components between production and assembly sites, spare parts logistics, supply in areas with difficult access) should benefit significantly from the following advantages of a UCA: The fast and flexible delivery of urgently needed goods and parts without being dependent on a road or rail network (Dauer et al. 2016). A premise for this would be that the cargo can be unloaded after landing and that the UCA can then possibly even carry out a turnaround. The system should depend on external supply as little as possible in order to be able to operate independently at airports and landing zones (Meincke et al. 2019).

4.2 Cargo Ground Handling in Case of no Infrastructure on Destination

After analyzing the integration of a UCA into the classic air freight supply chain—it is now necessary to focus on the cargo loading processes that can be handled at the destination due the lack of air freight infrastructure.

The cargo unloading of a UCA under the premise of maximum autonomy from the conventional airport infrastructure ("no cargo infrastructure at destination") can be divided into the following possible concepts (Meincke et al. 2019):

- Air cargo drop
- Click-out-and-go
- Use of appropriate equipment for unloading available at destination
- Loading equipment on board
- Mobile loading equipment taken previously to the destination (e.g. with pioneer-module)
- Unloading by automated guided vehicle (AGV)
- Unloading and delivery by autonomous delivery vehicle/robot
- Autonomous container-system included unit load device

In the following, these concepts are discussed with the restriction of no existing infrastructure at the destination.

4.2.1 Air Cargo Drop

Dropping cargo is a method especially practiced in military aircraft and in humanitarian relief operations when the ground infrastructure or security situation does not permit landing (Sadeck et al. 2009). For this purpose, goods are attached to parachutes and dropped over the destination. Dropping cargo is practiced; however, the effort for the preparation of the relief supplies and preparation of the drop zone is very high and are very application-specific. In addition, it cannot be guaranteed that the cargo arrives at the intended recipients. There are three main types of airdrop (Benney et al. 2005):

Low-Velocity Airdrop (LVAD) is the delivery of a load involving parachutes that are designed to slow down the load as much as possible.

High-Velocity Airdrop (HVAD) is the delivery of a load involving a parachute to stabilize its fall. One type of HVAD is the LAPES, a low altitude parachute extraction system, in which the aircraft almost completes a touch-and-go type maneuver (without actually touching the ground) (Fig. 7) and the load is ejected at an extremely low altitude (Desabrais et al. 2012). Utilizing modern UAS multicopters it becomes more of a "Click-out-and-go" concept (see next section).

Free fall air cargo drop is an airdrop with no parachute at all and a Low Cost Aerial Delivery System (LCADS) (Bonaceto and Stalker 2005) for a one-time use, stand-alone airdrop system: It consists of a Low Cost Parachute, LC Containers, and platforms. An example of a low-cost aerial system is the current Wings for Aid project "self-landing delivery cardboard box" (Fig. 8), where a simple folding box board box with flaps as a wing is used (Wings for Aid 2020).

The dropping of cargo must be kept in mind as an option for certain use cases, but a conceptual change of the current gyrocopter configuration of the ALAADy project (Fig. 1) would be necessary. However, it must be doubted whether it will be possible to integrate an airdrop as a regular mode of transport into the classic supply chain.

Fig. 7 Air cargo drop—LAPES (low altitude parachute extraction system)

Fig. 8 Free fall air cargo
drop with a low-cost aerial
system. Source: Wings for
Aid

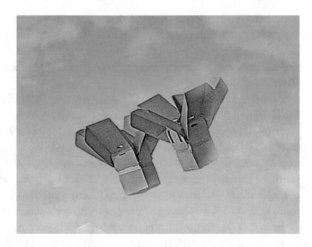

4.2.2   Click-Out-And-Go

In the following, a special type of cargo handling with an aircraft should be
mentioned, which will be defined as "click-out-and-go" (Fig. 9). In other words:
Release and deposit the cargo during flight (without direct landing). In this concept,
the cargo is usually attached with a fastening mechanism on the outside of the aircraft
and is disconnected or "softly" dropped off directly above at the landing point. This
method is increasingly being practiced in multicopters or drones in parcel deliveries

Fig. 9 Click-out-and-go with a cargo drone

(DHL Parcelcopter, DHL 2016; DPD Drone Parcel delivery, DPD 2017; Amazon Prime Air, Amazon 2017). Similarly, this is used for helicopters (including the Sikorsky S-64 Skycrane) to avoid direct landing, even engine shutdown or interruption of the flight phase. Since this case is not yet described in detail in the literature, it should be defined as "click-out-and-go", as it is clearly different from the "touch-and-go (includes a landing)" (Schmidt 2017) as well as the LAPES (see 4.2.1). Under "Click-out-and-go" there are various similar differentiated mechanisms as winch techniques, cf. Modular Aircraft Clip Air (EPF 2018) or with a full loading frame, cf. Airship cargo lifter (Aerospace engineering 2002). The benefits are obvious: Only a small landing zone is needed at the destination and specialist personnel for "unloading" is not required. At the same time, considerable time savings occur because no general landing of the aircraft is necessary, which eliminates the whole process of turnaround. However, the prerequisite is that the UCA is configured for this unloading technology, i.e. vertical take-off and landing is needed (Cakici and Leblebicioglu 2016).

4.2.3 Use of Appropriate Equipment for Unloading Available at Destination

Of all concepts listed, the possibilities of locally available means and the following concept with on-board resources are constructively the simplest to realize. Using appropriate equipment for unloading available at the destination means, for example, hand pallet trucks, goods carts or off-road vehicle winches which could be found at the destination (Fig. 10). It should be possible to handle this equipment even by low-skilled staff. This concept has the disadvantage that the dependence on local conditions is very high, since many influences could hinder a smooth freight handling. Therefore, the design of the UCA should be equipped with tools like roller system or a (lifting) ramp to support an effective unloading at the destination.

Fig. 10 Use of appropriate equipment for unloading available at destination

Fig. 11 Gyrocopter with roller system for cargo

4.2.4 Loading Equipment on Board

Similarly, simple tools or devices for loading could be integrated as on-board means into the UCA to achieve independence at the destination. These may be portable tools that are carried at the expense of the payload on each flight (e.g. truck-mounted forklift) or tools that are firmly integrated into the structure of the UCA. For example, the floor of the cargo deck in cargo aircraft consists of a combination of roller system and latches, which allow the moving of containers without machine assistance and can firmly lock them during flight. The advantage of this approach is a structurally simple and cost-effective implementation as well as high operational flexibility. In addition, all payloads whose size and weight allow manual unloading do not require any special equipment such as pallet trucks. The approach would be sufficient, for example, for the transport of medicine and other urgently needed items to people in remote areas. In the case of bulky and heavy items, however, this on-board-concept can be disadvantageous because suitable vehicles and devices must be available at the destination. There is also a need for ground personnel, who may need to be briefed to work safely on the aircraft. Necessary equipment on board in this case: Lifting ramp, roller system on the cargo hold floor (and/or rolls on the container Fig. 11), mobile winch and truck-mounted forklift (Fig. 12). The problem is obvious that this additional weight on board reduces the available payload significantly.

Fig. 12 Truck-mounted forklift, at the rear of a truck of the Austrian forwarding company Keimelmayr. Photo: Keimelmayr

4.2.5 Mobile Loading Equipment Taken Previously to the Destination

Alternatively, a pioneer module pre-transported by the UCA (Fig. 13) could set up a light infrastructure facility at the destination, for example, by carrying power generators and mobile fueling systems or equipment for a cargo handling (e.g. truck-mounted fork lifter).

This way, a destination could be prepared for the cargo handling of a UCA. Although this initially represents a significant financial and logistical overhead, various flights to the same destination would benefit from the existing facilities on the ground. After completion of the handling, or if no further flights are made to this destination, installed equipment must be loaded and transported again. If regular

Fig. 13 Mobile loading equipment (e.g. pioneer module) taken previously to the destination

flights are planned, it makes sense to build a minimum infrastructure. If, on the other hand, short-term deliveries to variable destinations are concerned, it is highly unlikely that the time required and the high outlay will make an economic operation possible. Furthermore, for a proper operation of the equipment trained personnel is necessary on site.

4.2.6 Unloading by Automated Guided Vehicle

The following two concepts have the advantage that no specialized personnel are required for unloading. For example, it is conceivable to perform an autonomous unloading of an UCA when an Automated Guided Vehicle (AGV) is carried on board in the form of a vertical or mid-level order picker to unload the freight units (Fig. 14).

Pickers are electrically powered trucks with two pallet forks, which are provided with an additional lift (Fig. 14). The speed is typically around one m/s in passenger traffic environments. Higher speeds can be achieved in completely automated areas. A major motivation for using automated guided vehicles, however, is generally the automation of work processes and thus the reduction of manual work. Other than a human resource, a driverless transport vehicle requires no breaks and is always ready to use (Lottermoser et al. 2017).

The navigation and spatial orientation is particularly important for AGV, because only if it works completely autonomous, safe and error-free, the system can be operated as desired. There are different approaches, although not all are suitable for the given case of operation in a terrain that has no infrastructure. Navigation with ground guidance, raster navigation, transponders embedded in the ground, magnets, optical grids or even laser navigation will be difficult to utilize. The challenge of navigation with ground guidance, grid navigation, transponders embedded in the floor, magnets, optical grids or even laser navigation must be solved (Vis 2006).

Fig. 14 UCA with an automated guided vehicle

Fig. 15 Automated guided
vehicle. Source:
Jungheinrich

Fig. 15 Automated guided vehicle. Source: Jungheinrich

The problem is the very high weight of the standard AGV, which would exceed the payload of the UCA construction. To solve this, a custom-made solution would be required.

4.2.7 Unloading and Delivery by Autonomous Delivery Vehicle

Another variant of unloading an UCA or aircraft without human resources would be an autonomous robotic vehicle, which leaves the aircraft independently and transports its cargo at the destination to the recipient: Autonomous Delivery Vehicle/Robots (ADV) or Self-Driving Vehicles (SDV) are autonomous robots used for transporting goods: They have the intelligence to make decisions when faced with new or unexpected situations. During the delivery, a variety of different sensors and scanners gather data that is used to avoid obstacles and/or continuously adjust the movement of the tug in order to follow the intended path more accurately. This kind of ADV

Fig. 16 Autonomous delivery vehicle e.g. Starship. Source: Mercedes-Benz Vans advance

(DHL Post Group 2019) could reduce further processes in the traditional air freight transport chain after the introduction of the UCA per se (Fig. 16). This would entail further time and cost potentials, as a more direct and straightforward delivery of the cargo to the recipient can be made. Starship Technologies' ADV and other competitors (e.g. Domino's/Marathon, Dispatch, Marble, Efficiency S.A.S.) are examples of this type of ADV.

Delivery robots require different features due to the variable fields of use. They are small in comparison to light trucks and hold only a few consignments, in extreme cases only one, and operate battery-electric. They will be tested with the aim of successfully delivering within 20 to 30 min. The vehicles seek the way to the receiver autonomously and avoid obstacles automatically. If there are situations in which the automatism of the machine cannot find a way, it passes the control to an operator who commands manually via remote control. The robot moves at 6 to 20 km/h and has a secured compartment inside, in which shipments with a total weight of 15 to 150 kg (Daimler 2018) can be transported in a radius of 5–20 km. According to the requirements of UCA in the ALAADy project, it should be considered to scale an existing robot up to a ton of weight (including load factor).

The UCA would need a ramp and the robot would have to be anchored (automatically) during the flight. At the destination, the autonomous robot could unload itself along with the freight and also carry out the delivery to the recipient. At present, this combination of unloading an aircraft and simultaneous delivery by a robot has not

Fig. 17 Mothership. Source: Mercedes-Benz Vans advance

yet been realized on the market (Meincke et al. 2019). Currently, only the cooperation of Mercedes-Benz Vans with Starship Technologies (Fig. 16) is known, where
a prototype of the so-called Mothership Concept is a Sprinter truck which serves as
a mobile loading and transport hub for eight Starship delivery robots (Punakivi et al.
2001).

The combination of airfreight unloading and delivery would have crucial time
and economic benefits and would be a significant solution to the "last mile" (DLR
2016) problem in airfreight logistics. This concept should have a high prioritization
in following research projects, as this autonomous variant would have a high degree
of future viability. Other solution combinations are possible, but it must be assessed
whether the benefits achieved outweigh a high design expenditure and decrease of the
payload. Disadvantages of this concept are clearly the high research and investment
costs for market readiness and for integration into the logistics chain of the companies.

4.2.8 Autonomous Robotic-Container-System

In addition to the infrastructure facilities, the load carriers in airfreight transport represent an important component. Because the use of containers has proven itself useful
(see Table 1 for the benefits and disadvantages), aviation-specific freight containers
are used in addition to conventional load carriers such as pallets. Since the existing
air cargo containers are based on the respective aircraft type, it is not possible to use
these in connection with the UCA of the ALAADy project.

Fig. 18 UCA with an autonomous robot container system

Nevertheless, to achieve the best possible landing space utilization for a UCA, it is necessary to develop a corresponding container system. In order to optimally integrate a UCA into existing logistics supply chain, it is important that a container system is based on already standardized units of measure of the load carriers. A reference point would therefore be a footprint of 1200 × 800 mm (EURO pallet) (Kazda and Caves 2011). In reference to the dimension of the load compartment (3000 × 1300 mm) of the gyrocopter configuration (Fig. 1) it would be possible to transport three container units per flight (Fig. 18).

5   Additional Challenges in the Cargo Topic Sector within the ALAADy Project

In the following some examples will illustrate how research on cargo transport within the ALAADy project contributes to the development of a UCA.

The basis for the studies is the configuration of a gyrocopter from the ALAADy project. For this purpose, the following dimensions or cargo hold sizes according to Table 10 were taken as a basis. These shall serve as a reference in the following analysis.

Table 10 Dimensions of the gyrocopter configuration

Dimensions of the gyrocopter configuration of the ALAADy project	
Fuselage width	180 cm
Hull height	180 cm
Hold height	130 cm
Cargo hold width	130 cm
Cargo hold length	300 cm
Inclination angle	6°

5.1 General Analysis of Cargo Doors for Freighters

In the first ALAADy animation (DLR 2016), the loading of the gyrocopter with two-part cargo doors was shown (Fig. 19) before the topic cargo door was analyzed again during the project.

For this purpose, possible variants of cargo access doors were checked for the gyrocopter configuration. Four main variants were identified (see Fig. 20):

- Cargo door at the side (a)
- Rear/tail access cargo door (b)
- Front/nose access cargo door (Upward-swinging hinged door) (c)
- Front/nose and rear cargo access door cargo (d)

A large number of cargo aircraft or aircraft used for cargo transport have cargo doors on the sides through which the cargo can be loaded. These also include

Fig. 19 Gyrocopter configuration of the ALAADy project—animation with two-part front access cargo doors

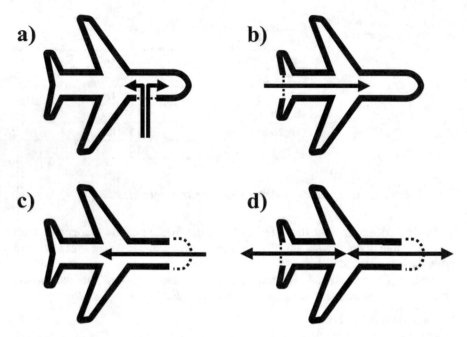

Fig. 20 Variants of aircraft cargo access doors

passenger aircraft that have been converted into cargo aircraft. These aircraft are equipped with a side cargo door or are fitted with a fuselage flap that can be opened upwards to allow extended loading. The side cargo door, however has an unfavorable effect due to the nonlinear loading, i.e. the cargo must always be loaded around the corner: The bottle neck is not only the dimensions of the door but also the width of the hull, so that long and bulky goods cannot be loaded. The Boeing 777F, Boeing 737-300F, McDonnell Douglas MD-11 and Airbus 310F, for example, have Cargo doors on the side.

Due to the space requirement for the cockpit in the front area of the aircraft, there is a general problem in finding a space for a door to load cargo into the aircraft. Loading over the tail is the obvious option if the length of the construction and the size of the body allow it. A rear cargo door has advantages for bulky loads and simplifies cargo access in general. Examples of loading via the tail are the following aircraft types: Antonov AN-12 and Lockheed L 1000 Hercules.

Cargo doors in the front area of commercial aircraft are difficult to realize as the cockpit interferes with the implementation. There are two common solutions: Either, the cargo doors are realized underneath the cockpit and the complete aircraft nose with the cockpit is an upward-swinging hinged door. (e.g. Boeing B747-100F/200F/400F). The exception is the Airbus A300-600ST "Beluga" or A330-700L Beluga XL, where the cockpit is below the cargo door or the nose flap can be opened above the cockpit. The disadvantage in this case is that the loading edge is significantly higher and no ramp for direct loading can be implemented. The general advantage of a frontal loading is that the door can be opened wide so that there is no bottleneck and no need to load "around the corner". Linear access saves time and makes it easier to load larger items. With the Boeing 747 freighter, for example, the large front cargo door has proven itself a beneficial option not only with oversized loads, but also makes loading and unloading faster (Stofels 2019).

Research on loading aircraft from the front has shown that only an upward folding door is common in current freight models and that two side-opening doors are not used at all. The only example of the latter is a construction of a propeller-driven cargo aircraft from 1945—the Bristol 170 Freighter. This propeller plane had two split doors at the nose below the cockpit, which could be opened to the left and to the right. From an aerodynamic and safety point of view, this configuration was probably not able to assert itself with higher speeds of the aero-engine age.

Another variant for a cargo access door on aircraft would be at both fuselage ends (front and rear): For quick loading and unloading the Antonov AN124 and AN225 do not only have a door in the rear, but also an extendable ramp integrated in the front. This allows trucks and forklifts to drive directly into the body. The nose with cockpit of the aircraft can be folded up completely for this purpose.

The Lockheed C-5 Galaxy has a front cargo door (the nose is folded upwards) as well as a rear loading opening, which consists of a combination of two lateral doors and a larger upper "sliding door". The cargo doors at the front and tail are equipped with ramps and winches, allowing both short turnaround times and complete trucks to be accommodated in their hold.

5.2 Transfer of the Results from Cargo Doors Analysis to the UCA

The results of the analysis of the loading door variants (Fig. 22) with the configuration of the gyrocopter from the ALAADy project are considered below. This check must take into account both the small size of the gyrocopter compared to a conventional cargo plane and the small distances between wings and propeller motors.

There is not enough space to construct a side door (Fig. 22a) that is large enough for loading access (e.g. when loading pallets). Both are disadvantages for timely and uncomplicated freight loading. It would also have to be analyzed whether the entire construction of the body would be stable enough at all. These characteristics make the alternative of side doors seem unusable for the gyrocopter.

A rear cargo door (Fig. 22b) is also likely to be ruled out due to the aerodynamic design and small size of the gyrocopter (Figs. 19 and 20). Even a complete opening of the tailgate in order to achieve unhindered access to the cargo area would severely affect the stability of the body.

Therefore, the most reasonable option seems to be to load the gyrocopter (Fig. 1) from the front (Fig. 22c), because one characteristic of aircrafts compared to a UCA is missing: No cockpit is needed and the loading of cargo from the front is not affected.

At this point it is now necessary to analyze the details of the door construction. When loading through an opening at the nose, the analysis showed that two split doors at the nose have no effect on the current aircraft designs.

Without initiating a security and aerodynamic investigation, this concept, which had been visualized in the DLR vision in 2016 (DLR 2016), was rejected as the project progressed. This decision matured further, as an electric opening mechanism for two doors would be more complex (e.g. because of the wiring) and heavier (e.g. twice the number of motors/hinges) than for a single door, reducing the payload. The symmetry of the opening mechanism and the good loading capacity of the UCA are in favor of a bow flap that opens upwards.

The variant of a combination of front and rear cargo door (Fig. 22d) was not pursued further at this point because the previous analysis was already negative for only a rear flap.

5.3   Challenges of the Gyrocopter Configuration During Loading

On the basis of the gyrocopter configuration of the ALAADy project, it is to be expected that challenges may occur during loading and unloading, because of the high loading edge and the inclination of the gyrocopter (Fig. 21).

The first obstacle to consider when loading the cargo is the height of the loading edge of 1.30 m. A possible solution to overcome the loading edge height could be an extendable ramp. In addition, the landing gear of the aircraft could be lowered (hydraulically) for loading in order to reduce the height of the loading edge (Fig. 22). A combination of both technical methods would probably be optimal. But it should be noted that the more technical aids are taken on board, the more this is at the expense of the payload.

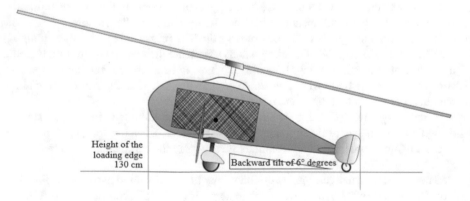

Fig. 21   Technical drawing of gyrocopter configuration of the ALAADy project

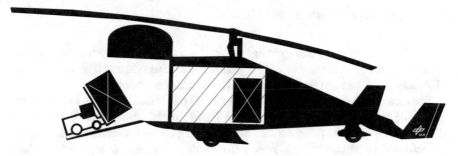

Fig. 22 UCA with ramp and lowering chassis

Fig. 23 Container dimensions for UCA of the ALAADy project

An additional challenge of the cargo loading is the 6° degree inclination of the gyrocopter. This rearward inclination also bears the risk of tipping over forwards during loading. As with the problem with the loading edge, the chassis of the gyrocopter could be lowered during the loading process and at the same time the UCA could be aligned horizontally to overcome the tilt angle. To avoid a forward inclination of the UCA in general, the gyrocopter should be anchored to the ground during handling. This would require further technical equipment (e.g. anchor) on board. However, anchoring the UCA would then increase the risk of hull fracture if the tilt is still present.

Of course, infrastructural conditions at the landing site at the start and destination could make it possible to anchor or balance the gyrocopter. However, this would considerably limit the flexibility of the gyrocopter's possible applications.

5.4   Container Solution for UCA

The analysis of air freight transport has shown that a container system offers far more advantages than disadvantages in a logistics chain. The original hypothesis of the ALAADy project was to use a EURO pallet for transportation.

The idea was that the 800 × 1200 mm EURO pallet would enable cost-effective and stable transport. The reusability of a EURO pallet also played a role in the planning decisions. The pallet enables easy handling when delivering goods: If goods are delivered to an end customer on a EURO pallet, the end customer takes back the same number of empty pallets. In addition, a wide variety of goods can be loaded onto a EURO pallets to save space. However, as the name suggests, the standards of the EURO pallet are of European origin. In international freight traffic, problems occur when containers are used. Since the container is an ISO-standardized transport container; i.e. developed according to American standards of measurement, the stowage volume of the container cannot be optimally utilized when using EURO pallets. Euro pallets are therefore not used in international container transport due to inadequate space utilization. This could be a disadvantage for a smooth entry of the UCA into the logistics chain (ISRA 2019).

For many industrial companies, packaging systems are standardized; e.g. by wrapping or shrink-wrapping EURO pallets with foil. When pallets are loaded, the foil provides protection for the transported goods. Whether this offers sufficient safety for transport in an aircraft such as a gyrocopter would have to be checked. But it is questionable whether EURO pallets with foil wrap can replace the safety aspects of a container.

EURO pallets consist of boards and squared lumber and weigh 20 to 24 kg (depending on the moisture content of the wood and material). There are also pallets made of plastic (approx. 5 kg), corrugated board (approx. 3 kg) and sheet metal (approx. 12 kg). The safe working load of a standard EURO pallet is 1,500 kg. The tare weight with a planned payload of 1,000 kg of the UCA is within the limits of the requirements. A ULD container for aircraft of comparable dimensions has a considerably higher basic weight of approx. 82 kg. The costs of manufacturing and the high tare weight for a container with all its advantages (Table 1), must be compared to the advantages from use of the lighter and cheaper EURO pallets (EPAL 2020).

According to the dimensions of the Gyrocopter ALAADy given in Table 10, the following indicators are available for containers:

- Height: 1.3 m; Width: 1.3 m and Depth: 3.0 m
- Payload of about 1,000 kg

In order not to reduce the payload of the UCA unnecessarily, a lightweight or foldable container is preferable to a container manufactured with ordinary material.

To make optimum use of the volume of the UCA, a container should be adapted to the internal dimensions of the UCA cargo hold (Fig. 22). (height 1.3 m, width 1.3 m and depth 3.0 m).

Since there are no suitable ULDs on the market with these dimensions, a special design would have to be constructed. For this purpose, the potential uses for ALAADy project should be considered (humanitarian logistics, internal transport of components between production and assembly sites, spare parts logistics, supply in areas difficult to access). The goods to be transported (e.g. dangerous or perishable goods; see Sect. 2.1.3) should also be considered.

6 Conclusion

We illustrated the significance of a UCA for the processes of an air cargo supply chain using the given configuration of a gyrocopter from the DLR project ALAADy. At the heart of this analysis was the freight handling within a turnaround process. It was required that the loading and unloading process takes place at a destination that has no (cargo) infrastructure. On the one hand, this premise was intended to promote the investigation of the automation of the cargo unloading process and, on the other hand, to enable coverage of versatile use cases.

When investigating the degree of automation of the process steps in the overall ground handling of a UCA during port-to-port operation, the following statements can be made: In fully automatic operation, the acquisition costs are very high, while the personnel costs are rather low. In manual operation, the conditions are exactly the opposite. A combination of both—the semi-automatic process sequence—offers advantages if the process steps that are most costly in fully automatic operation are replaced by manual steps. The decisive factor, however, remains the infrastructure on the destination site and how much personnel (e.g. for maintenance) will be required (on site).

Furthermore, it was shown that the type and arrangement of cargo doors and the choice of an ULD container can have a decisive influence on the project of an unmanned cargo aircrafts.

The following concepts were presented for cargo unloading: Air cargo drop, click-out-and-go, using equipment suitable for unloading at destination, loading equipment on board, unloading by automated guided vehicle, and unloading & delivery by an autonomous delivery robot and autonomous robot container-system. Among other things, it was shown that an autonomous delivery robot could eliminate further processes steps in the classic air freight transport chain, especially in the last mile. As a result, the introduction of an UCA (Fig. 3) into the air cargo supply chain will provide additional time and process reduction potential in daily operations by providing a more direct and straightforward delivery of the cargo to the recipient (Fig. 24).

At present, a combination of unloading an aircraft and simultaneous delivery has not been realized in the market yet, and thus would be trend-setting. However, the combination of air freight unloading and delivery would have significant time and economic benefits and would provide an important solution to last-mile problems in air cargo logistics. Further, the adaptation of a possible robotic container to the

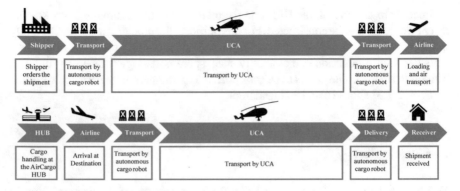

Fig. 24 Air cargo supply chain with UCA and an autonomous robot container system (inbound—outbound with HUB-connection)

usual units of measurement for ULDs seems indispensable. This could enable smooth handling between different transport systems within the supply chain. Other solutions and combinations are possible, but it is important to assess whether the benefits obtained outweigh the total of the investment. Another promising option could be to rethink the last-mile problem or the air cargo supply chain, in combination with the "click-out-and-go" concept described above and an autonomous delivery vehicle or robotic-container-system, which presents another promising option.

References

Aerospace-technology (2002) CargoLifter CL160. Kable Intelligence Limited, London. https://www.aerospace-technology.com/projects/cargolifter/. Accessed 3 June 2020

Amaruchkul K, Cooper WL et al (2011) A note on air-cargo capacity contracts. Prod Oper Manage 20 (1). Available via Wiley Online Libary. https://onlinelibrary.wiley.com/doi/full/10.1111/j.1937-5956.2010.01158.x, pp 152–162

Amazon (2017) Amazon Prime Air. Seattle/Washington, December 2016 https://www.amazon.de/p/feature/vwjfgns3yyjm9uo. Accessed 24 July 2020

Beier L, Einig J, Reich C (2016) Entwicklung und Validierung eines Unterstützungswerkzeuges zur Gestaltung von bodenseitigen Logistik- und Betriebsprozessen an Fracht-UAV. Project work DLR/BKTM, March 2016

Benney R, Barber J, et al (2005) The new military applications of precisi on Air drop systems. US Army Research, Development and Engineering Command, Natick, 26–29 Sep 2005

BMJV (2019a) LuftVG § 6. https://www.gesetze-im-internet.de/luftvg/__6.html. Accessed 18 24 July 2020

BMJV (2019b) LuftVG § 25. https://www.gesetze-im-internet.de/luftvg/__25.html. Accessed 24 July 2020

BMJV (2019c) LuftVG § 30 Abs.1. https://www.gesetze-im-internet.de/luftvg/__30.html. Accessed 24 July 2020

BMJV (2019d) LuftVG §1a Abs.1. https://www.gesetze-im-internet.de/luftvg/__1.html. Accessed 24 July 2020

Bonaceto B, Stalker P (2005) Design and development of a NewCargo parachute and container delivery system. In: Proceedings of 18th AIAA Aerodynamics Decelerator Systems Conference and Seminar, AIAA Paper 2005–1647, Munich, Germany, May 2005

Boonekamp T, Burghouwt G (2017) Measuring connectivity in the air freight industry. J Air Transp Manage 61:81–94

Cakici F, Leblebicioglu MK (2016) Control system design of a vertical take-off and landing fixed-wing UAV. IFAC-Papers Online 49(3):267–272

Collins MP (2016) The future market for Large Unmanned Cargo Aircraft in the national airspace system. Faculty of Lewis University, Aviation & Transportation, Illinois USA, December 2017

Daimler (2018) Mothership. http://media.daimler.com/marsMediaSite/de/instance/ko/Mercedes-Benz-Vans-investiert-in-Starship-Technologies-den-weltweit-fuehrenden-Hersteller-von-Liefer robotern.xhtml?oid=15274799. Accessed 13 July 2020

Dauer JC, Lorenz S, Dittrich JS (2016) Automated Low Altitude Air Delivery. Deutscher Luft- und Raumfahrtkongress. Brunswick 2016 http://www.dglr.de/publikationen/2017/420129.pdf. Accessed 06 July 2018

De Cerchio R, Riley C (2011) Aircraft systems cyber security. In: 2011 IEEE/AIAA 30th Digital Avionics Systems Conference (2011). IEEE, pp 2–3

Desabrais KJ, Riley J et al (2012) Low-cost high-altitude low-opening cargo airdrop systems. J Aircraft 49(1):349–354

DHL (2016) DHL parcelcopter. Deutsche Post DHL Group, Bonn 2016. http://www.dhl.com/en/press/releases/releases_2016/all/parcelecommerce/successful_trial_integration_dhl_parcelcop ter_logistics_chain.htm. Accessed 03 June 2020

DHL Post Group (2019) Neuer Zustell-Roboter unterstützt Postboten beim Austragen ihrer Sendungen. https://www.dpdhl.com/de/presse/pressemitteilungen/2017/neuer-zustell-rob oter-unterstuetzt-postboten.html. Accessed 11 July 2020

DLR German Aerospace Center (2016) Vision of a Drone—Automated Low Altitude Air Delivery, ALAADy. Video Animation 2016. https://www.youtube.com/watch?v=pU-4hrFA-Ec. Accessed 11 July 2020

DPD (2017) The DPDgroup Drone Parcel delivery 2.0. DPDgroup, Aschaffenburg 2016. https://www.dpd.com/home/insights/delivery_drones. Accessed 01 July 2020

EPAL (2020) Euro Pallet. www.epal-pallets.org. Accessed 26 July 2020

EPF (2018) Clip Air. Ecole polytechnique fédérale de Lausanne 2018. https://clipair.epfl.ch/home. Accessed 03 July 2020

EU regulation (1998) No. 2015/1998. https://eur-lex.europa.eu/legal-content/DE/TXT/?uri= CELEX%3A32015R1998. Accessed 03 July 2020

Feng B, Yanzhi L et al (2015) Air cargo operations. Literature review and comparison with practices. Available via Science Direct. https://www.sciencedirect.com/science/article/pii/S0968090X150 01175. Accessed 03 July 2020

Frye H (2011) Flächenbezogene Optimierung von Luftfrachtterminals. Dortmund 2011, p 61

Gomez F, Scholz D (2009) Improvements to ground handling operations and their benefits to direct operating costs. Hamburg University of Applied Sciences, Aero – Aircraft Design and Systems Group. Hamburg, 28 October 2009

Heerkens H (2017) Unmanned Cargo Aircraft—From anywhere to everywhere. Engineering & Technology Reference Online. Available via IET Digital Library. https://digital-library.theiet. org/content/reference/https://doi.org/10.1049/etr.2017.0009.2017. Accessed 14 June 2018

IATA International Air Transport Association (eds) (2005) Interview Giovanni Bisignani. Press release, 1 November 2005

IATA International Air Transport Association (eds) (2008) Fact Sheet IATA e-freight. http://www. iata.org/pressroom/facts_figures/fact_sheets/e-freight.htm. Accessed 14 Oct 2019

IATA International Air Transport Association (eds) (2013) Ground Handling, International Air Transport Association, October 2013

ISRA (2019) Freight & Logistics, Air Freight Container, https://www.isralogistics.com/gallery. html. Accessed 11 Oct 2019

Kazda A, Caves RE (2011) Airport design and operation. 2. ed. Emerald, Bingley, 2008 of the IEEE 99, 11 (2011), p 221

Kraus A (2001) Paradigmawechsel in der Luftfracht. VDI-Gesellschaft Fördertechnik, Materialfluss und Logistik, Airport Logistik Tagung (6), Stuttgart 4–5 December 2001

Lennane A (2017) Unmanned aircraft are the future for air cargo—but we need time to get it right. https://theloadstar.co.uk/unmanned-aircraft-future-air-cargo-need-time-get-right/. Accessed 08 September 2019

Lottermoser A, Berger C et al (2017) Method of usability for mobile robotics in a manufacturing environment. Procedia CIRP 62(2017):594–599

Meincke PA, Asmer L et al (2019) Concepts for cargo ground handling of unmanned cargo aircrafts and their influence on the supply chain. J Syst Manage Sci 2019

Meincke PA, Tkotz A (2015) Airports—types, functions, facilities, and accessibility. In: Wald A. (ed) Introduction to Aviation Management. Aviation Management. LIT, Berlin, 2015, pp 86–126

Mensen H (2001) Planung, Anlage und Betrieb von Flugplätzen. VDI-Buch. Springer-Verlag, Berlin, Heidelberg, 2007, pp 304–309

Merkert R, Van de Voorde E et al (2017) Making or breaking—key success factors in the air cargo market. J Air Transp Manage. Elsevier Publisher (eds), 2017. Available via ScienceDirect. https://www.sciencedirect.com/science/article/pii/S0969699717300455. 10 February 2017. Accessed 15 July 2020

NATO STANANG (1980) Standardization Agreement 3619. Helipad Marking – Helicopter Landing Zone. Available via GlobalSecurity.org, 10 July 1980. Accessed 30 Oct 2019

Nsakanda AL, Turcotte M et al (2004) Air cargo operations evaluation and analysis through simulation. WSC 2004: Proceedings of the 36th conference on Winter simulation December 2004, p 1791

Punakivi M, Yrjölä H et al (2001) Solving the last mile issue. Reception box or delivery box. International J of Physical Distribution & Logistics Management, Available via emerald insight. https://www.emerald.com/insight/content/doi/10.1108/09600030110399423/full/html#loginreload. Accessed 04 Feb 2019

Sadeck J, Riley J et al (2009) Low-cost HALO cargo airdrop systems. In: Proceedings of 20th AIAA Aerodynamics Decelerator Systems Conference and Seminar, AIAA Paper 2009–2010, Seattle, WA, May 2009

Sampigethaya K, Poovendran R et al (2008) Secure operation, control, and maintenance of future enabled airplanes. Proc IEEE 96(12):1992–2007

Sampigethaya K, Poovendran R et al (2011) Future-enabled aircraft communications and security—the next 20 years and beyond. Proc IEEE 99(11):2041

Schmidt M (2017) A review of aircraft turnaround operations and simulations. Munich Aerospace e.V., Faculty of Aerospace Engineering, Progress in Aerospace Sciences 92, July 2017, pp 25–38

Stofels F (2019) Airbus bietet mit A350 sehr gute Plattform für Frachter. Aero Telegraph Interview with Richard Forsen. www.aerotelegraph.com/interview-richard-forson-cargolux-airbus-bietet-mit-a350-sehr-gute-plattform-fuer-frachter, Accessed 11 Nov 2019

Vahrenkamp R (2007) Geschäftsmodelle und Entwicklungsstrategien von Airlines und Airports in der Luftfracht. Arbeitspapier zur Logistik. 2007:23

Vahrenkamp R, Kotzab H (2012) Logistik, Management und Strategien. Oldenbourg Verlag, Munich 2012, p 301

Valavanis K, Vachtsevanos GJ (eds) (2015) Handbook of Unmanned Aerial Vehicles. Springer Publishing 2015

Vis I (2006) Survey of research in the design and control of automated guided vehicle systems. Eur J Oper Res 170(3):677–709

Wings for Aid (2020) A self-landing delivery box. https://www.wingsforaid.org/news.html. Accessed 12 Jan 2020.

Warwick G (2017) Is there a commercial market for large unmanned aircraft? Available via Aviation week. http://aviationweek.com/technology/there-commercial-market-large-unmanned-aircraft. Accessed 7 Oct 2019

ZF Zukunftsstudie (2016) The last mile—Die letzte Meile. EuroTransportMedia Publisher. http://web-zf-zukunftsstudie-de.pixelpark.net/presse-zf-zukunftsstudie-letzte-meile/. Accessed 12 July 2020

Part III: System Components and Safe Autonomy

Cargo Drone Airspace Integration in Very Low Level Altitude

Niklas Peinecke and Thorsten Mühlhausen

Abstract Current initiatives to integrate UAS into airspace mostly target small or medium-sized drones. However, it is foreseeable that there will be a demand for cargo UAS with a payload in the range of about 1000 kg. It is proving to be a challenge to integrate these larger drones into a national airspace. The concept of choice for this problem is to use the EASA Specific Category. In this category it is possible that the reliability of the vehicles is not maximised by increasingly complex system components, but that the drones are guided on special trajectories so that any accident that occurs does not result in fatalities and only financial losses. The advantage is that the concept can be realized with existing technology and only moderate effort for new equipment on board and on ground.

However, this poses some further challenges to air traffic procedures. In Germany and Europe in general, the airspace is densely occupied, especially near airports. UAS missions should disturb this existing structure as little as possible. An integration concept must therefore ensure that the future air cargo system can strategically or at least tactically avoid any potential risk, regardless of whether it is a conurbation, an industrial infrastructure or other aircraft.

We present an integration concept for airspace structure, communication infrastructure and information management. Based on publicly available data on population and ground infrastructure, we present exemplary calculations that show the concept's feasibility.

Keywords UAS · Airspace integration · Airport integration · U-space · UTM

1 Introduction

N. Peinecke (✉) · T. Mühlhausen
German Aerospace Center (DLR), Institute of Flight Guidances, Lilienthalplatz 13, 38108 Braunschweig, Germany
e-mail: niklas.peinecke@dlr.de

T. Mühlhausen
e-mail: thorsten.muehlhausen@dlr.de

© Deutsches Zentrum für Luft- und Raumfahrt e. V. (DLR) 2022
J. C. Dauer (ed.), *Automated Low-Altitude Air Delivery*, Research Topics in Aerospace,
https://doi.org/10.1007/978-3-030-83144-8_10

247

Within the framework of the ALAADy (Automated Low Altitude Air Delivery) project, an integration concept for the vehicles used - automated transport aircraft - in the national airspace is to be developed. The idea is to realize a cargo drone that can carry a payload of up to 1000 kg and is highly automated. Such a relatively large cargo drone can cause major harm in the event of an accident and thus poses a considerable risk. The high degree of automation also implies that the UAS can operate without a remote pilot in some phases of the mission. A security concept is therefore required that avoids major risks for other air traffic, people on the ground and ground structures.

The concept of choice is to use the EASA Specific Category (European Comission 2019) together with an individual SORA (Specific Operations Risk Assessment). This means that the reliability of the vehicles is not maximised by increasingly complex system components, but drones are guided on dedicated trajectories, so that any accident that occurs does not result in fatalities and only monetary losses. The advantage of this approach is that the concept can be implemented with existing technology and with only moderate expenditure on new on-board or ground based technology.

However, this poses some challenges for air traffic procedures. In Germany, but also in Europe in general, airspace is densely populated, especially in the vicinity of airports. The UAS mission should disturb this existing structure as little as possible. An integration concept must therefore ensure that the future air delivery system can strategically or at least tactically avoid any potential risk, be it a populated area, industrial infrastructure or other aircraft.

2   Related Work

The integration of unmanned or remote-controlled aircraft has been studied for several years (Korn 2007; Dalamagkidis et al. 2008) and different concepts for different applications and scenarios have been proposed. Simulations have shown that, in principle, it is possible to integrate individual UAS into the existing ATM system, see Geister et al. (2013). Other concepts aim at adapting the airspace structure specifically for the integration of small drones, e.g. by introducing a specialised UAS traffic management (UTM) system. A well-known example is the UTM of NASA, see Kopardekar (2014). In Europe, for this concept the name *U-space* has been coined. The concept itself is targeted but not limited to urban airspace: The Warsaw Declaration, (EASA 2016) states that there is a need for "the development of the concept of the U-Space on access to low level airspace, especially in urban areas." This makes it clear that currently a demand is seen to solve the problem of integrating UAS in the airspace.

SESAR's study "European Drones Outlook Study – Unlocking the value of Europe" on the increasing number of drones predicts a European fleet of over 8 million by 2050. Among these more than 500,000 drones will be of commercial applications (SESAR Joint Undertaking 2016). The European Drone Investment - Advisory Platform projects the global number of mass-market drones to reach a volume of 35 million units by 2022 (EC 2019). The aforementioned UTM concepts – NASA UTM and U-space – provide infrastructure and some concepts for operating the airspace free of conflicts.

The concept presented in this paper has been first published in 2017 (Peinecke et al. 2017), predating the European U-space concept. Since then U-space has adapted and further defined its architecture and description of services in the publication of a concept of operations (SESAR Joint Undertaking 2019). In addition to our previous results we will analyse how our predating concept can be harmonized with SESAR's concept of X, Y and Z "airspace volumes" in Section "U-space Compatibility". It is shown that both concepts are compatible.

3   Airspace Structure

In Germany, the airspace is structured in airspace classes. These are usually designated by the letters A to G, each of which describes a specific purpose and a related set of rules that every air traffic participant must comply with. In Germany only classes C, D, E and G are used, see Fig. 1.

C, D and E are controlled airspaces with special requirements which a UAS in the Specific Category cannot normally fulfil. In addition, these airspace classes generally begin at higher altitudes, with the exception of the Class D control zone around an airport. ALAADy is designed to fly at very low altitudes to avoid possible risks

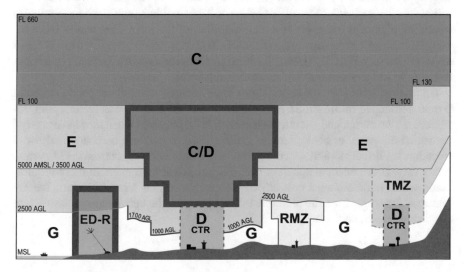

Fig. 1 Structure of the German airspace, Image source: https://commons.wikimedia.org/wiki/File: Luftraumstruktur_Deutschland.svg, Philipp Fischer CC BY 3.0

of collision with ground structures in an emergency. Therefore class G is the only remaining option for Germany.

On the other hand, class G normally means that the aircraft must follow the VFR rules. This means that the aircraft must be able to perform see-and-avoid or detect-and-avoid. Consequently, it needs to have the ability to avoid collisions with other aircraft on its own. For economic reasons, the ALAADy vehicle should not be equipped with expensive, heavy and error-prone sensor technology. Instead, in the event of an unrecoverable emergency, the vehicle should be terminated in such a way that it cannot damage people on the ground, ground structures or other aircraft. However, this should only be a last resort. Therefore, this concept should minimize the probability of an emergency as much as possible.

In Class G, a variety of other aircraft may potentially interfere with the ALAADy vehicle. These include:

- Regular General Aviation aircraft not equipped with transponders or ACAS (Airborne Collision Avoidance System) equipment.
- Ultralight aircraft not equipped with transponders or ACAS equipment.
- Gliders.
- Manned and unmanned balloons.
- Parachutes, sky-divers and paragliders.
- Other drones.

All of these possible vehicles may or may not be equipped with different forms of transponders, displays or collision warning systems, although the rules of Class G only require visual separation from other participants (Bundesrat 2015). Manned aviation follows the rules of the air which are documented in ICAO Annex 2 and recently transferred to SERA Part A (Standardized European Rules of the Air). The notion "well-clear" is used in ICAO Annex 2 but neither provides any minimum separation distances nor gives an exact definition of the term. This leaves room for interpretation of a safe separation distance. This is a major problem for a UAS or RPAS (Remotely Piloted Aircraft System) in situations when a human pilot is not available to judge about the safe separation distance.

Up to now, most integration strategies have been aimed at concepts that attempt to treat unmanned traffic similarly to manned aircraft. These led to various compromises, e.g. that unmanned aircraft must carry VHF equipment when they have to cross controlled airspace, or heavy sensors are required to perform detection and avoid procedures. Furthermore, it was the policy not to influence the current ATM system in any way. It has become apparent that it is not possible to meet all these requirements while maintaining a cost-effective and lightweight aircraft design. We are therefore departing from the principle of not changing the current airspace structure by introducing an additional class.

4 Class G+

As it is not possible to use airspace G in its current form, we propose the introduction of a modified class G with additional infrastructure and procedures. We call this class G+. In order to achieve a level of safety comparable to previous VFR regulation in the presence of drones not equipped with active sensor systems, we are adding communication infrastructure and some additional rules to airspace G and restricting access to the airspace that has traditionally been rather freely usable.

The main idea is to create an airspace-wide information network accessible to all airspace users. This will be implemented with relatively inexpensive, widely available hardware so that all participants can be equipped. In addition, the positions and intentions of all airspace users must be actively communicated by the aircraft themselves, similar to what would be done with an active transponder, e.g. ADS-B (Automatic Dependent Surveillance - Broadcast) or FLARM (Flight Alarm).

4.1 Airspace Location

In general, a Class G+ airspace can be located anywhere instead or as part of a Class G airspace. The purpose of a G+ airspace is to link points of interest, such as an airport and a distribution centre. A G+ airspace will therefore be tending to be shaped like a corridor connecting these points.

Furthermore, G+ airspaces must ensure that every airspace user can terminate safely at any time without endangering ground structures or people. The most likely location would therefore be along uninhabited, underdeveloped areas such as deserts, rivers, lakes, forests, undeveloped coastlines, etc. Outside Europe, such areas may be extensive, but especially in Germany this also leads to a corridor-like appearance.

In the vicinity of airports, class D/CTR is usually applied, i.e. the drones would enter a controlled airspace. In such cases, individual procedures would have to be defined to enable the UAS to land and take off safely even in the presence of other, possibly differently equipped aircraft, see the section on airport integration.

Figure 2 shows the airspace structure around Braunschweig. It can be seen that the two larger airports - Hannover and Braunschweig/Wolfsburg - are enclosed by a (purple) class CTR zone. Unmarked areas are usually class G.

Usually the ALAADy UAS are operated in the very low altitude below 500 ft. This decision was made to allow for a safe termination and as little impact on ground structures as possible in an emergency event. Further, since normal IFR/VFR operation takes place above this altitude, Class G+ can be restricted up to an altitude of 500 ft without significantly disturbing Class G airspace. As a further effect the containment volume of the UAS can be kept small.

Fig. 2 Airspace Structure around Braunschweig, Germany [image generated with Google Earth using data from skyfool.de]

Figure 3 shows the integration of Class G+ into the existing airspace structure. It should be noted that ground obstacles and buildings such as wind turbines or inhabited areas may be located in the airspace, so that occasionally a horizontal bypass is necessary.

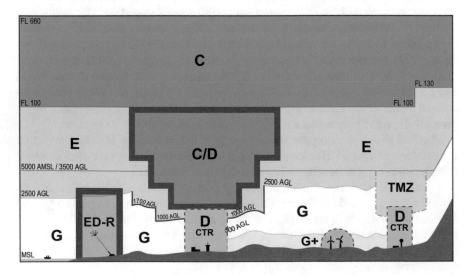

Fig. 3 Locating Airspace G+ in the German airspace, Image adapted from: https://commons.wikimedia.org/wiki/File:Luftraumstruktur_Deutschland.svg, CC BY 3.0

4.2 Communication Infrastructure

The establishment of an airspace-wide G+ communication infrastructure serves the following purposes:

- All participants have access to airspace-wide information management. This can be realized using well-known standards like SWIM (System Wide Information Management), but also more basic implementations based on ADS-B, AIS (Automatic Identification System) and similar standards are possible.
- Participants are obliged to submit their own position, attitude and intent.
- Participants have access to the positions, attitudes and intentions of all other (relevant) participants.
- There is no need to install on-board sensor equipment for airspace users. This is particularly useful for small, light-weight UAs.

To achieve all these aspects, a mobile radio network is planned to be used. This could be based, for example, on a 3G communications network, which would have the advantage of low-cost hardware for both the distribution network and the receivers. Newer standards like 4G LTE (Long Term Evolution) and 5G are also possible. Another option would be to use ad-hoc networks using D2D (drone to drone) communications, forming a network that is relayed by all airspace participants.

Access to the network can be implemented using SIM (Subscriber Identity Module) card technologies to achieve a high level of security and avoid spoofing to a certain extent. When utilizing D2D solutions exist to ensure secure and reliable transmission even via relay nodes.

To ensure secure communication everywhere, in case of mobile networks access nodes have to be installed that cover the entire airspace. In contrast to terrestrial mobile radio, the antennas of these nodes must also be directed upwards so that airspace users can establish connections at all necessary altitudes. Since the focus of the concept is at lower altitudes, the technical effort required for this is limited. As G+ airspaces are most likely shaped like corridors (see section "Airspace location"), a chain of individual access nodes would be the most likely configuration. Figure 4 shows a typical antenna distribution.

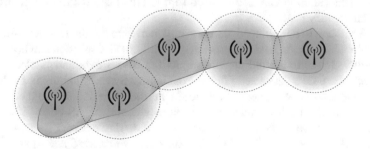

Fig. 4 Antenna distribution along a corridor

In case of D2D communication no further ground installations are necessary. However, it might be advisable to have at least some ground receivers (D2X, drone to infrastructure) installed, to relay information received from other information channels, e.g. weather, NOTAMS (Notices to Airmen), geo-fences or ADS-B and FLARM, see below considerations about cooperative DAA (Detect and Avoid).

If airspace use grows beyond a certain level, it may not be necessary or possible to communicate the status of each airspace user to all other participants. In this case, the individual access nodes can pre-filter these respective data packets and forward them to neighbouring nodes only. In this way, data traffic and necessary on-board processing can be kept at a reasonable level.

In detail, the following data have to be exchanged via the ALAADy network:

1. **Position of the aircraft**: This position is usually determined by means of GNSS equipment on board. However, since the UAS will operate exclusively in airspace G+, the own transmitter cells of the ALAADy network could also be used for positioning, e.g. by multi-lateration solutions based on mobile cells. If using D2D a relative positioning solution can also be used to further enhance the accuracy of the position.

2. **Altitude of the aircraft**: This includes both the barometric altitude and the altitude above MSL and, if possible, above ground. The height above ground could be calculated based on an altitude database if no suitable on-board sensor is available. However, the source of the altitude information must be specified in the data. Again, the accuracy of the measurements can be augmented using multi-lateration and differential GNSS.

3. **Flight attitude of the aircraft:** The inclination of the three principal axes of an airplane in flight can be assessed by on-board equipment and should be submitted.

4. **Speed over ground**

5. **Accuracy values** should be given for all parameters to allow the calculation of safety margins.

Additionally to the regular function of an airspace-wide network for DAA and capacity management purposes there is the possibility that degraded airspace users can be located by the network as long as there is a functioning subscription to the network. This function can also be used to track the last position of an aircraft in case of emergency procedures.

All aircraft are to be equipped with low-cost hardware similar to a standard mobile phone into which an ALAADy network SIM card or D2X solution can be installed. The hardware solution could contain inputs for GNSS data and is also intended to implement A-GPS (Assisted Global Positioning System) solutions for location determination based on the nearest cell nodes. Furthermore, inputs could be provided for connection to FLARM, radar or laser altimeters and ACAS boxes, if available. Another solution would be to equip dedicated ALAADy network ground antennas with FLARM, ACAS or TCAS (Traffic Collision Avoidance System) and ADS-B, so that the known aircraft positions can also be used for intruders not equipped with ALAADy network receivers, see Fig. 5.

Fig. 5 ALAADy network
integrating different airspace
participants

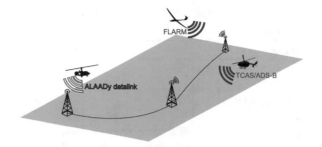

4.3 U-space Compatibility

EASA together with the European Commission has developed a preliminary draft regulatory framework containing high-level safety requirements on the establishment of U-space. This draft regulation published in 2019 already contains recommendations for implementing U-space services, competent authorities and general requirements for aircraft operators and U-space service providers. The classifica-tion of the future U-space airspace and its integration into the present airspace classification is still not completely solved. Therefore, different concepts for fu-ture airspace volumes exist (e.g. SESAR CORUS, EASA and EUROCONTROL). The compatibility of the ALAADy concept with U-space will be described in the following paragraphs based on the concept presented by CORUS (SESAR Joint Undertaking 2019).

The basic idea of U-space is to have "services" and "capabilities": Services describe a type of task a provider can carry out for airspace participants in order to increase for example safety or efficiency. A capability is the ability of a participant to carry out tasks or parts thereof by themselves. Sometimes services require certain capabilities in order to become usable. U-space is organized in four timely stages known as "U1" to "U4", containing the different services. In each stage a sub-set of services is thought to become available.

Given the notion of "airspace volumes" introduced with latest U-space CORUS CONOPS (Concept of Operations), it is necessary to revisit the earlier introduced concept of airspace G+.

CORUS devises three volumes of airspace: X, Y and Z, with Z being subdivided into Za and Zu. The operations possible according to CORUS in different volumes are summarized in the Table 1 (SESAR Joint Undertaking 2019).

Reviewing the type of operations possible in the different volumes, it becomes evident that in G+ no unknown drone operations should be present. These cannot be excluded from X and Y. In G+ the drone itself operates automated. Thus, airspace G+ should be in principle identical to CORUS airspace volume Zu, except when landing on an airport which results in Za. This means that the concept presented here can be implemented in line with CORUS.

Table 1 Characterisation of U-space airspace volumes

Operation		X	Y	Z
Drone	VLOS	Yes	Yes	Yes
	Follow-me	Yes	Only be undertaken with reasonable assessment of the risk involved	
	Open	Yes	Yes, provided access requirements are met	
	Specific	Yes. However, the risk of unknown drone operations must be considered, evaluated and mitigated appropriately	Yes	Yes
	Certified		Yes	Yes
	BVLOS		Yes	Yes
	Automated		As for X	Yes in Zu
Crewed	VFR		Yes, but the use of U-space services by VFR flights is strongly recommended	Yes. However, type Za is controlled airspace. Crewed flights in Za will need to behave as such
	IFR	No	No	

4.4   Airport Integration

It has been demonstrated in the past that UAS can be integrated into normal airport operations without interfering with current ATC procedures; see Geister et al. (2013), Korn (2007). When taking off and landing at a major airport, UAS usually enter Class D or CTR airspace. In this situation, the UAS must follow the guidance and vectorisation information provided by ATC. Since the UAS is not usually equipped for VHF operation, voice radio has been relayed over the UAS in the past. In the present case, we recommend the solution of communicating directly with the operator of the UAS via the telephone or - if applicable - via the dedicated ALAADy network. Thus, the UAS datalink does not need any additional voice capacity. The pilot would then have to respond to the instructions of the ATC.

In the case of a fully automated flight, the company that operates the UAS could, for example, provide ground crew specifically for landing at the airport. If this is not possible, an alternative solution would be to have a service remote pilot (SRP), which could be hired from a third party provider. This SRP could be located at the destination airport and be there on request when the UAS lands and takes off. Figure 6 shows the assignment of the optional SRP when landing at an airport. Take-off and flight in airspace G+ would normally take place automatically.

Fully automated cargo distribution centres would typically not require SRPs, as there would be no interference with manned aviation.

Entering a controlled airspace would in terms of CORUS mean to enter airspace volume Za. In Za it is mandatory to "behave like manned aviation". In the context of U-space this would be realized by using the service called "Procedural interface with ATC" that is available from U2 onwards.

In practice, three types of integration concepts can be distinguished:

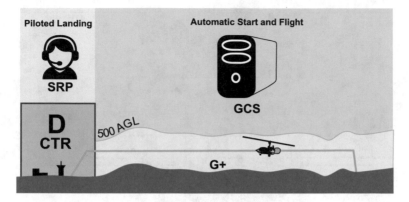

Fig. 6 ALAADy mission including landing in a controlled airspace

1. Landing on an airport with the airport in regular operation. In this case a drone pilot or SRP must be involved.
2. Landing on an airport without SRP. If no drone pilot or SRP can be provided an automatic landing on the airport is still possible. However, this would mean to close down the regular operation of the airport during the drone approach and landing. Usually this also means larger safety margins and thus costly airport fees.
3. Landing on a dedicated UAS hub near the airport. A dedicated hub can be separated sufficiently far from the airport. On the one hand, this would mean that the cargo carried by the drone would have to be transferred between the airport and the hub by other means. On the other hand one minimizes the disturbance of regular airport operations. This way, airport fees can be avoided altogether.

Figure 7 illustrates the three types from top (type 1) to bottom (type 3). The "H"-symbol marks an optional dedicated UAS hub. The red area symbolizes a possible existing zone that needs to be avoided, for example, an area restricted for reasons of safety or noise emissions.

5 Risk Mitigation

5.1 Avoidance of Inhabited Areas

Along the path of a UAS and in the vicinity of airports a number of areas exist that needs to be avoided. Most prominently this includes inhabited areas. As an example Fig. 8 shows the surroundings of Braunschweig. The area can serve as a typical example of European settlement structures. Land use data are shown, with the red areas representing urban settlement and thus possible ground risks.

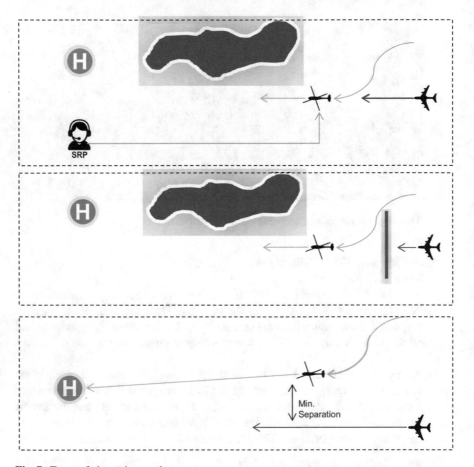

Fig. 7 Types of airport integration

These data are based on CORINE Landcover data; see CORINE Landcover (2012). Furthermore, the controlled airspace of Braunschweig/Wolfsburg Airport (EDVE, darker area, center of the image) and the airspaces belonging to Hannover Airport (dark structures to the top-left) are shown.

Suppose that a UAS is to be routed from Hannover on the left to the automobile plant in Wolfsburg on the top-right of the image. It is obvious that a hypothetical airspace G+ would have to avoid the more densely populated areas. A possible solution in this case would therefore be to create an airspace G+ over agricultural and similar areas, bypassing the inhabited areas. In any case, however, the airspace would have to cross major roads, such as motorways.

Fig. 8 Airspace structure and ground risks near Braunschweig/Germany

5.2 Mid-Air Collision

Collisions in the air must be taken into account between the ALAADy UAS and other airspace participants. Depending on the flight phase, possible intruders may be regular airspace G users such as gliders or balloons, general aviation during an airport approach or rescue helicopters in all flight phases.

Our concept and also the requirements of CORUS volume Zu require the integration of airspace G users into the information network when they enter the airspace. This would mean that the position and flight attitude of such aircraft would be known to the ALAADy UAS and the usual separation procedures could be used. As the airspace G+ is intended for use by the ALAADy UAS, in this particular case we recommend that the other aircraft should give way. This can be regulated locally, especially since current regulation drafts recommend that UAS should give way in any case. This makes sense especially if one expects other airspace participants, e.g., balloons that have limited manoeuvrability. Separation during an airport approach must be ensured by ATC in cooperation with the pilot of the UAS or SRP (see section "Airport Integration").

Similar considerations apply to the avoidance of rescue helicopters. Generally, the UAS pilot should be notified of the helicopter emergency operation. In case of an automated flight, it would be sufficient to inform the ALAADy UAS about the planned helicopter flight path, so that the UAS can avoid a possible loss of separation. If this is not possible due to planning constraints, at least the position and flight attitude of the helicopter must be communicated via the ALAADy network. This can be done by feeding ADS-B data from the helicopter's transponder into the network using ground-based receivers. However, since such information broadcast can be prone to delays and inaccuracies, a safety margin between the UAS and the helicopter should be included.

As a last resort, if all the above separation methods fail, the flight of the UAS can be terminated.

The U-space services that are involved here are "Tracking and Position Reporting", "Strategic Conflict Resolution" and "Emergency Management", that are available from U2 onwards. More advanced services involved are "Tactical Conflict Resolution" and possibly "Dynamic Capacity Management" as scheduled from U3 onwards.

5.3   Loss of Control Due to Turbulence

Turbulence is most likely to occur on approach to the airport, as otherwise the ALAADy UAS would not encounter larger aircraft in airspace G+. In case of an airport approach, the UAS should be treated by ATC as "Light" ICAO Wake Turbulence Category (see section "Airport Integration") and the respective separation to manned aircraft should be applied.

5.4   Loss of Control Data Link

In the case of a fully automatic flight, the ALAADy UAS does not have to maintain a constant control data link. As long as the current traffic situation awareness is ensured via the ALAADy network, the UAS can operate automatically.

In flight phases in which a Control Data Link is absolutely necessary, e.g. during operation in airspaces other than G+ , the Data Link function is safety critical. It must be ensured that a permanently functioning data link is in place before entering a non-G+ airspace. In the event of a permanent loss or if the proper function cannot be guaranteed, the flight must not enter the controlled airspace and should be re-routed or terminated.

5.5 Loss of the ALAADy Network

The presence and function of the ALAADy network is essential for all the above mentioned separation requirements. Therefore if a loss of connection to the ALAADy network cannot be compensated by other means it is a reason for termination of the flight.

5.6 Loss of Position Solution

The presence and function of a working position solution is essential for all the above mentioned separation requirements. If possible, the position solution should be designed redundantly. A complete loss of position awareness is a reason for termination of the flight.

6 Inter-operation with Manned Air-Traffic

A rare case can occur when a manned aircraft, e.g. a rescue helicopter or a fixed-wing aircraft, crosses G+ airspace during an emergency landing. In this scenario, all other unmanned aircraft would have to make way for the intruder, even if this would mean terminating the drone flight when there are no other options.

Class G+ manned traffic will not normally carry ADS-B or even ACAS equipment. Therefore, in this case the nodes of the ALAADy network must be equipped with ADS-B and ACAS-in receivers. In this way, the network itself can locate the incoming manned traffic and inform all unmanned airspace users of the intruder's aircraft status.

In U-space this situation should be handled by the "Emergency Management" service.

7 Conclusion and Outlook

We have presented a concept for the integration of larger cargo drones into the airspace. The focus was on the densely populated German area, which places special demands on safety. In a world-wide context especially in largely uninhabited areas lesser demands may apply. Thus we expect the general concept to be applicable in a variety of applications and environments. Our concept is based on the EASA Specific Category with an individual SORA leading to the use of a dedicated airspace G+. Being first published in 2017, we have revisited and updated the concept taking latest CORUS CONOPS for U-space into consideration.

Open questions that need to be clarified include the safety margins for typical corridors. The dimensions of an airspace G+ corridor depend on the performance characteristics of the aircraft, e.g. the minimum turn radius or the glide path angle in case of engine failure. These parameters must be taken into account in future calculations.

Acknowledgements The authors would like to thank Andreas Volkert for insights and discussions about U-space and flight regulations in general.

References

Bundesrat (2015) Verordnung zur Anpassung nationaler Regelungen an die Durchführungsverordnung (EU) No. 923/2012, Drucksache 337/15, Bundesanzeiger Verlag: §40

CORINE Landcover (2012) CLC 2012. http://land.copernicus.eu/pan-european/corine-land-cover. Accessed 8 Jan 2020

Dalamagkidis K, Valavanis KP, Piegl LA (2008) On unmanned aircraft systems issues, challenges and operational restrictions preventing integration into the national airspace system. Prog Aerosp Sci 44(7):503–519

EASA (2016) Warsaw Declaration: Drones as a leverage for jobs and new business opportunities. High Level Conference on Drones, Warsaw, Poland. https://www.easa.europa.eu/system/files/dfu/Warsaw%20Declaration%20on%20Drones_24%20Nov%202016_final_EN.PDF. Accessed 8 Jan 2020

EC: European Drone Investment - Advisory Platform (2019), European Investment Bank Handout. https://ec.europa.eu/transport/sites/transport/files/drone_investment_advisory_platform_hand_out.pdf. Accessed 8 Jan 2020

European Comission (2019) COMMISSION IMPLEMENTING REGULATION (EU) 2019/947 of 24 May 2019 on the rules and procedures for the operation of unmanned aircraft, Official Journal of the European Union, 11 June 2019

Geister D, Korn B, Tittel S, Edinger C (2013) Operational integration of UAS into the ATM system. In: AIAA Infotech@Aerospace (I@A) Conference, AIAA 2013-5051, Boston, MA. https://doi.org/10.2514/6.2013-5051

Kopardekar PH (2014) Unmanned Aerial System (UAS) Traffic Management (UTM): Enabling Low-Altitude Airspace and UAS Operations. NASA Technical Memorandum 2014-218299. https://ntrs.nasa.gov/citations/20140013436. Accessed 4 Sep 2020

Korn B (2007) Operational Procedures for Integration of UAVs in Controlled Airspace. DLR–USAF Workshop on UAS, 25–27 April 2007, Braunschweig, Germany

Peinecke N, Volkert A, Korn B (2017) Minimum risk low altitude airspace integration for larger cargo UAS. In: 17th Integrated Communications, Navigation and Surveillance Systems Conference, ICNS 2017. IEEE Press. Integrated Communications Navigation and Surveillance Conference (ICNS 2017), 18–20 April 2017, Washington DC, USA. ISBN 978-150905375-9

SESAR Joint Undertaking (2016) European Drones Outlook Study: Unlocking the value for Europe. https://www.sesarju.eu/sites/default/files/documents/reports/European_Drones_Outlook_Study_2016.pdf. Accessed 4 Sep 2020

SESAR Joint Undertaking (2019) SESAR Concept of Operations for U-space, vol 1, 2 & 3, September 9. https://www.sesarju.eu/node/3411. Accessed 8 Jan 2020

System Architectures and Its Development Efforts Based on Different Risk Classifications

Daniel Rothe and Florian Nikodem

Abstract This work develops a set of high-level system architectures for an Unmanned Aircraft System (UAS) for the low altitude transport of one ton payload within the Automated Low Altitude Air Delivery (ALAADy) project. These architectures are further analyzed regarding their development effort. Based on the Specific Operation Risk Assessment (SORA) as Acceptable Means of Compliance (AMC) for the Specific Category of the new EU regulation for UAS, requirements are derived and processed into the architectures. A reference architecture for a certified system is developed for comparison as well. Overall, four major architectures are found. Within SORA, regarding the use case of a large cargo drone, only flights over sparsely populated areas are possible. Consequently, for flights over populated areas a certified system is needed. Requirements in SORA are based on the intrinsic risk of an operation. The risk can be modified by applying mitigations. Therefore, different levels of requirements have to be used for the same mission depending on the mitigations used. Within the analysis of the development effort a nondimensional relative effort factor is found. As a result an ALAADy mission within SORA with limited mitigations and high-risk requirements has little difference to a certified system. Furthermore, a gap is found between two adjacent requirement levels. It is shown that the development effort ranges from the least demanding to the most demanding architecture in half an order of magnitude.

Keywords UAS · ALAADy · Cargo · Transport · System architecture · JARUS SORA · Automated flight · Low altitude · Hazard assessment · Effort analysis · Cost analysis

D. Rothe (✉) · F. Nikodem
Institute of Flight Systems, German Aerospace Center (DLR), Lilienthalplatz 7, 38108 Braunschweig, Germany
e-mail: daniel.rothe@dlr.de

F. Nikodem
e-mail: florian.nikodem@dlr.de

© Deutsches Zentrum für Luft- und Raumfahrt e. V. (DLR) 2022
J. C. Dauer (ed.), *Automated Low-Altitude Air Delivery*, Research Topics in Aerospace,
https://doi.org/10.1007/978-3-030-83144-8_11

1   Introduction

The Automated Low Altitude Air Delivery (ALAADy) project develops concepts
to transport medium-size payloads in an automated vehicle which has short takeoff
and landing capabilities. The aim is to find a solution which is not significantly more
expensive than a transport on road but much faster. Within the project this work
focusses on the development and effort analysis of a set of system architectures
which fulfill different regulatory levels. The different levels evolve from the new EU
regulation for Unmanned Aircraft Systems (UAS). While an Open Category operation covers harmless operation of for example small camera drones, the Certified
Category is comparable to manned aviation. A transition is given by the Specific Category. An Acceptable Means of Compliance (AMC) for the Specific Category is the
Specific Operation Risk Assessment (SORA). Within ALAADy possible scenarios
reach from Specific to Certified Category and are covered by the analysis in this
work.

The different architectures result in different economic characteristics. First of
all the development and operating costs differ. Furthermore the reliability varies and
even the feasibility of a mission can be dependent on the resulting requirements. With
the developed architectures and the effort analysis it shall be possible to find the best
economic approach for the intended operation. This work focusses on technical issues
and does not cover an operational perspective. A first estimation on the development
effort for the architectures is given to find tendencies which can be further considered
in future work.

The development of mission and vehicle characteristics were important steps
within the ALAADy project. They are covered closer in Dauer and Dittrich (2021)
and Hasan and Sachs (2021). The important vehicle characteristics are summarized
in the following. Three possible aircraft configurations were found: *Twin Boom, Box
Wing and Gyrocopter*. All vehicles have a payload capacity of 1 ton. The total mass
of the configurations differs within an insignificant range for this study and can be
assumed to be about 2.5 tons. The cruise speed is 200 km/h for all configurations.
The flight altitude shall be below 150 m. The aerodynamic steering is conducted
by actuating aerodynamic control surfaces respectively a rotor head for the *Gyrocopter*. The flight can immediately be ended by a parachute ejection for the *Twin
Boom* and *Box Wing Configuration* or an autorotation landing for the *Gyrocopter*.
All configurations have multiple engines.

The structure of the remaining sections is as follows: First the related work is
analyzed. In *Sect.* 3 the legal demands are outlined and the resulting technical requirements are derived. Following in *Sect.* 4 important decisions for the structure of the
architectures are made. *Section* 5 conducts a hazard assessment, sets up the architecture and analyzes the development effort. Finally a perspective on future work and a
conclusion is given.

2 Related Work

Papers about the necessity of regulation in the field of UAS missions are Weibel and Hansman (2004), Hayhurst et al. (2006) and Loh, Bian and Roe (2009). The implementation of regulations came much later once a drone market had developed after 2010. The Commission Implementing Regulation (EU) 2019/947 of 24 May 2019 on the rules and procedures for the operation of unmanned aircraft in Europe was adopted in 2019. An AMC to this regulation for the specific operation is SORA. This work uses SORA to develop system architectures. SORA was recently published so there was no comparable work found. Other work which analyzes or applies SORA is Martin, Huang and McFadyen (2018) as an analysis on air risk for UAS missions, Capitán et al. (2019) on the application of SORA for a cinematography UAS and Terkildsen and Jensen (2019) on a tool for checking SORA compliance. The development of system architectures apart of SORA however is conducted in Casarosa et al. (2004). A risk based approach for protecting adjacent areas with geofences is presented in Schirmer, Torens & Adolf (2018). Safety analyses on UAS are described in Hammer, Murray and Lowman (2017) and Clothier and Walker (2015).

3 Regulatory Framework

Commission Implementing Regulation (EU) 2019/947 on the rules and procedures for the operation of unmanned aircraft defines three categories for UAS: open, specific and certified. The open category allows nonhazardous operations to be conducted without further oversight. Due to the size of the ALAADy configurations the missions will not fit the conditions for an open operation. It is the goal of the ALAADy project not to fall into the certified category. Therefore, a risk assessment is conducted with SORA to fit into the specific category which is closer described in Nikodem et al. (2021). A short summary is given in the following. It is shown that within ALAADy missions a flight over sparsely populated areas is in the scope of SORA. Flights over populated areas are out of scope of SORA. Due to the fact that ALAADy missions focus on flights over sparsely populated areas the following legal demands also focus on SORA. Furthermore, the requirements dictated by SORA are summarized. The information given in this section refers to SORA 2.0. A detailed set of requirements for certified UAS is not available yet. The approach to model the certified operation is described in *Sect.* 5.2.

3.1 SORA Summary

In this section the general SORA process is briefly summarized. For a more detailed description see Nikodem et al. (2021).

SORA aims to reduce the risk of an unmanned flight operation by evaluating the severity of risks and giving instructions to reduce the probability of failure conditions happening in a sequence of steps. This is done by categorizing the risk into a Ground Risk Class (GRC) and an Air Risk Class (ARC). Both risk classes are defined by an initial risk class and applying mitigations.

In Step #1 of SORA, the relevant information has to be gathered in a concept of operation description. In Step #2, the initial GRC is found by the vehicle characteristics of mass or impact energy, the overflown areas and whether the vehicle is flown in Visual Line Of Sight (VLOS) or Beyond Visual Line Of Sight (BVLOS). In Step #3, there are three mitigations which can be applied to lower the GRC. They cover the reduction of the number of people at risk, the reduction of the impact dynamics and the emergency response.

The initial Air Risk Class (ARC) is found by the characteristics of the used airspace in Step #4. It is possible to reduce the ARC by applying a strategic mitigation in Step #5. In this mitigation, it has to be shown that the used airspace is less frequented than generally assumed for this type of airspace. Resulting from the final ARC in Step #6, Tactical Air Risk Mitigations (TARM) have to be applied to lower the probability of a midair collision. These cover different robustness levels of detect and avoid capabilities.

In Step #7, the two risk classes are combined into one of six Specific Assurance and Integrity Levels (SAIL). Resulting from the SAIL, multiple Operational Safety Objectives (OSO) have to be met, which is described with Step #8. OSO exist on three different robustness levels: low, medium and high. Which of the levels has to be used is depending on the SAIL. Some OSO are called optional for low SAIL.

Additionally Step #9 defines criteria to avoid endangering adjacent areas by containing the operation. Finally, the compliance with the requirements has to be shown in Step #10.

A semantic model is used to describe the operation: The flight is conducted within a *Flight Geography*. It is surrounded by an optional *Contingency Volume* to apply contingency procedures. *Flight Geography* and *Contingency Volume* form the *Operational Volume*. The ground area around the *Operational Volume* is the *Ground Risk Buffer* to protect the *Adjacent Area*. Volumes are generally understood as a 2.5D-volume as a ground area combined with altitude information.

3.2 Possible Operational Scenarios

SORA enables multiple approaches for a project by varying the mission scenario or mitigations. The following section shows which possibilities are given by conducting a risk assessment according to the SORA process.

The GRC is evaluated by overflown areas and vehicle characteristics. It is scored by a positive integer which rises with the risk. The ALAADy configurations are classified into the most dangerous vehicle class because of the possible impact energy resulting from the mass of about 2.5 tons. In addition, all flights will be conducted beyond visual line of sight. Still two different initial GRCs can be set for ALAADy missions depending on flying over populated areas or avoiding them. Within the context of the ALAADy missions some mitigations can be used. By applying the three mitigations within their possible range the initial GRC changes to the final values. A closer review of the applicability of the mitigations can be found in Nikodem et al. (2021). The final values provide a range between best- and worst cases depending on the used mitigations and are presented in *Table* 1.

SORA can only be applied when the final GRC is 7 or lower. Otherwise the Certified Category has to be applied. Therefore, flights over populated areas are out of the SORA scope.

The ARC is evaluated by the characteristics of the used airspace. It is scored by a (low risk) to d (high risk). ARC-a is reserved for restrictive airspace. Because ALAADy missions are planned to use nonrestrictive airspace the ARC can ranges from b, when avoiding frequented airspace, up to d, when using public airfields and their control zones.

The SAIL evolves from GRC and ARC by *Table* 2. The possible SAIL are marked green and reach from 3 to 6.

3.3 Resulting Technical Requirements

Technical requirements evolve from three sections of SORA. These are Step #9 in the main part, Annex D with requirements for TARM and Annex E with OSO.

Table 1 Possible GRC depending on mission scenario and applied mitigations

Value	Sparsely populated	Populated
Initial GRC	6	10
Change by mitigations in worst case	+1	
Change by mitigations in best case	− 2	
Highest possible GRC	7	11
Lowest possible GRC	4	8

Table 2 SAIL determination. The green marked SAIL are the possible range within SORA for ALAADy missions

Final GRC	Residual ARC			
	a	b	c	d
<=2	I	II	IV	VI
3	II	II	IV	VI
4	III	III	IV	VI
5	IV	IV	IV	VI
6	V	V	V	VI
7	VI	VI	VI	VI
>7	Certified Category			

Step #9 defines requirements for containment. The resulting safety requirements which affect the system architecture are on the one hand:

- No probable failure of the UAS or any external system supporting the operation shall lead to operation outside of the *Operational Volume*.

On the other hand the following requirements have to be considered when there are gatherings of people in the adjacent area:

- The probability of leaving the *Operational Volume* shall be less than 10^{-4} per Flight Hour (FH).
- No single failure of the UAS or any external system supporting the operation shall lead to operation outside of the *Ground Risk Buffer*.
- Software (SW) and Airborne Electronic Hardware (AEH) whose development error(s) could directly lead to operations outside of the *Ground Risk Buffer* shall be developed to an industry standard or methodology recognized as adequate by the competent authority.

According to the SORA terms a *Ground Risk Buffer* is a zone following the *Operational Volume* to protect adjacent areas of enhanced vulnerability. The width of the *Ground Risk Buffer* has to be at least the same number as the flight altitude. Due to the large *Flight Geography* and the possible range of the used vehicle it is assumed that ALAADy missions will have to consider gatherings of people in adjacent areas. Therefore, all four requirements above will have to be applied.

Requirements for a TARM System (TARMS) are defined in Annex D of SORA. This mandatory mitigation defines capabilities of a detect and avoid system. For flights beyond visual line of sight each ARC has different requirements. The Annex gives recommendations for systems to be used, recommendations for agility and requirements for reaction times. For ARC-d a system meeting RTCA SC-228 or EUROCAE WG-105 Minimum Operational Performance Standards (MOPS)/ Minimum Aviation System Performance Standard (MASPS) or similar is required. Requirements for detection rate and reliability are given in *Table* 3.

Table 3 Requirements for TARMS from Annex D regarding detection and failure rate

Value	ARC-b	ARC-c	ARC-d
Detection rate	~50%	~90%	See RTCA SC-228 or EUROCAE WG-105 or similar
Failure rate	$<10^{-2}$/FH	$<10^{-3}$/FH	$<10^{-5}$/FH

General requirements are provided by Annex E of SORA by OSO. Not all OSO contain technical requirements. The requirements which contain technical regulations are described below. Only the level of integrity is summarized because the level of assurance does not affect the system architecture. Still, the level of assurance can be important for the development effort. It is assumed that this influence is covered with the assigned Design Assurance Level (DAL). For closer details see Annex E of SORA. The technical OSO are:

- OSO #4: UAS developed to authority recognized design standards
- OSO #5: UAS is designed considering system safety and reliability

  - On a medium level of integrity a strategy is required to detect, alert and manage any malfunction or failure which would lead to a hazard.
  - On a high level of integrity given probabilities of different failure conditions need to be met.

- OSO #6: C3 link characteristics

  - This OSO prescribes that the command, control and communication (C3) link has to be appropriate for the mission and that it has to be monitorable. The used methodology in this work does not further process this requirement.

- OSO #10: Safe recovery from technical issue
- OSO #12: The UAS is designed to manage the deterioration of external systems supporting UAS operation
- OSO #10 and #12 are described commonly in one paragraph. They only apply when flying over populated areas. Therefore, it is not further considered.
- OSO #13: External services supporting UAS operations are adequate to the operation

  - This OSO prescribes that the external services have to be appropriate for the mission. The used methodology in this work does not further process this requirement.

- OSO #18: Automatic protection of the flight envelope from human errors
- OSO #19: Safe recovery from Human Error
- OSO #20: A Human Factors evaluation has been performed and the HMI found appropriate for the mission

  - This OSO prescribes that the Human Machine Interfaces (HMI) has to be appropriate for the mission. The used methodology in this work does not further process this requirement.

Table 4 Assignment of robustness levels to SAIL

Requirement	SAIL III	SAIL IV	SAIL V	SAIL VI	Certified
Step #9	Required in general				SORA requirements not applicable
TARMS	Depending on ARC flown				
OSO #4	-	L	M	H	
OSO #5	L	M	H	H	
OSO #18	L	M	H	H	
OSO #19	L	M	M	H	

- OSO #24: UAS designed and qualified for adverse environmental conditions

  - This OSO prescribes that UAS has to be appropriate for the environmental conditions during the mission. The used methodology in this work does not further process this requirement.

The robustness level to be used is defined by the SAIL. The assignment is summarized in *Table* 4.

4 System Definition

The different system architectures are based on assumptions which are outlined in this section. First a Monitoring System, which is indirectly suggested by SORA, is characterized. Then general assumptions are described which the system architectures are derived from. Finally the used system terms are defined.

4.1 Monitoring System

Step #9 in the SORA process dictates requirements to protect adjacent areas. A detailed description is found in *Sect.* 3.3. Step #9 requires the Unmanned Aircraft (UA) to leave the *Operational Volume* just with a certain probability of 10^{-4}/FH. In addition no single failure is allowed for the UA to operate outside the *Ground Risk Buffer*. When aiming for a simple UAS without redundancy, a possibility to satisfy these requirements is to implement a Monitoring System which is able to end the flight immediately by a controlled crash when certain criteria are violated. This flight termination shall be conducted by deactivating the propulsion system and triggering the Impact Dynamics Reduction System (IDRS) which is described in *Sect.* 4.3. According to the SORA terms a flight termination is an emergency procedure and not an operation.

When using a Monitoring System there are two ways to fulfill the requirements by Step #9:

Table 5 Requirements for systems and mission depending on the used termination variant. "Flight System" can be interpreted as all systems excluding the monitoring system

Subject	Variant 1	Variant 2
Verified failure rate for the monitoring system	No requirement	$<10^{-4}$/FH (combined with Flight System)
Verified rate for failures that lead to operation outside the *operational volume* for the flight system	$<10^{-4}$/FH	$<10^{-4}$/FH (combined with monitoring system)
Size of *ground risk buffer*	1 to 1 (reference: altitude)	1 to 1 (reference: altitude)
Size of *contingency volume*	No requirement	Depending on possible range after termination

1. The Monitoring System terminates when leaving the *Contingency Volume*. The controlled crash is conducted inside the *Ground Risk Buffer*.
2. The Monitoring System terminates when leaving the officially defined *Flight Geography*. The controlled crash is conducted inside the *Contingency Volume*.

According to the SORA terms the *Operational Volume* consists of a *Flight Geography* where the mission is conducted normally and a *Contingency Volume* to apply contingency procedures. The size of the *Contingency Volume* is not prescribed in general. Without substantiated evidence of the reliability of the pilot, the technical systems alone have to fulfill the requirements dictated by Step #9. SORA does not require to adjust the size of the *Ground Risk Buffer* to the potential range of the vehicle in case of a termination. Variant 2 allows setting internal contingency procedures to prevent breaching the *Flight Geography* without terminating the flight. The two variants are summarized in *Table 5*.

The flight paths of the two variants are shown in *Fig. 1*. It can be seen, that Variant 1 allows for a larger *Flight Geography*.

Both variants have a specific advantage. They are:

- Advantage Variant 1: Larger *Flight Geography*
- Advantage Variant 2: Lower complexity challenges due to decreased reliability requirements

The selection of one variant does not affect the system architecture but the effort to realize the system because of a changing DAL. Therefore, the effort of each Variant is evaluated in *Sect. 5.4*.

4.2 General Assumptions

The setup of the system architectures follows general assumptions which are outlined in the following. The ALAADy aircraft configurations differ in certain aspects like the principle of aerodynamic steering or the setup of the propulsion elements. The

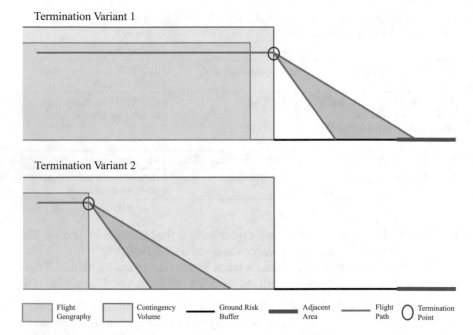

Termination Variant 1

Termination Variant 2

| Flight Geography | Contingency Volume | Ground Risk Buffer | Adjacent Area | Flight Path | Termination Point |

Fig. 1 Visualization of flight paths depending on the termination variant

system architectures are therefore described on an abstract level to cover all aircraft configurations in one view. Consequently, it is not possible to evaluate the configurations in this context. It is assumed that redundant systems are much more expensive to develop than non-redundant systems. Therefore, redundancies are only applied when necessary because of legal demands. Single failures are avoided by using a Monitoring System when possible. Redundant systems are implemented when failure conditions need to be considered and a general failure in a system would lead to a failure condition which is hazardous or higher. The type of necessary redundancy, like the number of components or dissimilarities, is not considered further.

All Systems could cause a loss of the vehicle and shall be developed to a DAL of at least D (DAL as defined in ARP4754A). This shall be done independent of a lack of legal demands to ensure a minimum reliability of an ALAADy mission.

The difference of the requirements for SAIL V and VI is small. In the considered requirements it covers only OSO #4 and #19. OSO #19 differs only in the level of assurance. The difference in OSO #4 is a soft description about the compliance with standards used. These differences are not further considered to reduce the complexity of this work. This leads to the simplification that there are no differences in the effort to develop a system for SAIL V or VI.

The solutions found generally try to describe the simplest architecture for a set of requirements. There is always the possibility to increase the complexity of an architecture for non-safety reasons.

4.3 Terms

To describe the system architectures independent of specific solutions it is described on a logical level. The terms used are defined in the following.

4.3.1 Central Avionics Elements

CS – Communication System. Covers the elements in the UA to establish a C2 link and communication avionics like transponders. Communication avionics are commercially available. To focus on the systems which have to be developed for an ALAADy configuration they are not further modelled.

FCS – Flight Control System. Is able to translate operator inputs and position, attitude and velocity information from the PFAS into steering commands. Operator inputs can be direct control or waypoints. It therefore includes an autopilot as well.

FCELS – Flight Control and Emergency Landing System. Contains a FCS with the additional capability of conducting an emergency landing and further emergency procedures to be able to operate over populated areas in a certified system.

FEHEP – Flight Envelope and Human Error Protection. Monitors steering commands to prevent the vehicle from exceeding certain limits by giving feedback to the FCS. Limits are for example maximum velocities or loads. It can also protect the vehicle from breaching the allowed *Flight Geography*. Satisfies OSO #18 and #19.

GS – Ground Station. Covers the elements to establish a C2 link on ground, interfaces for the operator and related computational elements.

Link – Data Connection between CS and GS.

MDS – Malfunction Detection System. Is able to monitor critical functions and elements of the vehicle. The feedback can be used by the FCS autonomously or the operator. Satisfies medium level of integrity of OSO #5.

PFAS – Position and Flight Condition Acquisition System. Gathers information like position, altitude, velocities and attitude.

TARMS – Tactical Air Risk Mitigation System. SORA requires Tactical Air Risk Mitigations (TARM) when flying an ARC-b or higher. A closer review of the requirements for the TARM can be found in *Sect.* 3.3. Architectures on the same SAIL differ only in the required TARMS. Therefore, it is represented by the same block to reduce the number of visualized architectures. In the cost estimation the difference for each ARC is considered.

4.3.2 Actuation Elements

PAS–Primary Actuation Systems. For aerodynamic steering the actuation of flaps or rotors is necessary. The PAS cover the elements required for controlled flight.

SAS–Secondary Actuation Systems. SAS cover all actuated elements which are not necessarily needed for control of flight like drag flaps.

4.3.3 Propulsion Elements

FSS–Fuel Supply System. To propel the vehicle a supply of fuel is needed. This function is covered with the FSS. It covers the supply of internal combustion engines as well as fuel cells if implemented in the configuration.

PS–Propulsion Systems. The aerodynamic lift generates drag which needs to be compensated by Propulsion Systems. The Propulsion Systems also enable the vehicle to lift off and climb. The ALAADy configurations generally have multiple propulsion systems. PS covers internal combustion engines and electrical engines if implemented in the configuration. It is assumed that the engine will be bought. This component therefore focusses on the link between the control system and the engine.

4.3.4 Other Elements

MS–Monitoring System. As described in *Sect.* 4.1. The MS is completely independent from other aircraft systems regarding position and attitude acquisition to avoid common failures. If not independent in power supply the system has to be deployed when losing the power supply. Contains elements to calculate the validity of the flight state and the elements to bring the vehicle into the terminated flight state, e.g. a parachute. The Monitoring System is implemented to satisfy Step #9 for low SAIL.

IDRS–Impact Dynamics Reduction System. For ALAADy missions a termination of the flight shall be possible. The responsible system can be triggered as an emergency procedure. For the *Box Wing* and *Twin Boom Configuration* the termination system relies on a parachute ejection. The *Gyrocopter Configuration* shall land in autorotation. In addition the Propulsion System is deactivated. These systems are modelled as an Impact Dynamics Reduction System (IDRS).

Power. Contains all elements to provide power in the UA. It also covers the supply of electrical engines, if implemented in the configuration.

5 System Variants

In this section the different system architectures are developed and analyzed. The system architectures are defined for each system variant found. The range of possible variants is defined by *Table* 2 and if a Monitoring System is applied the termination variant as described in *Sect.* 4.1. In addition, a system architecture for a certified system is developed to give a comparison to high SAIL as well. The naming scheme for the architectures based on SORA includes the specific Assurance and Integrity

Level, the Air Risk Class and, if applied, the Termination Variant (TV). An example is SAIL IV, ARC-b, TV 1.

5.1 Required Components

To perform a mission, most of the defined systems have to be implemented. There is either a Flight Control System (FCS) or a Flight Control and Emergency Landing System (FCELS). FCELS is only implemented in a certified system to be able to apply autonomous emergency procedures to avoid crashing into populated areas. Flight Envelope and Human Error Protection (FEHEP) and Malfunction Detection System (MDS) are implemented depending on the SORA requirement. FEHEP is required from SAIL III on and therefore for all variants. MDS is required from SAIL IV on.

A MS is only implemented for SAIL III and IV. From SAIL V on OSO #5 dictates that a loss of the vehicle is a hazardous failure condition and must not occur with a higher probability than 10^{-7}/FH as described more detailed in *Sect.* 5.2. This however is a stricter requirement than not to leave the *Operational Volume* with a probability of 10^{-4}/FH dictated by Step #9. Because MS is implemented to avoid redundancies and with OSO #5 on high robustness level redundancies are necessary anyway, MS is not implemented for higher SAIL and a certified system.

The Impact Dynamics Reduction System (IDRS) is not implemented for a certified system, because failure is not an option. A summary of these information is listed in *Table* 6.

5.2 System Hazard Assessments

A hazard assessment is conducted to find the DAL and required redundancies of the components. The established fault trees follow ARP 4761. The target numbers evolve from Step #9 and OSO #5 on high level robustness. Step #9 is closer described in *Sect.* 3.3. These requirements are only applied to the ground risk, because it is assumed that as long as the vehicle does not leave the *Operational Volume* the air risk is already considered within the TARM. Leaving the *Operational Volume* is covered with Step #9 though.

For the SAIL III and IV variants OSO #5 on high level of robustness does not need to be considered and only Step #9 is relevant. In SAIL III and IV a monitoring system is to be used as described in *Sect.* 4.1. Therefore, only the requirement of not leaving the *Operational Volume* with a probability of 10^{-4}/FH is evaluated here.

For each of the termination variants as listed in *Table* 5 a different fault tree has to be prepared. Without substantiated evidence of the reliability of the pilot, the technical systems alone have to fulfill the requirements dictated by Step #9. Therefore, it is assumed for both termination variants that the flight path will leave

Table 6 Implemented components for different architectures

Component	SAIL III, ARC-b, TV 1/2	SAIL IV, ARC-b/c, TV 1/2	SAIL V or VI, ARC-b/c/d	Certified
CS	x	x	x	x
FCS	x	x	x	
FCELS				x
FEHEP	x	x	x	x
GS	x	x	x	x
Link	x	x	x	x
MDS		x	x	x
PFAS	x	x	x	x
TARMS	x	x	x	x
PAS	x	x	x	x
SAS	x	x	x	x
FSS	x	x	x	x
PS	x	x	x	x
MS	x	x		
IDRS	x	x	x	
Power	x	x	x	x

the *Operational Volume* and the *Ground Risk Buffer*. *Figure* 2 shows the related fault tree for termination variant 1 when a termination is performed just before leaving the *Operational Volume*.

It can be seen that the Flight Envelope and Human Error Protection (FEHEP) component is responsible for preventing the leaving of the *Operational Volume*. Therefore, it has to fulfill the reliability of 10^{-4}/FH. Although a DAL cannot be set

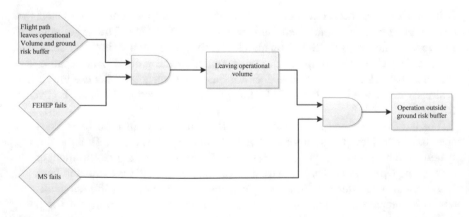

Fig. 2 Fault tree for SAIL III or IV, TV1

Table 7 Required DAL and redundancies for SAIL III or IV, TV 1 architectures

Component	DAL	Redundancy
CS	C	No
FCS	C	No
FEHEP	C	No
GS	C	No
Link	C	No
MDS (only SAIL IV)	D	No
PFAS	C	No
TARMS	Depending on ARC flown	
PAS	D	No
SAS	D	No
FSS	D	No
PS	D	No
MS	D	No
IDRS	D	No
Power	D	No

in direct relation to reliability, it is assumed that the general DAL D of safety critical systems, as described in *Sect.* 4.2, will not meet the required reliability. This statement is supported by AMC RPAS. 1309 which is also included in *Table 9*. As consequence the FEHEP and all components that could interfere have to be developed as DAL C. All other systems could cause a loss of the vehicle and are therefore developed as DAL D though it is not necessary by legal demands. No outage of any component can directly lead to a violation of the Step #9 requirements. Therefore, no redundancies are mandatory. The results are summarized in *Table 7*.

The fault tree for termination variant 2 is similar to that of termination variant 1 and can be seen in *Fig.* 3. Here, the termination occurs before leaving the *Operational Volume*.

In this variant the FEHEP component and the monitoring system are responsible for preventing the leaving of the *Operational Volume*. Thus the two systems combined have to fulfill the reliability of 10^{-4}/FH. As consequence the FEHEP and all components that could interfere as well as the monitoring system have to be developed as DAL D. All other systems could cause a loss of the vehicle and are therefore developed as DAL D as well for own interest. No outage of any component can directly lead to a violation of the Step #9 requirements. Therefore, no redundancies are mandatory. The results are summarized in *Table 8*.

For SAIL V or VI Step #9 and OSO #5 on a high level of robustness have to be considered. OSO #5 requires in the category of the high level of integrity:

- Major Failure Conditions are not more frequent than Remote
- Hazardous Failure Conditions are not more frequent than Extremely Remote
- Catastrophic Failure Conditions are not more frequent than Extremely Improbable

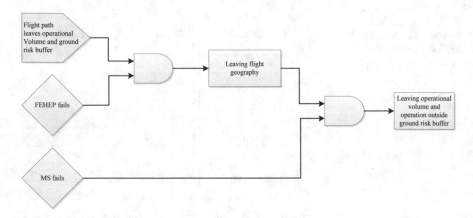

Fig. 3 Fault tree for SAIL III or IV, TV2

Table 8 Required DAL and redundancies for SAIL III or IV, TV 2 architectures

Component	DAL	Redundancy
CS	D	No
FCS	D	No
FEHEP	D	No
GS	D	No
Link	D	No
MDS (only SAIL IV)	D	No
PFAS	D	No
TARMS	Depending on ARC flown	
PAS	D	No
SAS	D	No
FSS	D	No
PS	D	No
MS	D	No
IDRS	D	No
Power	D	No

Table 9 Required quantitative failure probability and DAL for different failure conditions from AMC RPAS. 1309

Failure condition	Allowable quantitative probability	Required design assurance level
Major	$<10^{-5}$/FH	C
Hazardous	$<10^{-7}$/FH	B
Catastrophic	$<10^{-8}$/FH	B

Specific definitions of the terms are given in AMC RPAS. 1309 with an interpretation for the current issue in the following. A malfunction of any component with exception of the IDRS is considered as a major failure condition. A loss of the vehicle or an emergency landing is considered as a hazardous failure condition. One or more fatalities are considered as a catastrophic failure condition. The allowed probabilities for the given failure conditions evolving from a multi reciprocating or turbine engine of less than 6000 lbs with complexity level II are shown in *Table* 9.

From the given information the following consequences evolve: Leaving the *Operational Volume* is considered to be a hazardous failure condition, because it will likely result in an emergency landing or the loss of the vehicle. The operation outside the *Ground Risk Buffer* is considered to be a catastrophic failure condition, because it is likely that an emergency landing or a crash will result in fatalities here. As a consequence, the requirements of Step #9 are exceeded by the requirements resulting from OSO #5 on a high level of robustness and are not further considered.

The fault tree of a SAIL V or VI architecture can be seen in *Fig.* 4.

Most systems can cause catastrophic failure conditions and arc therefore DAL B. An outage could lead to a failure condition which is at least major. Therefore, redundancies are applied to meet the quantitative probabilities. However, some exceptions can be made.

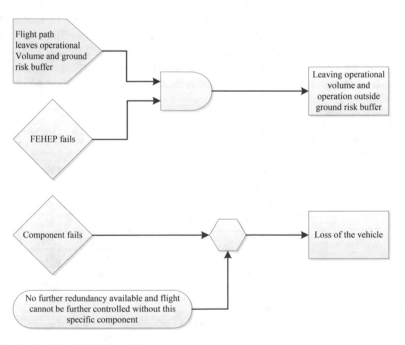

Fig. 4 Fault tree for SAIL V or VI

DAL B is necessary for the MDS, because of possible interferences with other systems. A redundancy does not seem to be necessary because a loss of this component would just result in a safety landing at the next possibility which is not a major failure condition.

The SAS cannot lead to hazardous failure conditions and are therefore DAL C without redundancy. However, in a specific system this could change for certain components like a retractable gear whose outage in the retracted position could cause a loss of the vehicle.

The PS is assumed to base on commercial off the shelf engines. Because multiple engines are implemented the PS does not require a redundancy for each engine.

The IDRS does not lead to major failure conditions so the minimum DAL of D without redundancy is assumed. However, this decision needs to be closer reviewed if the IDRS is able disturb the flight significantly e.g. with the ejection of a parachute.

ARP4754A allows to reduce the DAL, when having independent redundancies. So the DAL B systems could be reduced to DAL C systems, when having independent redundancy anyway. However, it was not further examined, in which cases interdependencies would forbid the use of this possibility, so it was not used within this work. Future analysis should consider this option.

A summary of the results can be seen in *Table* 10.

Requirements for a certified operation are not available yet. To find a comparative architecture, assumptions were made on how a certified system will be structured.

For SAIL V and VI, OSO #5 dictates that certain failure probabilities have to be met for certain failure conditions. These derive from the AMC RPAS. 1309 which also applies for the certified category. It is assumed that the development effort mainly evolves from the process to attain required reliabilities. When the same failure rates

Table 10 Required DAL and redundancies for SAIL V or VI architectures

Component	DAL	Redundancy
CS	B	Yes
FCS	B	Yes
FEHEP	B	Yes
GS	B	Yes
Link	B	Yes
MDS	B	No
PFAS	B	Yes
TARMS	Depending on ARC flown	
PAS	B	Yes
SAS	C	No
FSS	B	Yes
PS	B	No
IDRS	D	No
Power	B	Yes

Table 11 Required DAL and redundancies for a certified system architecture

Component	DAL	Redundancy
CS	B	Yes
FCELS	B	Yes
FEHEP	B	Yes
GS	B	Yes
Link	B	Yes
MDS	B	No
PFAS	B	Yes
TARMS-d	C	Yes
PAS	B	Yes
SAS	C	No
FSS	B	Yes
PS	B	No
Power	B	Yes

Table 12 Required DAL and redundancies for TARMS

ARC	DAL	Redundancy
b	none	No
c	D	No
d	C	Yes

have to be met, equal processes are necessary which build on similar requirements. Therefore, the evolving architecture for SAIL V or VI is taken with the highest ARC to find the certified architecture. Some modifications have to be made to be applicable for flights over populated environments. These cover the Flight Control System that should be able to conduct emergency landings and the possibility to terminate a flight is discarded.

The differing systems used result in *Table* 11.

The DAL and Redundancies for the TARMS evolve from Annex D which is closer described in *Sect.* 3.3. The results are listed in *Table* 12. For TARMS-b no DAL is necessary. This component can probably not result in a loss of the vehicle. Therefore, no DAL is applied for reliability reasons.

5.3 System Architectures

The visualized architectures evolve from the required components listed in *Sect.* 5.1 and the redundancies listed in *Sect.* 5.2. The ARC or Termination variants are not considered here. To visualize the architectures, the legend shown in *Fig.* 5 is applied.

Fig. 5 Key for interpreting the architectures shown in Fig. 6, 7, 8 and 9.

Components are illustrated with boxes. Multiple components like actuators for the different control surfaces are illustrated with a bar beyond the box. Redundant components are illustrated with a second box behind the main box.

The simplest architecture is for SAIL III. Only ARC-b is possible here. The architecture is shown in *Fig.* 6.

For SAIL IV a MDS is added. In addition, missions with ARC-c are possible in this variant. The architecture is shown in *Fig.* 7.

The SAIL V or VI architecture differs mainly in the necessary redundancies. In addition, no monitoring system is implemented. With this architecture all ARCs can be flown if the sufficient TARMS is implemented. The architecture is shown in *Fig.* 8.

The architecture for a certified system is similar to the SAIL V or VI architecture. The FCS is replaced with a FCELS and the IDRS has been removed. The architecture is shown in *Fig.* 9.

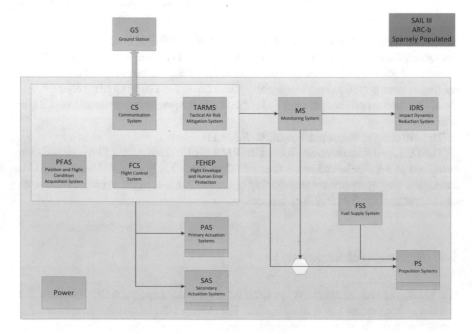

Fig. 6 Architecture for SAIL III

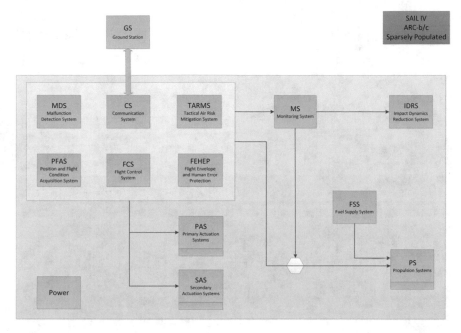

Fig. 7 Architecture for SAIL IV

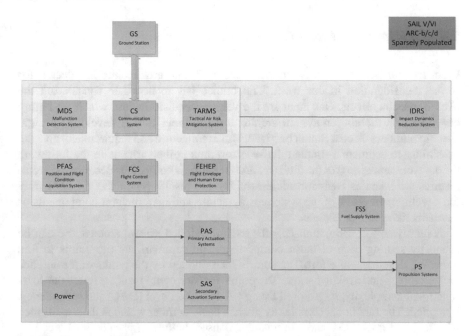

Fig. 8 Architecture for SAIL V

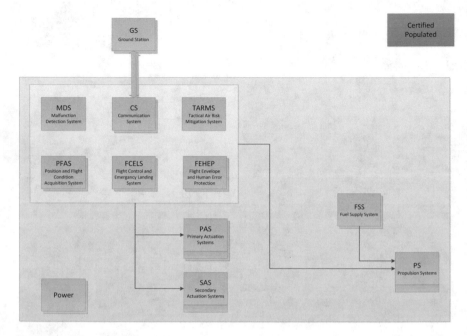

Fig. 9 Architecture for a certified system

5.4 Development Effort Estimation

From the number of system architectures found, one architecture is selected for
further consideration in this work. The choice has to be made considering the
overall mission effects. One input are the vehicle acquisition cost, which are highly
depending on the chosen system architecture. Without further developed design,
specific numbers on cost cannot be given. But architecture and requirements give the
possibility to estimate a simple relative effort among the variants. In the following
the development effort for the different architectures is compared. Because the architectures are taken as basis for this estimation the resulting numbers describe only
the development effort of the system architecture and no structural components or
elements like gears or engines.

A relative effort estimation E_{rel} for each developed system architecture can be
found by scoring each of the n components and summarizing the points to find
a relative development effort. The sum is then divided by the points of a reference
architecture to find a relative value. The points for each component are found due to its
functional complexity, its required Design Assurance Level (DAL) and redundancy,
if applied. The functional complexity gives a basic number which is then multiplied
with factors for the DAL and redundancy. Consequently E_{rel} is calculated by.

$$E_{rel} = \sum_{i=1}^{n} \frac{B_i \cdot f_{a,i} \cdot f_{DAL,i} \cdot f_{r,i}}{P_{ref}}.$$

Table 13 Points for the basic functional complexity of each component

Component	B_i	Component	B_i
CS	2	TARMS-c	4
FCS	5	TARMS-d	8
FCELS	7	PAS	4
FEHEP	4	SAS	2
GS	4	FSS	3
Link	3	PS	2
MDS	4	MS	6
PFAS	4	IDRS	3
TARMS-b	2	Power	3

The base for the analysis is the basic functional complexity of each component B_i. The estimation used evolves from the number of general tasks of each component to be conducted and can be found in *Table* 13. Components which contain a high amount of commercial off the shelf products were considered less sophisticated.

The factor $f_{a,i}$ describes if a component is used in the architecture. It is 1 if it is and 0 if it is not. The usage of a component in a certain architecture can be found in *Sect.* 5.1.

The factor for the complexity depending on the DAL $f_{DAL,i}$ was found by reviewing RockwellCollins (2009). The different approaches were averaged to a factor of 1.5 for each following DAL. This simplified factor disregards important but unknown influences like team experience. It is assumed that this factor can be applied for all subjects. The factor for each DAL can be found in *Table* 14.

The factor for the complexity depending on the redundancy $f_{r,i}$ is assumed to be at least the double effort of a simplex system. It is assumed that the additional testing required for verifying interdependencies increases this factor to 3.

The resulting numbers are divided by a reference value P_{ref} to find a relative value. The number of SAIL IV, ARC-c, TV 1 is chosen as reference value because it is the preliminary preferred variant within SORA as described below. The final relative values can be found in *Fig.* 10.

Figure 10 shows that the architectures can be categorized into two groups: Architectures evolving from SAIL III or IV and architectures evolving from SAIL V or IV or a certified system. The step between the two groups is mainly driven by OSO#5.

Table 14 Effort factors for different DALs

DAL	$f_{DAL,i}$
None	1
D	1.5
C	2.25
B	3.38

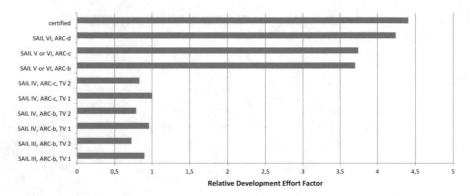

Fig. 10 Relative Development Effort Factor for different architectures

This OSO requires failure conditions to occur with a certain probability which leads to an increased DAL and required redundancy.

The development effort is more dependent on the SAIL than the ARC. An exception is ARC-d which increases the development effort significantly. If a termination is possible, the termination variant has significant impact on the development effort. The challenging requirements of SAIL V and VI differ by just a small increase of the development effort to a certified system, if ARC-d is to be used.

For further analysis it seems useful to reduce the number of considered architectures. If a complex and highly reliable system shall be developed, it seems beneficial to choose a certified system. The development effort is just slightly higher than a SAIL V or VI system, but has no restrictions for operation over populated areas, which will shorten many routes or enable scenarios entirely. If a simple system shall be developed, it seems valuable to aim for SAIL IV and ARC-c. Further requirement reduction by using more mitigations will not reduce the development effort significantly, but restrict the flown missions. It is assumed that the use of termination variant 2 could restrict the route and this restriction does not justify the lower development effort. Therefore, SAIL IV, ARC-c, TV 1 is chosen.

The given results can be used as a support for architectural decisions. However, it is important to note that many simplifications were made to find these numbers and that these numbers can significantly differ in a specific project. A big influence is the team experience in developing systems with design standards and DALs. Some systems may also react sensible to additional mass from redundant systems which was not covered in this analysis and could have an impact on the feasibility. Finally, the input parameters cannot be attested as free from subjectivity. Nevertheless, the given tendencies and especially the statement that a SORA system should stay within SAIL IV can be used for further development.

6 Future Perspective

The developed architectures and effort estimations give the possibility to take them as an input for further economic models for the targeted operation. This information can be used to iterate payload mass, velocity and mission scenarios. In a feedback loop information about necessary reliability for an economic operation could be considered. The results indicate that the ALAADy aircraft configurations seem to be operated best in either SORA SAIL IV with the highest possible Air Risk Class or as a certified system. These two possibilities should be further examined.

Furthermore, the results in this work are based on SORA 2.0. Eventually, there will be more Annexes published or further versions which will have to be incorporated into the existing work. The comparison to a certified system can be evaluated in more detail when a certification specification for certified UAS is available.

7 Conclusion

This work developed system architectures and estimated related development efforts for ALAADy configurations based on safety requirements derived from SORA. Possible architectures were diversified depending on the Specific Assurance and Integrity Level (SAIL), the Air Risk Class (ARC) and, if applied, the Termination Variant (TV). Two TVs were found to satisfy the SORA requirements which generate slightly different requirements for the architectural components. While the first variant allows a larger *Flight Geography*, the evolving requirements are more sophisticated. The different architectures were analyzed regarding their crucial components, including the components Design Assurance Level (DAL) as well as necessary redundancies. The related development effort was estimated by scoring the individual components.

There was a big discrepancy found between SAIL IV and V, which indicates that the costs of applying additional mitigations will be much smaller than the increased cost of a system for a higher SAIL. This evolves from the requirement to fulfill certain probabilities of failure condition which will be equal to the ones of a certified system. This explains the small difference between SAIL VI and a certified system, too. It is likely that assurance processes for certified systems will be adapted to fulfill this requirement within SORA. Consequently, this could lead to a situation where certified systems will be preferred over systems of SAIL V and VI.

SAIL IV with a high ARC or a certified system is found to be best suitable for ALAADy missions. Which of these different approaches is the best suited has to be examined in further considerations which are highly integrated between economical, technical and operational perspectives. This could change some statements within this work because they evolve from limiting the perspective to safety requirements and not economically driven reliability.

References

Capitán C et al (2019) Risk assessment based on SORA methodology for a UAS media production application. Paper presented at International Conference on Unmanned Aircraft Systems (ICUAS). Atlanta, GA, USA, 11–14 June 2019

Casarosa C et al (2004) Impact of safety requirements on the weight of civil unmanned aerial vehicles. In: Aircraft engineering and aerospace technology, vol 76, issue 6. Emerald, pp 600–606

Clothier R, Walker R (2015) Safety risk management of unmanned aircraft systems. Handbook of Unmanned Aerial Vehicles. Springer Science+Business Media, Dordrecht, pp 2229–2275

Dauer JC (2021) Unmanned aircraft for transportation in low-level altitudes: a systems perspective on design and operation. In: Dauer JC (ed) Automated low-altitude air delivery - towards autonomous cargo transportation with drones. Springer, New York, Berlin, Heidelberg

Hammer J, Murray A, Lowman A (2017) Safety analysis paradigm for UAS: development and use of a common architecture and fault tree model. Paper presented at the 36th Digital Avionics Systems Conference (DASC), St. Petersburg, 17–21 Sept 2017

Hayhurst K et al (2006) Unmanned aircraft hazards and their implications for regulation. Paper presented at 25th digital avionics systems conference, Portland, OR, USA, 15 Oct 2006

Hasan YJ, Sachs F (2021) Performance-based preliminary design and selection of aircraft configurations for unmanned cargo operations. In: Dauer JC (ed) Automated low-altitude air delivery - towards autonomous cargo transportation with drones. Springer, New York, Berlin, Heidelberg

Loh R, Bian Y, Roe T (2009) UAVs in civil airspace: safety requirements. IEEE A E Syst Mag

Martin T, Huang Z, McFadyen (2018) Airspace risk management for UAVs: a framework for optimising detector performance standards and airspace traffic using JARUS SORA. Paper presented at the 37th Digital Avioncs Systems Conference (DASC), London, 23–27 Sept 2018

Nikodem F, Rothe D, Dittrich J (2021) Operations risk based concept for specific cargo drone operation in low altitudes. In: Dauer JC (ed) Automated low-altitude air delivery - towards autonomous cargo transportation with drones. Springer, New York, Berlin, Heidelberg

RockwellCollins (2009) Certification cost estimates for future communication radio platforms. https://docplayer.net/13974967-Certification-cost-estimates-for-future-communication-radio-platforms.html. Accessed 23 September 2020

Schirmer S, Torens C, Adolf F (2018) Formal monitoring of risk-based geo-fences. Paper presented at the AIAA Information Systems-AIAA Infotech at Aerospace, 2018 (209989). https://doi.org/10.2514/6.2018-1986

Terkildsen K, Jensen K (2019) Towards a tool for assessing UAS compliance with the JARUS SORA guidelines. Paper presented at International Conference on Unmanned Aircraft Systems (ICUAS), Atlanta, GA, USA, 11–14 June 2019

Weibel R, Hansman J (2004) Safety Considerations for Operation of Different Classes of UAVs in the NAS. Paper presented at AIAA 3rd "Unmanned Unlimited" Technical Conference, Workshop and Exhibit, Chicago, Illinois, 20–23 Sept 2004

ARP4754A: guidelines for development of civil aircraft and systems. SAE Aerospace

ARP4761: Guidelines and methods for conducting the safety assessment process on civil airborne systems and equipment. SAE international, 400 commonwealth drive, Warrendale, PA 15096–0001

AMC RPAS.1309 Safety assessment of remotely piloted aircraft systems. Issue 2, Joint authorities for rulemaking of unmanned systems

Commission implementing regulation (EU) 2019/947 of 24 May 2019 on the rules and procedures for the operation of unmanned aircraft

SORA: JARUS guidelines on specific operations risk assessment. Edition 2.0, Joint authorities for rulemaking of unmanned systems

Human Machine Interface Aspects of the Ground Control Station for Unmanned Air Transport

Max Friedrich, Niklas Peinecke, and Dagi Geister

Abstract The GCS (Ground Control Station) is an elementary part of the UAS (Unmanned Aircraft System). It provides the connection between the human pilot and the airborne part of the UAS, the drone. While early GCS have been merely more than simple remote controls, a modern GCS can do more than just relaying steering commands to the drone. Instead, it provides an important contribution to the safety of the UAS operation. This is achieved by presenting the pilot pre-processed information from the drone and secondary data sources. Utilizing design principles developed in human machine interface (HMI) theory, these information can be shown to the pilot without distraction and making optimal use of the data. This chapter summarizes design principles and challenges for HMIs in aviation use. The focus is on challenges related to UAS operations and UAS GCS in particular. The design process for the GCS *U-FLY* will be described, and some results of a real-world test in the context of two application projects will be presented.

Keywords Unmanned aircraft systems · Human machine interface · Ground control station · Human machine interface design · Human factors

1 Introduction

A UAS (Unmanned Aircraft System) consists not only of the flying vehicle, the UA (Unmanned Aircraft) but also of other components that need to be taken into consideration. At least the pilot needs to be considered as a further element of the UAS. Another important element is the GCS (Ground Control Station) that provides the

M. Friedrich (✉) · N. Peinecke · D. Geister
German Aerospace Center (DLR), Institute of Flight Guidance, Lilienthalplatz 13,
38108 Braunschweig, Germany
e-mail: max.friedrich@dlr.de

N. Peinecke
e-mail: niklas.peinecke@dlr.de

D. Geister
e-mail: dagi.geister@dlr.de

© Deutsches Zentrum für Luft- und Raumfahrt e. V. (DLR) 2022
J. C. Dauer (ed.), *Automated Low-Altitude Air Delivery*, Research Topics in Aerospace,
https://doi.org/10.1007/978-3-030-83144-8_12

connection between pilot and UA. In order for the entire system to operate safely and fulfil its purpose the HMI (human–machine interface) of the GCS needs to be properly designed. In this chapter the basic principles of a GCS HMI design will be described. Further, some of these principles will be illustrated by presenting their implementations in the *U-FLY*, the GCS and mission planning software developed at DLR (German Aerospace Center). In detail, the use of *U-FLY* in the ALAADy (Automated Low Altitude Air Delivery) project as a ground control station for mission control and supervision will be described. The scope of ALAADy is described in Dauer and Dittrich (Dauer and Dittrich 2021).

2   Human Factors Challenges in UAS Control

From a human factors' perspective, controlling a UAS is a challenging task. In the history of unmanned aviation higher accident rates have been observed than for conventional, manned aviation (2004). These accidents are oftentimes attributable to shortcomings in the design of the HMI for control and system safety monitoring.

Major design problems of HMIs for UAS guidance include:

- The proliferation of displays and pop-up windows obscuring displays presenting safety critical information, or the textual instead of graphical presentation of critical information (Hobbs, Lyall (Hobbs and Lyall 2016),
- Overloading the UAS pilot with raw data instead of aggregated information requires engagement in additional cognitive processing (Tvaryanas, Thompson (Tvaryanas and Thompson 2008),
- Insufficiently thought-through concepts for color use, warning philosophies and information management.

These design deficiencies could be attributed to situations in which safety critical information and warnings were not accurately recognized or understood leading to inadequate responses of the pilot to the situation (Hobbs 2010; 2006). Many of the shortcomings in HMI design are a consequence of insufficient consideration of human factors engineering principles in the systems' early development phase.

Since safety is the highest priority in aviation, it is necessary to identify the unique human factors challenges of operating UAS in controlled airspace and close to populated areas, as it is envisioned in the operational concept of ALAADy. Based on an analysis carried out in the project, design requirements have to be established and taken into consideration in the development of the HMI. Furthermore, HMIs for UAS control should be validated and evaluated extensively before using them in actual operation. As the authors in Hobbs, Shively (2013) point out, the approach of identifying design deficiencies of immature systems through the investigation of accidents and incidents, also referred to as "tombstone safety" won't be tolerated anymore due to the increased community expectations of safety and reliability. Thus, it is inevitable to put as much effort as possible into human factors issues already in the design phase of an unmanned system.

Therefore, a major aim in the development of the human machine interface of the ground station used in ALAADy was the application of empirically validated human factors methods in order to design an HMI that adequately supports pilots in their tasks. With the aid of cognitive work analysis methods and findings from scientific literature on HMI design approaches, a novel HMI was designed for the supervision and guidance of multiple highly automated UAS operated in low-altitude airspace.

A major challenge in the design of the HMI was to account for the compensation of the reduced accessibility of sensory information due to the remote position of the pilot, which was found to be a critical human factor in remote aircraft control (2008). Therefore, the next section will deal with the topic of sensory isolation and possible impacts on safety.

2.1 Sensory Isolation

As a consequence of the GCS being located on the ground the UAS pilot is deprived of possibly safety critical sensory cues which a pilot on board of an aircraft has access to. A pilot especially lacks the vestibular and olfactory senses, which are important as they can inform about, for example, unusual attitudes (2008). Additionally, there are latencies in up-link of commands and down-link of responses due to delays in transmission caused by the data link characteristics.

A usual method for compensating at least for the lack of visual information is the installation of cameras on board of the aircraft. However, even if the UAS pilot has access to visual information from the aircraft, there will always remain a lack of information sources due to the fact that the pilot is not present in the actual situation on board. This lack of possibly safety critical information may impact the quality of the pilot's decisions. As such, detailed analyses of information sources should be conducted in order to identify situation specific information important for decision making (2018). The results of such analysis should serve as input for the design of HMIs for remote control and supervision of aircraft. Concepts should be developed that allow for efficient comprehension and recognition of safety critical situations despite reduced access to sensory information. In the course of the HMI development for the GCS in ALAADy, detailed analyzes of critical information were carried out, some of which are published in (2018).

One approach to overcome the challenges of sensory deprivation in UAS control and supervision is the use of highly sophisticated automation that detects abnormal situations and informs the pilot. However, the introduction of automation bears a variety of human factors challenges which are discussed in the following section.

2.2   System Autonomy Classification

From a human factors point of view, one of the major challenges of introducing automation into a safety critical system is the allocation and distribution of system functions between the human and the automation (Feigh (Feigh and Pritchett 2014),Pritchett (Feigh and Pritchett 2014). In the course of the years different classifications in terms of system automation levels have been introduced. The first classification of automation levels was published in 1978, when Sheridan and Verplank distinguished between ten automation levels ranging from *manual* to *complete autonomous system control* (Sheridan 1987). In the year 2000 another classification, namely the PACT (Pilot authorization and control tasks) framework (Bonner (2000) was introduced which differentiates between six levels of system autonomy. If a system exceeds an automation level of four on the PACT scale, the pilot's main task becomes to monitor the safe execution of system functions by the automation. Only in case of abnormal behaviour the pilot needs to react and initiate appropriate counter measures. This form of system control is also referred to as supervisory control (Sheridan 1987). If the system's autonomy reaches level four on the PACT scale, system designers should assure that the system informs the pilot constantly about its actions and intends, otherwise a human intervention becomes extremely difficult in case of a system failure.

More recently *Drone Industry Insights* lists five levels of drone autonomy, ranging from Level 0 with no automation, over Level 3 with conditional automation where the pilot acts merely as a fall-back-option, up to Level 5 with full automation including AI (artificial intelligence) assisted flight planning (DRONEII 2019).

3   Human Machine Interface Development

The first step in designing an HMI is to develop a well-defined function allocation concept. The function allocation defines the scope of functions that need to be executed by the pilot with the devices integrated in the HMI of the GCS. A clearly defined allocation of functions forms the basis for defining information requirements for the HMI, i.e. which information to be displayed on the display screens of the GCS.

In order to determine the functions that the pilot will execute and thus specify the actual information needs, it is important to get an understanding about the underlying work domain of the pilot. A well-known method for this purpose is CWA (cognitive work analysis); (Jenkins (Jenkins 2009),Rasmussen (Rasmussen et al. 1994; Vicente 1999). CWA is a framework, developed to model how work could progress within complex sociotechnical work systems. The term sociotechnical work system refers to a system within which humans and technical elements work together to achieve a common purpose (Jenkins 2012). CWA follows the so-called ecological approach to system design, meaning that it focuses on the constraints and capabilities of the work system (Vicente 1999). The focus on constraints separates CWA from techniques that intent to describe how work actually is or should be done by identifying the tasks and actions the pilot is required to perform when interacting with a system in order to achieve a certain system goal (Kirwan (Kirwan and Ainsworth 1992). In case of not yet existing systems, it is difficult to define specific tasks that have to be executed since these systems cannot be observed. Furthermore, workers often develop new ways of working while they gain experience with the new systems (Elix (2008). CWA offers a solution for the problem, as it models how work could be done instead of how it should be done. Thus, as opposed to designing systems based on predefined tasks, basing the design on constraints imposed by operational conditions and system resources the CWA approach leaves a space of possibilities for actions the pilots can take (Jenkins 2012). Since actions are highly dependent on the situation, this space of possibilities permits the system to better account for unanticipated situations than systems that were designed based on predefined tasks (Vicente 1999). Further, in complex sociotechnical systems there is usually not only one right way to accomplish a certain purpose. Instead, workers can adopt different strategies of how to proceed, depending on the specific situation. Through a CWA the workspace within which workers can operate is determined and information requirements as well as human–machine interface requirements can be derived.

The goal of the first phase of CWA, the WDA (work domain analysis), is to gather a thorough understanding about the systems' architecture in terms of systems, actors and information flow but representing the work system independent of specific events, tasks, activities and actions. During the WDA, the whole system is decomposed into sub-systems and further into smaller units. Furthermore, purposes and functions of the system are derived and it is extracted which subsystem is responsible for fulfilling which specific sub-function in order to fulfil the purpose of the system as a whole. The abstraction hierarchy is a commonly used tool for summarizing a WDA (Vincente (Vicente and Rasmussen 1992). An abstraction hierarchy consists of five different levels of abstraction ranging from functional purposes of the system to processes related to the actual physical objects of the system. Figure 1 depicts an extraction of the abstraction hierarchy for the work domain of a pilot controlling a UAS in controlled airspace.

In the vertical axis the five levels of abstraction can be seen. Each level comprises a number of objects of different nature depending on their abstraction level. For example, on the topmost level there are the overall functional purposes of the system, while the bottom level consists of the actual physical objects involved. Relations are represented by edges that always connect objects of directly adjacent levels. The relations are characterized by means-end relationships (Vicente and Rasmussen 1992). This means that the respective upper abstraction level describes why a function is being executed and the level below describes how it is executed.

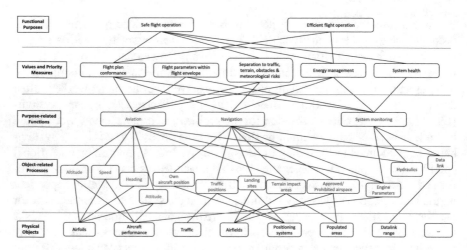

Fig. 1 Abstraction hierarchy for the work domain of a UAS

3.1 Derivation of Information Requirements

The abstraction hierarchy depicted in Fig. 1 allows deriving information requirements for the HMI. On the object related processes level, various functions can be found which the pilot needs to be aware of. An example for an information requirement is the question of whether the HMI should show the position of the controlled aircraft and the positions of other traffic. Further information that can be be displayed include the positions of suitable landing sites, approved and prohibited airspaces and possible pre-defined areas where the vehicle could be purposely crashed without endangering persons or infrastructure. These terrain impact areas are especially important in the case of an engine failure with subsequent loss of thrust. Figure 2 illustrates a possibility how these information requirements are implemented in an HMI. The example shown is the main screen the GCS *U-FLY* that was developed at DLR, the German Aerospace Center. The figure shows a situation in which a pilot is supervising the flight of a UAS-swarm consisting of three UAS. One UAS experienced a loss of thrust and consequently needs to divert to the closest suitable airfield or crash into a nearby terrain impact area as quickly as possible. The HMI visualizes the closest suitable airport as well as the closest terrain impact area together with the estimated flight trajectory (visualized as arrows). Furthermore, prohibited airspaces over populated areas are shown in red. In such a situation the GCS would suggest two options to choose from: Gliding to the closest airfield (visualized by the green arrow) or gliding into a terrain impact area (visualized in orange). It is necessary to inform the pilot properly of these options so that an informed choice can be made.

The presented abstraction hierarchy and the derived information requirements formed a basis for designing an HMI which allows for safety supervision and guidance of UAS while aiming to serve the needs of the pilot. In DLR's GCS *U-FLY* several of the aforementioned concepts have been integrated. The *U-FLY* and its

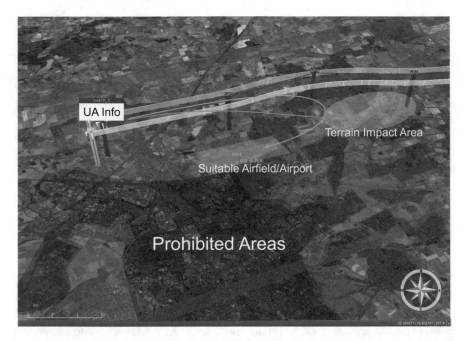

Fig. 2 Exemplary information requirements. Different types of areas can be represented by color in a map display: Prohibited Areas (dark red), Suitable Airports (green) and potential Terrain Impact Area (yellow)

functionalities are presented in detail in the next section followed by a description of the HMI elements that were used in the context of the ALAADy project.

4 The Ground Control Station *U-FLY*

At the Institute of Flight Guidance at DLR an HMI of the GCS *U-FLY* for supervision and guidance of UAS is currently developed. In the design concept of the HMI cognitive work analysis was applied to derive information requirements. The HMI was validated and evaluated within the scope of two studies. Positive results regarding the usability and suitability of the HMI and the design concept were obtained. The results of both studies can be found in Friedrich (2019).

The *U-FLY* supports flight mission planning and execution as well as the technical supervision of several UA, regardless of the aircraft category. This means that the *U-FLY* can be used to plan and execute missions with small multicopters, VTOL (vertical take-off and landing) systems as well as MALE and HALE (medium altitude long endurance, resp. high altitude long endurance) systems. The *U-FLY* enables 4D trajectory planning, i.e. trajectories composed of 3D positions and timing information. The planning can be based on the performance parameters of the respective aircraft. Beyond the scope of a pure GCS, the *U-FLY* supports a number of missions planning and surveillance functions, e.g. a scan pattern creation optimized for the performance parameters of the aircraft and its sensors in order to be able to efficiently surveil areas using several UA. During mission planning and execution, the *U-FLY* also enables strategic conflict detection and avoidance with, for example, other air traffic participants or dynamic airspace closures. Furthermore, the *U-FLY* enables flight panning that considers flight paths inducing low ground risks. For this *U-FLY* allows to import areas with high ground risk from external sources to identify the flight route with the lowest risk. The aim is to reduce the overall mission risk according to SORA (Specific Operational Risk Assessment) methodology (2019). In the specific instance of ALAADy the lowest risk flight path is derived by avoiding populated areas. A connection to the envisioned European framework of U-space services is also provided (SESAR (SESAR Joint Undertaking 2017).

4.1 UAS Supervision Concept for the ALAADy Demonstration

Some of the concepts for a GCS HMI as developed in ALAADy were implemented in a technology demonstrator. This offered the opportunity to test these parts of the HMI under real-word conditions. Details on the demonstration can be found in (Lorenz et al. 2021). In this paragraph the aspects of the GCS relevant for ALAADy will be briefly described.

In the ALAADy demonstration the *U-FLY* is primarily used as a mission supervision station. This is seen as a first iteration to avoid regulatory details not connected to the project aims. The actual task of steering the UA is then carried out by a dedicated GCS operated by the pilot. In a later iteration a fully integrated control of the UA by *U-FLY* is envisioned.

From the perspective of the *U-FLY* HMI three roles are involved in controlling and supervising the UAS during the flight mission. Note that these represent only certain aspects of the roles in the actual mission, thus the role names differ from those given in (Lorenz et al. 2021). The *pilot in charge* of actually controlling the aircraft used a GCS in form of a hand-held remote control. In (Lorenz et al. 2021) this role is identified as *Pilot Flying*. A dedicated *flight test engineer* monitors safety critical technical parameters and is also in charge of deconfliction in case of an unexpected encounter with another aircraft as well as flight termination, if needed. This role is identified in (Lorenz et al. 2021) as *Flight Termination System Operator*. The third role is the *mission controller* who is in charge of planning the flight and supervising mission status and progress using the functions integrated in the *U-FLY*. This role is called *Aircraft System Operator* in (Lorenz et al. 2021). Both, the mission controller and the flight test engineer have access to the technical parameters of the aircraft (Fig. 3). Other roles mentioned in (Lorenz et al. 2021) are related to higher order processes or systems not related to the *U-FLY* GCS and are therefore not documented here.

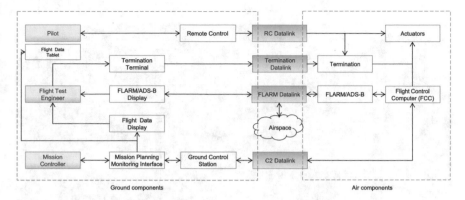

Ground components   Air components

Fig. 3 Control and system architecture of the UAS used in ALAADy

Figure 3 details the three roles (beige boxes) in connection to technical modules (white boxes). Note that apart from the described GCS *U-FLY* (here Ground Control Station) and the steering GCS Remote Control additional equipment was necessary for regulatory reasons, e.g. a FLARM (Flight Alarm) and ADS-B (Automatic dependent surveillance–broadcast) display. Several radio data links connecting ground and air components are shown as blue boxes. More than one data link is necessary due to redundancy and safety reasons, e.g. to separate the C2 (Command and Control) datalink from the flight termination link. Arrows represent the principle data flows without detailing nature and content of these data.

4.2   The U-FLY HMI

An overview of the HMI of the *U-FLY* in ALAADy can be seen in Fig. 4. It consists of three panels or screens (Fig. 5):

Fig. 4 The *U-FLY* HMI showing a planned route for air transport

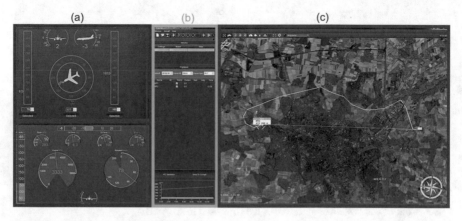

Fig. 5 Panels of the *U-FLY* HMI

- The technical supervision screen (Fig. 5, a).
- The flight plan widget (Fig. 5, b).
- The map display showing the position of the aircraft and flight mission critical areas including geo-fences on the map (Fig. 5, c).

Each of these panels includes a number of display elements that represent object related processes derived in the abstraction hierarchy (Fig. 1). Cross-checking that each object related process is present in the design ensures the basic functional purposes of the HMI. Further, there are a number of higher order requirements that arise from the values and priority measures level of the abstraction hierarchy. An example is the flight plan widget (Fig. 5, b). It is not directly linked to one of the object related processes but links several of them to additional information – the current flight plan – in order to measure the flight plan conformance.

The technical supervision screen presents important technical parameters of the vehicle:

- On its upper part (Fig. 5, d) data is shown that normally would be displayed on a primary flight display in a manned aircraft. The parameters presented in this screen are altitude, attitude, heading and airspeed. These parameters were considered as especially important in the context of the project.
- On its lower part (Fig. 5, e) the display shows further technical parameters, mainly regarding the engine. The parameters shown in this screen are rotor and engine RPM, as well as oil pressure, oil temperature and cylinder head temperature.
- In the centre of the display, the flight plan widget is located (Fig. 5, b). In this widget the flight plan of the UAS and a vertical profile display are shown.

To the right side of the flight plan widget, the map display is located (Fig. 5, c). This display shows the position of the UAS, represented as a symbol on a map. The planned flight route is shown in yellow, and restricted and prohibited areas are marked in red.

Figure 4 presents an exemplary situation of a low altitude air delivery use case. The aim is to transport goods from the western part of the city of Braunschweig, Germany to the east without crossing any restricted or prohibited areas. As such, the flight route was planned to avoid these areas. The map display may also be tilted and rotated, to enable a three-dimensional understanding for the user.

5 Flight Test Campaign

A prototype of the ALAADy vehicle has been tested during an extensive campaign involving several flight tests (Lorenz et al. 2021). For this, *U-FLY* was integrated in a mobile truck station (Fig. 6). During the flight tests *U-FLY* provided the mission controller with information about the planned mission, the current situation and derivations of both. These tests served as a proof that a display of relevant mission data is possible in a real-world set-up. A detailed final evaluation of the functionality for the mission controller has not yet been carried out.

Fig. 6 Mobile truck station with *U-FLY* used during the test campaign

6 Conclusion and Outlook

We have presented a brief overview of design principles and challenges for human machine interfaces in aviation use with a main focus on the challenges related to UAS operations and UAS GCS in particular. A design process for DLR's GCS *U-FLY* was detailed, and the results in the context of project ALAADy were presented.

During ALAADy the *U-FLY* was already used in the technology demonstration. Here, already some of the concepts could be tested under real-world conditions. For future developments it is planned to further develop and utilize the *U-FLY* in follow-up projects, especially in the application areas of humanitarian aid and logistics.

References

Bonner M, Taylor R, Fletcher K (2000) Adaptive automation and decision aiding in the military fast jet domain. In: Proceedings of the conference on Human Performance, Situation Awareness and Automation: User centered design for the new Millenium, Savannah, Georgia, pp 154–159

Dauer JC, Dittrich JS (2021) Automated cargo delivery in low altitudes: concepts and research questions of an operational-risk-based approach. In: Dauer JC (ed) Automated Low-Altitude Air Delivery - Towards Autonomous Cargo Transportation with Drones. Springer, New York, Berlin, Heidelberg

DRONEII (2019) The 5 Levels of Drone Autonomy. https://www.droneii.com/project/drone-autonomy-levels. Accessed 8 Sep 2020

Elix B, Naikar N (2008) Designing safe and effective future systems new approach for modelling decisions in future systems with cognitive work analysis. In: Proceedings of the 8th international symposium of the Australian Aviation Psychology Association. Australian Aviation Psychology Association Sydney, NSW, Australia

Feigh KM, Pritchett AR (2014) Requirements for effective function allocation: a critical review. J Cognit Eng Decis Mak 8(1):23–32

Friedrich M, Papenfuß A, Hasselberg A (2018) Transition from conventionally to remotely piloted aircraft – investigation of possible impacts on function allocation and information accessibility using cognitive work analysis methods. In: Advances in Human Aspects of Transportation: Proceedings of the AHFE 2017 International Conference on Human Factors in Transportation, 17–21 July 2017, The Westin Bonaventure Hotel, Los Angeles, California, USA, N.A. Stanton, Editor. Springer International Publishing, Cham, pp 96–107

Friedrich M, Lieb J (2019) A novel human machine interface to support supervision and guidance of multiple highly automated unmanned aircraft. In: 38th Digital Avionics Systems Conference. 8–12 Sep 2019, San Diego, USA

Hobbs A (2010) Unmanned aircraft systems. Human factors in aviation. Elsevier, San Diego, CA, USA, pp 505–531

Hobbs A, Shively RJ (2013) Human factor guidelines for UAS in the national airspace system. In: Association for Unmanned Vehicle Systems International (AUVSI). Washington DC

Hobbs A, Lyall B (2016) Human factors guidelines for unmanned aircraft systems. Ergon Des 24(3):23–28

Hopcroft R, Burchat E, Vince JP (2006) Unmanned Aerial Vehicles for Maritime Patrol: Human Factors Issues. No. DSTO-GD-0463. Defence Science and Technology Organisation Edinburgh (Australia) Air Operations DIV

Jenkins DP et al (2009) Cognitive work analysis: coping with complexity. Ashgate Publishing Ltd., Farnham, UK

Jenkins DP (2012) Using cognitive work analysis to describe the role of UAVs in military operations. Theor Issues Ergon Sci 13(3):335–357

Kirwan B, Ainsworth LK (1992) A guide to task analysis: the task analysis working group. CRC Press, Boca Raton, FL, USA

Lorenz S, Benders S, Goormann L, Bornscheuer T, Laubner M, Pruter I, Dauer JC (2021) Design and flight testing of a gyrocopter drone technology demonstrator. In: Dauer JC (ed) Automated Low-Altitude Air Delivery - Towards Autonomous Cargo Transportation with Drones. Springer, New York, Berlin, Heidelberg

Murzilli L, et al. (2019) JARUS guidelines on specific operations risk assessment (SORA), Public Release Edition 2.0, Joint Authorities for Rulemaking of Unmanned Systems, pp 11–38. http://jarus-rpas.org/sites/jarus-rpas.org/files/jar_doc_06_jarus_sora_v2.0.pdf. Accessed 21 Sep 2020

Rasmussen J, Pejtersen AM, Goodstein LP (1994) Cognitive systems engineering. Wiley, New York

SESAR Joint Undertaking (2017), "U-space blueprint", Luxembourg. https://doi.org/10.2829/335092, https://www.sesarju.eu/sites/default/files/documents/reports/U-space%20Blueprint%20brochure%20final.PDF. Accessed 21 Sep 2020

Sheridan TB (1987) Supervisory control. Handbook of human factors. Wiley-Interscience, Hoboken, pp 1243–1268

Tvaryanas AP, Thompson WT (2008) Recurrent error pathways in HFACS data: analysis of 95 mishaps with remotely piloted aircraft. Aviat Space Environ Med 79(5):525–532

Vicente KJ (1999) Cognitive work analysis: Toward safe, productive, and healthy computer-based work. CRC Press, Boca Raton, FL, USA

Vicente KJ, Rasmussen J (1992) Ecological interface design: theoretical foundations. IEEE Trans Syst Man Cybern 22(4):589–606

Williams KW (2004), A summary of unmanned aircraft accident/incident data: Human factors implications. F.A.A.O.C.O.C.A.M. INST., Technical Report Publication No. DOT/FAA/AM-04/24

Williams KW (2008) Documentation of Sensory Information in the Operation of Unmanned Aircraft Systems. F.A.A.O.C.O.C.A.M. INST., Technical Report Publication No. DOT/FAA/AM-08/23

Data Link Concept for Unmanned Aircraft in the Context of Operational Risk

Lukas Marcel Schalk and Dennis Becker

Abstract Long-distance cargo transport is one of the manifold applications envisaged for highly automated unmanned aircraft (UA), which operate in the very low-level airspace below 150 m. Despite the level of automation, remote pilots need an interface for command and control of UA, which is typically provided by a digital data link. In this contribution, we investigate the impact of certification based on operational risk, which the European Aviation Safety Agency (EASA) introduced for UA within the specific category on the data link concept. In order to reach the required level of safety, which is the outcome of the Specific Operation Risk Assessment (SORA), the data link concept has to fulfill certain safety objectives. At the same time, the data link concept should be as cost-efficient as possible from an economical point of view. Keeping these boundary conditions in mind, we present an operation risk-based data link concept based on commercial off-the-shelf technologies. In dependence of the mission profile and the required safety level, cellular networks, satellite networks, as well as a combination of both, are viable options. Specifically, we evaluate by simulations the suitability of 4th generation Long Term Evolution (LTE) cellular networks for long-distance cargo UA. With regard to the concepts of operation, the goal of our simulations is estimating low-altitude cellular coverage in rural areas and we factor in radio propagation effects, link budgets, and the distribution of cellular base stations. The results of our system simulations with realistic flight routes show that LTE cellular networks offer good rural coverage in Germany and ensure operational safety in general. Nevertheless, connection disruptions in the order of some seconds can occur in valleys that are not served well by LTE and have to be considered in the risk assessment process.

Keywords Channel modelling · Cellular coverage · Command and control · Long term evolution · Unmanned aircraft · Unmanned aircraft systems

L. M. Schalk (✉) · D. Becker
German Aerospace Center (DLR), Institute of Communications and Navigation, Münchener Str. 20, 82234 Weßling, Germany
e-mail: Lukas.Schalk@dlr.de

D. Becker
e-mail: Dennis.Becker@dlr.de

© Deutsches Zentrum für Luft- und Raumfahrt e. V. (DLR) 2022
J. C. Dauer (ed.), *Automated Low-Altitude Air Delivery*, Research Topics in Aerospace,
https://doi.org/10.1007/978-3-030-83144-8_13

1 Introduction

Command and control (C2) of unmanned aircraft (UA) inherently requires a bidirectional digital data link, which provides an interface between the UA and its remote pilot in control. On the one hand, commands are sent from the remote pilot to the UA in order to steer the UA, especially for low level of UA automation. On the other hand, status information is sent from the UA to the remote pilot in order to monitor the UA and its surroundings, especially if the UA operates beyond visual line of sight (BVLOS).

Today, popular data link candidates for BVLOS C2 of civilian low-altitude UA, which are foreseen to travel over distances of several hundreds of kilometers, are cellular and satellite networks. In the last years, several satellite service providers started to offer new services with regard to civilian UA. For example, Inmarsat launched a data link service for civilian UA already in 2016. Moreover, existing cellular networks for mobile phone users have been tested for UA at low altitudes. Especially, Long Term Evolution (LTE) cellular networks have been proven suitable for C2 of low-altitude UA in urban areas but their suitability outside urban areas is still an open question, especially due to network coverage holes in rural areas.

In this chapter, we discuss the question which data link technologies are suitable for C2 of the investigated long-distance cargo UA in the context of operational risk based certification. At first, we discuss the requirements, which depend on the top-level aircraft requirements (TLAR) introduced by Dauer and Dittrich (2021) and the specific operation risk assessment (SORA) discussed by Nikodem et al. (2021). Afterwards, we present the architecture of the C2 data link concept. Eventually, we investigate the operational safety of the data link concept in simulations. In order to reduce the complexity of our simulations, we only focus on the suitability of LTE networks for low-altitude cargo UA and do not discuss satellite networks any further. Our simulation framework allows comprehensive system simulations of cellular LTE networks in combination with effects of radio propagation and realistic UA flight routes in low-altitude environments. The results of our system simulations show that cellular LTE networks in Germany offer good coverage and throughput for C2 in general, but there can be connection disruptions in the order of some seconds in valleys that are not served well by LTE.

2 Related Work

Digital data links for C2 or payload communications of commercial UA are in the focus of several research and standardization groups. For example, the Radio Technical Commission for Aeronautics (RTCA 2016) published Minimum Operational Performance Standards (MOPS) for L-band and C-band terrestrial data link networks in 2016, including specification and test procedures for airborne and ground radios. Furthermore, the team of Mielke at the German Aerospace Center (DLR) is working towards the development of the C-band digital aeronautical communications system (CDACS), which makes use of modern wireless communication technologies known from other aeronautical communication standards like LDACS and AeroMACS. As of 2020, the initial design concept was presented (Mielke 2017; Mielke 2018) and a flight measurement campaign was performed (Mielke 2019) to support the data link development.

Nevertheless, a data link network for UA has neither been standardized nor deployed so far. Moreover, building up a dedicated terrestrial C2 communications network exclusively for UA with ubiquitous coverage would cost hundreds of millions of dollars (Ponchak 2016). Therefore, the use of non-dedicated communications networks came into focus of researchers and industry in order to save money and speed up UA integration into the national airspace. Especially, the suitability of the fourth-generation cellular network standard Long Term Evolution (LTE) has been assessed in several studies.

In 2017, Qualcomm and the 3rd Generation Partnership Project (3GPP) released studies containing initial results of flight measurements and simulations concerning the suitability of commercial LTE networks as a data link for C2 of small UA (Qualcomm 2017; 3GPP 2017). The goal was to provide information that enhances understanding of the applicability and performance of ground-based cellular networks for providing connectivity to low-altitude UA (120 m above ground level and below). The key result of the trial is that commercial LTE networks are capable of providing initial connectivity to UA without any modifications. However, no flight trials have been conducted in rural areas. Therefore, the suitability of commercial LTE for long-distance cargo UA remains unanswered.

Similarly, the research group around Amorim et al. (2017; 2019), Nguyen et al. (2017) and Stanczak et al. (2018) of the Aalborg University in Denmark investigated the suitability of LTE networks for UA above urban rooftops and in rural areas by measurements and in simulations. Most importantly for our work, they investigated the LTE coverage in rural areas for BVLOS operations by simulations. Their first results show that LTE is capable of delivering a throughput of 60 kbps to airborne UA. Unfortunately, the simulation area was limited to 70 km × 70 km, which is smaller than the range of our cargo UA. Hence, the question how LTE performs in our scenario remains unanswered. Furthermore, they discussed the benefit of dual-network connectivity. Their measurement results show, that simultaneous connections to at least two independent LTE networks increase the reliability of the connection without modifying the network itself. In our work, we want to verify these results with respect to our scenario in simulations.

3   Requirements for the Data Link Concept

The operation risk-based data link concept has to meet the following requirements:

1. Requirements imposed by the operational boundary conditions described in the Concept of Operations (ConOps).
2. Requirements imposed by the assigned Specific Assurance and Integrity Level (SAIL), which are the outcome of the SORA process.

   In general, requirements are reduced if a low SAIL is applicable and vice versa. Especially at low SAIL, this allows the use of cost-efficient commercial off-the-shelf data link technologies, which might only provide a limited reliability. In the following, we summarize the operational boundary conditions defined in the TLARs discussed by Dauer and Dittrich (2021) as well as defined in the JARUS SORA document (JARUS 2019).

3.1   Operational Boundary Conditions and Requirements

The investigated cargo UA are foreseen to operate BVLOS at a maximum altitude of 150 m above ground with a maximum speed of 200 km/h. Hence, the data link concept has to provide connectivity at the given altitude and speed.

   Furthermore, the data link concept has to provide sufficient bandwidth to enable a low latency data exchange between remote pilots and UA in order to allow remote pilots to safely control, monitor, and manage their UA during all phases of flight, i.e. taxi, takeoff, en route, and landing. Specifically, mission tasks, e.g. waypoints, have to be transmitted from remote pilots to UA (forward C2 data). The other way, status and position information has to be transmitted from UA to the ground to allow remote pilots a thorough monitoring (reverse C2 data). Subsequently, we derive bandwidth requirements for forward and reverse C2 data.

   Typical data packets containing mission tasks, which are sent to the UA, include among others: unique identifiers, 4D waypoint coordinates, speed limits, flight modes, and avionics commands. An overview of possible forward C2 data and their size is given in Table 1. Moreover, we assume that forward C2 commands are sent with an update rate of 1 Hz to the UA. As a result, the required bandwidth amounts to roughly 1 kbit/s if the data packet size is 128 byte.

   Typical data packets containing status information, sent from UA to remote pilots, include among others: unique identifiers, 4D position, flight state, airspeed, orientation, GNSS status, engine status, and DAA information. An overview of possible reverse C2 data and their size is given in Table 2.

   Regarding the DAA concept, which is introduced by Schalk and Peinecke (2021), the data link concept has to support the transmission of target information from UA to remote pilots. In other words, as soon as sensors onboard the UA identify and track an intruder aircraft, the collected information has to be made available

Table 1 Forward C2 commands and size in bytes for a cargo UA

Command	Size [byte]
UA-ID	16
Mission-ID	16
4D waypoint coordinates (Latitude, Longitude, Altitude, Time)	32
Speed	8
Flight Mode	8
Lights	8
Engine and landing gear commands	8
Landing gear	8
Break	8
Commands for avionics	16

to the remote pilot. As discussed by Schalk and Peinecke (2021), we assume that a FLARM transponder is installed onboard the UA to detect cooperative intruders. FLARM transponders can receive signals of up to 50 aircraft within range according to the manufacturer. Assuming an update rate of 2 Hz and a data packet size of 146 byte, if 50 intruder aircraft are tracked, we estimate a maximum bandwidth of 116.8 kbit/s for DAA data.

For reverse C2 data packets, which do not contain DAA data, we assume an update rate of 1 Hz and the data packet size is calculated as 232 byte. Hence, we obtain a required bandwidth of 2 kbit/s for reverse C2 data without DAA data. In total, we expect a required bandwidth of roughly 120 kbit/s for reverse C2 data including DAA data.

Table 2 Status information

Status information	Size [byte]
UA-ID	16
Mission-ID	16
Flight state	8
Position (latitude, longtiude, altitude, relative altitude, orientation, time)	48
Next waypoint (latitude, longitude, altitude, time)	32
Airspeed and speed over ground	16
Roll, pitch, yaw	24
Data link status	8
GNSS status	8
GNSS accuracy	24
Engine temperature and RPM	16
Fuel fill level and battery voltage	16
DAA information per aircraft	146

All in all, the data link concept for cargo transport UA has to fulfill several normative requirements, which are summarized below:

1. The data link concept has to provide a bi-directional interface between UA and remote pilots for command and control.
2. In support of the first requirement, the data link concept has to fulfill the data rate and latency requirements of the interface. In the forward direction, the data link concept has to support a data rate of 1 kbit/s per UA. In the reverse direction, the data link concept has to support a data rate of about 120 kbit/s per UA.
3. The data link concept has to support addressed communications to allow remote pilots to communicate with specific UA.
4. Cargo UA usually travel hundreds of kilometers. Hence, remote pilots should be able to communicate with their UA BVLOS as well as beyond the radio line of sight.
5. The data link concept has to support mobility of up to 200 km/h, which is the maximum speed of the investigated low-altitude cargo UA.
6. The data link concept has to support flight altitudes from 0 m up to 150 m.
7. The interface between UA and remote pilots has to be protected against attackers.
8. The remote pilot should be able to monitor the status of the interface in order to detect a loss of a data link.

3.2 Digital Data Links in the Context of JARUS SORA

Within the SORA process, a Specific Assurance and Integrity Levels (SAIL) is determined with respect to the ConOps. Then, the SAIL is used to identify operational safety objectives (OSOs) which are associated with different robustness levels. A high SAIL requires a high OSO robustness, whereas a low SAIL requires a low or even no OSO robustness. More information on OSOs in JARUS SORA in the context of this project is given in Nikodem et al. (2021).

In JARUS SORA (2019), the OSO#06 is defined as shown in Table 3.

Instead of the term data link, JARUS uses the equivalent term C3 link. In our understanding, it comprises all previously mentioned functions, which are necessary to ensure the safety of a mission. The OSO#06 is the only OSO that affects the communications between UA and remote pilots.

As an outcome of the SORA process, we expect a SAIL between III and VI for the investigated ConOps. Hence, the required level of robustness for OSO#06 ranges

Table 3 Jarus SORA Operational Safety Objective #06 levels of robustness

OSO number		SAIL					
		I	II	III	IV	V	VI
OSO#06	C3 link performance is appropriate for the operation	Optional	Low	Low	Medium	High	High

Table 4 Criteria for integrity

Technical issue with the UAS		Level of integrity		
		Low	Medium	High
OSO#06	Criteria	• (…) performance, RF spectrum usage and environmental conditions for C3 links are adequate (…) • (…) means to continuously monitor the C3 performance (...)	Same as Low	Same as Low. In addition, the use of licensed bands for C2 Link is required

from low to high in dependence of the SAIL. For example, a SAIL III is associated with a low level of robustness for OSO#06. In order to achieve the desired level of robustness, an equivalent level of integrity and level of assurance has to be achieved. By definition, the achieved level of robustness cannot be higher than the lowest level of either integrity or assurance. The criteria for integrity and assurance of OSO#06 according to JARUS SORA Annex E are given in Table 4 and in Table 5 respectively.

First, we take a look at the criteria for integrity with respect to the data link concept. Please note that the criteria for low and medium integrity of OSO#06 are the same. Therefore, in case of a SAIL III to IV, the applicant has to determine that performance, RF spectrum usage, and environmental conditions are adequate. Main performance parameters are availability, continuity, integrity, and transaction expiration time as defined by ICAO (2017). Moreover, remote pilots have to be able to monitor the data link performance. In case of SAILs equal or higher than SAIL V, the same criteria as for lower SAIL have to be fulfilled. In addition, only licensed frequency bands may be used for the data link (Table 4). This excludes the use of industrial, scientific, and medical (ISM) bands, like the 2.4 GHz band, which is typically used by Wi-Fi networks, or the 868/900 MHz bands for ZigBee networks. Data links in licensed frequency bands are for example cellular networks, like GSM, UMTS or LTE.

The criteria for assurance of OSO#06 increases for each level of robustness as shown in Table 5. In case a low level of assurance is required, the applicant has to declare that the required level of integrity has been achieved. In case of a medium level of assurance and above, the applicant has to demonstrate that the data link performance is in accordance with standards. At the moment, standards have not been defined yet, but may be defined by National Aviation Authorities in the future.

4 Data Link Architecture

The current non-existence of dedicated data link networks for C2 of long-distance cargo UA leaves only the use non-dedicated networks if a fast and cost-efficient

Table 5 Criteria for Assurance

Technical issue with the UAS		Level of assurance		
		Low	Medium	High
OSO#06	Criteria	The applicant declares that the required level of integrity has been achieved	Demonstration of the C3 link performance is in accordance with standards considered adequate by the competent authority and/or in accordance with means of compliance acceptable to that authority	Same as medium. In addition, evidence is validated by a competent third party

deployment is desired. In particular, we identified two data link technologies, which are suitable for the investigated cargo UA: cellular (terrestrial) networks and satellite networks.

Both technologies are in general able to provide connectivity to airborne UA for the use cases identified by Pak (2021) and both fulfill the bandwidth requirements derived in Sect. 3.1 as well. For example, Inmarsat's SB-UAV service provides global coverage except for extreme polar regions and a throughput of 200 kbps (Inmarsat 2020). Cellular networks, e.g. LTE networks, are capable of transmitting several hundreds of Mbps for a single user in theoretically. However, the total bandwidth has to be shared among all users in a cell, which reduces the available bandwidth for each user. However, tests performed by the Bundesnetzagentur showed that 90% of all mobile users in Germany receive at least a throughput of 700 kbps (Bundesnetzagentur 2019).

However, the question whether these technologies are sufficiently robust for the intended missions remains an open question, which we address later on. In Nikodem et al. (2021) and Sect. 3.2, the SAIL and its implication on the data link have been discussed. Due to the ConOps, SAILs from III to VI can apply to the same UA depending on the mission profile. The data link architecture could be designed to fulfil the requirements of the highest SAIL, i.e. SAIL VI. However, this would unnecessarily increase complexity and costs of SAIL III missions, as discussed by Rothe and Nikodem (2021). Instead, we propose the use of a flexible data link architecture, which can be easily adapted to the operational risk. In the following, we describe an architecture, which combines both technologies into a comprehensive data link concept within the operational risk framework.

The required robustness of the data link is closely related to the risk the mission poses to the environment and the safety impact of the remote pilot's role. The risk to the environment and the safety impact of the pilot's role are discussed in detail Nikodem et al. (2021). If the risk to the environment is low, the robustness of the data link can be low as well and vice versa. Similarly, if the safety impact of the remote pilot's role is low, the required robustness of the datalink can be low as well. Only if the risk to the environment is high and the safety impact of the remote pilot

Table 6 Required data link
~robustness

		Risk to environment	
		Low	High
Safety impact	Low	Low	Low
	High	Low	High

is high as well, the required robustness has to be high inevitably. The four cases are summarized in Table 6. Nevertheless, the operator might want to achieve a higher robustness than the risk assessment requires in order to cut expenses due to numerous lost aircraft.

In case non-dedicated data link networks are used, it is hard to make changes to the network itself, e.g. increase transmit power, to increase robustness of the data link. Instead, we propose to increase data link robustness by using multiple independent data links simultaneously to interconnect the same UA and remote pilot. In case one of the data links fails, the remaining links can be used to exchange data without interruption.

Our proposed operation risk-based data link architecture is depicted in Fig. 1. In principle, the UA and the remote pilot are part of the same virtual private network (VPN). VPNs are a viable means to secure public network connections by authentication and encryption. The remote pilot is connected to the VPN gateway as well either through a local network connection or through the Internet. In the latter case, the remote pilot can be in principle located anywhere on the globe as long as an Internet connection is available. From the remote pilot's point of view, the functionality of his ground control station should not differ due to the used data link network significantly. Independently of the data link network, the remote pilot sees the UA as a client in the VPN.

The UA is foreseen to use either a single or multiple cellular (terrestrial) networks to connect to the VPN. In addition, a satellite network is used if coverage, redundancy, or both needs to be increased. VPN bonding allows the aggregation of multiple network connections to a single network interface to increase stability and speed. In addition, VPN bonding enables seamless handovers between individual links. It is commonly used in case redundant links, fault tolerance or load balancing networks are needed. For example, VPN bonding can create a 2 Mbps data link out of two 1 Mbps links. If one of the links fails, a 1 Mbps connection is still available. VPN bonding includes as well an automatic monitoring of the individual links quality and recovery detection.

Our data link architecture does not prescribe any particular communications standard and each individual network can be operated by a different provider. In principle, the data link architecture is able to work with a connection to a single cellular or satellite data link network only. Depending on the mission profile, the risk assessment, and the individual network coverage, additional connections to other cellular or satellite networks might be needed to cover the whole mission area and to increase reliability to the required level.

For example, let us suppose that the cellular network of provider A covers only the area around the departure airport. Then, the cellular network of provider B, which covers the remaining area around the arrival airport, has to be used in addition to provider A's network to guarantee cellular coverage throughout the mission. Alternatively, the aircraft could connect to a satellite network, which covers the whole operational area. For seamless operations, reliable handovers between the individual networks have to be ensured without any interaction by the remote pilots.

We propose for the investigated cargo UA the usage of at least one and up to N cellular networks in combination with a single satellite network, if needed. A view of the potential on-board setup is given in Fig. 2. For each connection to a cellular network, a cellular modem and at least one dedicated antenna per cellular modem has to be installed. Additionally, a satellite modem and a satellite antenna have to be installed. The antennas of the cellular modems have to be faced downwards, whereas the satellite antenna has to be faced upwards. Each modem is connected to the aircraft computer, which handles the VPN connections.

The number of cellular modems is limited by the number of providers, which operate cellular networks in the mission area. More than one modem per provider adds redundancy but does not increase the coverage. To reach maximum diversity, all available providers should be used. However, for each provider an additional usage contract is needed and costs will roughly multiple by a factor of N. Similarly, we expect that the use of satellite networks increases costs significantly.

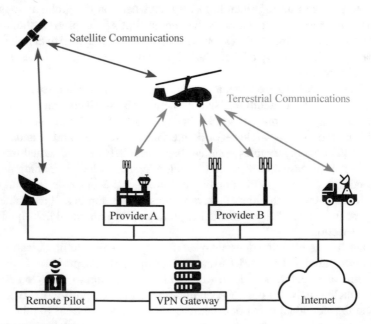

Fig. 1 The data link concept

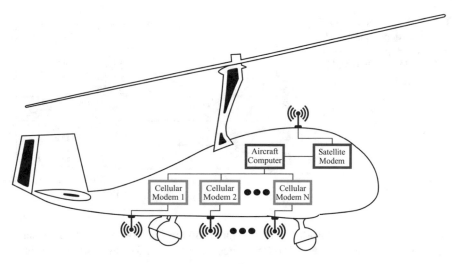

Fig. 2 Onboard setup of the data link concept

5 Prerequisites for the Evaluation

In this section, we provide background information regarding our simulation framework. In order to evaluate the suitability of LTE networks (or cellular networks in general) as a data link for low-altitude UA, it is of primary importance to characterize the air-to-ground radio channel at first. Subsequently, we introduce an LTE link model and calculate the LTE link budget, which is necessary to estimate the range of an LTE base station. Finally, we analyze the coverage of commercial LTE networks for flight routes in Germany discussed by Schopferer and Donkels (2021).

5.1 Air-to-Ground Radio Channel Model at Low Altitudes

In digital wireless communications systems, data bits are modulated on electromagnetic waves, which radiate towards receivers. Naturally, electromagnetic waves are affected in various ways, e.g. by the earth surface, objects on ground or the airframe. Severe distortions can lead to loss of the data bits, if no countermeasures are taken. Therefore, the understanding and modelling of radio channels play an essential role in system design and performance evaluation of any wireless communication systems. Such models are used for example to estimate the signal attenuation due to free space path loss (FSPL) or multipath propagation and to assess the system performance by means of software simulations, prior to implementing the system in hardware. For the evaluation as well as specification of terrestrial wireless communications systems for UA in Very Low Level (VLL) airspace, an air-to-ground (AG) radio channel model

Fig. 3 Distance to the radio
horizon in dependence of UA
altitude over ground and base
station (BS) antenna height

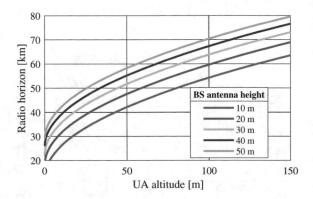

is inevitably needed. Unfortunately, adequate models based on real-world measurements have not been presented so far. Hence, we use an AG channel model based on geometrical considerations.

For the modelling of large scale fading effects over large distances we choose the well-known curved earth two-ray (CE2R) ground-reflection model as introduced by Matolak and Sun (2017). In comparison to a flat-earth model, the CE2R model takes the blockage of the transmitted signal due to the earth's curvature into account and limits the LOS range to the radio horizon. For the modelling of small-scale fading effects in LOS conditions we combine the CE2R model with the well-known Rice model. In the following, we briefly recapitulate the model derivation.

In Fig. 3, we show the distance to the radio horizon in relation the base station (BS) antenna height and the UA altitude if the terrain is not considered. Our results show that the maximum range of the BS heavily depends on the UA altitude and the BS antenna height. Even if the BS antenna is installed at 50 m height and the UA is at an altitude of 150 m, the maximum range does not exceed 80 km. In the flat earth case, the LOS range would be unlimited. When looking on the number of possible mobile radio stations that are in radio range in rural areas or evaluating how the mobile radio stations are affected by many low-altitude UA, this radio range is important and motivates the use of the CE2R model.

Beside large-scale fading effects, we also consider small scale fading effects by applying a Rice channel model that suitable for environments when receiving a strong line of sight signal component compared to the signal reflections. For our simulations, we assume that a valid connection between UA and a mobile base station can only be established when being in line of sight, because otherwise the signal attenuation is too strong. With the help of the CE2R model and the Rice model, which are recapitulated in the next, we are able to estimate the overall path loss or total attenuation of a AG data link. The total attenuation L_{total} of the received signal in dB is the sum of the attenuation by the CE2R model L_{CE2R} and the attenuation by the Rice model L_{Rice} given as

$$L_{total,dB} = L_{CE2R,dB} + L_{Rice,dB}. \tag{1}$$

Likewise, L_{CE2R} is composed of the attenuation of the LOS ray and the attenuation of the ground-reflected (GREF) ray as

$$L_{CE2R,dB} = 20\log_{10}\left[\frac{4\pi}{\lambda}d_{LOS}\left|1 + \frac{d_{LOS}}{d_{GREF}}\Gamma(\psi)e^{-j\Delta\varphi}\right|^{-1}\right] \qquad (2)$$

where λ is the wavelength, d_{LOS} is the length of the LOS ray, d_{GREF} is the length of the GREF ray, Γ is the reflection coefficient, ψ is the reflection angle, $\Delta\varphi$ is the phase difference between the two rays and j is the imaginary unit.

The curved earth geometry is shown in Fig. 4. Earth's effective radius R_{eff} is used instead of the true earth radius to account for atmospheric refraction. It is calculated as

$$R_{eff} = k \, R_e, \qquad (3)$$

where k is the refractive coefficient and R_e is the true earth radius. Throughout we assume an altitude independent coefficient of $k = \frac{4}{3}$ which is commonly used for near earth surface communications and a location independent earth radius of $R_e = 6{,}370$ km.

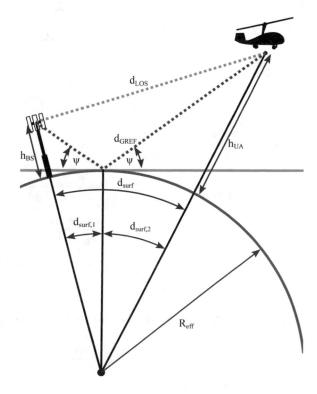

Fig. 4 Curved earth geometry for the two-ray radio channel model

The LOS ray length d_{LOS} is computed by the law of cosines with known base station antenna height h_{BS}, UA antenna height h_{UA} and earth surface distance d_{surf} by

$$d_{LOS}^2 = (k + h_{UA})^2 + (k + h_{BS})^2 - 2(k + h_{UA})(k + h_{BS})\cos(q) \qquad (4)$$

with

$$q = \frac{d_{surf}}{k} \qquad (5)$$

The length difference of the LOS ray and the GREF ray Δ d is needed in order to calculate the phase difference between the LOS component and the GREF component. It is given by

$$\Delta d = 2\frac{d_{surf,1}d_{surf,2}\Psi^2}{d} \qquad (6)$$

with the earth surface distances

$$d_{surf,1} = \frac{d_{surf}(1+b)}{2} \qquad (7)$$

$$d_{surf,2} = d_{surf} - d_{surf,1} \qquad (8)$$

and angle

$$\psi = \frac{h_{UA} + h_{BS}}{d_{surf}}(1 - m(1 + b^2)) \qquad (9)$$

and variables

$$m = \frac{d_{surf}^2}{4k(h_{UA} + h_{BS})} \qquad (10)$$

$$u = \frac{h_{UA} + h_{BS}}{h_{UA} + h_{BS}} \qquad (11)$$

$$b = 2\sqrt{\frac{m+1}{3m}}\cos\left(\frac{\pi}{3} + \frac{1}{3}\cos^{-1}\left(\frac{3u}{2}\sqrt{\frac{3m}{(m+1)^2}}\right)\right). \qquad (12)$$

The phase difference $\Delta\varphi$ between the LOS path and the reflected path is given by

$$\Delta\varphi = \frac{2\pi\,\Delta d}{\lambda}. \qquad (13)$$

The reflection coefficient for a vertically polarized wave depends on the permittivity of the reflecting medium. The vertical reflection coefficient is given by

$$\Gamma = \frac{\sin\psi - \frac{1}{\bar{\varepsilon}_r}\sqrt{\bar{\varepsilon}_r - \cos^2\psi}}{\sin\psi + \frac{1}{\bar{\varepsilon}_r}\sqrt{\bar{\varepsilon}_r - \cos^2\psi}}, \tag{14}$$

where ψ is angle between the ground and the reflected ray and $\bar{\varepsilon}_r$ is the complex relative permittivity of the medium given by

$$\bar{\varepsilon}_r = \varepsilon_r - j60\gamma\lambda. \tag{15}$$

The attenuation L_{Rice} due to Rician fading is modeled using a random variable X as

$$L_{Rice,dB} = 20\log_{10}(X). \tag{16}$$

Its probability density function $f_X(x)$ is the Rice distribution, which is defined as

$$f_X(x) = \frac{x}{\sigma^2}\exp\left(-\frac{x^2 + v^2}{2\sigma^2}\right)I_0\left(\frac{xv}{\sigma^2}\right), x \geq 0 \tag{17}$$

where v and σ denote the power of the specular components (LOS and GREF) and the reflected, non-dominant signals respectively, and I_0 is the modified Bessel function of the first kind with zero order. Typically, the relation between both is called the Rician K-factor given by

$$K = \frac{v^2}{2\sigma^2}. \tag{18}$$

An example of the AG path loss with respect to the surface distance is given in Fig. 5. Here, the frequency is 2 GHz, the UA altitude is 150 m, the base station height is 20 m and the K-Factor is 10 dB. In comparison to the FSPL model, high attenuation peaks are significantly reducing the received signal, which can lead to a data link loss for several kilometers.

5.2 LTE Model and Link Budget

Next, we need to model the LTE data link performance in order to estimate the throughput in relation to the distance between LTE base station, called eNodeB, and UA. Typically, throughput reduces with distance due to the reduced received signal power. The distance at which zero throughput, or a minimum throughput, is

Fig. 5 Air-ground (AG) path loss of the proposed radio channel model (CE2R + Rice) in comparison to the free-space path loss (FSPL)

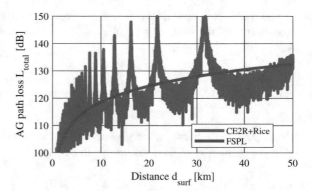

reached is defined as the maximum range of the LTE eNodeB. Usually, throughput is estimated in extensive simulations. To reduce complexity, we choose a simplified yet accurate model, which is briefly described in the following.

Mogensen et al. (2017) showed that the well-known Shannon capacity bound for additive white Gaussian noise (AWGN) channels can be modified in a way so that it accurately matches the LTE throughput. We adopt this approach and extend it towards a multi-user scenario.

The Shannon capacity formula for the maximum channel spectral efficiency C with respect to the Signal-to-Noise Ratio (SNR) is known as

$$C = \log_2(1 + SNR). \tag{19}$$

Similarly, we express the LTE user throughput T_{user} as a function of the allocated bandwidth per user B_{user} as

$$T_{user} = \xi B_{user} \log_2(1 + SNR) \tag{20}$$

with a correction factor ξ, $0 \le \xi \le 1$ to model the offset between the Shannon capacity bound and the LTE data link, e.g. due to imperfect channel coding. Mogensen et al. (2017) proved in simulations that $\xi = 0.8$ provides a good fit in most scenarios.

If the LTE eNodeB allocates the same amount of frequency bandwidth to each user, then the allocated user bandwidth B_{user} is given by.

$$B_{user} = \frac{N_{RB}}{N_{user}} N_{SC} B_{SC}, \tag{21}$$

where N_{user} is the number of users which the base station serves and N_{RB} is the number of totally available resource blocks. N_{RB} depends on the system bandwidth. In the following, we consider only LTE systems with 20 MHz frequency bandwidth, which is the maximum configuration for LTE Release 8 systems. In that case, 100 resource blocks are available. N_{SC} is the number of subcarriers per resource block and B_{SC} is the subcarrier spacing. Typically, $N_{SC} = 12$ and $B_{SC} = 15$ kHz in LTE

Fig. 6 LTE User
Throughput in dependence
of SNR according to the
presented model

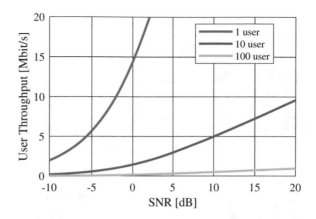

systems. In Fig. 6, we show the results of our LTE model, namely the achievable user
throughput in relation to SNR and number of users. Since LTE is a shared medium,
the total throughput of an LTE eNodeB is split up among all users.

Now, we have a look at the LTE link budget calculations, which are needed to
map SNR values to the distance between LTE eNodeB and UA. In the end, we are
able to determine the range of an LTE eNodeB and we use this result to estimate the
coverage of LTE networks.

As the name indicates, the SNR at the receiver always is the ratio between the
received signal power and the noise level. In addition, we also account for interference
and banking losses. Hence, the SNR in dB at the receiver is given as

$$SNR = P_{TX} - PL - L_{int} + G_{RX} - L_{RX} - N_{RX} - NF_{RX} \qquad (22)$$

where P_{TX} is the effective isotropic radiated power of the transmitter, PL is the
path loss due to the communications channel, L_{int} is the loss due to interference
of adjacent cells, G_{RX} is the receive antenna gain, L_{RX} is the cable loss in the
receiver, N_{RX} is the receiver noise power and NF_{RX} is the receiver noise figure.
Thereby P_{TX} includes the transmitter antenna gain. The actual receive antenna gain
G_{RX} for a certain angle of sight is dependent on the antenna radiation pattern and
the maximum receiver antenna gain in direction of the maximum radiated signal
strength. We discuss the antenna radiation patterns in more details in Sect. 5.4. In
Table 7, we show our assumed parameters for LTE downlink (DL) and uplink (UL).

Since the transmission power of the UA is much lower than the transmission power
of the eNodeB, we consider the uplink to be crucial for the connection between remote
pilot and UA. In other words, if the uplink connection breaks, no downlink connection
can be established either. Hence, we use only the uplink link budget for the range
estimation. Furthermore, we assume the same LTE link model for the uplink and the
downlink.

Using our CE2R channel model and the LTE link model, we show the result of the
range analysis in Fig. 7. Here, we assume a carrier frequency of 1.8 GHz, a required
user throughput of 1 Mbit/s and 10 users per LTE cell. The LTE eNodeB height is

Table 7 Assumed LTE Link budget parameters

Parameter		Unit	DL	UL
Transmitted power (EIRP)	P_{TX}	dBm	44	24
Interference margin	L_{int}	dB	3	3
RX antenna gain max	$G_{Rx_{max}}$	dB	3	17
RX cable loss	L_{RX}	dB	1	3
RX noise power	N_{RX}	dBm	−101	−101
RX noise figure	NF_{RX}	dB	9	2

Fig. 7 Visualization of the SNR in dependence of the distance between UA and a mobile base station for the given LTE Link Budget.When applying the CE2R model, the SNR drops below the SNR limit and starts becoming unreliable at around 5 km distance

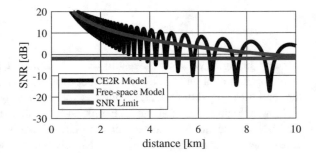

20 m and the UA altitude is 150 m. According to our results, we can expect a first loss of the LTE link around 4 km, when the SNR drops below the minimum required SNR and multiple losses after 5 km. In the following section, we assume that the maximum LOS range of an LTE eNodeB is 5 km, since the link becomes unreliable at greater distances and the connection will very likely break down for the positions of the UA when it flies in the shown distances to a connected mobile radio station. Therefore, we simplify our simulations by applying this maximum range.

5.3 LTE Coverage Analysis for Cargo UA in Germany

Cellular network coverage is currently a highly discussed topic in research, industry and politics. Although, ubiquitous coverage is demanded by public authorities, providers cannot close all coverage holes due to massive costs. Since every cellular base station can only cover a limited surrounding area, coverage mainly depends on the number of base stations. In principle, coverage is easily increased by placing more base stations at appropriate locations. However, missing infrastructure, missing grand authorizations, and resistance of local residents often slow down network expansion. Moreover, providers tend to increase coverage only in urban areas, because of the higher number of paying users compared to rural areas. Nevertheless, network coverage in urban areas is not relevant for long-distance operations of cargo UA, since these UA are foreseen to operate only above non-populated or sparsely populated areas. Please see Nikodem et al. (2021) for more information.

Three pieces of information are required to estimate the LTE network coverage for UA at altitudes up to 150 m:

1. Locations of the LTE eNodeBs
2. Information about the terrain elevation
3. LTE link model and link budget calculations

The LTE link model and link budget calculations have been presented in the previous section of this chapter and account for all influences from interference, noise, link utilization and antenna radiation patterns. To obtain the locations of the LTE eNodeBs, we used the open-source cellular base station database Open-CelliD (2019). This database provides location estimates based on radio signal strength measurements by volunteers. The database allows filtering of the locations by country, provider, technology, and measurement date. It is updated daily and we used the database update from 1st of August, 2019. Information of the terrain elevation was obtained from the publicly available NASA Shuttle Radar Topography Mission (SRTM) database with 3-arc second resolution.

Especially at low-altitudes, terrain plays a significant role with respect to coverage. For example, even low hills between base station and UA can block the LOS signal, which can lead to loss of connectivity. Exemplarily, we show the LOS coverage of a theoretical base station at the airport Oberpfaffenhofen (EDMO) in Germany in Fig. 8 with and without terrain obstruction. The base station is fixed at 20 m height and coverage was evaluated for UA altitudes of 50 m, 100 m, and 150 m above ground.

The three circles indicate the particular LOS range without terrain influence. In that case, the only limiting factor is the earth curvature. The filling of the circles shows the LOS range due to terrain obstruction. Please note that we only consider the first link breakage point. Naturally, a smaller area is covered by a particular base station if UA fly at lower altitudes. But even at the highest altitude of 150 m, the LOS range does reach the radio horizon only occasionally. Of course, this example is taken from the southern part of Germany, which is known to be hillier than the

Fig. 8 LOS Range with and without terrain obstruction for a base station with 20 m height at Oberpfaffenhofen (EDMO) airport in Germany. Circles indicate the LOS range without terrain influence. Filling shows the LOS range with terrain obstruction

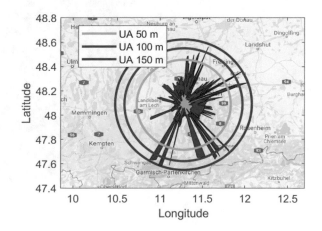

northern part. In the northern part, the coverage of a single base station is likely to be increased due to the flat terrain. In turn, this means that more base stations have to be placed in hillier terrain to cover an area of the same size.

Taking into account the LTE eNodeB locations of the OpenCellid database, we are able to estimate the coverage of LTE networks in Germany. In Fig. 9, we show the coverage of the LTE network belonging to the Deutsche Telekom. Here, we assume that UA fly at an altitude of 150 m and that each LTE eNodeB has a maximum range of 5 km, if it is not obstructed by any terrain features. This assumption was the outcome of our analysis of the LTE link budget in the previous section of this chapter. Coverage is ubiquitous around big cities like Berlin, Hamburg or Munich, where more than 3 LTE eNodeBs can be usually received at the same time. On the contrary, many areas without network coverage can be identified apart from cities and any other infrastructure, e.g., motorways.

In Fig. 10, we show the coverage if more than just one provider is used. In Germany, three LTE providers with distinct networks exist in total: Deutsche Telekom, Vodafone, and O2. Of course, coverage of the individual providers is somehow correlated, since providers use identical locations for their base stations from time to time to reduce costs. Nevertheless, the use of all three providers improves coverage evidently due to the much higher number of deployed base stations. In this case, less and smaller red areas without coverage can be identified in comparison to single provider case. More detail will be provided in the next section of this chapter, where we take a look at the coverage with realistic flight routes.

Unfortunately, our approach has some limitations, which we point out in the following. Since the OpenCellid database is open source and we know nothing about the contributors, we have to assume that it is neither complete nor accurate enough to

Fig. 9 Deutsche Telekom LTE network coverage in Germany for UA at 150 m altitude and eNodeB height of 20 m

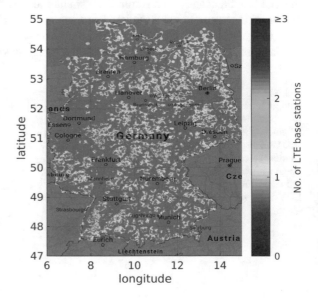

provide a trustworthy coverage estimate. Moreover, it does not provide information about the used frequency, antenna beam direction, and antenna height, which are important factors for coverage estimation. Intuitively, the OpenCellid database seems to be more accurate near cities and streets since participants mainly measure the radio signals in these locations. For example, the measurement equipment is mounted into a car and is driven along motorways. Hence, the database is not complete in rural areas with less streets. Eventually, we are not able to distinguish the reason for missing network coverage. It could be either because no LTE eNodeB is installed or because it is not listed in the database. In the future, cooperation with providers could increase the quality of our simulations, if more accurate databases are provided.

5.4   Antenna Radiation Patterns

LTE base station antennas do not radiate omnidirectional and are optimized for ground users. Thus, airborne UA's might be served only by the sidelobes of the down tilted antennas or experience antenna nulls during flight at some angles relative to the connected base station. Therefore, we consider this with more realistic antenna radiation patterns for the channel model of the LTE data link emulator. Figure 11 and Fig. 12 illustrate the simulated 3D antenna radiation patterns for a LTE base station, used in Lin X (2018), and a half wave dipole for the UA looking vertically down to the ground. The patterns are plotted for a certain range starting from the maximum gain that make it easy to visualize the shape. Basically, we can see for the base station several sidelobes in vertical direction beside the mainlobe. As we do

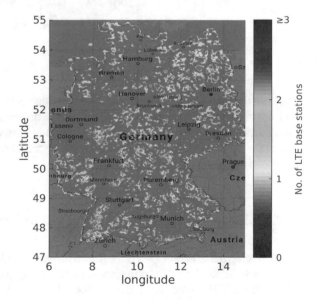

Fig. 10 Deutsche Telekom, Vodafone and O2 LTE network coverage in Germany for UA at 150 m altitude and eNodeB height of 20 m

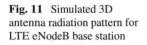

Fig. 11 Simulated 3D antenna radiation pattern for LTE eNodeB base station

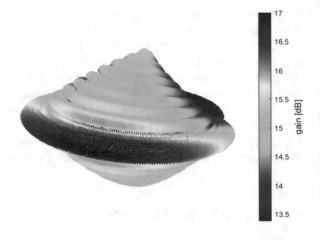

Fig. 12 Simulated 3D antenna radiation pattern for half wave dipole antenna at UA

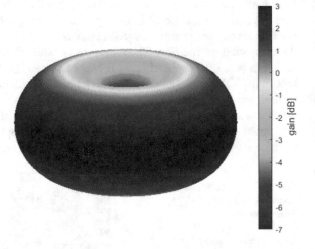

not know the directions of the base station antennas in the OpenCellId database and there are usually more antennas mounted to serve different directions, we assume an omnidirectional pattern in the horizontal plane and in the vertical plane the presented pattern for the base station's antenna. For the aircraft's antenna we do not consider influences on the radiation by the air frame.

6   LTE Data Link Emulator and System Simulations

For evaluating the suitability of LTE cellular networks for UA in low-altitude environment, we simulate the radio propagation effects using the presented channel model with realistic flight routes for a cargo drone, LTE link budgets and distributions of cellular base stations. By this, we can estimate the availability of a LTE data link for such flight scenarios and provide information about the expected latencies for typical C2 commands for a cargo drone. Additional information like expected distances between UA and the cellular base stations or handovers between base station can be analyzed as well. We first describe the building blocks of our simulation and show necessary steps in order to achieve real-time capability. Then, we evaluate three realistic flight routes at two different heights with the LTE data link emulator.

6.1 LTE Data Link Emulator Description

The LTE data link emulator emulates the datalink between the UA and the ground control station and is implemented in Matlab Simulink for simulations. Basically, it includes a LTE model, an environmental model and a channel model. For the LTE model, described in more details in Scct. 5.2, we consider cellular base station distributions and calculate link budgets as well as transmission latencies and simulate base station handovers. With the environmental model, we determine which base stations are in line of sight of the aircraft or obstructed by terrain, which is explained in Sect. 5.3. Thereby we do not consider diffraction or reflection effects. If a base station is obstructed, no connection can be established. The channel model defines the actual link quality by estimating the received signal to noise ratio (SNR) with the model described in Sect. 5.1.

Figure 13 visualizes the three implemented main building blocks and the basic signal flow for the data link emulator. The LTE data link emulator requires the position and the orientation of the UA as input in addition to the data messages that are sent through the simulated data link. In the figure one message stream *dataA2G_in1* is exemplarily depicted for all messages of same type that are sent through the downlink channel (A2G). If there are concurrent messages, they are modeled with additional message streams. For each message stream either in the downlink or the uplink channel there is one dedicated *DataChannel* block, that affect the messages by delaying or erasing them depending on the simulated data link quality. Thus, the output on *dataA2G_out1* is a delayed version of the input *dataA2G_in1* or the input message gets lost, if the data link connection breaks or the data link quality is poor during transmission. The delay accounts for the time the message needs to travel the distance between transmitter and receiver and the achieved and the transmission time that is dependent on the achieved data rate and the packet size. The data link quality is determined by estimating the received SNR in the *A2GChannelControl* block for the downlink and the *G2AChannelControl* block for the uplink. They interact with the assigned *DataChannel* blocks as shown exemplarily in the figure for one message stream in the downlink. These blocks controlling the behavior of the data link channels are in turn controlled by the block *DatalinkControl*, which takes care of connecting to a LTE base station eNodeB and determines the viewing angles between the base station antenna and the UA's antenna in order to calculate the antenna gains by considering the antenna radiation patterns described in Sect. 5.4.

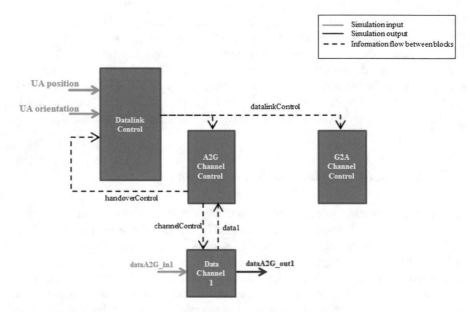

Fig. 13 Main building blocks and structure of LTE datalink emulator

For the connection to a certain eNodeB the block considers the current position of the UA, the base stations that are in line of sight for that position and the current data link quality for the current connected eNodeB. If the link quality drops below a certain SNR, a handover to another eNodeB is initiated and a new connection is established if a new eNodeB is found in line of sight with sufficient link quality. Thus, the connection is disrupted for the times when the link quality is poor or there is even no base station in line of sight, because the next one is too far away or obstructed by terrain.

6.2 LTE Data Link Emulator Preprocessing for Real-time Capability

The LTE data link emulator is one major building block of the whole simulation framework that defines a real-world time that is simulated for each simulation step. Therefore, in order to achieve real-time for an online simulation, the processing time of the LTE data link emulator must be considered and minimized until reaching reatime or at least a minimum. For the data link emulator the bottleneck is the process of finding the base station eNodeB for a valid data link connection.

In order to calculate the LTE link quality for each processing step in the simulation, the distance between the UA and the connected base station eNodeB must be known. When connecting to a new base station, the nearest base station in visual line of sight that fulfills the link quality requirements is used. Because there might be hundreds of base stations in line of sight that must be evaluated, this can be computationally expensive and therefore time-consuming. In order to provide real-time capability during simulation we preprocess the information about the number of base stations in line of sight and sort them by their distances to the possible UA's position. In order to accelerate the computation, we therefore defined the maximum LOS range of an LTE eNodeB to be 5 km in Sect. 5.2 as the link becomes unreliable and it minimizes the number of possibilities. The horizontal resolution is defined by the given terrain data with 3-arc second resolution and we define a vertical resolution of 50 m. Figure 14 illustrates the preprocessed grid with the amount of base stations in line of sight for a certain flight height. This results to the LTE coverage plots shown in Fig. 10 and Fig. 11.

In addition to the mentioned sorting as preprocessing step, we further proof for each simulation step, if the underlying geometry has sufficiently changed and only run the building block *DatalinkControl* mentioned before in Sect. 6.1, if necessary. Then, the base station selection process, as the bottleneck, is not executed and the overall processing time is minimized.

Fig. 14 Excerpt showing exemplarily the preprocessed grid for the amount of base stations in line of sight depending on the UA's position at a certain flight height

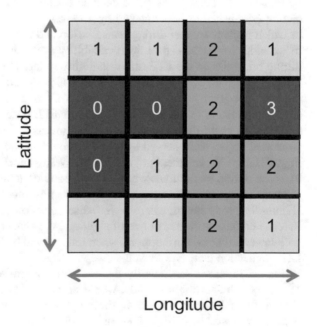

6.3   LTE Data Link Emulator Evaluation

In order to evaluate the suitability of a LTE data link outside urban areas by simulation, we used exemplarily the three different ALAADy mission scenarios, Hamm-Neubrandenburg, Hamm-Schoenenberg and Hamm-Waldkappel at a lower and a higher altitude below 150 m. These flight routes are also used for demonstration of the risk-based trajectory planning in Schopferer and Donkels (2021) and we utilize them as they provide feasible trajectories for an cargo UA. During simulation 232 byte status messages in the downlink A2G channel and 128 byte command messages in the uplink G2A channel are sent every second like defined in Sect. 3.1. First, we investigate the LTE data link availability and explain link failures occurred in the simulation. Then, we provide information about the expected latencies and discuss the overall simulation results. The LTE link is not available, if no base station is connected or if the link quality is bad because the SINR drops below a defined threshold of -6dBm during transmission. No base station is connected if there is none in line of sight or if the link quality to all visible base stations is not sufficient, thus a handover was initiated but no new base station with sufficient link quality can be found. In this case, a packet is delayed for transmission until a new connection can be established or gets updated by a newer message. We distinguish the cases in which packets are lost during transmission and in which no connection is established at all for the LTE data link availability evaluation. In terms of packet losses during transmission the LTE link was highly reliable in all simulation cases. Only for the flight Hamm-Schoenenburg at higher altitude one packet loss occurred, when the UA experienced an antenna null due to disadvanteous view angles leading to strong attenuation of the signal. In terms of LTE base station connection the UA experienced some disruptions. Table 8 summarizes the simulation results for all simulated flights when served by all LTE providers and by Telekom alone. For the flights Hamm-Neubrandenburg and Hamm-Waldkappel the aircraft was connected to a base station for all times when flying the routes at higher altitudes. When flying at lower altitudes the times without a connection increase in all cases except for Hamm-Neubrandenburg, where no link disconnection occurred at all. This happens because no base station is in line of sight and the chances for obstructions increase with lower altitudes. Furthermore, when only served by Telekom, the times without LTE connection increases as well.

Figure 15 shows exemplarily the case for Hamm-Schoenenberg with all LTE providers at higher flight altitude, where no base station could be found in line of sight at the indicated red spots. In these spots the UA flies through a valley which is not served well by LTE. In comparison Fig. 16 shows the same flight when only served by Telekom LTE. The link outages occur at the same region but get worse as there are even less base stations in line of sight.

For all packet transmissions the latency was relatively low due to high SNR values leading to high achievable data rates and because of the relatively small packet sizes. Only for the view cases without LTE connection the latencies raise, because the packets are delayed until a new connection can be established. Furthermore, every

Table 8 Simulation results for LTE data link availability for different flight routes. Times were no connection to a base station can be established for the simulation with all LTE providers in first case and only Telekom in second case

| | Hamm-Neubrandenburg | | Hamm-Schoenenberg | | Hamm-Waldkappel | |
	Higher altitude	Lower altitude	Higher altitude	Lower altitude	Higher altitude	Lower altitude
Max time [s]	0/0	0/0	4.7/9.6	9/12.1	0/0	3.8/6.6
Mean time [s]	0/0	0/0	3.15/4.13	3.66/4.45	0/0	2.7/2.24
Median time [s]	0/0	0/0	– /3	2.25/2.5	0/0	– /2
Amount	0/0	0/0	2/4	10/11	0/0	2/7

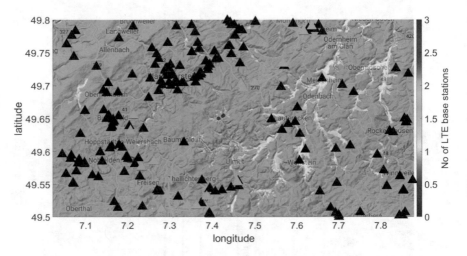

Fig. 15 Flight route Hamm-Schoenenberg in green with red spots when the LTE link was disrupted. Black triangles show all **assumed LTE base stations of all providers** and green triangles show the stations the UA was connected to during flight. The map reveals some valleys which are not served well by LTE

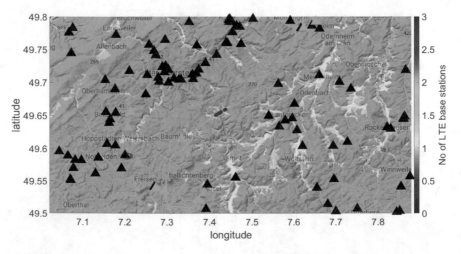

Fig. 16 Flight route Hamm-Schoenenberg in green with red spots when the LTE link was disrupted. Black triangles show **only assumed Telekom LTE base stations** and green triangles show the stations the UA was connected to during flight. The map reveals some valleys which are not served well by LTE

handover from one to another base station increases the latency. The average distance between the UA and a connected base station was about 2 to 3 km in all cases, which indicates a relatively good LTE coverage as well.

7 Conclusion

In this contribution, we discussed data link technologies for command and control of long-distance cargo UA in the context of certification based on operational risk. In dependence of the mission profile and the required safety level, we propose the use of cellular networks, satellite networks, as well as a combination of both. In detail, we evaluated the suitability of 4th generation Long Term Evolution (LTE) cellular networks by simulations using realistic flight routes of a cargo UA. In this context, we evaluated the low-altitude cellular coverage in rural areas. Our results show that cellular LTE networks offers high data link availability and throughputs in Germany, but connection disruptions in the order of some seconds occur in valleys that are not served well by LTE. Moreover, we showed that the simultaneous use of multiple, distinct LTE networks, which are operated by different providers, is a viable option to increase coverage and robustness. We evaluated the LTE coverage only for flight routes in Germany, but the same approach can be applied worldwide. In the future, the impact of connection disruptions on operational safety has to be further investigated and how areas with insufficient coverage can be identified before the missions in order to circumfly them.

References

3GPP (2017) Enhanced LTE support for aerial vehicles (TR 36.777 v15.0.0). Valbonne, France. https://portal.3gpp.org/desktopmodules/Specifications/SpecificationDetails.aspx?specif icationId=3231. Accessed 13 May 2019

Amorim R et al (2017) Radio Channel modeling for UAV communication over cellular networks. IEEE Wirel Commun Lett 6(4):514–517. https://doi.org/10.1109/LWC.2017.2710045

Amorim R, et al (2019) Improving drone's command and control link reliability through dual-network connectivity. In: 2019 IEEE 89th vehicular technology conference (VTC2019-Spring), Kuala Lumpur, Malaysia, 28 April–1 May 2019

Bundesnetzagentur (2019) Breitbandmessung Aktueller Jahresbericht 2018/2019. https://www.bun desnetzagentur.de/SharedDocs/Downloads/DE/Sachgebiete/Telekommunikation/Verbraucher/ Breitbandmessung/Breitbandmessung_Jahresbericht_%202018_2019.pdf. Accessed 01 June 2020

Dauer JC, Dittrich JS (2021) Automated cargo delivery in low altitudes: concepts and research questions of an operational-risk-based approach. In: Dauer JC (ed) Automated low-altitude air delivery—towards autonomous cargo transportation with drones. Springer, New York, Berlin, Heidelberg

Inmarsat (2020) SB-UAV aviation communication. https://www.inmarsat.com/service/sb-uav. Accessed 01 June 2020

International Civil Aviation Organization (2017) Doc 9869: performance-based communication and surveillance (PBCS) Manual

Joint Authorities for Rulemaking on Unmanned Systems (2019) JARUS guidelines on Specific Operations Risk Assessment (SORA) Ed 2.0

Lin X et al (2018) The sky is not the limit: LTE for unmanned aerial vehicles. IEEE Commun Mag 56:204–210. https://doi.org/10.1109/MCOM.2018.1700643

Matolak DW, Sun R (2017) Air-ground channel characterization for unmanned aircraft systems—part I: methods, measurements, and models for over-water settings. IEEE Trans Veh Technol 66:26–44. https://doi.org/10.1109/TVT.2016.2530306

Mielke D (2017) C-band digital aeronautical communication for unmanned aircraft systems. In: 2017 IEEE/AIAA 36th digital avionics systems conference (DASC), St. Petersburg, FL, 17–21 September 2017

Mielke D (2018) Frame structure of the C-band digital aeronautical communications system. In: 2018 Integrated communications, navigation, surveillance conference (ICNS), Herndon, VA, 10–12 April 2018

Mielke D, Schneckenburger N (2019) Towards a data link for unmanned aviation: DLR flight measurement campaign for C2 data link development. In: 2019 Integrated communications, navigation and surveillance conference (ICNS), Herndon, VA, USA, 9–11 April 2019

Mogensen P, et al (2017) LTE capacity compared to the shannon bound. In: 2007 IEEE 65th vehicular technology conference—VTC2007-spring, Dublin, 22–25 April 2007

Nikodem F, Rothe D, Dittrich JS (2021) Operations risk based concept for specific cargo drone operation in low altitudes. In: Dauer JC (ed) automated low-altitude air delivery—towards autonomous cargo transportation with drones. Springer, New York, Berlin, Heidelberg

OpenCelliD (2019) Unwired labs, Hyderabad. https://www.opencellid.org. Accessed 01 Aug 2019

Pak H (2021) Use-cases for heavy lift unmanned cargo aircraft. In: Dauer JC (ed) Automated low-altitude air delivery—towards autonomous cargo transportation with drones. Springer, New York, Berlin, Heidelberg

Rothe D, Nikodem F (2021) System architectures and its development efforts based on different risk classifications. In: Dauer JC (ed) Automated low-altitude air delivery—towards autonomous cargo transportation with drones. Springer, New York, Berlin, Heidelberg

Schalk LM, Peinecke N (2021) Detect and avoid for unmanned aircraft in very low level airspace. In: Dauer JC (ed) Automated low-altitude air delivery—towards autonomous cargo transportation with drones. Springer, New York, Berlin, Heidelberg

Schopferer S, Donkels A (2021) Trajectory risk modelling and planning for unmanned cargo aircraft. In: Dauer JC (ed) Automated low-altitude air delivery—towards autonomous cargo transportation with drones. Springer, New York, Berlin, Heidelberg

Stanczak J, et al (2018) Mobility challenges for unmanned aerial vehicles connected to cellular LTE networks. In: 2018 IEEE 87th vehicular technology conference (VTC Spring), Porto, 3–6 June 2018

RTCA (2016) DO-362 with Errata—Command and Control (C2) Data Link Minimum Operational Performance Standards (MOPS) (Terrestrial), Washington, DC, USA

Qualcomm (2017) LTE Unmanned Aircraft Systems Trial Report v.1.0.1. San Diego, USA. https://www.qualcomm.com/documents/lte-unmanned-aircraft-systems-trial-report. Accessed 5 June 2017

Ponchak D, et al (2016) A summary of two recent UAS command and control (C2) communications feasibility studies. In: 2016 IEEE Aerospace Conference, Big Sky, MT, USA, 5-12 March 2016.

Detect and Avoid for Unmanned Aircraft in Very Low Level Airspace

Lukas Marcel Schalk and Niklas Peinecke

Abstract Regular aviation does usually not utilize the very low level airspace below 150 m. Even so, this altitude range cannot be exclusively reserved for the operation of unmanned aircraft. Therefore, measures have to be implemented to avoid mid-air collisions, especially with manned aviation. Primarily, the mid-air collision risk is supposed to be reduced on a strategic level preflight by thorough flight route planning and making use of advanced unmanned aircraft system traffic management concepts. Besides, the residual mid-air collision risk is supposed to be reduced on a tactical level through the use of detect and avoid systems, which support remote pilots especially during flights beyond visual line of sight. However, an accepted technical solution for unmanned aircraft does not exist at the moment. On that account, the contribution of this chapter is three-fold. First, we identify suitable detect and avoid system architectures. Second, we review suitable cooperative sensor technologies. Third, we determine the minimum required sensor range in simulations. As a result, we propose a solution for operation in European airspace that is already available: FLARM cooperative sensors onboard the UA surveille the surrounding airspace and the remote pilot tracks intruders and commands avoidance maneuvers.

Keywords ADS-B · Collision avoidance · Cooperative sensors · Detect and avoid · FLARM · Self-separation · Unmanned aircraft · Unmanned aircraft systems

1 Introduction

The evaluation of the intrinsic risk of mid-air collisions, the so-called air risk, is part of the specific operation risk assessment (SORA) process described in JARUS (2019).

L. M. Schalk (✉)
German Aerospace Center (DLR), Institute of Communications and Navigation, Münchener Str. 20, 82334 Weßling, Germany
e-mail: Lukas.Schalk@dlr.de

N. Peinecke
German Aerospace Center (DLR), Institute of Flight Guidance, Lilienthalpl. 7, 38108 Braunschweig, Germany
e-mail: Niklas.Peinecke@dlr.de

© Deutsches Zentrum für Luft- und Raumfahrt e. V. (DLR) 2022
J. C. Dauer (ed.), *Automated Low-Altitude Air Delivery*, Research Topics in Aerospace,
https://doi.org/10.1007/978-3-030-83144-8_14

333

It is used to determine in conjunction with the ground risk the specific assurance an integrity level (SAIL). In order to reach a low SAIL, the air risk class can be lowered by applying strategic mitigation means in advance of the takeoff, e.g. by operating only within certain boundaries. The residual risk is reduced by tactical mitigation means during flight, e.g. by the use of Detect and Avoid (DAA) systems.

The lack of human pilots' eyesight onboard unmanned aircraft (UA) requires enhanced technical solutions compared to manned aviation. Especially during flights beyond visual line of sight (BVLOS), remote pilots are not able to surveille the airspace around UA from ground using only their eyesight. DAA systems are viable technical solutions to enable remote pilots, UA, or both, to detect stationary obstacles, e.g., cranes, as well as moving obstacles, e.g., other aircraft, in reasonable time in order to initiate appropriate maneuvers for collision avoidance. DAA systems rely on onboard or on ground sensors, which operate either in a cooperative or non-cooperative manner. In this contribution, we focus primarily on cooperative sensors for DAA.

Recently, Europe's SESAR has released the concept of operations for the European unmanned aircraft system traffic management (UTM) initiative U-space, see SESAR (2019). In the document released by the U-space project CORUS the overall U-space concept and architecture is detailed. While the first two U-space stages, U1 and U2, are envisioned to be carried out without further means of DAA, from U3 onward it is foreseen that DAA will play a crucial role within U-space. The CORUS concept of operations vol. 2 state that it will mainly rely on "collaborative DAA", defining collaborative DAA as an environment where "all participating hazards (aircraft) make themselves detectable in some agreed way". This definition matches our notion of cooperative DAA.

In the following, we first present possible DAA system architectures for BVLOS flights of cargo UA in VLL airspace. Second, we review existing cooperative sensor technologies and select appropriate solutions considering the average equipment of the majority of aircraft operating in the mission area. Third, we determine the required range of DAA sensors for tactical mitigations to further evaluate the suitability of the selected cooperative sensor using simulations.

2 Related Work

Hottman et al. (2009) reviewed literature on DAA sensor technologies which are available and useful for UA operating in the national airspace system. The authors included cooperative as well as non-cooperative sensor technologies and rated their suitability with respect to different requirements. The main advantage of non-cooperative sensors over their cooperative counterparts is their ability to detect targets without requiring any action from the target. In other words, non-cooperative sensors operate independently from the targets' equipment. Generally, non-cooperative sensors are class-divided in active and passive sensors. Characteristic for active sensors is the need to transmit energy in order to detect obstacles in their field of view. The authors state that a single sensor solution might not be sufficient since cooperative as well as non-cooperative airspace users have to be tracked. Instead, multiple, independent sensors should be used especially if redundancy is required to reach a certain level of safety.

Furthermore, research on DAA systems for UA has to include topics such as DAA architectures and DAA algorithms. A comprehensive survey was given by Fasano et al. (2016) for all kinds of UA, not only in VLL airspace. The authors identified a broad range of possible solutions for changing requirements and operational environments. Also, numerous open issues, which should be resolved before UA are equipped, were pointed out.

Recently, Opromolla et al. (2019) presented an overview and analysis of non-cooperative DAA concepts for small UA in VLL airspace. Recent advances in sensor technology allow the use of formerly heavy and energy-intensive equipment, e.g., radar and LIDAR, without any reduction of detection probability or detection range.

3 DAA System Architectures for BVLOS Missions

Typically, DAA systems consist of three interconnected functional components: the sensing component, the processing component, and the decision-making component. Each component is related to a specific task, which can be either executed by autonomous functions or by human pilots. The task of the sensing component is the surveillance of a certain airspace volume around the UA. Raw sensing data is fed from the sensing component to the processing component which identifies and tracks targets. The decision-making component subsequently gets the identified targets, e.g., intruder aircraft, as input. Its output is control actions in case at least one of the identified targets has to be avoided. Naturally, pilots are capable of performing the tasks of all three components, which means that all components are located onboard the manned aircraft. This is commonly known as "see and avoid". Moreover, the pilot's capabilities can be enhanced by technical solutions in order to monitor the surrounding airspace and process information. For example, the Traffic Alert and Collision Avoidance System (TCAS) enables the detection of similar equipped aircraft and is even able to issue advisories in case of conflict. Nevertheless, the pilot still has to take appropriate actions in accordance with the TCAS output.

In contrast to manned aviation, in unmanned aviation no pilot onboard can perform the tasks necessary for DAA. Instead, either the remote pilot or onboard autonomous functions have to sense, process, and decide. In other words, the functional components can be placed either onboard the UA or off-board on the ground. The separate placement of functional components inherently requires a digital data link for interconnection.

Subsequently, we refer to any specific combination of onboard and on ground placement of components as the DAA system architecture and identify suitable architectures for BVLOS flights in VLL airspace.

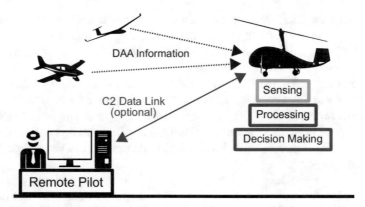

Fig. 1 Onboard DAA architecture. All components are placed onboard the UA. Hence, the C2 data link is optional for DAA

3.1 Onboard Architecture

The onboard DAA system architecture implies the placement of all three components onboard UA. This architecture is most similar to DAA in manned aviation, where onboard pilots use their own eyesight or technical equipment to surveille the environment, identify targets, and directly decide and take action if necessary.

Similarly, UA are equipped with one or multiple cooperative or non-cooperative sensors, which enable the aircraft to sense its surrounding. Sensor outputs are processed and targets are identified by an onboard computer directly on the UA itself. Moreover, the UA is capable of deciding if and what kind of maneuver has to be executed to avoid identified targets without any human input. Hence, a command and control (C2) data link is not necessary but optional to allow the remote pilot to observe the DAA process and to overwrite actions if necessary. In Fig. 1, we depict the onboard DAA architecture.

The main limitations of this architecture are size, weight and power (SWaP) constraints onboard the UA. Especially for small and lightweight UA, sensor and computer equipment can easily violate the constraints. Moreover, a high level of autonomy is required onboard the UA, which may raise regulatory issues.

3.2 Ground-based Architecture

A fully ground-based DAA architecture implies that all components are placed on ground, but not that all components are necessarily gathered in the same place. This specific architecture may be employed especially if SWaP requirements imposed on the UA by the DAA hardware have to be relaxed.

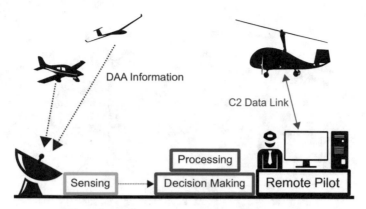

Fig. 2 Ground-based DAA architecture with centralized processing

The cooperative or non-cooperative sensors are placed suitably on the ground, from where they are able to observe the UA and their surrounding airspace. Hence, the sensors should not be placed near the ground control station (GCS) but rather in the vicinity of the BVLOS flight path. If the sensors are placed in a remote location, raw sensor data has to be transferred to the processing component, e.g., via an on-ground data link. On the one hand, multiple processing components can be distributed and send the processed data to the GCS individually. On the other hand, all data from the remote sensors can be transferred to a central processing unit, which can be located near the GCS. In any case, a ground network is needed for the data exchange. In Fig. 2, we show a ground-based DAA architecture with centralized processing.

Typically, remote pilots decide on the avoidance maneuvers if this architecture is employed. In that case, no autonomy is required onboard the UA. Instead, a data link for C2 is required to transfer the avoidance commands to the UA. Nevertheless, the decision-making of the remote pilot can be additionally supported by GCS functions.

Especially for BVLOS operations, individual ground sensors need to cover a wide area or a network of ground sensors has to be used. Moreover, ground sensor networks for cooperative as well as non-cooperative VLL airspace users are not existent as of now. Ground-based sensor networks for the surveillance of UA are under investigation in the scope of the U-space project TERRA, see SESAR (2017).

3.3 Hybrid Architecture

The term hybrid architecture refers to any possible combination of the three components' placement. However, not all combinations make sense especially in the context of UA. For example, if the sensing component and the processing component are not located nearby, a huge amount of raw sensing data has to be transferred from one component to the other which is usually inefficient over large distances via wireless data links. We therefore highlight only one combination in the following. First of all,

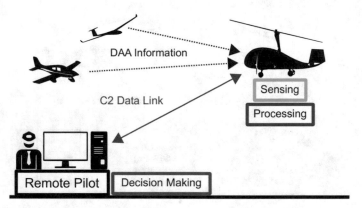

Fig. 3 Hybrid DAA architecture

we assume that the autonomy of the UA is rather low. Thus, the remote pilot has to be in the decision loop.

As already mentioned before, the placement of the sensing component on the UA is favorable especially for BVLOS operations since no network of ground sensors is needed. Consequently, the processing component should be placed on the UA as well to avoid the transfer of raw sensor data via the C2 data link. Instead, only target information is transmitted to the remote pilot who decides on the avoidance maneuvers. Then, control actions are transmitted back to the UA via the C2 data link. Please note that additional latencies introduced by C2 link have to be taken into account. Further, this architecture includes the possibility for redundant approaches, e.g., with UA implementing a full onboard architecture while mirroring some components on ground for safety reasons (Fig. 3).

4   Cooperative DAA Sensors for UA

Since sensors are the first entity of the DAA processing chain, they are crucial for the avoidance of collisions. If the sensors do not sense the target in the surveillance volume around the UA, targets can neither be identified nor avoided. Therefore, in particular cooperative sensors have to be carefully chosen with respect to the surrounding airspace since, by design, only similarly equipped aircraft can be detected. All other aircraft remain undiscovered. In this section, we present the most important cooperative sensor technologies for UA. Apart from cooperative sensor the term transponder is often used synonymously.

The working principle of cooperative sensors is based on the exchange of cooperative information, e.g. position and intent information, among nearby airspace users. The structure is depicted in Fig. 4. Typically, the position and intent information is obtained by a global navigation satellite system (GNSS) receiver and broadcast as a cooperative message via a digital data link.

Fig. 4 Working principle of cooperative sensors

The performance of cooperative sensors is naturally limited by the range and capacity of the data link and by the accuracy of the cooperative information. First of all, if the broadcast position does not match the true position of the intruder aircraft because of GNSS failures, conflicts are not detected or avoidance maneuvers are ineffective. Additionally, the range of any non-cooperative or cooperative sensor has to be large enough so that DAA information can be processed and the avoidance maneuver can be completed, especially at high relative velocities. Finally, if the capacity of the broadcast data link is exceeded, i.e., the aircraft density exceeds a certain maximum per airspace volume, cooperative messages may be lost. Intuitively, message loss directly leads to undetected traffic conflicts and reduced safety. This issue will be further discussed with respect to the individual systems in the following.

4.1 ADS-B

Automatic Dependent Surveillance – Broadcast (ADS-B) is considered to be a key cooperative sensor technology for enhancing safety and efficiency in controlled airspace operations since it allows long range, omni-directional intruder detection, and self-separation with low SWaP and cost requirements. The primary purpose of ADS-B is to provide surveillance information to other aircraft and air traffic controllers. ADS-B is automatic since cooperative messages are broadcast periodically without any external input, e.g., by the pilot, to everyone within the transmission range. It depends the position and velocity data from the aircraft's navigation system which is obtained for example from a GNSS receiver. Without precise position information, the system does not work.

Two systems on designated frequencies are currently in operational use for the transmission of ADS-B messages on designated frequencies: 1090 MHz Extended Squitter (1090ES) described in RTCA (2009) for commercial aircraft and 978 MHz Universal Access Transceiver (UAT) described in RTCA (2011) for general aviation and airport vehicles. 1090ES is the most widespread technology and the only one used internationally whereas UAT is permitted only for use in the United States.

1090ES uses the standard secondary surveillance radar (SSR) 1090 MHz reply frequency and adds the ADS-B information for other aircraft and ground stations to the SSR reply data stream. Therefore, the ADS-B function can be integrated into existing Mode S transponders. In 2010, the Federal Aviation Administration (FAA) mandated ADS-B Out equipage of aircraft which fly in certain airspace by January

2020. More specifically, all aircraft flying in class A airspace above Flight Level (FL) 180 are required to employ 1090ES. Below FL180, aircraft have to be equipped with either 1090ES or UAT. Similarly, the Commission Regulation No 1207/2011 mandates that in European airspace all aircraft weighing more than 5,700 kg or cruising at more than 250 knots need to be equipped with 1090ES.

ADS-B technology, especially UAT, for UA has been the focus of several studies during the last years. The authors Stark et al. (2013) underlined the remaining implementation and regulatory issues concerning ADS-B for small UA. The relation between the UA density, the UAT transmission power, and the UAT performance was investigated in simulations by Guterres et al. (2017). The authors state that the transmission power has to be reduced in very dense scenarios to keep the required level of performance. Inspired by this work, Matheou et al. (2018) further investigated the UAT performance in mixed airspace scenarios. The authors extended previous models and additionally implemented a DAA algorithm to show that ADS-B is suitable for the investigated scenario. In 2018, Goossen et al. (2018) designed and tested a lightweight implementation of UAT which can be used on small UA. Furthermore, Duffy and Glaab (2019) tested the suitability of a reduced-power UAT transmitter for small UA. The aim of their research was to find an adequate output power that reduces the congestion of the frequency channel without reducing the reliability of the conflict detection.

However, several other studies outline that the capacity of ADS-B systems is insufficient for future air traffic scenarios. For example, Strohmeier et al. (2014) pointed out that high aircraft densities lead to congestion on the 1090 MHz frequency channel which causes severe message loss. In turn, conflicts cannot be reliably detected any more. The situation on the 978 MHz channel is less severe, but this may be used only in the US. Therefore, the use of current ADS-B technology for a large number of UA is not recommended without carefully monitoring the impact on manned aircraft.

4.2 FLARM

FLARM ("flight alarm") is a cooperative sensor especially designed for the requirements of small aircraft, e.g., gliders, in today's VLL airspace. It was launched in 2004 and more than 40,000 aircraft have been equipped with FLARM sensors so far. It is not an ADS-B system, but it is approved by the European Union Aviation Safety Agency (EASA) for installation in certified aircraft. Nevertheless, FLARM cannot be used for ATC communications since it is incompatible with ADS-B systems and its radio communications range is also far smaller when compared to ADS-B systems. Typically, classic FLARM sensors reach up to 3 km whereas PowerFLARM sensors reach up to 10 km.

The FLARM communication protocol is proprietary, patented, and copyright protected. Therefore, no official documentation is accessible. In the following, we summarize the publicly available information. FLARM sensors do not use aeronautical frequency bands for the radio communications as ADS-B systems do.

Instead, the free-to-use SRD860 or the Industrial, Scientific and Medical (ISM) band is used. Due to national regulations, the operational frequency band changes for different areas. FLARM transponders are capable of automatically tuning to the correct frequency based on GNSS information. Since the FLARM frequency bands are not protected, other devices are likely to interfere with FLARM. For example, WiFi networks use ISM bands as well. Compared to ADS-B systems, the low range and the use of open frequency bands may lead to a reduced reliability of the system.

The unique feature of FLARM compared to ADS-B systems is the use of symmetric encryption to guarantee the integrity of the system. Only licensed FLARM sensors are able to encrypt and decrypt messages correctly which protects the system from abuse. On the one hand, encryption increases security and privacy of the users. On the other hand, the encryption limits the interoperability and protects a monopoly position which may endanger airspace safety. The process itself needs computational power and time which reduces safety even further. Moreover, once the symmetric encryption key is found, the whole system is compromised. This issue could be circumvented by the use of asymmetric encryption.

4.3 Cyber Security Considerations

In this section we survey the vulnerabilities of cooperative sensors, namely ADS-B systems (1090ES and UAT) and FLARM with respect to UA in VLL airspace. The cyber security of DAA systems is very important for UA since no pilot onboard can mitigate attacks.

A wide variety of attacks on cooperative sensors can be carried out if the attacker is able to inject, modify or delete any ADS-B or FLARM message at will. In the following we describe the most important attacks on cooperative sensors. These attacks have also been presented by Schäfer et al. (2013) with respect to 1090ES. Nevertheless, we point out that these attacks can be carried out against UAT and FLARM systems as well.

- **Reconnaissance**
  Reconnaissance is a very simple, passive attack which enables the attacker to gain information about a UA's position and direction of flight by listening to unsecured broadcast messages. This information can be used for example to intercept the UA or to support further cyber security attacks.
- **Jamming**
  Attackers can severely disturb the cooperative sensors by transmitting an arbitrary wideband signal with high power on the same frequency channel. The attacker's signal jams any nearby aircraft or ground-based cooperative sensors. If no other sensors, either cooperative sensors on other frequency bands or non-cooperative sensors, are onboard the UA, a jamming attack is comparable to blindfolding the UA.

- **Ghost Target Injection**
  Any attacker with knowledge of the message structure can compile and broadcast fake messages to create ghost targets. Since other aircraft, pilots, or ATC usually cannot distinguish ghost targets from real targets, attackers can easily use ghost targets to create confusion among airspace participants. For example, attackers can use ghost targets to create fake conflict warnings which can be used to reroute aircraft. UA are vulnerable in particular since no pilot onboard can verify the existence or nonexistence of the ghost aircraft by his eyesight.
- **Aircraft Disappearance**
  Highly critical situations arise if the cooperative sensors cannot detect an intruder on a collision course. Especially, UA are affected by malfunctioning sensors since no pilot onboard can see the intruder. Attackers can exploit this by deleting all broadcast messages of a specific aircraft to make this aircraft disappear for the cooperative sensor. Again, UA are vulnerable due to the missing pilot onboard.
- **Virtual Trajectory Modification/False Alarm Attack**
  Instead of injecting ghost targets, attacks can modify the broadcast position and trajectory of real aircraft. The effect is comparable to the injection of ghost targets but the attack is harder to recognize since the ghost targets do not appear suddenly.
- **Aircraft Spoofing**
  Every broadcast message contains a unique ID which identifies the aircraft. Attackers can change this ID to an arbitrary one. Moreover, the ID can also be changed to an ID of a trusted aircraft to disguise the own aircraft and to enter restricted areas.

In Table 1, we evaluate whether the cooperative sensors 1090ES, UAT or FLARM are protected against the presented cyber security threats. The main result of our analysis is that both ADS-B systems are vulnerable to all threats. Only FLARM is protected against 4 out of 6 threats since it uses symmetric message encryption. FLARM is invulnerable against reconnaissance, ghost target injection, virtual trajectory modification, and aircraft spoofing. Nevertheless FLARM is still vulnerable to jamming and aircraft disappearance attacks.

Table 1 Evaluation of vulnerability against cyber security attacks on cooperative sensor systems

Attack	1090ES	UAT	FLARM
Reconnaissance	No	No	Yes
Jamming	No	No	No
Ghost target injection	No	No	Yes
Virtual trajectory modification	No	No	Yes
Aircraft disappearance	No	No	No
Aircraft spoofing	No	No	Yes

5 Evaluation by Simulations

In this section, we determine the minimum required range of DAA sensors with respect to the investigated cargo UA and to the surrounding airspace. Only if a cooperative sensor like FLARM is able to provide the minimum range, we consider it suitable for the intended application. First of all, we present the assumptions for our evaluation, i.e. the characteristics of typical aircraft in VLL airspace, the characteristics of the investigated cargo UA, and the required sensor processing and remote pilot reaction times. Subsequently, we simulate various aircraft encounters and check whether the distance requirements for collision-avoidance and self-separation can be satisfied. In the end, we obtain the required DAA sensor range.

5.1 Assumptions

5.1.1 Aircraft in VLL Airspace

Today, a broad range of airspace users is present in the German airspace class G as well as in the adjacent airspace classes. The users can be divided into five distinct classes (see column "Class" in Table 2). Within the specific operations risk context,

Table 2 Parameters of typical aircraft in VLL airspace

Class	Type	V_{max} [km/h]	V_{min} [km/h]	$V_{vertical}$ [m/s]	$a_{vertical}$ [m/s^2]	$a_{lateral}$ [m/s^2]	$a_{longitudinal}$ [m/s^2]
Business jet	Learjet 60	840*	170	22,9	7,6	4	0,6
Sports plane	Glasair III	480	80	15	3,5	4	0,4
Glider	ASW 20	280	60	10	5	7	0,9
Multicopter	DJI Phantom 4	72	0	6	6	5	5
Military jet	Eurofighter	1270*	203	255*	25	28	2,8
Parachutist	-		198	0	55	20	4	0,4

*Below 10,000 ft (3048 m) only 250 kts (460 km/h; 128 m/s) allowed

especially commonly occurring intruder aircraft with high cruising speeds must be considered. Therefore, we are particularly interested in aircraft with outstanding performance specifications in their class. Nevertheless, we only consider aircraft which are produced and sold in moderate to large quantities, since encounters with these aircraft are more likely. Although we believe that the assumptions shown are typical for general airspace G, there may be different traffic compositions in countries other than Germany.

Comments on Table 3:

Table 3 Sensor processing and remote pilot reaction time

Action	Actor	Duration
Sensor rate	Sensor system	1 s
Intruder tracking and alerting	Sensor system	1 s
Decision to turn left or right	Pilot	4 s
Muscular reaction	Pilot	0.4 s
Aircraft lag	UA	1 s
Total (autonomous UAS)		3 s
Total (RPAS)		7 s

- The specified vertical velocity $v_{vertical}$ is always the maximum of the particular aircraft.
- The specified accelerations $a_{vertical}, a_{lateral}, a_{longitudial}$ are always the maximum of the particular aircraft.
- The maximum vertical acceleration $a_{vertical}$ of the parachutist is the braking acceleration when opening the glider. The vertical velocity $v_{vertical}$ corresponds to the normal speed during a jump, record attempts reach higher speeds.

5.1.2 Cargo UA Configuration

Several possible configurations for cargo UA have been investigated by Hasan and Sachs (2021), from which the concept of a gyrocopter turned out to be a promising candidate. Within the scope of this chapter, the performance variables of the gyrocopter are used as a basis for the ownship's performance characteristics.

- Maximum acceptable rate of climb at 200 km/h → 6.5 m/s

  - The airspeed decreases by 2 m/s during climbing.

- Descent rate without speed increase at 200 km/h → 4.5 m/s
- Descent rate without speed increase at 100 km/h → 4.5 m/s
- Max. yaw rate at 200 km/h due to maximum rolling angle of 45° → 6.8°/s
- Max. yaw rate at 100 km/h due to maximum rolling angle of 45° → 11°/s
- Max. yaw rate during climb at 200 km/h at climb rate of 5 m/s → 5.5°/s
- Max. yaw rate during descent at 200 km/h at descent rate of 5 m/s → 7.4°/s
- Limit maximum dynamically induced load multiplier by pulling or pushing to ± 0.3 g (subjective estimate - conservative and not determined).

5.1.3 Sensor Processing Time and Remote Pilot Response Time

Directly after an intruder enters the range of DAA sensors, an additional period of time is required before the actual avoidance maneuver can be carried out. We assume that this time span is composed of different parts. On the one hand, we account for the time the sensor system needs to process raw sensor data. On the other hand, we account for the time the remote pilot needs to act according to the displayed information. In a fully autonomous environment, this decision time can be assumed to be close to zero as it is already included in the tracking and alerting process. However, we assume that the remote pilot always is in the decision loop.

Estimates for the pilot's response are found in EASA (2017), based on FAA (1983). In accordance with RTCA DO-365, we call this parameter PIC (pilot in command time). In Table 3 we list the assumed components of the PIC.

In many publications it is common to refer to the older estimates presented in FAA (1983). Therefore, we also use a higher value of 12 s here, leaving an additional buffer of 5 s for maneuver times ("set-on time") resulting from the flight dynamics of the aircraft during a change of flight direction.

5.2 Results

By means of a series of simulations, the necessary ranges of DAA sensors have been determined. In this context it does not matter how the DAA sensors are technically implemented, the values determined apply to non-cooperative sensors (radar, lidar) as well as cooperative sensors (ADS-B, FLARM), since the specifics of actual sensor models are not taken into account in the simulation yet.

Two cases are distinguished:

- **Collision-Avoidance (CA):** An aircraft tries to avoid an impending collision that could not be avoided by previous measures. Considerations of economy and efficiency are set aside in favor of the safety of all participants. TCAS MOPS defines a near mid-air collision (NMAC) as a situation where the aircraft are closer together than 100 ft laterally. Thus, in such a situation, we will consider a separation distance of about 210 m, being roughly 7 times the distance, still acceptable for the design of a DAA system.
- **Self-Separation:** Aircraft attempt to maintain a safe but efficient distance from all surrounding airspace users (and obstacles) for normal, unimpaired operation. The applicable aviation standards must be taken into account. ICAO Doc 4444 demandes a separation of 5 NM, i.e. 9260 m, when using a cooperative sensor such as ADS-B for DAA.

The results presented in this chapter were obtained using a simulation tool called AirES (Aircraft Encounter Simulator). AirES was developed by DLR as part of the KoKo project (collision detection/collision avoidance) in 2010. Its functionality was significantly extended during the course of the MasterUAS project until 2019.

The simulation takes the initial aircraft positions, attitude, speeds and minimum curve radii of both the ownship and the intruder as an input. Basic parameters of the DAA sensor, namely field of view in azimuth and elevation direction, maximum detection range and leveling capability of the sensor must be specified. Finally, the reaction time of the remote pilot can be determined before executing an avoidance maneuver. Furthermore, we can specify whether the original course should to be resumed after reaching the Closest Point of Approach (CPA), and whether the intruder executes avoidance maneuvers, too. Based on these parameters, AirES is used to calculate a minimum separation distance at the CPA, the visibility of the intruder for the sensor and the safety status for all flight path points. As an option, an avoidance maneuver can be triggered at the point of first detection. In the results presented here, the intruder should not perform an avoidance maneuver.

In Fig. 5, we show an exemplary avoidance maneuver calculated by AirES. In this example, the ownship and the intruder fly on a straight course and in opposite flight direction towards the point of the first DAA sensor detection (marked by the blue triangles) at a distance of 3,216 m. From there, a delay of PIC equal to 25 s is considered before the avoidance maneuver is started. The maneuver itself is a constant turn with 8°/s angular velocity until an avoidance angle is reached that maximizes the distance at the CPA. Black dots mark the actual CPA at a distance of 1,436 m (red circle). The trajectory of the ownship is marked in four colors:

- Green means well clear.
- Yellow means that the DAA sensor has detected the intruder, but the situation is still well clear.
- Orange indicates a Well Clear Violation (WCV) without sensor detection.

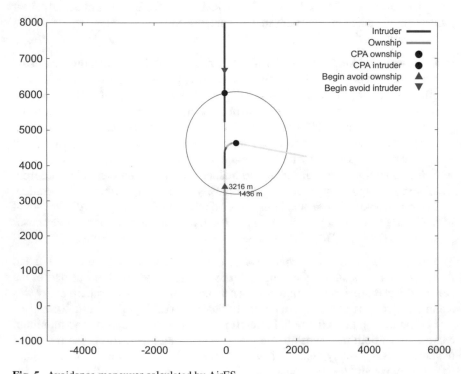

Fig. 5 Avoidance maneuver calculated by AirES

- Red indicates that the situation is not well clear and the DAA sensor has detected the intruder.

It should be noted that instead of well clear, any separation criterion can be used here, e.g., whether the horizontal separation at the CPA was greater than 210 m.

Next, we vary individual parameters of the simulation and convert the results into angle-distance plots. In detail, the procedure was as follows:

1. The ownship was set on a straight course with a fixed speed of 200 km/h.
2. The intruder is also on a straight course with fixed speed.
3. The intruder was set to a defined point with relative coordinates (x,y) and relative course α with respect to the ownship.
4. The ownship carries out the avoidance with a turn rate of 6.8°/s, the reaction time is PIC = 12 s.
5. The avoidance maneuver is executed either in left or right direction, depending on which direction allows the larger minimum separation distance.
6. The DAA sensor range was set to ∞, i.e., the ownship immediately initiates an avoidance maneuver.
7. We define 210 m as the minimum separation distance. Note that no temporal separation component is required. In TCAS terminology this means that TAUMOD is set to 0 s, see Muñoz (2015).

For all courses α, it was calculated, whether the ownship can reach a course, which keeps the minimum separation, or whether the minimum separation is inevitably violated.

5.2.1 Collision-Avoidance

The resulting requirements for the collision-avoidance application with respect to the DAA sensor range are shown in Fig. 6. We require that a minimum separation distance of 210 m could be maintained at the CPA. The ownship moves at a speed of 200 km/h, which corresponds to 55.55 m/s.

The cyan-coloured area represents the results for a fixed intruder, e.g., stationary obstacles such as wind turbines or construction cranes. It turns out that a minimum DAA sensor range of 1,176 m in flight direction is sufficient to maintain a safe separation. To the sides, smaller DAA sensor ranges are sufficient, for example 932 m at 10°, 658 m at 20°, 260 m at 60° and 226 m at 90°.

The yellow area represents the results for an intruder which is approaching at 200 km/h, the same velocity as the ownship. Here, the minimum detection range of the DAA sensor must be 2,238 m at 0° to 6°. At 60°, 1,184 m are sufficient, at 90° 626 m.

The red area represents the results for an intruder approaching at 480 km/h. This corresponds to the sports plane listed in Table 3. In the lower airspace below 10,000 feet the maximum speed is limited to 128 m/s, therefore 133.33 m/s is a valid worst

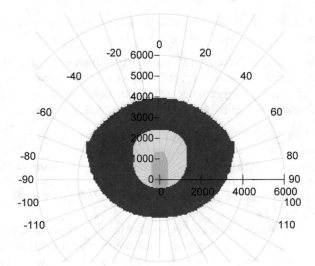

Fig. 6 Simulation results for collision-avoidance

case assumption. The required DAA sensor range here is 3,779 m at 0° to 10°. Up to 64° it is still 3,687 m. At larger angles, only then it reduces to 2,926 m at 90°.

5.2.2 Self-Separation

In ICAO Doc 4444 when using a system such as ADS-B a separation distance of 5 NM, i.e., 9260 m is demanded. Similarly, the airspace integration concept presented by Peinecke and Mühlhausen (2021) requires a safety distance to inhabited areas. Initially, the safety distance was set to 400 m in order to allow safe termination of the UA even in situations without control by the remote pilot. It seems reasonable to use a value of 400 m (or a different value which may be determined) also as a separation distance for regular operation, since in our opinion the ICAO regulation is not suited for the VLL airspace with its associated speed restrictions. In order to verify this assumption, the simulations from Sect. 5.2.1 were repeated with a required minimum separation distance of 400 m at the CPA instead of 210 m. The results are shown in Fig. 7.

Again, the results for fixed intruders are represented by the cyan colored area. It is shown that a minimum DAA sensor range of 1,376 m in a straight flight direction (0°) is sufficient. To the sides, shorter DAA sensor ranges are sufficient, for example, 1,234 m at 10°, 1,036 m at 20°, 434 m at 60° and 425 m at 90°.

The yellow area represents the results for an intruder approaching at 200 km/h. Here, the minimum range of a DAA sensor must be 2,626 m at 0° to 6°. At 60°, 1,473 m are sufficient, at 90° 876 m.

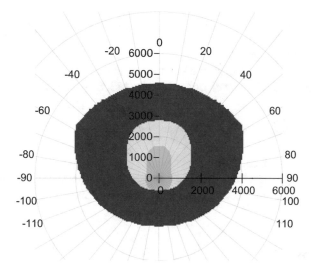

Fig. 7 Simulation results for self-separation

The red area represents results for an intruder approaching at 480 km/h. The minimum required sensor range here is 4,511 m at 58° to 10°. At 0° it is 4,426 m. At 90°, it is 3,475 m.

5.3 Discussion

The results of our simulations show that every DAA sensor that is used for DAA in VLL airspace with respect to the characteristics of the cargo UA should provide a minimum sensor range of 3,779 m for the collision-avoidance application and a range of 4,511 m for the self-separation application. One should note that the minimum sensor range given here is independent of the sensor technology. Both, cooperative and non-cooperative sensors could be used to fulfill the requirement.

In this contribution, we focus on cooperative sensors solely and in Sect. 4 we identified FLARM as a suitable cooperative DAA sensor for the European VLL airspace, since many VLL airspace users are already equipped with FLARM sensors as of today. Nevertheless, we still have to validate whether FLARM is able to provide the minimum required DAA sensor range.

With this in mind, we have to take the update rate of the cooperative message broadcasts into consideration. Since the ownship is unable to detect the intruder if it is not broadcasting cooperative messages, the gaps between individual broadcasts reduce the maximum range of the cooperative DAA sensor. Let us assume that the distance d between ownship and intruder is just a little bit larger than the maximum range of the DAA sensor, the relative speed is v_{rel}, and the update rate of the coopera-tive messages is f. In the worst case, the intruder can still approach to $d' = d - \frac{1}{f}v_{rel}$ before detection is triggered.

FLARM claims a range of $d = 5$ km and an update rate of $f = 2$ Hz. Hence, the effective range for a relative velocity of $v_{rel} = 133, 33\frac{m}{s}$ is calculated by

$$d' = 5000\,\text{m} - \frac{133, 33\,\text{m}}{2\,\text{Hz}\,\text{s}} = 4933, 33\,\text{m}. \tag{1}$$

The required DAA sensor range for collision-avoidance was 3,779 m and 4,511 m for self-separation. Therefore, FLARM cooperative sensors support the requirements even in worst case scenarios, i.e., a fast intruder on a frontal collision course.

6 Conclusion

The use of cooperative sensors onboard UA is a viable approach to enable collision avoidance on a tactical level during flight. For BVLOS flights of the investigated cargo UA, we propose a hybrid DAA architecture. In detail, cooperative sensors and processing units are placed onboard UA, whereas decisions are made by a remote pilot on ground and transferred via a C2 data link.

The selection of cooperative sensors for the electronic detection of intruder aircraft should be made considering the average equipment of the majority of aircraft oper-ating in the area. A viable solution at low altitudes in European airspace is the use of FLARM transponders. Elsewhere, other systems like UAT or web-based tracking services would have to be used.

Since the cooperative sensor range is crucial for the effectiveness of the DAA system, we determined the minimum required sensor range for UA in VLL airspace using simulations. Our results show that the range of FLARM is sufficient for the application.

References

Duffy B, Glaab L (2019) Variable-power ADS-B for UAS. In: 2019 IEEE/AIAA 38th digital avionics systems conference (DASC), San Diego, CA, 8–12 September 2019
EASA (2011) Research Project EASA.2011/07: Scoping improvements to 'see and avoid' for general aviation (SISA)
FAA (1983) Advisory Circular 90-48C, Pilot role in collision avoidance
Fasano G et al (2016) Sense and avoid for unmanned aircraft systems. IEEE Aerosp Electron Syst Mag 31(11):82–110
Guterres R et al (2017) ADS-B surveillance system performance with small UAS at low altitudes. https://www.mitre.org/sites/default/files/publications/16-4497-AIAA-2017-ADS-B.pdf. Accessed 01 Sep 2019

Goossen J et al (2018) Development of a versatile ADS-B communications system. In: 2018 Texas symposium on wireless and microwave circuits and systems (WMCS), Waco, TX, 5–6 April 2018

Hasan YJ, Sachs F (2021) Performance-based preliminary design and selection of aircraft configurations for unmanned cargo operations. In: Dauer JC (ed) Automated low-altitude air delivery - towards autonomous cargo transportation with drones. Springer, New York

Holtmann SB et al (2009) Literature review on detect, sense, and avoid technology for unmanned aircraft systems, Washington, DC

Joint authorities for rulemaking on unmanned systems (2019) JARUS guidelines on Specific Operations Risk Assessment (SORA) Ed 2.0

Matheou K et al (2018) ADS-B mixed SUAS and NAS system capacity analysis and DAA performance. In: 2018 integrated communications, navigation, surveillance conference (ICNS), Herndon, VA, 10–12 April 2018

Muñoz C et al (2015) DAIDALUS: detect and avoid alerting logic for unmanned systems. In: 2015 IEEE/AIAA 34th digital avionics systems conference (DASC), Prague, 2015, 13–17 September 2015

Opromolla R et al (2019) Conflict detection performance of non-cooperative sensing architectures for small UAS sense and avoid. In: 2019 integrated communications, navigation and surveillance conference (ICNS), Herndon, VA, USA, 9–11 April 2019

Peinecke N, Mühlhausen T (2021) Cargo drone airspace integration in very low level altitude. In: Dauer JC (ed) Automated low-altitude air delivery - towards autonomous cargo transportation with drones. Springer, New York

RTCA (2009) DO-260B - Minimum operational performance standards for 1090 MHz extended squitter automatic dependent surveillance - broadcast (ADS-B) and Traffic Information Services - Broadcast (TIS-B), Washington, DC, USA

RTCA (2011) DO-282B with corrigendum 1 - minimum operational performance standards for universal access transceiver (UAT) Automatic Dependent Surveillance – Broadcast, Washington, DC, USA

RTCA (2017) DO-365 - minimum operational performance standards (MOPS) for detect and avoid (DAA) Systems, 2017 Edition, May 31, Washington, DC, USA

Schäfer M et al (2013) Experimental analysis of attacks on next generation air traffic communication. In: Applied cryptography and network security, pp 253–271

SESAR (2017) Technological European research for RPAS in ATM – TERRA. https://www.sesarju.eu/projects/terra. Accessed 05 Oct 2020

SESAR (2019) SESAR Concept of Operations for U-space, vol 1, 2 & 3. https://www.sesarju.eu/node/3411. Accessed 5 Oct 2020

Stark B et al (2013) ADS-B for small unmanned aerial systems: case study and regulatory practices. In: 2013 International conference on unmanned aircraft systems (ICUAS), Atlanta, GA, 28–31 May 2013

Strohmeier M et al (2014) Realities and challenges of nextgen air traffic management: the case of ADS-B. IEEE Commun Mag 52(5):111–118

Trajectory Risk Modelling and Planning for Unmanned Cargo Aircraft

Simon Schopferer and Alexander Donkels

Abstract A key challenge for low-altitude unmanned air transportation is to minimize operational risks by all means. Besides many other measures to be considered, the aircraft's trajectory must be planned carefully and optimized as there are inevitable remaining risks which should be minimized when flying over sparsely populated areas. The risk may be mitigated by a safe termination of the flight if circumstances permit. Also, the probability of violating any operational constraints that would lead to a flight termination should be reduced as much as possible. Adequate risk models and efficient algorithmic risk assessment techniques are required to perform such optimizations. Furthermore, the aircraft may have to react to certain events such as high-priority traffic by changing its trajectory online during flight. As command and control (C2) links may have limited reliability, it must be possible to perform trajectory re-planning onboard with limited computational resources. This poses high demands on the runtime efficiency of the planning algorithms. In this work, we present conceptual approaches to risk modeling and assessment based on geospatial datasets and aircraft dynamic models. We further present the design and experimental results of a software framework for onboard and online trajectory planning. Our results demonstrate that risk-based motion planning for unmanned aircraft can be performed with limited onboard computational resources allowing for safe autonomous flight.

Keywords Motion planning · Path planning · Sampling-based planning · Risk-based planning · Trajectory generation · Flight termination

S. Schopferer (✉) · A. Donkels
German Aerospace Center (DLR), Institute of Flight Systems, Lilienthalplatz 7 , 38108
Braunschweig, Germany
e-mail: simon.schopferer@dlr.de

A. Donkels
e-mail: alexander.donkels@dlr.de

© Deutsches Zentrum für Luft- und Raumfahrt e. V. (DLR) 2022
J. C. Dauer (ed.), *Automated Low-Altitude Air Delivery*, Research Topics in Aerospace,
https://doi.org/10.1007/978-3-030-83144-8_15

1 Introduction

The concept developed in the scope of the ALAADy (Automated Low Altitude Air Delivery) research project envisions highly automated low-altitude flight over long distances (Dauer and Dittrich 2021). In contrast to many established civil applications of unmanned aircraft, this implies flying beyond the visual line-of-sight of a pilot on the ground. Consequently, the aircraft has to be controlled via a command-and-control (C2) link. While in the military sector, drones are commonly remotely piloted via dedicated satellite links, this would be unfeasible for a civil unmanned cargo drone, or a fleet thereof due to cost and safety concerns. From a safety perspective, risk mitigations must not depend on the C2 link to be perfectly reliable. Contingency procedures will, therefore, have to be executed autonomously in case of a link loss. At the very least, this includes a safe flight termination in case operational constraints are violated. However, from a use-case and economic perspective, it would be beneficial if the aircraft is able to autonomously mitigate any operational risks before a termination must be triggered. The likely event of a temporary loss or drop-out of the C2 link should be mitigated in other ways. With high degrees of autonomy, the dependence on the C2 link reliability can be reduced effectively. Also, the workload of the ground crew and requirements for ground station equipment can be reduced if the aircraft is able to guide, navigate and control its flight autonomously.

Even when flying over sparsely populated areas, certain risks remain that cannot be mitigated. For example, overflying streets or infrastructure such as powerlines may be considered an operational risk that cannot be avoided completely. However, such inevitable risks can be minimized, for example by crossing small streets rather than busy highways and by overflying these streets perpendicularly to reduce the exposure time. In certain situations, remaining risks may be reduced by a trade-off. For example, in strong winds it may be advisable to reduce altitude in order to ensure that the aircraft would stay within its designated airspace. The airspace may be surrounded by a safety buffer, which is ensures sufficient distance between the aircraft and to be protected people or property. The immediate grounding of the vehicle or flight termination is a mitigation measure to contain the vehicle within the safety buffer in case of unintended flight behavior. Although, terminating from lower altitudes may lead to higher impact velocities and increase the harm or damage potential. Peinecke et al. (2017) further analyze risks and challenges involved for low-altitude unmanned flights as envisioned in the ALAADy project.

In general, all knowledge available a priori (e.g. airspace structure, terrain and known obstacles, street maps etc.) should be considered for pre-flight trajectory planning. However, certain information may not be available before flight and certain events may occur unexpectedly during flight. For example, a change in wind conditions may have to be accounted for by increasing safety buffers along the trajectory. Another example is high-priority traffic crossing the flight path ahead which may have to be encountered with a mitigation procedure involving a change of the planned trajectory. The ability to re-plan the trajectory online and during flight would provide alternative means of mitigation before having to resort to contingencies.

Planning and optimizing trajectories is a complex task. Automating this task requires equally complex mathematical models and databases that describe the aircraft and its environment as well as efficient algorithms. In the literature, a broad variety of approaches have been described to solve this problem. In this work, we focus on a prevalent concept for runtime efficient motion planning: The combination of a global sampling-based planning algorithm and a trajectory generation algorithm acting as a local planner. Figure 1 shows a simplified functional architecture of this approach: For a given mission, a global planning algorithm plans and optimizes waypoints along the route to the destination. A trajectory generation algorithm connects consecutive waypoints with safe and feasible trajectory segments based on an aircraft performance model. These precisely planned trajectory segments can be assessed in regard to their clearance towards terrain and obstacles as well as operational risks involved using an environment model. The result of the safety and risk assessment is fed back into the global planning algorithm, which may compare and select alternative waypoints or generate new ones to find an optimal route.

A key component of the risk-based planning procedure is the cost function used to assess and compare candidate trajectories. The risk assessment algorithm allows for quantifying specific operational risks. However, these risks have to be balanced and compared to each other. Also, a trade-off may have to be found between minimizing remaining risks and achieving the mission objectives in time. In this work, we use a multi-objective weighted cost function in order to be able to consider different risks in

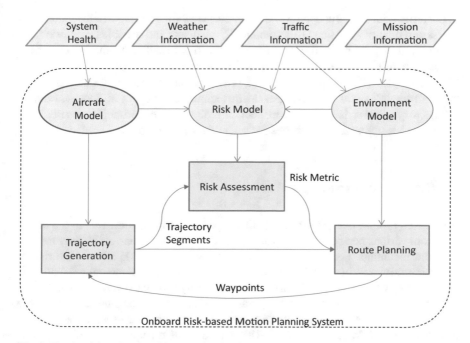

Fig. 1 Top-level functional architecture for risk-based motion planning as an onboard component of an autonomous unmanned aircraft

the planning procedure. In the context of the ALAADy, only two risks are considered: The *ground risk*, which is the risk of causing harm to people or infrastructure on the ground, and the *air risk*, which is the risk of causing harm to other airspace users. The mathematical definition of the risk-based cost function used in this work is based on the interpretation of risk as the product of the probability of an undesired event to happen and the harm that this event would inflict. Consequently, all risk cost are integrated over the flight time as the probability of system failures or other unexpected events that may result in uncontrolled flight is assumed to be proportional to the flight time. In order to be able to design the trade-offs mentioned above, the flight time is used as the minimum cost value which is increased based on risks values varying along the trajectory. Equation 1 shows the mathematical definition with the weighting factors for ground risk and air risk, K_{GR} and K_{AR}, and the quantified ground and air risk g_{GR} and g_{AR} which are normalized to the range [0, 1]. This formulation allows to design weighting factors in terms of multiples of the flight time. For example, by setting the ground risk weighting factor to 2, the additional cost due to ground risk would be limited to twice the flight time cost. In other words, a detour to avoid ground risks could triple the flight time and would still be considered more optimal with regard to the risk-based cost function

$$g(t_f) = \int_{t_0}^{t_f} \left(1 + K_{GR} \frac{g_{GR}(t)}{g_{GR,max}} + K_{AR} \frac{g_{AR}(t)}{g_{AR,max}} \right) dt. \tag{1}$$

In the remainder of this chapter, we give an overview over the algorithms required to assess and quantify risks and to automatically plan safe and feasible trajectories that are optimal or near-optimal with respect to the risk-based cost function described above. The following section gives an overview over related and previous work. In Sect. 3, we describe the processing and provisioning of geospatial datasets which lay the foundation for the risk assessment techniques laid out in Sect. 4. Finally, a trajectory planner is presented in Sect. 5 along with experimental results from planning risk-optimized trajectories for a set of ALAADy mission scenarios.

2 Related and Previous Work

Risk-based motion planning for unmanned aircraft has been addressed in the literature from various perspectives. Depending on the mission scenario, risk may take on different meanings. For example, Goerzen and Whalley (2011) address the risk of a mission failure in general. The risk of uncertainty in environmental perception is addressed in Chaves (2015). Nahrwold et al. (2019) and Babel (2013) consider the risk of a hostile threat. In this work, operational risks such as causing harm to people or infrastructure on the ground or endangering other airspace users are addressed. This is in line of Primatesta et al. (2018) or Rudnick-Cohen et al. (2016), however, we focus on the envisioned mission scenarios of the ALAADy concept.

The concepts and experimental results presented in this work are based on the authors' previous works on related subjects, especially the risk assessment approach described in Donkels (2020) and the sampling-based path planning framework presented in Schopferer and Benders (2020). The various aspects are brought together in this work and the algorithmic approaches are demonstrated in scenarios representative for the ALAADy vision. The remainder of this section gives a brief literature overview over the two key disciplines addressed in this work: risk modelling and sampling-based path planning.

2.1 Risk Modelling

Washington et al. (2017) give a broad overview of existing models to calculate the ground risk of UAS and introduce a way to decompose the different aspects of ground risk into several sub-models. Many available risk models can be associated with one or more of these sub-models, which allows for their categorization. Also, each risk model is associated with a level of uncertainty. Uncertainty results from the fact that an a priori prediction of the exact cause and consequences of the vehicle malfunction as well as the influence of environmental factors (e.g. wind, terrain influence) is impossible. The level of uncertainty allows classifying, if and how detailed uncertainty is recognized in assumptions and calculations.

Many risk models described in the literature, e.g. Bertrand et al. (2017), Aalmoes et al. (2015), la Cour-Harbo (2017), decompose the ground risk into three aspects: The fallibility of the system causing the malfunction, the spatial probability of hitting property or people at impact location, and the likely consequences of the damage on property or people as the severity of the impact. All three aspects can be modeled by a chain of dependent probabilities, which can be calculated separately and accumulated to a quantified risk. The following three paragraphs give an overview of literature for each aspect.

Fallibility of the system—This refers to the identification of possible failure modes and their influence on the system performance as well as their time-related probability of occurrence. The ability for failure recovery may also be reviewed. The majority of related literature assumes a constant failure rate for the overall UAS. Popular methods to assess the failure rate are:

- Analysis of available data obtained from historical events in manned aviation (see e.g. Aalmoes et al. (2015) and Lum and Waggoner (2011)) or tailored on unmanned aircraft (Belcastro et al. 2017). Also, data generated from hardware-in-the-loop simulations and experimental results can be used (Lum et al. 2011).
- Tailoring of available failure-mode analysis techniques on UAS (e.g. FMECA in Burke and Hall (2011)), System Engineering approaches as in Denney et al. (2012) or the use of Bayesian Networks to compute failure probabilities (Allouch et al. 2019).

- Target Level of Safety (TLS) approaches, which recognize failure rate requirements of available regulations and standards (Melnyk et al. 2014) or assume a
commonly accepted Fallibility (Wu and Clothier 2012). An accepted failure rate
may be defined by an expert commission, government decisions, or the society's
risk perception.

Impact location—This aspect covers the spatial part of the risk and models the probability for collision of the vehicle with a person or property on the ground. The
nature of the failure condition and the vehicle state in the moment of malfunction are
unknown a priori, so that the impact location is highly uncertain. Bertrand et al. (2017)
and la Cour-Harbo (2017) model the impact uncertainty using the bivariate normal
distribution to describe the impact probability. In Bertrand et al. (2017) the distribution is obtained from the vehicle's performance parameters, such as the maximum
gliding distance. La Cour-Harbo (2017) introduces normally distributed parameters
in simplified kinematic equations for a given flight termination event to create a
normal distribution around the termination point. The kinematic equations include
wind vectors which allow consideration of wind influence.

Rudnick-Cohen et al. (2016) and Lum et al. (2011) use Monte Carlo simulations
without wind influence to generate an impact distribution from a starting point, which
can be adapted to every point along a flight path to determine the impact probability
distribution.

The *impact location* is strongly connected to the *impact severity* as it determines
the probability of hitting a person or property and their exposure to the vehicle.
The probability of hitting a person or property naturally increases with the population density of the impact region and its land use. The connection with the population density is assumed in Harmsel et al. (2017) where the detailed distribution of
New York's population density is used to find risk minimal flight paths for small
drones. The proposed risk algorithm transforms the distribution into a weighted
graph, which can be searched by a path planning algorithm. The transformation is
done by multiplying the assumed area for safe landing with the local area-related
population density.

The exposure of people on the ground varies with the type of the area and reaches
from relatively low exposure in residential areas to high exposure in open landscape.
Castelli et al. (2016) use accurate datasets such as Google Maps, to identify safe and
unsafe flight zones for a small UAS. The flight zones are categorized according to
the exposure of people to the UAS. Thus, walkways, paths, and roads are considered
unsafe while buildings and free (unpopulated) spaces are treated to be safe.

Impact severity—The *impact severity* covers the actual consequence in terms of
damage created when the vehicle impacts on the ground. A common way of modelling
the Impact Severity is through the introduction of an impact fatality probability or
lethality. Here, the damage to property is neglected and fatal injury of a person as a
result of blunt trauma from direct collision, penetration of impact debris, or contact
with spinning parts like rotors is considered. As mentioned for the Impact Location,
exposure and density of people depend on the land use of the region in which the
impact occurs and have an influence on the impact fatality probability. Clare et al.

(1975) and the Range Safety Group of the Range Commanders Council (2001b) introduce functions for the probability based on the kinetic energy of the vehicle at impact and the existence of sheltering objects for the people. Objects like houses and walls are able to absorb kinetic energy and thus lower the impact fatality probability.

Another way of obtaining the fatality probability is through the calculation of the area around the impact location, in which lethal injuries may occur. The relation of this area to the population density yields the fatality probability. Lum et al. (2011) calculate the affected area from the geometry and the gliding angle of a gliding vehicle to determine the area in which a collision with a person is possible. Stevens et al. (2012) provide a fatality probability obtained from a statistical analysis of crash area data points. The probability results form an empirically determined correlation between impact area and the weight of the impacted vehicle. Here, the probability is a ratio of the total number of fatalities and the number of persons present in the crash area.

Melnyk (2014) assess the fatality probability by introducing an event tree and mapping the ends of the tree with probability numbers obtained from applicable literature. For instance, the vehicle impact into a house is modeled based on an earthquake fatality analysis.

2.2 Sampling-Based Path Planning

Planning low-altitude flight paths that are optimal in respect to a given cost function is a complex optimization problem. At low altitudes, terrain and obstacles form an unstructured environment which can hardly be modeled mathematically. Also, risk-based cost functions that depend on geospatial data have a complex structure which makes it difficult to apply gradient-based optimization approaches. Under such conditions, sampling-based planning algorithms have proven to be of great value. The key concept is to place random or quasi-random samples in the configuration space, i.e. the manifold spanned by the degrees of freedom to be considered. Connections between neighboring samples are inspected to analyze the local connectivity of the configuration space and to evaluate environment models and cost functions locally. The set of samples and connections resemble a graph or network. This graph provides an abstract and discretized view on the solution space of the optimization problem. It allows solving the problem efficiently using graph search algorithms.

Although sampling-based planning algorithms have proven to be useful for solving complex motion planning problems, they are inherently suboptimal and incomplete, i.e. they may fail to find a solution although one exists or they may return spatially suboptimal solutions. This is due to the limited number of samples and the resulting limited resolution. To account for this, the terms resolution or probabilistic completeness and asymptotic optimality are used to identify algorithms that converge to the optimal solution with increasing number of samples and which are complete in respect to the limited sampling resolution. There are two main groups of sampling-based algorithms, some of which provide such guarantees: The rapidly-exploring random tree (RRT) and variants thereof on one hand and algorithms based on probabilistic roadmaps (PRM) on the other hand. The RRT algorithm is based on the concept of exploring the configuration space by incrementally building a search tree. The first variant proposed by LaValle (1998) was shown to be resolution complete. A well-known variant named RRT* was proposed by Karaman and Frazzoli (2010) and is asymptotically optimal. The PRM algorithm is based on the idea of a persistent cyclic search graph, also referred to as *roadmap*. It was first proposed by Kavraki et al. (1996) and has since been applied to numerous motion planning problems. An asymptotically optimal version, referred to as PRM*, was proposed by Karaman and Frazzoli (2011). In contrast to RRT-based algorithms, PRM algorithms allow for multi-query planning, i.e. the search graph built in a pre-processing phase may be re-used to solve multiple planning problems. For mostly static environments, where the graph does not have to be updated as a whole, this may result in significant runtime benefits compared to tree-based approaches because the evaluation of cost function and collision checks are performed only once.

Another aspect to consider for sampling-based planning is how kinematic or dynamic constraints are incorporated into the planning problem in order to ensure that the aircraft can safely follow the planned trajectory. With RRT-based algorithms, new samples may be produced by integrating a dynamic system with random inputs, thus, resulting in inherently feasible connections between samples (LaValle and Kuffner 2001). When building cyclic search graphs, a local planner is required that is capable of solving the two-point boundary value problem of connecting two given samples to each other. For PRM-like algorithms, respecting kinematic and dynamic constraints is the responsibility of the local planner. Independent of the algorithmic approach, increasing the dimension of the planning space demands for graphs with higher degree which results in an increase of computational complexity. For this reason, sampling-based planning is often restricted to 2 or 3 dimensions with samples representing waypoints that are connected by linear segments as described in Pettersson and Doherty (2006) and Adolf and Andert (2011). Smooth and dynamically feasible trajectories are then generated in a post-processing step based on the path of linear segments. Although computationally more efficient, this approach also requires a local planner that is capable of connecting consecutive waypoints with dynamically feasible trajectory segments.

Generating smooth and feasible trajectory segments based on simplified kinematic models has received attention in the past due to the low algorithmic complexity and good runtime efficiency. Most works are based on the 2D path planning scheme based on the Dubins vehicle model (Dubins 1961). Its extension to the 3D case with flight path angle constraints has been referred to as the Dubins airplane model (Chitsaz and LaValle 2007). In Niendorf et al. (2013) the vertical profile is calculated based on a double integrator system to allow for vertical acceleration constraints. In Techy and Woolsey (2009) the 2D case is extended to use trochoid curves instead of circular segments in order to account for the prevailing wind. Both extensions have been adopted in Schopferer and Pfeifer (2015) for trajectory generation in a sampling-based path planning approach. Even dealing with uncertainties and disturbances can be achieved with simple kinematic models. In Al-Sabban et al. (2013), uncertain wind information is incorporated into the planning procedure as a stochastic problem. In Wolek and Woolsey (2015), margins towards the turn rate constraint are defined that allow for rejecting disturbances and unexpected changes in wind condition. A similar approach, however, based on an adaptive safety distance is presented in Schopferer et al. (2018).

The PRM approach has been applied for runtime efficient path planning for unmanned aircraft in the past. In Adolf and Andert (2011) it is used to plan paths for an unmanned rotorcraft in a-priori unknown environments. In Niendorf et al. (2013) the same approach is applied to fixed-wing UAV path planning using trajectory generation with a Dubins-based kinematic model. This line of work is continued in Schopferer and Pfeifer (2015) and Benders and Schopferer (2017). The applicability of the PRM approach to long-range mission scenarios as considered in the ALAADy concept was shown Schopferer and Benders (2020). Building on this work, we present experimental results of a sampling-based minimum-risk path planning for exemplary mission scenarios in Sect. 5.

3 Geospatial Database for Trajectory Planning

Planning and optimizing trajectories requires knowledge about the mission environment. For example, to calculate the clearance towards terrain and obstacles, machine-readable maps or, more general, geospatial data is required with terrain elevation as well as position and extent of obstacles. Also, in order to assess the risk of a given trajectory, information regarding the harm potential of overflown areas and objects on the ground has to be known. Geospatial data forms the basis for both the risk model and the environment model as depicted in Fig. 1. In this section, we describe how a geospatial database for trajectory planning may be composed and how efficient interfaces to be used by planning and optimization algorithms are designed. The design and implementation of a database is based on previous work presented in Schopferer and Benders (2020).

3.1 Processing Geospatial Datasets

The geospatial datasets used in this work are freely available as so-called *open data*. The *EU DEM*[1] (digital elevation model) and the *CORINE Land Cover*[2] maps used in this work are provided by the *Copernicus* program. The land use maps provide a basis

---

[1] EU - Digital Elevation Model, Copernicus Land Monitoring Service, https://www.eea.europa.eu/data-and-maps/data/copernicusland-monitoring-service-eu-dem.

[2] CORINE Land Cover, Copernicus Land Monitoring Service, https://land.copernicus.eu/pan-europaean/corineland-cover.

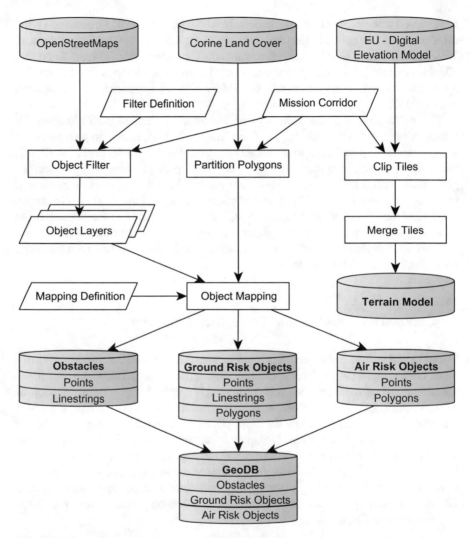

Fig. 2 Processing geospatial datasets into a database for trajectory planning (Schopferer and Benders 2020)

for assessing the ground risk with a set of 44 land use classes such as forests, fields, residential areas and more. The *OpenStreetMap*[3] dataset is a versatile, community-driven geospatial dataset from which a rich set of objects, for example power lines, wind turbines, towers, streets, waterways, airfields, and heliports, are extracted and stored in the database. Figure 2 depicts the process of database creation from input datasets. One of the key steps in this process is the mapping from objects of certain semantic classes to objects useful for trajectory planning, i.e. obstacle, ground risk or

---

[3] OpenStreetMap, https://www.openstreetmap.org/.

air risk objects. Required parameters that are not available in the input dataset must be derived based on the semantic class of the object. For example, the height of a wind turbine may not be specified and is set to a conservative value, i.e. the maximum height of wind turbines of the given type. Another example for a parameter that may have to be derived is the width of a highway for which only the number of lanes is given. These examples illustrate the challenges in working with geospatial datasets in the context of trajectory planning.

3.2 Calculating Horizontal and Vertical Clearance

In order to calculate the clearance of trajectories against terrain and obstacles and the distance to ground and air risk objects, interfaces have to be implemented to efficiently query the database and process the result. Traditionally, common vector data formats supported by geospatial information systems (GIS) are comprised of objects of three basic geometric types: points, line strings, and polygons. However, for trajectory planning, a 3D environmental model is required in order to represent the geometries of obstacles or air space boundaries. While 3D collision checking and distance calculation has been widely adopted in computer graphics and games, it is a computationally intensive task. Therefore, we propose a hybrid approach that allows storing objects as 2D geometric primitives as supported by GIS software frameworks while still having objects represent 3D geometries using additional attributes. The approach is based on the concept that 3D geometries may be extruded vertically from their 2D geometric primitives by a height attribute and buffered horizontally by a radius or width attribute. Table 1 shows the mapping from 2 to 3D geometries using additional attributes. Figure 3 shows some examples of objects represented by 3D geometries.

Instead of calculating the minimum 3D distance to objects in the database, a horizontal and vertical clearance metric is assessed as follows:

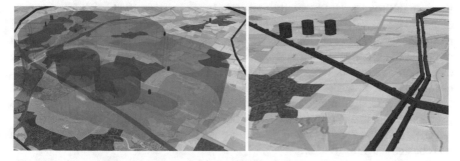

Fig. 3 3D representation of objects in the geospatial database: Cylindric obstacles (e.g. wind turbines), wall-shaped obstacles (e.g. power lines), extruded polygons (e.g. restricted airspaces)

Table 1 2D geometries and
3D representations for objects
in the geospatial database

2D Geom.	Attributes	3D representation
Point	Radius, height	Axis-aligned cylinder
Line string	Width, height	Extruded wall
Polygon	Height	Extruded polygon

1. The database is queried for objects within a search radius from a given position. This query is 2D only. Therefore, the horizontal extension of objects beyond their corresponding 2D geometric primitives has to be added to the effective search radius.
2. The horizontal clearance to each object is calculated as the distance in the horizontal plane. If the given position lies within the horizontal projection of the object, the distance is set to zero.
3. The vertical clearance to each object is calculated based on the local terrain elevation, the object's height and the given position's altitude.

If either the vertical clearance or the horizontal clearance is above the required thresholds, then the position is defined as having sufficient clearance towards the set of objects considered. In contrast to calculating the 3D minimum distance to objects, this approach is less computationally intensive with only a 2D Euclidean distance calculation in the horizontal plane and scalar operations for the vertical axis. Furthermore, the notion of distinct horizontal and vertical clearances is meaningful in the context of path planning for unmanned aircraft as flying above objects may have different implications than flying around obstacles.

3.3 Assessing the Harm Potential of Overflown Ground

A vital part of risk assessment is to create knowledge about harm arising from mishaps or system failures during the vehicle's operation. In general, risk can be defined as the probability of a damage event multiplied with its quantified harm (Christensen et al. 2003; SAE 2010; International Programme On Chemical Safety 2004). Typical manifestations of harm are injury or death of people or costs for repair or replacement of damaged property. The harm is depending on the nature or severity of the damage event. For example, the repair costs of a house may differ greatly for different damage events like flooding, fire, or stormy wind, but are correlated with the value of the house. For this simple example, three aspects of risk need to be addressed: The probability of damage occurrence (e.g. floodings per year), the damage severity (e.g. water level of the flooding), and the harm potential (e.g. the value of the house and its robustness against flooding).

Within the proposed risk cost functions, the determination of damage probability and severity is done by the used risk models. The harm potential shall be provided by the database. A major challenge is to find and transform available information about the overflown area into semantic data for risk aware planning. We define semantic data as quantified information, which can serve different kinds of requests, so that different risk models are able to work with the same data. In other words, it shall be avoided to build a database for fire damage, flooding damage, and storm damage, which all contain the same house. Instead, the house shall be modeled generally enough, so that fire model, flooding model, and storm model can calculate the harm potential based on a single dataset.

In a first approach, available land use data from *CORINE land cover* maps was extended to hold information about the local harm potential. The maps consist of polygons, which represent the land use of the covered area. Each polygon consists of at least three non-collinear points and is assigned with a specific land use class. Based on the class, a damage value was added to quantify the potentially endangered property values or human lives on the ground. For proof of concept, a normalized damage value is assigned, based on logical assumptions. The assignment of values to the land cover classes is displayed in Table 2.

Urban areas with high populations and close distribution of property, like houses or critical infrastructure were rated with high damage values. The damage value decreases over rural to uninhabited areas. These areas still have a significant damage value, which results from the resolution of the land cover data (single houses in a farming area are considered farmland) and the fact that hikers or workers may temporarily occupy a natural area. Possible ways to overcome this may be to increase the map resolution, for instance through AI-processing of satellite images or the use of population density distributions, which allows more accurate estimations of the damage value for inhabited areas. Outside inhabited areas, groups of people might

Table 2 Assignment of normalized damage value to available land cover classes

Land cover class	Damage value
Urban areas	100
Urban green areas and sports ground	85
Industrial and traffic areas	80
Construction sites and landfill	70
Permanent crops	30
Inland areas of water	25
Open spaces	22
Heterogeneous agricultural land	20
Inshore moisty areas	12
Arable areas	10
Marine waters	8
Forests	7
Inland moisty areas	6
Grassland	5
Scrub vegetation	4
All other classes	0

Table 3 Assignment of normalized damage value to different transport routes

Transport routes	Damage value
Motorway	80
Trunk road	80
Primary road	70
Secondary road	50
Canal	40
Railway	30
River	25

be recognized through image processing onboard the vehicle or the tracking of cell phone position data.

The *CORINE land cover* maps provide gapless coverage of entire countries in Europe, but do not consider transport routes, such as roads or rivers, which lie inside and lead across the land cover polygons. The transport routes in between densely populated regions are usually very busy and create a local and very sharp increase of assets (cars, trucks, ships) or human lives (drivers and passengers). Thus, transport routes from *OpenStreetMap* were reclassified into the seven groups shown in Table 3 and also assigned with a damage value according to the expected utilization and traffic density. Utilization and traffic density may vary during each day; thus, a possible improvement would be the introduction of a daytime depending factor to the assignment of damage values.

Another important aspect of harm potential is the exposure of the people or property to the damage event. People in urban areas inside their homes might be less exposed to a crashing vehicle, than on the open land without any sheltering objects. The risk cost function for flight termination, which is presented in Sect. 4.5, is able to recognize the exposure of the people on the ground. Thus, each polygon was assigned with values describing the local exposure of people as well as the presence of sheltering objects and their protective ability. The values are chosen to be the same for all polygons and future work will focus on ways to adjust the values for each polygon individually.

4 Risk Assessment for Safe Flight Termination

4.1 Definition of Flight Termination

The main focus of science, engineering, rulemaking, and certification efforts lies usually on the continuous improvement of safety and reliability of aircraft and spacecraft systems (Oster Jr. et al. 2013). Thus, it is the final aim to keep flying under the worst imaginable circumstances and even beyond them. The term of flight termination seems to seek for the complete opposite, by finding the easiest and most robust way to end a vehicles flight, which embraces the destruction and total loss of the vehicle. This seemingly contradiction can be solved by viewing the air transportation as a whole system. Then the system embraces damage done to parts other than the aircraft itself that needs to be considered when talking about the system's fallibility. Flight termination systems (FTS) are well known and widely used as an integral part of flight safety systems (FSS) of rockets or spacecraft. Their sole purpose lies on containment of the vehicle's threat to the environment as a last resort option, e.g. by self-destruction of a rocket to prevent it from carrying its highly explosive fuel or payload towards inhabited areas, if a malfunction causes it to leave its pre-planned trajectory. To ensure containment of an uncontrolled vehicle is only one reason to terminate a vehicle's flight. Also, an otherwise not avoidable mid-air collision may be mitigated by flight termination. The ability to sacrifice the unmanned vehicle for the sake of protection of people on the ground or seated in other aircraft is not only applicable on rockets, but it also provides a protection function for the unmanned air transport (Range Safety Group, Range Commanders Council 2001b; Sgobba et al. 2013).

The activation of flight terminating measurements can be divided into two main functions of the system. The first function is the ability of fast and reliable detection of abnormal flight situations, followed by activation or deployment of flight termination measurements. Abnormal flight behavior is defined by the European Aviation Safety Agency (2017) and may be the consequence of wrong sensor signals or sensor signal loss, such as deteriorated GNSS, or the loss of control over critical flight control functions affecting the controllability of the vehicle. The delay between first occurrence of abnormal behavior and activation of termination measurements, the termination reaction time, should be kept as short as possible.

The detection of abnormal behavior may involve human factors like a remote pilot or operator, who monitors transmitted data from the vehicle. In case of abnormal behavior, flight termination is activated manually from a control station via datalink. Systems used for manual flight termination are commonly available and follow design standards like IRIG Standard 331–01 (Range Safety Group 2001a). The human factor may be replaced by the use of Autonomous FTS (AFTS), capable of automated monitoring onboard the vehicle. AFTS may allow shorter termination reaction times and increased rates of detection of abnormal flight behavior. AFTS are current research topics as presented by Bull and Lanzi (2008) and also in Schirmer and Torens (2021).

A vehicle may be equipped with one or more flight termination measurements, according to its design and fitted to its mission and performance parameters. Sachs (2021) and Schirmer and Torens (2021) give examples of flight termination measurements investigated in ALAADy. The following subsection complements the examples with a list of superficial design and functional requirements for these measurements.

4.2   Design and Function of Flight Termination Systems

NASA technical standard STD-8719.25, Range Safety Commonality Standard RCC 319-19 (see National Aeronautics and Space Administration (2018), and Range Safety Group, Range Commanders Council (2019) respectively) provide up-to-date FTS requirements and design specifications for unmanned aircraft. The little available experience with FTS in aircraft systems and their rulemaking contrasts the wide availability of documents and systems within the rocket and space domain. Contrary to rockets, the majority of aircraft are not designed to be expendable. The aim to protect the lives of passengers and crew aboard naturally excludes the use of FTS. Air frame parachute systems, such as the Cirrus Airframe Parachute System (CAPS), may provide FTS capabilities but are designed to protect or rescue the passengers on board. So far, there is no comparable FTS framework of standards and rulemaking for aviation. However, the holistic risk assessment approach SORA (Specific Operations Risk Assessment) by the JARUS (Joint Authorities for Rulemaking on Unmanned Systems) defines the term flight termination as a system, procedure, or function that aims to immediately end the flight (JARUS 2017).

A deeper understanding of FTS function gives the following set of top-level performance requirements obtained from RCC 319 and suited to ALAADy.

The FTS shall perform the following:

- Ensure the flight-terminated vehicle's debris impact, resulting from residual lift or drift under worst-case wind conditions, will not endanger any protected area.
- When termination is initiated, it shall be irrevocable.
- Render each propulsion system that has the capability of reaching a protected area. This includes each stage and any strap-on motor or propulsion system that is part of any payload.
- Result in aerodynamic control surface manipulation or deployment of aerodynamically acting surfaces (e.g. parachutes) that make a vehicle unable to glide.
- FTS termination action shall reduce the size and severity of damage to the vehicle's impact area
- FTS termination action shall not detonate solid or liquid fuel or propellant.

The FTS aboard unmanned aircraft or spacecraft are commonly used to ensure spatial separation of the vehicle to inhabited areas on the ground and to other aircraft in the air. Spatial separation may be implemented by defining an allowed zone for the unmanned aircraft, the *flight geography*, which separates the airspace used by the unmanned vehicle from the environment to be protected. Within the *flight geography*, flight trajectories may be planned and executed by the vehicle. The delay caused by the termination reaction time and the generally not instantaneous effect of flight termination measurements may not prevent the vehicle from leaving the *flight geography*. Thus, a buffer zone between the *flight geography* and the protected area becomes necessary. According to the two stage process of FTS activation as introduced in Sect. 4.1, the buffer zone may be divided into a contingency buffer and

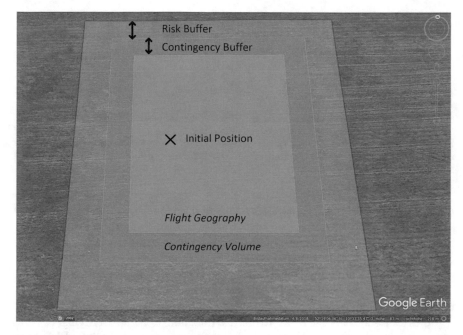

Fig. 4 Geofencing for rectangular flight geometry

a risk buffer. The detection and activation of termination measurements is performed within the contingency buffer leaving the risk buffer as a space for the measurements to act and to ground the vehicle. The dimensions of risk and contingency buffer are generally depending on the vehicle's specifications and the missions performed within the *flight geography*, but may be designed to fit other criteria. For instance, an advice introduced in (European Aviation Safety Agency 2018) for the scaling of buffer zones is the so called 1:1 rule. The 1:1 rule states that the horizontal distance to protected areas must equal or exceed the height of the unmanned aircraft above ground level. This rather simple rule does not include wind conditions or the aircraft's ability to glide, which could cause the vehicle to leave the buffer zone.

Figure 4 visualizes the typical setup of flight volumes for a delimited and restricted area, like missile or unmanned aircraft test ranges. These test ranges are usually areas of uncultivated ground with restricted access to the public. The limits of the *flight geography* and buffer zones are usually fitted to the size of the restricted area. The size of the area itself is usually chosen to allow proper testing of missiles or aircraft.

4.3 Flight Termination Within ALAADy

The ALAADy concept of unmanned air transportation seeks for operation across the country to obtain flexible and fast connections between shipper and recipient of the

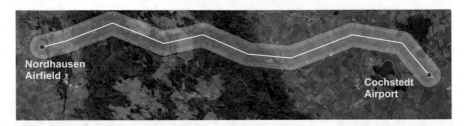

Fig. 5 Exemplary Flight Geometry (green area) and buffer zones (orange and white areas) are defined by the transport route (white line), which connects the airfields Nordhausen and Cochstedt in central Germany

transported goods. Thus, the aircraft is operated outside restricted areas and follows a path planned above sparsely populated areas (villages, farms, isolated houses). Figure 5 visualizes this difference and reuses the same coloring from Fig. 4 to indicate the flight volume and safety buffers around the portrayed example path. The path is planned to ensure minimal interference between the vehicle and the people on the ground or within other aircraft, which share the same airspace. Due to the dense transport infrastructure of most countries, it might occur that the aircraft frequently overflies transport routs like roads, rivers, and train tracks. While overflying, vehicles and vessels containing people and valuable freight are exposed to the vehicle for a limited time. FTS activation could cause the vehicle to impact on or close to the transport route causing direct damage at the impact location and indirect damage through startled drivers or compensations for delayed traffic.

The main function of FTS aboard the ALAADy vehicle is to ensure that the vehicle does not leave the sparsely populated area, thus threatening a densely populated area or an overly exposed group of people, such as occurring in stadiums or open air festivities. The flight termination measurements in ALAADy consist of a cutoff of thrusters and propulsion and a secondary measurement, which depends on the vehicle configuration. The fixed and box-wing configurations use airframe parachutes, while the gyrocopter performs an autorotation sequence (see Sachs 2021). The parachute and autorotation measurements were chosen, because both measures allow significant reduction of impact energy.

The major challenge of FTS activation arises when the aircraft approaches or passes a road or the outskirts of a village. In this case, the buffer zones from Fig. 5 might overlap with the road and the inhabited area. Now, the immediate grounding of the vehicle protects more densely populated regions, but produces the chance of harm for people living or driving within the buffer zones. The chance for FTS activation connected with the severity of consequences from the vehicle's impact on the ground form the risk to be assessed by the risk assessment for flight termination.

The risk assessment presented in this work focuses on the case of successful FTS activation only. Thus, only the system failures directly leading to this case will be considered separately from other failures like loss of transport goods, unsuccessful FTS activation or mid-air and terrain collision. The damage event of FTS activation is the impact of the vehicle on the ground leading to consideration of ground risk only. The term "impact" means that a vehicle of several hundred kilograms or several tons, as considered in ALAADy, will always lead to damage of people or property on the ground, even if the impact velocity is very small. The aim of ALAADy is to reduce the probability for damage and the severity of the damage to a minimum or at least to an acceptable level. The here addressed way is the introduction of a risk metric to quantify the probability and severity of the damage. The metric requires a description of the covered flight termination event and its consequences from which a risk model can be derived. The risk model can quantify the risk when applied to a concrete flight termination event. As part of a risk assessment, the risk model can be used to quantify the risk of a concrete flight path or within the path planning stage of a mission to identify risk-minimal paths.

4.4 Heuristic Risk Assessment

In order to quantify the risk of a given trajectory, a heuristic approach considers only the most prominent aspects affecting the risk. It allows for quickly approximating the risk of candidate trajectories with limited computational resources as required for runtime efficient planning. Given the assumption that any loss of control of the aircraft will result in a flight termination, the key aspects to consider for the risk quantification are the extent and the harm potential of the impact area.

Without further differentiating the probability of impact locations and corresponding harm potentials within the impact area, as described in the subsequent section, an approximation of the risk is obtained by considering only the outer bounds of the impact area and the maximum harm potential within it. In order for this approximation to produce conservative risk values, the extent of the impact area must also be over-estimated. The most important aspects affecting the reachability of impact locations from a given position along a flight path are the wind conditions, the aircraft's altitude, airspeed and heading and the dynamics of the aircraft during the flight termination.

The aforementioned 1:1 rule described by the European Aviation Safety Agency (2018) represents a simple and straight-forward approach to calculate the extent of buffer zones depending on the height above ground. It can be directly applied for approximating impact areas along a given flight path as shown in Fig. 5. The 1:1 rule is based on the assumption of a minimum angle-of-descent of 45°. While this may be a good approximation in many cases, it certainly cannot be considered a general worst-case approximation. By reducing the assumed minimum angle-of-descent, the resulting area will increase and the approximation of the impact area will become more conservative, i.e. it will cover more cases of wind conditions, initial flight states, aircraft configurations and flight termination systems. Hence, with this simple method, the problem of acquiring a sufficiently conservative over-estimation of the impact area can be reduced to selecting an appropriately small angle-of-descent.

It is important to emphasize, that the heuristic risk assessment described here only provides rough estimates of the true risk. The first systematic error lies in neglecting the probability of reaching specific impact locations within the impact area, i.e. the impact locations are assumed to be equally distributed. A second systematic error results from considering only the maximum harm potential within the impact area in order to over-approximate the risk. Despite these systematic errors, the heuristic approach may be useful for runtime efficient risk approximation, as it may be implemented in terms of simple geometric calculations and efficient database queries. This facilitates integration with sampling-based path planning algorithms. A more accurate but also computationally more intensive approach to risk modelling is described in the following section.

4.5 Bayesian Based Risk Modelling

This section introduces a newly developed cost function for risk aware path planning with flight termination. The risk model of the cost function recognizes the three aspects of risk *fallibility of the system, impact location*, and *impact severity* which were introduced and discussed in Sect. 2.1. The *fallibility of the system* as well as the *impact severity* are modeled after established methods from Wu and Clothier (2012), and the Range Commanders Council (2001b) respectively. For the aspect of *impact location*, a new method was developed with the aim to use the information provided by the geospatial database. The interaction of the *impact location* model with the database is shown in Fig. 6. The new *impact location* model is described after the following introduction of the cost function.

Risk Cost Function
Assuming a flight trajectory τ segmented into a set of N points, a point $n \in [1 \dots N]$ can be evaluated by a cost function $c(n)$ to obtain the risk for flight termination for this point. The risk costs $C(\tau)$ of the considered trajectory are the aggregation of the point costs

$$C(\tau) = \sum_{n=1}^{N} c(n). \tag{2}$$

The risk cost function $c(n)$ is obtained from

$$c(n) = P_{CF} T_n D_{Impact}(n) \tag{3}$$

With P_{CF} as the time related *fallibility of the system*, T_n the flight time to pass point n, and $D_{Impact}(n)$ as the damage consequence of flight termination at point n. The model for *impact severity* from the Range Commanders Council (2001b) recognizes the sheltering or exposure of people on the ground (parameters α, β, and p_s) and calculates the lethality of people hit by a moving object. The exposure and

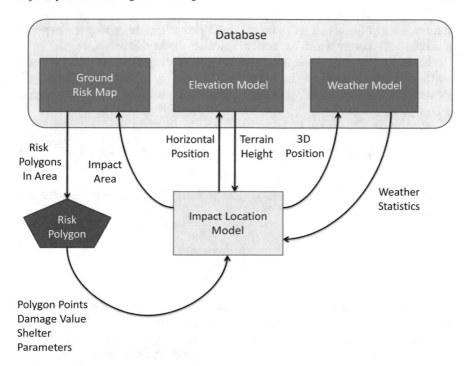

Fig. 6 Interaction of the impact location model with the database

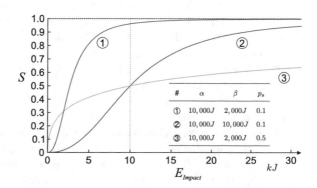

Fig. 7 Impact Severity for different shelter and exposure parameters

#	α	β	p_s
①	$10,000J$	$2,000J$	0.1
②	$10,000J$	$10,000J$	0.1
③	$10,000J$	$2,000J$	0.5

sheltering varies with the type of the overflown area and is depending on the *impact location*. Thus, the *impact severity* and the *impact location* are calculated together and combined within $D_{Impact}(n)$.

The exposure as well as the damage potential of the overflown area can be obtained from the geospatial database, which holds a polygon map of the planning area (see Fig. 6). Polygons are defined by a set of at least three non-collinear points within a geographic or local Cartesian coordinate system. Each polygon is assigned with a normalized damage value w_{Damage}, which represents the damage potential. In addition, each polygon can be associated with parameters describing the exposure of people and property which are used to calculate the *impact severity*.

The uncertainty of the *impact location* is modeled as spatial probability which is used to calculate and assign an impact probability p_{Impact} to each available polygon. For each polygon the individual damage consequence is calculated and aggregated to $D_{Impact}(n)$ using

$$D_{Impact}(n) = \sum_{Polygons} w_{Damage}\, p_{Impact}\, S(\alpha, \beta, p_s, E_{Impact}), \tag{4}$$

with S as the *impact severity* function (Range Safety Group, Range Commanders Council 2001b)

$$S = \frac{1}{1 + \sqrt{\frac{\alpha}{\beta}\left(\frac{\beta}{E_{Impact}}\right)^{\frac{1}{4p_s}}}}. \tag{5}$$

Here, α, β, and p_s are parameters for exposure and shelter of people on the ground and E_{Impact} is the kinetic energy of the vehicle at impact. From the Eq. 5, p_{Impact} and E_{Impact} are unknown and are calculated by the *impact location* model. Figure 7 shows the *impact severity* function for different shelter and exposure values.

The shelter parameter p_s refers to the average sheltering of people to the impacting vehicle (no shelter: $p_s = 0$, maximum shelter $p_s = 1$). The parameters α and β quantify the minimal kinetic energy to lethally hit an unprotected person (β) and the kinetic energy for a lethality of 50% if the person is sheltered by a factor of $p_s = 0.5$. Thus, Graph 1 shows the impact severity value for a rather exposed person. The impact of β on the severity value is shown in Graph 2, where the robustness of a person was increased (e.g. by wearing protective gear). Graph 3 contrasts Graph 1 through an increased shelter value.

Impact Location Model

The *impact location* model bases on two functions necessary to model the kinematical behavior of the terminated vehicle and the vehicle's height above ground to determine the time of impact of the vehicle. To describe the vehicle's kinematics after FTS activation, a time-discrete nonlinear equation system f is assumed. The difference equation f calculates the vehicle state $\vec{x}$ for the next time step t from the previous state at $t-1$ as well as control inputs $\vec{u}$ and uncontrollable inputs $\vec{z}$ (e.g. wind):

$$\vec{x}^t = f(\vec{x}^{t-1}, \vec{u}^{t-1}, \vec{z}^{t-1}). \tag{6}$$

The vehicle state $\vec{x}$ must at least contain the spatial position of the vehicle as local Cartesian coordinates (x_g, y_g, z_g), and the velocity of the vehicle $\vec{V}$, which is needed to calculate the kinetic energy E_{Impact}.

The height function H is used to obtain elevation data from the geospatial database and returns the elevation h (see Fig. 8) of the terrain below the vehicle

$$h = H\left(x_g^t, y_g^t\right). \tag{7}$$

In the following, the vehicle kinematics are used to propagate a normally distributed covariance $\boldsymbol{P}$ of the vehicle state $\vec{x}$ from the point of flight termination to impact. With the position of the vehicle as part of the vehicle state, it is possible to obtain the spatial probability of the vehicle position from the covariance $\boldsymbol{P}$. The spatial probability of the vehicle at impact is then used to calculate the impact probability p_{Impact} for the polygons.

For a time-discrete linearized system, Bayes theorem states that the covariance of the next state solely depends on the previous state

$$P\left(\vec{x}^t\right) = \frac{P\left(\vec{x}^t \mid \vec{x}^{t-1}\right) P\left(\vec{x}^{t-1}\right)}{P\left(\vec{x}^{t-1} \mid \vec{x}^t\right)}. \tag{8}$$

Bayes theorem applied on f, which, if f is time-variant, is linearized to

$$\vec{x}^t = A^{t-1}\vec{x}^t + B^{t-1}\vec{u}^t + L^{t-1}\vec{z}^t \tag{9}$$

with

Fig. 8 An example vehicle with parachute terminates the flight around evaluation point n around the trajectory and follows a termination path. The picture shows the position of the vehicle and the terrain elevation with respect to a local cartesian reference frame

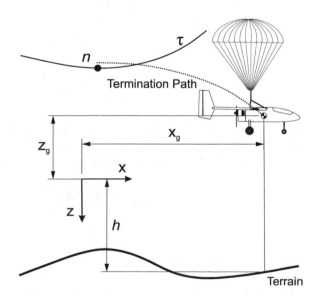

$$A^{t-1} = \left.\frac{\partial f}{\partial x}\right|_{\vec{u}^{t-1},\vec{z}^{t-1}} \qquad B^{t-1} = \left.\frac{\partial f}{\partial u}\right|_{\vec{x}^{t-1},\vec{z}^{t-1}}$$

$$L^{t-1} = \left.\frac{\partial f}{\partial z}\right|_{\vec{x}^{t-1},\vec{u}^{t-1}}$$

(10)

yields

$$P^t = A^{t-1}P^{t-1}A^{t-1^T} + B^{t-1}U^{t-1}B^{t-1^T} + L^{t-1}Q^{t-1}L^{t-1^T}.$$

(11)

This time-variant formulation is also used in the prediction step of an extended Kálmán filter (Kálmán 1960). The matrix U represents uncertainty of control inputs after flight termination and matrix Q represents uncertainty of uncontrollable inputs, such as external disturbances like wind and turbulence. The design of Q offers possibilities to use statistical weather data of the planning area, which enables planning with different wind situations.

The initial covariance P^0 contains the assumed normally distributed uncertainty of the vehicle's state around $\vec{x}^0$ in the moment of flight termination. The *impact location* model does not consider specified termination reasons. Instead it is assumed that violation of the Flight Geography or violation of operating constraints would always lead to FTS activation. Thus, the exact vehicle position in the moment of FTS activation is assumed to be around the evaluation point n on the trajectory and inside the Flight Geography. Thus, the initial state $\vec{x}^0$ is obtained from the evaluation point n. Other kinematic variables, such as the velocity, are assumed to be inside the operating constraints. If kinematic variables are modeled to be independent and uniformly distributed within the operating constraints, they can be easily transformed using

$$\sigma_{normal} = \frac{(Max - Min)^2}{12}$$

(12)

With σ_{normal} as the variance and *Min* and *Max* as the lower and upper limit of the variable, respectively.

The covariance P contains the covariance for the vehicle's position P_p, which is

$$P_p = \begin{pmatrix} \sigma_{xx} & \sigma_{xy} & \sigma_{xz} \\ \sigma_{xy} & \sigma_{yy} & \sigma_{yz} \\ \sigma_{xz} & \sigma_{yz} & \sigma_{zz} \end{pmatrix}.$$

(13)

The position obtained from $\vec{x}^t$ and the covariance P_p^t represent the expected position of the vehicle and the uncertainty distribution at time step t. With the forward calculation of time steps, the vehicle's height above ground h will decrease, because of the FTS action, while the uncertainty of the position increases. The forward calculation is stopped when the vehicle is considered to be impacted. Because of the uncertainty of the position, the criteria $z_g \geq h$ is not sufficient, as z_g refers to the most likely position. Thus, there is a chance that the vehicle is still airborne. Instead, the covariance of the position is evaluated in z-direction for the interval $[h, \infty]$. The according probability P_I is calculated by the multivariate cumulative distribution function:

$$P_I^t = \int_{h^t}^{\infty} \int_{-\infty}^{\infty} \int_{-\infty}^{\infty} P^t(x, y, z) dx dy dz. \tag{14}$$

If P_I satisfies a defined threshold, the vehicle is considered to be impacted. The horizontal part of the position covariance at the time step at impact and the horizontal vehicle position obtained at $z_g \approx h$ form the impact uncertainty distribution[4]

$$\vec{\mu} = \left(x_g y_g \right)^T \tag{15}$$

$$\Sigma = \begin{pmatrix} \sigma_{xx} & \sigma_{xy} \\ \sigma_{xy} & \sigma_{yy} \end{pmatrix}, \tag{16}$$

with $\vec{\mu}$ as the expected value and Σ as the covariance matrix.

The impact uncertainty distribution can now be used to calculate the impact probability p_{Impact} for each polygon, using the method from DiDonato and Hageman (1980). For the calculation of E_{Impact} the vehicle's expected velocity at $z_g \approx h$ is used.

The variances from covariance Σ and the expected value $\vec{\mu}$ can be interpreted as an ellipsoid. The eigenvalues are the major and minor axes and the corresponding eigenvectors represent the orientation of the axes. For $\sigma_{xy} = 0$ the ellipsoid is axis aligned. The axes of the ellipsoid can be scaled by factor k to transform the ellipsoid into a confidence region according to a given confidence level p_L. The confidence level p_L is the probability for the impact of the vehicle inside the confidence region. The scaling factor k is calculated according to Hoover (1984)

$$k = [-2ln(1 - p_L)]^{\frac{1}{2}}. \tag{17}$$

The confidence region allows the visualization of simulation results as well as the selection of relevant polygons to be recognized by the cost function. For the polygon selection, only polygons which intersect with the confidence region are selected and all others neglected, which decreases the computational effort.

Implementation and Simulative Results
For demonstration and validation of the method, a simplified example aircraft with parachute termination system was assumed and modeled and used for the planning of an example transport mission. The setup is further described in Donkels (2020). The terminated vehicle is modeled as point mass m of 25 kg without thrust or control

---

[4] The most likely impact spot and impact energy E_{Impact} are reached when the propagated vehicle position touches the ground.

surfaces with deployed parachute. The parachute influence is modeled with time-variant drag-surface coefficient $(C_D S)_{Para}$, which recognizes the parachute opening delay and air density ρ. The parachute model is mainly influenced by the wind, which is recognized by the variables $u_{W,g}$, $v_{W,g}$, and $w_{W,g}$. The state vector and the state equations are:

$$\vec{x}^t = \left(x_g^t, y_g^t, z_g^t, V_{Aa}^t, \gamma_A^t, \chi_A^t \right)^T \tag{18}$$

$$\dot{x}_g = V_{Aa} \cos(\gamma_A) \cos(\chi_A) + u_{W,g} \tag{19}$$

$$\dot{y}_g = V_{Aa}\cos(\gamma_A)\sin(\chi_A) + v_{W,g} \tag{20}$$

$$\dot{z}_g = -V_{Aa}\sin(\gamma_A) + w_{W,g} \tag{21}$$

$$\dot{V}_{Aa} = -g \sin(\gamma_A) - \frac{\rho}{2m}(C_D S)_{Para}(t)V_{Aa}^2 \tag{22}$$

$$\dot{\gamma}_A = \frac{g}{V_{Aa}}\cos(\gamma_A) \tag{23}$$

$$\dot{\chi}_A = 0. \tag{24}$$

Benchmark Scenarios and Demonstration

Several benchmark scenarios have been created to evaluate the behavior of the new cost function. Two of these scenarios are discussed in the following. The planning task of the first scenario is to fly over a highway, which is modeled as a line of 60 m width (including safety buffers) and a risk value of 100%. The surrounding environment is assumed to have no risk, thus $w_{Damage} = 0\%$. The task is to fly from an initial position to a diagonally oriented target across the highway. For the benchmark scenarios and the following demonstration, the path planner introduced in Sect. 5 was used. As a reference to the risk aware path, a time optimal path was also created. Contrary to the time-optimal path which leads from start to goal in a direct line, the risk aware path crosses the highway at an almost perpendicular angle.

Also, the chosen altitude of the risk aware path is much higher, which allows the parachute to open and slow down the vehicle. The costs for the risk aware path are 4.6×10^{-9}, while the time optimal path, evaluated with the risk cost function, yields 1.9×10^{-7} and is more than forty times lower.

In a second benchmark scenario, a square shaped region of 100% damage value has been placed between initial and target position. Again, the surrounding environment has a damage value of 0%. In this scenario, three paths were planned, which are displayed in Fig. 10. For each path, the wind condition has been altered. The very left yellow colored path was planned for calm wind. The orange path was planned for horizontal wind of 10 m/s blowing from the right to the left, and the red path was planned for an increased wind velocity of 17 m/s. The displayed confidence intervals are scaled for an impact probability inside the highlighted area of $p_L = 99.9\%$ and belong to the middle segment of the red path.

All paths have an altitude of about 140 m above ground, but shift towards the wind for increasing wind velocities. Main reason for this behavior is the drift of the parachute with the wind to the left side. Thus, the detour taken by the yellow path can be reduced, which decreases the flight time or timely exposure of the square shaped region to the vehicle.

The risk cost function was demonstrated with a simulated path planning task for a transport mission across the Bavarian Tegernsee lake in the southern part of Germany. The task is to fly the vehicle from a starting point on a small hill on the western shore of the lake to a 6.1 km distant goal at the eastern shore (see Fig. 11) with a cruising speed of 25 m/s. The planning space was designed to be 12 km long in the northern direction, 10 km wide, and 2 km high above the lake surface to allow for many different trajectories. It was assumed that the mission takes place on a hot summer day with calm wind and many boats and swimmers on the lake. The latter increases the damage value of the lake. The planning algorithm created a trajectory graph with 1427 knots connected with each other via 5973 feasible trajectories and found the green path displayed in Fig. 9. The path risk value is 2.58×10^{-8}, and the flight time is 6.5 min. The blue flight path, planned for a minimal flight time, has a flight duration of 3 min and a path risk value of 5.29×10^{-8}.

Fig. 9 Benchmark scenario "street". The time optimal path is colored blue and the risk aware path is colored yellow

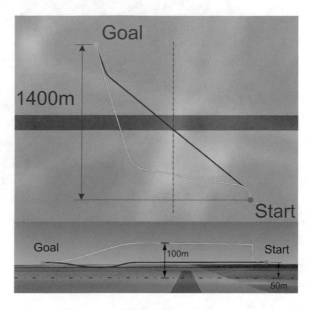

Fig. 10 Planned risk aware
paths for different wind
conditions. Yellow path:
calm wind, orange path:
10 m/s horizontal wind from
right to left, red path: 17 m/s
horizontal wind from right to
left. Confidence intervals
shown in blue for middle
segment of red path

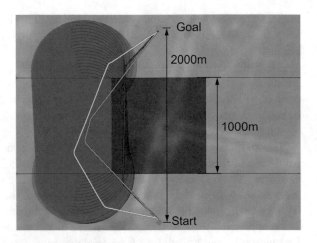

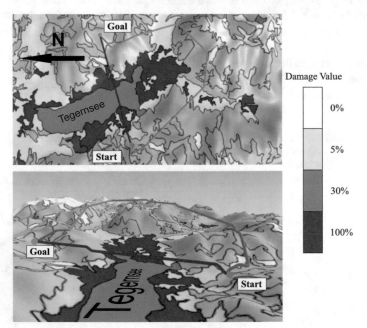

Fig. 11 Demonstration scenario "Tegernsee": The top view (upper picture) and the south-looking view. The shading of the areas corresponds to their assumed damage value

5 Sampling-Based Motion Planning

In this section, an algorithmic framework for planning long-range and low-altitude flight paths with limited onboard computational resources is described and evaluated. The framework is based on Schopferer and Benders (2020). As the focus of this work is targeted towards runtime efficient online planning, the heuristic risk assessment described in Sect. 4.4 is used to evaluate a risk-based cost function as described in this chapter's introduction. Furthermore, the sampling-based planning procedure is limited to the three spatial dimensions and a flight path consisting of linear segments. A trajectory generation algorithm using a simplified kinematic aircraft model is applied in a post-processing step to compute safe and dynamically feasible trajectories. We present experimental results for exemplary ALAADy mission scenarios and evaluate the impact of weighting factors of the risk-based cost function.

5.1 Search Graph Initialization

The first step of building a search graph is to generate samples which cover the required portions of the configuration space. In general, a mission area or corridor has to be defined in order to limit the extent of the search graph. Within this mission corridor, the samples should be distributed effectively to limit the number of required samples. Without any further prior knowledge of the environment, an even distribution of samples is most desirable. Also, the sampling algorithm used should be deterministic in order to be able to exactly reproduce the results of the planning procedure. As described in LaValle (2006), quasi-random number sequences may be used to deterministically generate sets of samples with low-discrepancy. In this work, the Halton sequence was chosen as it has been widely adopted for sampling-based motion planning. As the admissible altitude range is restricted to ca. 500 ft above ground level for the considered low-altitude scenarios, the quasi-random sampling is limited to the horizontal plane and samples are arranged in layers of constant altitude above ground. An example of this sampling procedure is shown in Fig. 12.

Once the samples have been generated, each sample is connected to its nearest neighbors to build a directed cyclic search graph. Each node connection or edge in the graph resembles a linear corridor which extends horizontally and vertically in order to allow maneuvering within this corridor, i.e. to ensure that transitions with a bounded turn radius and constrained vertical acceleration from one linear corridor to the next one will be safe within the corridors. Each edge is discretized into segments of fixed length and for each segment the clearance towards terrain and obstacles is assessed. Furthermore, the heuristic risk assessment is conducted for each segment by querying ground and air risk objects from the database and calculating a risk cost value. The edge cost are then calculated by adding up flight time and risk cost over the edge's segments and applying the risk-specific weighting factors. The result of this procedure is a cyclic, directed and weighted search graph. Figure 12 shows an example.

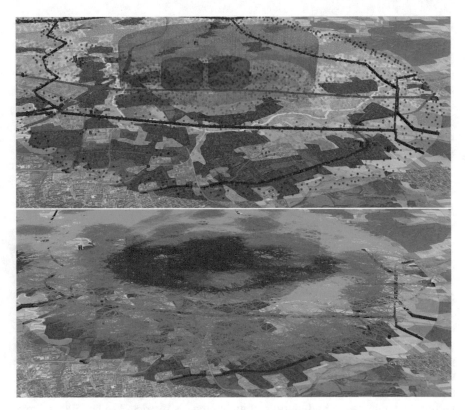

Fig. 12 Visualization of search graph samples (top) and edges (bottom)

5.2 Route Planning

In order to plan a flight path from a given start waypoint to a destination waypoint, the two waypoints have to be connected to the search graph first. This requires finding their closest neighbor nodes and calculating new edges connecting the waypoints to the graph. Then, a graph search algorithm can be applied to find an optimal path through the graph from the start to the destination. We use A*-algorithm which is a depth-first search that is guided towards the destination by a cost heuristic. In order for the heuristic cost function to be admissible, it must not underestimate the cost-to-go. Regarding the risk-based cost function described by Eq. 1, it can be seen that the flight time resembles the minimum cost or, in other words, any risk-related cost are added to the flight time cost. This allows to use the direct line-of-sight flight time to the destination as the cost heuristic of the graph search. The better the cost heuristic matches the true cost, the quicker the search will find a solution and terminate. Consequently, the runtime performance of the graph search must be expected to depend on the harm potential and especially harm potential gradients in the considered area.

Regarding the use-case of online and inflight planning, the runtime of the route planning step is critical. While the search graph can be computed before flight and the path smoothing can be performed incrementally as the aircraft flies along the path, the route planning resembles an automated decision making that directly attributes to the reaction time of the autonomous system. If, for example, the aircraft is to be redirected to the closest landing site, the route planning has to be finished before the aircraft can execute this request.

5.3   Trajectory Generation and Smoothing

The approach used to calculate trajectory segments is based on Schopferer and Pfeifer (2015). Between two given waypoints with fixed flight path azimuths and inclination angles, a trajectory segment is calculated using mathematically independent parametric curves for the horizontal and the vertical profile. The airspeed is assumed to be constant over each segment, but may vary between consecutive segments. Turn rate, flight path inclination angle as well as the vertical acceleration are constrained to feasible values.

In order to process the result of the route planning, a path of linear segments, into a dynamically feasible trajectory, we apply an iterative smoothing procedure. Each segment knot along the linear path is assigned flight path azimuth and inclination angles according to the previous segment and the subsequent segment's direction. Following a greedy algorithm strategy, from each waypoint the best next waypoint is searched for within a limited number of segments ahead taking into account the different possible path azimuth and inclination angles at each waypoint. To evaluate candidates, trajectory segments are calculated and assessed with regard to safety and risk cost, following the same procedure as with linear edges of the search graph. Finally, the trajectory segments are stitched together to form a C1-continous trajectory from start to destination.

5.4   Experimental Results

To demonstrate online-capable risk-based trajectory planning, we selected a set of three exemplary ALAADy mission scenarios in the field of spare parts logistics. As discussed in more detail in Pak (2021), the objective is to deliver urgently required spare parts from a warehouse in Hamm, Germany to local dealerships in different regions of the country. The scenarios are listed with distances from start to destination in Table 4. The mission corridors were confined to segments of 100 km in width.

The planning algorithm was executed on a desktop PC with an Intel i7 multi-core processor running at 3.4 GHz and with 8 GB RAM. While the search graph was built using up to 8 parallel threads over timespans ranging from 1 to 12 h, the planning and

Table 4 Exemplary spare parts logistics mission scenarios

Destination	Distance	Samples
Waldkappel	150 km	500,000
Schönenberg	265 km	750,000
Neubrandenburg	418 km	1,000,000

smoothing procedures were executed on a single core in order to emulate reduced onboard computational resources.

In order to assess the impact of the risk cost weighting factors, the factors for ground risk and air risk, K_{GR} and K_{AR} are varied in the range from 0 to 20 for all scenarios. For brevity, these factors are referred to as the *risk factors* below. Two runtimes were measured: The *query runtime*, i.e. the time to find a minimum cost route consisting of linear segments as described in Sect. 5.2, and the *smoothing runtime*, which is the time to calculate a smooth and dynamically feasible trajectory along the route (see Sect. 5.3). Also, the relative flight time increase compared to the case with zero risk cost was calculated for each planned trajectory. The results are shown in Fig. 13, 14, 15, 16 and 17.

Hamm—Waldkappel
As shown in Fig. 13, the flight time increases with the risk factor. As can be seen from the visualized trajectories in Fig. 14, the higher the risk factor, the further the trajectories tend to deviate from the minimum flight time trajectory. Considering that a risk factor of 20 may theoretically favor low risk paths that are 20 times longer than alternative paths with high risk, the corresponding relative flight time increase

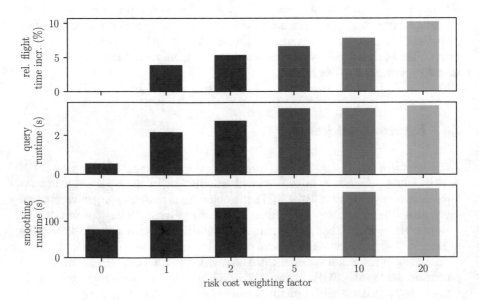

Fig. 13 Scenario Hamm-Waldkappel: Relative flight time increase and planning times

Fig. 14 Scenario Hamm-Waldkappel: Top-view of trajectories from start to goal (top) and close-up of departure from airfield Hamm (bottom)

of ca. 10% is quite low. Although the environment within the mission corridor may seem challenging, low-risk areas such as farmland and forests are more prevalent and allow trajectories to circumvent cities and villages.

The query and smoothing runtimes both increase with the risk factor, however, for different reasons. The main portion of the query runtime is attributed to the graph search. As discussed in Sect. 5.2, the error margin of the heuristic cost function

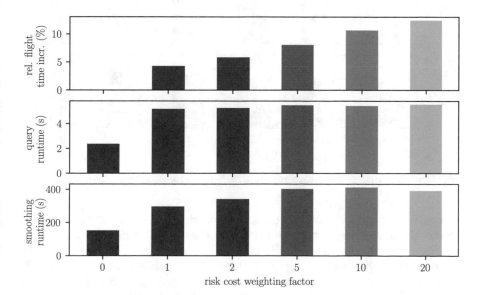

Fig. 15 Scenario Hamm-Schönenberg: Relative flight time increase and planning times

Fig. 16 Scenario Hamm-Schönenberg: Top-view of trajectories

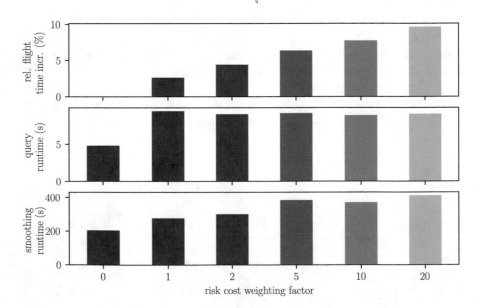

Fig. 17 Scenario Hamm-Neubrandenburg: Relative flight time increase and planning times

increases with the risk factor. Consequently, the graph search is guided less effectively towards the destination leading to a larger number of nodes to be expanded, i.e. processed by the graph search algorithm. With zero risk cost only 56,780 nodes are expanded and with risk cost weighting factor of 20, the number increases to 383,770 which is 77% of the search graph. Still, the query runtime lies within the order of a few seconds. The smoothing runtime increases with the risk factor due to the longer flight times and trajectory lengths. The runtime is significantly higher compared to the query runtime, which is due to the iterative generation and risk cost evaluation of new trajectory segments along the route.

Hamm—Schönenberg

The results shown in Fig. 15 are similar to the preceding scenario. The maximum relative flight time increase is 12.4%. The increase in query runtime with increasing risk factor is not as high in this scenario. With a risk factor of 1, 72% of the search graph nodes are expanded. This value increases to 86% for a risk factor of 20. The difference in runtime behavior for each scenario can be explained with differences in size and shape of the mission corridors and the structure and complexity of the environment within the mission corridors.

Hamm—Neubrandenburg

In the largest scenario, with a search graph of 1 million nodes, the maximum query runtime is 9.4 s. Compared to the Hamm-Schönenberg scenario, the trajectory leads through a more rural region of Germany. Consequently, the risk-optimized trajectories deviate less from the minimum flight time trajectory with a maximum flight time increase of 9.6%. Figure 18 shows a close-up example of how the risk-optimized trajectories tend to cross streets perpendicularly in order to decrease the exposure time to the street as a region of high damage potential.

Fig. 18 Scenario Hamm-Neubrandenburg: Top-view of trajectories

6   Conclusions and Future Work

We presented an algorithmic framework for risk-based motion planning targeted towards online and inflight trajectory planning with limited computational resources onboard. As the ALAADy concept envisions a risk-based approach to realizing low-altitude air delivery, the focus of this work is on risk assessment techniques and application of risk-based cost functions in sampling-based motion planning. We presented experimental results from planning trajectories for long-range transportation missions. Our results show that minimum risk trajectories can be planned within seconds using sampling-based algorithms and runtime efficient heuristic risk assessment.

The Bayesian based risk model for flight termination offers possibilities to deal with uncertainty within the vehicle state and environmental influence. Both uncertainties could be complemented by uncertainty in control inputs through adding a linearized input matrix. The risk model is freely adaptable in its uncertainty and level of detail. The ability to increase the level of detail also grows with more available data in the geospatial database and the use of real-time data.

An interesting aspect to be addressed in future work is the integration of the presented probabilistic risk assessment and the sampling-based planning algorithm. This would alleviate the need for the heuristic risk assessment used in this work. A first step towards this goal would be to increase the runtime performance of the risk assessment in order to maintain acceptable planning times. Another aspect to consider in future work is planning in higher dimensional state spaces. As of now, the result of the sampling-based planning is a 3D flight path corridor of linear segments from which a smooth trajectory is generated in a post-processing step. A more detailed planning, e.g. taking into account the aircraft's heading and airspeed, would improve the quality of the risk assessment's result.

Further research topics regarding the introduced risk models are their validation and comparison with other available risk models. Validation should focus on proving of the underlying assumptions made for the uncertainty and probability distributions.

Almost every risk model comes with its own risk quantification, which makes their direct comparison rather difficult. One way is the integration of different risk models and their cost function into the available risk planning framework and to compare the planning results for benchmark missions. This should allow further minimization of the remaining risks and contribute towards the realization of safe and efficient automated low altitude transport.

References

Aalmoes R, Cheung Y, Sunil E, Jacco H (2015) A conceptual third party risk model for personal and unmanned aerial vehicles. In: International conference on unmanned aircraft systems (ICUAS), Denver, CO, USA, 9–12 June 2015

Adolf FM, Andert F (2011) Rapid multi-query path planning for a vertical take-off and landing unmanned aerial vehicle. J Aerosp Comput Inf Commun 8(11):310–327

Allouch A, Koubaa A, Khalgui M, Abbes T (2019) Qualitative and quantitative risk analysis and safety assessment of unmanned aerial vehicles missions over the internet. IEEE Access 7:53392–53410

Al-Sabban WH, Gonzalez LF, Smith RN (2013) Wind-energy based path planning for unmanned aerial vehicles using Markov decision processes. In: IEEE international conference on robotics and automation (ICRA), Karlsruhe, Germany, pp 784–789

Babel L (2013) Three-dimensional route planning for unmanned aerial vehicles in a risk environment. J Intell Robot Syst 71:255–269

Belcastro C et al (2017) Hazards identification and analysis for unmanned aircraft system operations. In: 17th AIAA aviation technology, integration, and operations conference, Denver, CO, USA, 5–9 June 2017

Benders S, Schopferer S (2017) A line-graph path planner for performance constrained fixed-wing UAVs in wind fields. In: International conference on unmanned aircraft systems (ICUAS), Miami, FL, USA, pp 79–86

Bertrand S, Raballand N, Viguier F, Muller F (2017) Ground risk assessment for long-range inspection missions of railways by UAVs. In: International conference on unmanned aircraft systems (ICUAS), Miami, FL, USA, 13–16 June 2017, pp 1343–1351

Bull J, Lanzi R (2008) An autonomous flight safety system. In: AIAA missile sciences conference 2008, Monterey, CA

Burke DA, Hall CE (2011) Cook, SP (2011) System-level airworthiness tool. J Aircr 48(3):777–785

Castelli T et al (2016) Autonomous navigation for low-altitude UAVs in urban areas. CoRR, Volume abs/1602.08141

Chaves SM, Walls JM, Galceran E, Eustice RM (2015) Risk aversion in belief-space planning under measurement acquisition uncertainty. In: 2015 IEEE/RSJ international conference on intelligent robots and systems (IROS), Hamburg, 28 September–02 October 2015, pp 2079–2086

Chitsaz H, LaValle SM (2007) Time-optimal paths for a Dubins airplane. In: 2007 46th IEEE conference on decision and control, New Orleans, LA, USA, 12–14 December 2007, pp 2379–2384

Christensen FM, Andersen O, Duijm NJ, Harremoes P (2003) Risk terminology - a platform for common understanding and better communication. J Hazard Mater 103(3):181–203

Clare VR, Mickiewicz AP, Lewis JH, Sturdiven LM (1975) Blunt Trauma Data Correlation. Aberdeen Proving Ground, Aberdeen, MD, USA

Dauer JC, Dittrich JS (2021) Automated cargo delivery in low altitudes: concepts and research questions of an operational-risk-based approach. In: Dauer JC (ed) Automated low-altitude air delivery - towards autonomous cargo transportation with drones. Springer, Heidelberg

Denney E, Pai G, Ippolito C, Lee R (2012) An integrated safety and systems engineering methodology for small unmanned aircraft systems. In: Infotech@Aerospace conference, Garden Grove, CA, USA, 19–21 June 2012

DiDonato A, Hageman R (1980) Computation of the Integral of the Bivariate Normal Distribution Over Arbitrary Polygons. Naval Surface Weapons Center, Dahlgren, VA

Donkels A (2020) Trajectory risk evaluation for autonomous low-flying air transport. J Guidance Control Dyn 43(5):1026–1033

Dubins L (1961) On plane curves with curvature. Pac J Math 11(2):471–481

European Aviation Safety Agency (2017) Guidance Material (GM) to Annex I – Definitions for terms used in Annexes II to VIII of Commission Regulation (EU) 965/2012 on air operations, March 2019. https://www.easa.europa.eu/sites/default/files/dfu/Consolidated%20AMC-GM_Annex%20I%20Definitions_March%202019.pdf

European Aviation Safety Agency (2018) Flying a Drone. EASA Poster. https://www.easa.europa.eu/sites/default/files/dfu/217307_EASA_DRONE_POSTER_2018%20final.pdf

Goerzen C, Whalley M (2011) Minimal risk motion planning: a new planner for autonomous UAVs in uncertain environments. In: AHS international specialists meeting on unmmaned rotorcraft, Tempe, Arizona, USA, 25–27 January 2011

Harmsel AJ, Olson IJ, Atkins EM (2017) Emergency flight planning for an energy-constrained multicopter. J Intell Robot Syst 85:145–165

Hoover WE (1984) Algorithms for confidence circles and ellipses. NOAA Technical report NOS 107 C. U.S. Department of Commerce, National Oceanic and Atmospheric Administration, Rockville, MD, USA

International Programme on Chemical Safety (2004) Risk Assessment Terminology - Part 1 and Part 2. WHO, Geneva, Switzerland. http://www.inchem.org/documents/harmproj/harmproj/har mproj1.pdf

Joint Authorities for Rulemaking of Unmanned Systems (2017) JARUS Specific Operations Risk Assessment – Annex I - Glossary, Public Release Edition 1.0. http://jarus-rpas.org/sites/jarus-rpas.org/files/jar_doc_06_jarus_sora_annex_i_v1.0.pdf

Kálmán RE (1960) A new approach to linear filtering and prediction problems. J Basic Eng 82(1):35–45

Karaman S, Frazzoli E (2010) Incremental sampling-based algorithms for optimal motion planning. In: Robotics: science and systems VI, Zaragoza, Spain, 27–30 June. MIT Press

Karaman S, Frazzoli E (2011) Sampling-based algorithms for optimal motion planning. Int J Robot Res 30(7):846–894

Kavraki LE, Svestka P, Latombe JC, Overmars MH (1996) Probabilistic roadmaps for path planning in high-dimensional configuration spaces. IEEE Trans Robot Autom 12(4):566–580

la Cour-Harbo A (2017) Quantifying risk of ground impact fatalities of power line inspection BVLOS flight with small unmanned aircraft. In: International conference on unmanned aircraft systems (ICUAS), Miami, FL, USA, 13–16 June

LaValle SM (1998) Rapidly-exploring random trees: a new tool for path planning. Technical report TR. 1998, Computer Science Dept, Iowa State University, pp 98–11

LaValle SM (2006) Planning algorithms. Cambridge University Press, Cambridge

LaValle SM, Kuffner JJ (2001) Randomized kinodynamic planning. Int J Robot Res 20(5):378–400

Lum CW, Waggoner B (2011) A risk based paradigm and model for unmanned aerial systems in the National Airspace. In: Infotech@Aerospace, St. Louis, Missouri, USA, 29–31 March 2011

Lum CW et al (2011) Assessing and estimating risk of operating unmanned aerial systems in populated areas. In: 11th AIAA aviation technology, integration, and operations (ATIO) conference, Virginia Beach, VA, USA, 20–22 September 2011

Melnyk R, Schrage D, Volovoi V, Jimenez H (2014) A third-party casualty risk model for unmanned aircraft system operations. Reliab Eng Syst Saf 124:105–116

Nahrwold A, Petit PJ, Plorin J (2019) Analysis, Assessment and Development of Online Path Planning Algorithms for Helicopters under Threat. DGLR, Darmstadt, Germany, 30 September–02 October 2019

National Aeronautics and Space Administration (2018) NASA-STD-8719.25 Range Flight Safety Requirements

Niendorf M, Schmitt F, Adolf FM (2013) Multi-query path planning for an unmanned fixed-wing aircraft. In: AIAA guidance, navigation, and control conference, Boston, MA, USA, 19–22 August 2013

OpenStreetMap (2019) https://www.openstreetmap.de. Accessed 20 July 2020

Oster CV Jr, Strong JS, Zornc CK (2013) Analyzing aviation safety: problems, challenges opportunities. Res Transp Econ 43(1):148–164

Pak H (2021) Use-cases for heavy lift unmanned cargo aircraft. In: Dauer JC (ed) Automated low-altitude air delivery - towards autonomous cargo transportation with drones. Springer, Heidelberg

Peinecke N, Volkert A, Korn BR (2017) Minimum risk low altitude airspace integration for larger cargo UAS. In: Integrated communications, navigation and surveillance conference (ICNS), Herndon, VA, USA, 18–20 April 2017

Pettersson PO, Doherty P (2006) Probabilistic roadmap based path planning for an autonomous unmanned helicopter. J Intell Fuzzy Syst 17(4):395–405

Primatesta S, Cuomo LS, Guglieri G, Rizzo A (2018) An innovative algorithm to estimate risk optimum path for unmanned aerial vehicles in urban environments. Transp Res Procedia 35:44–53

Range Safety Group, Flight Termination System Committee, Range Commanders Council (2001a) IRIG Standard 313-01 - Test Standards for Flight Termination Receivers/Decoders. Secretariat Range Commanders Council U.S. Army White Sands Missile Range, NM, USA, May 2001

Range Safety Group, Range Commanders Council (2001b) Range Safety Criteria For Unmanned Air Vehicles: Rationale And Methodology Supplement. Secretariat Range Commanders Council U.S. Army White Sands Missile Range, NM, USA, April 2001

Range Safety Group, Range Commanders Council (2019) Commonality Standard 319-19 - Flight Termination Systems. Secretariat Range Commanders Council U.S. Army White Sands Missile Range, NM, USA, June 2019

Rudnick-Cohen E, Herrmann JW, Azarm S (2016) Risk-based path planning optimization methods for unmanned aerial vehicles over inhabited areas. J Comput Inf Sci Eng 16(2):021004

SAE (2010) ARP4754A/EUROCAE ED-79A - Guidelines for Development of Civil Aircraft and Systems

Sachs F (2021) Configurational aspects and vehicle specific investigations for future unmanned cargo aircraft. In: Dauer JC (ed) Automated low-altitude air delivery - towards autonomous cargo transportation with drones. Springer, Heidelberg

Schirmer S, Torens C (2021) Safe operation monitoring for specific category unmanned aircraft. In: Dauer JC (ed) Automated low-altitude air delivery - towards autonomous cargo transportation with drones. Springer, Heidelberg

Schopferer S, Benders S (2020) Minimum-risk path planning for long-range and low-altitude flights of autonomous unmanned aircraft. In: AIAA Scitech 2020 forum, Orlando, FL, USA, 6–10 January 2020

Schopferer S, Lorenz JS, Keipour A, Scherer S (2018) Path planning for unmanned fixed-wing aircraft in uncertain wind conditions using trochoids. In: International conference on unmanned aircraft systems (ICUAS), Dallas, TX, USA, 12–15 June 2018, pp 503–512

Schopferer S, Pfeifer T (2015) Performance-aware flight path planning for unmanned aircraft in uniform wind fields. In: International conference on unmanned aircraft systems (ICUAS), Denver, CO, USA, 9–12 June 2015, pp 1138–1147

Sgobba T, Wilde PD, Rongier I, Allahdadi FA (2013) Safety design for space operations. Butterworth-Heinemann, Oxford

Stevens J et al (2012) Helicopter Safety Related Research at NLR: a Multi-Disciplinary Task. National Aerospace Laboratory NLR, Amsterdam, The Netherlands, NLR-TP-2012-412

Techy L, Woolsey CA (2009) Minimum-time path planning for unmanned aerial vehicles in steady uniform winds. J Guidance Control Dyn 32:1736–1746

Washington A, Reece C, Silva JM (2017) A review of unmanned aircraft system ground risk models. Prog Aerosp Sci 95:24–44

Wolek A, Woolsey C (2015) Feasible Dubins paths in presence of unknown, unsteady velocity disturbances. J Guidance Control Dyn 38(4):782–787

Wu P, Clothier R (2012) The development of ground impact models for the analysis of the risks associated with unmanned aircraft operations over inhabited areas. In: Proceedings of the 11th international probabilistic safety assessment and management conference and the 2012 annual European safety and reliability conference, Helsinki, Finland, 25–29 June 2012. IAPSAM & ESRA, New York, pp 5222–5234

Safe Operation Monitoring for Specific Category Unmanned Aircraft

Sebastian Schirmer and Christoph Torens

Abstract Future unmanned aircraft systems are allowed to incorporate operational aspects for flight approval due to the new EASA "specific" category. Incorporating operational aspects offer new possibilities for the verification and validation of complex functions used especially in highly automated vehicles. For these functions, verification and validation can focus on predefined operational aspects prior to flight. In-flight, limits of the operation are monitored to assure the correct working environment for these functions resulting in a safe operation. In this paper, we present the notion of safe operation monitoring and depict operational limits to be supervised. One prominent example for such an operational limit is geofencing. Geofencing prevents an unmanned aircraft from entering a forbidden airspace by using virtual fences. Specifically, in this paper, we present an algorithm and describe parameters for the buffer distance used for the geofence boundary values. The algorithm can be highly parallelized which is important when considering realistic geofences of future operations. Further, we highlight the use of a formal specification language and simulation results which support the verification and validation of geofencing, respectively. The chosen specification language is not limited to geofencing, other operational limits can be expressed and monitored in-flight to assure the safe operation.

Keywords Unmanned aircraft system · Formal methods · Safe operation · Monitoring · Geofencing · Risk-buffer · Simulation

S. Schirmer (✉) · C. Torens
German Aerospace Center (DLR), Institute of Flight Systems, Lilienthalplatz 13, 38108 Braunschweig, Germany
e-mail: sebastian.schirmer@dlr.de

C. Torens
e-mail: christoph.torens@dlr.de

© Deutsches Zentrum für Luft- und Raumfahrt e. V. (DLR) 2022
J. C. Dauer (ed.), *Automated Low-Altitude Air Delivery*, Research Topics in Aerospace,
https://doi.org/10.1007/978-3-030-83144-8_16

1   Introduction

There are three categories of operations for unmanned aircraft systems (UAS), introduced by EASA in 2015. These categories are "open", "specific", and "certified", and
are based on the intrinsic risks involved in operating unmanned aircraft, see EASA
(EASA, 2015, 2015-10, 2017). The open category is intended for operations requiring
no or minimal regulation. The certified category is intended for high risk operations,
comparable to manned aviation. With the specific category it is possible to scale the
requirements for regulation and certification between the open and certified category.
For the specific category, the certification is not based solely on the safety assurance
of the aircraft, but also on the overall risk caused by the operation. A so-called Specific
Operation Risk Assessment (SORA) is used to determine the required level of rigor
for UAS development and operation (JARUS 2016). Details on the categories of UAS
operation and on the severity can be found in Nikodem et al. (2021). The involvement of operation risk is new to the process of receiving a permission to operate an
UAS. This risk-based adaptation to the requirements of the unmanned aircraft development and operation opens up new possibilities in aircraft design and approaches
to ensure safety. An important factor for the assessment of the operation is the so-
called ConOps document, the document that describes the Concept of Operations,
the details and also boundary conditions of the operation. With the safe operation
monitoring, the German Aerospace Center (DLR) is researching a safe and reliable
methodology to supervise the safe operation of unmanned aircraft in the specific
category. With the safe operation monitoring, the safety of the operation is ensured
and enforced by supervision of the boundary conditions of the ConOps description,
that are the basis of the SORA and the flight approval. If the safe operation monitor
detects a violation of the ConOps description, contingency or emergency procedures
can be activated to cope with abnormal or emergency situations, respectively.

The project Automated Low Altitude Air Delivery (ALAADy) fits in well here,
since it focuses on the identification of the limits of the SORA approach. Some
key objectives for the project are the identification of important boundary operation
conditions for the system's supervision as well as certification concepts for complex,
i.e. hard to certify, functions. Aspects of a flight termination as an emergency procedure in the context of ALAADy are discussed in Dauer et al. (2021), Sachs et al.
(2021), and Schopferer et al. (2021).

Since the system monitoring is intended to ensure that the system is only operated
within pre-defined operational limits, it represents a key safety-critical function and
must therefore be development with the greatest possible effort. A common way
for implementing such a monitoring function would be to implement it using hand-
written code. Written code is traced to requirements, i.e. operational limits here, and
tested rigorously. Also, static analyses can be used to show bounds on desired properties like memory or runtime. The disadvantage of this approach is that code for such a
complex function becomes quite extensive which makes it difficult to understand and
consequently difficult to maintain. Additionally, the performance of static analyses
can be impacted. A different way which addresses this problem is runtime monitoring

based on a formal specification language. The focus of such a specification language is to describe *what* should be monitored instead of *how*, which is typical the focus of hand-written code. As a consequence, the specification of operational limits is more concise and easier to understand and to maintain. Also, since the language handles the *how*, i.e. the actual implementation, errors for instance in the context of memory management are directly avoided during development. Further, since the language is formally fully-defined, analysis of memory consumption and runtime as well as consistency checks can be directly inferred by analysis of the specification. As specification language we chose Lola which is described in later on in more detail. Besides mentioned advantages, Lola offers modularity and high-expressiveness which makes it the perfect fit for specifying operational limits.

This chapter is structured as follows: after introducing the motivation and background of this research in this section, related work on monitoring and specifically geofencing is presented in Sect. 2. Section 3 describes the approach of safe operation monitoring in more detail, discussing the impact on the overall risk of the operation and the aircraft safety requirements. Furthermore, an overview of supervision properties is given. For the remaining chapter, geofencing is exemplified as a supervision property and the specifics of the geofence functionality for safe operation in the context of the specific category are introduced. Section 4 describes the case study that was used for the ALAADy project and its simulation studies. It gives details on how the geofence functionality is implemented using the formal language Lola. Furthermore, it shows the experimental results of the simulation studies of the safe operation monitoring. Section 5 describes how the geofence monitoring is realized on a Field Programmable Gate Array (FPGA), a programmable hardware circuit. Benefits and limitations of an FPGA implementation as well as certification aspects of FPGA are discussed in this context briefly. Finally, Sect. 6 concludes the chapter on safe operation monitoring and the possibilities that this methodology could enable.

2 Related Work

The concept of monitoring a system component is commonly used in safety critical development, and guidelines for development of civil aircraft mention the concept (S.A.E. 2010). In addition to that, automated monitoring on an aircraft level, i.e. aircraft components, and even on a mission level, i.e. components involved in the mission, is getting more research interest recently. ASTM has published a standard that uses the concept of monitoring to bind the behavior of unmanned aircraft containing complex functions. The standard is intended to increase the use of functions that are difficult to verify with traditional verification methods. An example for such a complex function is a deep neural network. The basic idea behind the approach is to switch from a complex and hard to verify function to a simple and verified function in case of a monitored misbehavior. A revision of the standard is currently being developed, incorporating lessons learned of first use cases (ASTM 2017).

NASA is utilizing the monitoring concept in several projects. The NASA Safeguard technology is an assured safety net for UAS implemented as independent system component. Safeguard has the ability to trigger mitigation tactics upon violations of geo-limitations, e.g. terminate flight (Gilabert et al. 2017). Used algorithms, code, and mathematical models have been formally verified (NASA 2017). The more complex the trigger conditions are, the harder the verification of handwritten code gets. Our approach is based on a formal specification language, which offers a higher level of abstraction for specifying system monitor properties. This approach promises high flexibility due to automatic monitor code generation, and overall improved maintainability due to a clear and concise specification of the system properties.

A similar approach based on formal methodology has been utilized with the R2U2 framework (Geist et al. 2014). This tool has been used to ensure safety as well as security properties. Furthermore, it was synthesized on an FPGA (Rozier et al. 2017). The approach differs from our approach in utilizing a mission time linear temporal logic and a Bayesian Network to monitor and reason about the system, respectively. In contrast to this, due to numerical support of the specification language, a single formalism is used for both monitoring and reasoning.

Another comprehensive framework for runtime monitoring is the NASA EVAA RTA network architecture (Skoog et al. 2020). This approach utilizes a moral compass to select the best behavior when conflicting violations occur at the same time. The moral compass can be seen as an implementation of the pilot's moral. Finally, the NASA System-wide Safety Initiative uses monitoring as one component to guarantee safety of the future airspace in combination with airspace services (Ellis et al. 2019).

For the specific category concept of EASA, the conditions being monitored are described in the ConOps document. Formalizing the ConOps description can help with the automated transformation of ConOps to monitoring properties (Torens et al. 2018, 2019).

Geofencing is a concept that is and is becoming more and more important with the upcoming use of unmanned aircraft. Aircraft that pass-through SORA have to show compliance with safety requirements associated with technical containment design features required to stay within the operational volume define in ConOps (Step #9: Adjacent Area/Aircraft Considerations). For the general problem of deciding if a point is inside of a polygon, several algorithms exist. A detailed analysis of the mathematical approach and a comparison in the context of the geofencing has been done (Stevens et al. 2018). Utilizing Lola for the use case of geo-fencing has been proposed in (Schirmer et al. 2018). There is also increasing interest specifically for runtime monitoring utilizing FPGAs (Solet et al. 2016), (Nguyen et al. 2016), (Stamenkovich et al. 2019) (Pellizzoni et al. 2008).

3 Safe Operation Monitoring

To motivate the idea behind safe operation monitoring, it is necessary to fully understand the idea of the specific category and its implications. The general idea behind the specific category is that the risk of the operation can be incorporated in the safety argumentation, and with this also the requirements to the safety of the unmanned aircraft can be adjusted. This is different for manned aviation, as well as the certified category in a general sense. There, the risk of the operation is always at a maximum level since lives are always at risk. This results from the fact that this is obviously true for any manned flight due to the onboard pilot. Similarly, this risk must also be assumed for any unmanned flight in the general airspace, since there is always a risk of third-party airspace participants or population on the ground.

Now, let us assume that a specific operation with restrictions written in the ConOps has reduced risk. This creates a new implicit risk to be focus of this work, which is that the specific operation violates the operational limitations defined by the ConOps. For example, imagine the simplistic scenario where an unmanned aircraft flies over an unpopulated area. For this operation, there is a reduced risk of injuring people. There is no pilot onboard and also it is unlikely to hit people in case of a crash. If this mission is performed as intended, the operation is safe. Even in the case of a technical failure and the crash and subsequent loss of the unmanned aircraft, the overall operation would still be safe. This holds true, because the area of operation is unpopulated and the vehicle is unmanned. However, if the aircraft is about to leave the unpopulated and flies over populated area, the risk of injuring people would increase drastically. As a result, it is of great importance to assure and enforce the operational limitations of the specific operation. This is where the safe operation monitoring comes into play. The safe operation monitoring supervises all operational limitations and triggers mitigations like a safe flight termination to ensure that the operational limitations are not violated. The safety requirements of the safe operation monitoring are in fact always at a high level, since the failure of this safe operation monitoring could lead to leaving the operational limits and thus leaving the risk reducing factors of the operation.

The safe operation monitor should supervise the overall state of the aircraft as well as the operational limitations, defined by the concept of operations. With respect to safe operation monitoring, the following supervision properties have been identified in the categories of Mission, Aircraft, Component and Sensors as well as Cargo, Environment, and Hardware. Table 1 depicts some more detailed properties on the mentioned different supervision levels. However, some of these properties might be easier to monitor for safety than others. Additionally, for some of these property limits will be known, for others the limits might be difficult to explicitly specify. A topic for future research will be how to define the limits of normal operation for such properties, e.g. boundary values for sensor readings.

Although it is the goal to supervise all properties, one of the most important property for safe operation is geofencing, i.e. containment in the context of SORA. In the following, this property will be exemplified for the remainder of this chapter.

Table 1 Description of the
flight volumes used in the
SORA Semantic Model.
Nominal flight is intended in
the Flight

Supervision level	Properties
Mission	• Operational limitations, e.g. geofencing, altitude, or airspeed • Fuel consumption • Progress • Trajectory deviation
Aircraft	• Overall system state • Performance metrics • Aircraft vibrations
Component and Sensor	• Signal frequency • Signal quality • Cross checks • Boundary values • Module responsiveness • State transitions
Cargo	• Temperature • Acceleration/load factor • Max bank angle
Environment	• Ground risk/population density • Air risk/intruders • Weather conditions • Availability of U-space[a] services

Hardware	• CPU load
	• Memory consumption
	• Temperature
	• Utilization of message buffer and system bus
	• Communication link
	  – Quality
	  – Load

[a]https://www.sesarju.eu/U-space

3.1 Example of Safe Operation Monitoring: Geofencing

Geofencing deals with ensuring the separation of an aircraft from a predefined area, a so called no-fly zone. Hence, the pre-defined area is virtually fenced and, upon exceeding, mitigation actions can be triggered. Different mitigation actions do exist. In this paper, we differentiate between contingencies and safe flight termination. The main difference between the two is that contingencies allow re-entering of the nominal flight volume while the activation of a safe flight termination is final. In SORA, the nominal flight volume is called *Flight Geography* and the contingency flight space is referred to as *Contingency Volume*. Both flight volumes are illustrated in Fig. 1. According to SORA, the available time and space for a contingency maneuver

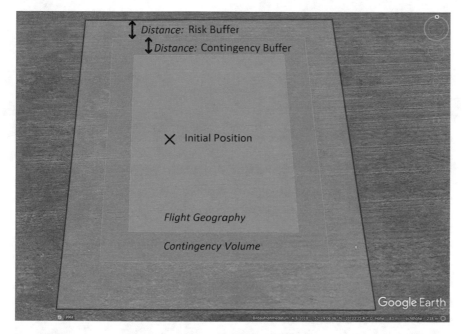

Fig. 1 Description of the flight volumes used in the SORA Semantic Model. Nominal flight is intended in the Flight Geography, contingency mitigations are triggered in the Contingency Volume, and the Risk Buffer is used for emergency procedures to assure that the UAS never exceeds the boundary

or a safe flight termination is defined by the *Contingency Buffer* and *Risk Buffer*, respectively.

Geofencing Algorithm in the Context of Safe Operation Monitoring

For the automatic detection of geofence violations, we describe the flight geometry as polygonal chain and check for line crossing between the polygonal chain and the flown trajectory. We assume a fixed height limit, and that we fly over or around obstacles, but not underneath them. Hence, we can project the 3D geofencing to 2D with an additional height check: if height $> h_{max}$ then trigger mitigation.

In Fig. 2, the polygonal chain is depicted as red line with points $p_1, \ldots p_5$. Due to the sampling of a position sensor, the flown trajectory is also given discretized by $t_1, \ldots t_6$. As mentioned, we compute line crossing between trajectory lines $t := [t_i, t_j]$, and lines of the polygonal chain $p := [p_k, p_l]$.

The algorithm depicted in the following consists of the computation of the trajectory line (1) and the intersection of the trajectory line with the polygon lines (2).

(1) The trajectory line computations are, where "p.x" and "p.y" refer to the x-coordinate and y-coordinate of the point p, respectively:

- $\Delta_x := t_j.x - t_i.x$, x-difference
- $\Delta_y := t_j.y - t_i.y$, y-difference
- $dst_t := \sqrt{\Delta_x^2 + \Delta_y^2}$, trajectory line distance

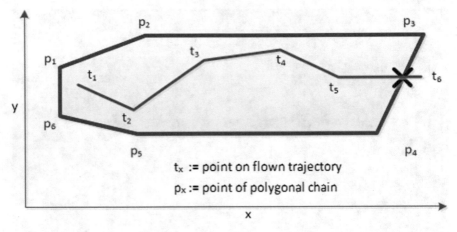

Fig. 2 UAS trajectory violating the predefined geofence described as polygonal chain

- $ori_{t_x} := \Delta_x < 0$, x-orientation of trajectory line
- $ori_{t_y} := \Delta_y < 0$, y-orientation of trajectory line
- $m_t := \frac{\Delta_y}{\Delta_x}$, slope of the trajectory line
- $b_t := t_j.y - m_t \cdot t_j.x$, y-intersect of the trajectory line

(2) For each p in the polygonal chain, we compute an intersection point and check whether the point is valid, i.e. lies on both: the current trajectory line and the considered polygonal chain line. Further, we use the Jordan Curve Theorem to check for point-in-polygon. The computations are:

- m_p (slope of p) and b_p(y-intersect of p) are computed similar to m_t and b_t, but using points p
- $i_x := \frac{b_t - b_p}{m_p - m_t}$, x coordinate of intersection point
- $i_y := m_p \cdot i_x + b_p$, y coordinate of intersection point
- $\Delta_{ix} := i_x - t_i.x$, intersection point x-difference
- $\Delta_{iy} := i_y - t_i.y$, intersection point y-difference
- $ori_{valid} := \left((\Delta_{ix} < 0) = ori_{t_x}\right) \bigwedge \left((\Delta_{iy} < 0) = ori_{t_y}\right)$, checks whether both lines t and [t_i, (i_x, i_y)] have the same orientation
- $dst_{valid} := \sqrt{\Delta_{i_x}^2 + \Delta_{i_y}^2} \leq dst_t$, checks whether the intersection point is between t_i and t_j by comparing the distance of t with [t_i, (i_x, i_y)]
- $bound_{valid} := \min(p_k.x, p_l.x) < i_x < \max(p_k.x, p_l.x)$, checks whether the intersection point is on the polygonal chain line
- $violation_p := ori_{valid} \,\&\, dst_{valid} \,\&\, bound_{valid}$, if all checks are valid, then an intersection point exist, i.e. a contingency violation.

At its core, geofencing targets the question: "is the UAS within or outside of an area". Several algorithms exist and are compared in Stevens et al. (2019). We choose the previously presented algorithm due to several reasons: First, its simplicity. In geometry, computing line intersections is a basic and well-known problem. Second, the distance computation combined with the velocity of the aircraft offers a good metric on "how far/how long until a line violation". Third, the algorithm can be highly parallelized. Each intersection point between the trajectory line and the polygonal chain lines can be computed independently. Therefore, they can be run very efficiently on hardware solutions like GPU or FPGA. Forth, at runtime, polygonal changes can be dynamically updated more easily compared to algorithms which rely on precomputed data structures. Last, we are not limited to e.g. convex polygonal chains, but instead allow arbitrary ones.

Geofencing Extensions Incorporating Risk
In Schirmer et al. (2018), *risk areas* and a position prediction are added to geofencing. These extensions are required to allow e.g. road crossings based on the current confidence in the UAS. In the paper, the *confidence* of a system is based on parameters like the current and past flight performance, e.g. how well could the commanded trajectory be followed, or the navigation performance, e.g. GPS performance is high. Based on the current confidence of the system, transitions from one risk area into another are allowed or restricted. Risk areas are adjacent geofences with an attached risk. The risk of such an area reflects its criticality of the area, i.e. risk areas incorporating road crossings or cities have a higher value compared to ones covering forests. With this formalism it is possible to model situation where a decision has to be made.

3.2 Buffer Calculation for Safe Termination

In geofencing, the polygonal chain lines are set according to the pre-flight calculated buffers. The term buffer is kept relative general, since it refers to both: the risk buffer and the contingency buffer, as depicted in Fig. 1. These buffer calculations guarantee that when activating mitigation such as a safe flight termination, the touch down happens within the buffer bounds. Hence, harm for people on the ground can be reduced by putting sufficiently large buffers around cities or infrastructures. When calculating the buffer value, the following parameters should be incorporated:

- Position uncertainty – geofencing is based on the position estimation which can deviate from ground truth.
- Reaction time – computing or manually activating the correct mitigation strategy takes time while the UAS continues its flight.
- Wind – the UAS can be carried away by wind during the maneuver.
- Maximal height difference buffer – reflects the difference between the maximal and minimal terrain height of the operation area, see Fig. 3.
- Maneuver—is depending on the planned maneuver, e.g. loitering radius. There are contingency and emergency maneuver.
- Safety Margin—other operation specific parameters limitations, e.g. based on experience.

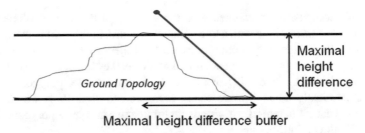

Fig. 3 Ground height profile with UAS depicted as red dot. In case of termination, the gliding from highest to lowest point in area assumes a 1-to −1 rule, i.e. 45°. The maximal height difference buffer incorporates the possible height differences which would be neglected when assuming a flat surface

In the following, as a showcase, we compute the buffers for the safe flight termination of a gyrocopter and a box wing in case of breaching the risk buffer bound, i.e. activation of an emergency procedure. The gyrocopter uses as termination maneuver its passive autorotation to safely land on ground. As termination maneuver for the box wing, we consider the use of a parachute. We assume the scenarios depicted in Table 2. For more details on the safe flight termination, we refer to Dauer et al. (2021) and Sachs et al. (2021). For the simulation presented in Sect. 4, we use these buffer distances for geofencing and show how simulation results under different environmental conditions can be used to validate and refine geofence bounds.

For the gyrocopter, based on simulation results we assume that descending from 50 m to ground takes 15 s during a light breeze of 10 km/h. Hence, wind carries the UAS 40 m away. Further, we assume maximal height difference of 200 m. Also, we consider an automatic mitigation. Therefore, the decision takes around 30 m (200 km/h · 0.5 s). The gyrocopter has a lift-to-drag ratio of six, implying a maneuver buffer distance of 900 m (150 m · 6). In total: 1180 m = 10 m (position uncertainty) + 30 m (reaction time) + 40 m (wind) + 200 m (maximal height difference buffer) + 900 m (maneuver) + 0 m (safety margin).

For the box wing, the general conditions remain the same. Simulation results of (Russell 2018) indicate that it takes 15 s upon parachute activation at 150 m until touch down and that it takes 250 m until the parachute completely slowed down the UAS. Termination at the maximal height difference of 200 m takes 20 s and, therefore, wind carries the UAS 60 m away. In total: 350 m = 10 m (position uncertainty) + 30 m

Table 2 Vehicle specific characteristics used to calculate the buffer distance

Parameter	Gyrocopter	Box wing
Cruise speed [km/h]	200	200
Maximal Wind [km/h]	10	10
Reaction time [s]	0.5	0.5
Position uncertainty [m]	10	10
Maximal height [m]	50	50

(reaction time) + 60 m (wind at maximal height difference) + 250 m (maneuver) + 0 m (safety margin).

The sample computations show that the buffer values highly depend on the vehicle and its planned maneuvers. The box wing achieved a smaller buffer value than the gyrocopter. In the presented example, to derive the buffer calculation solely on values coming from simulation results is highly desirable, since it promised significant cost reduction and time savings. However, certification of the simulation environment is very hard to achieve since physical UAS models are complex.

4 Case-Study

So far, we have presented an algorithm for geofencing and parameters for the buffer calculation. In this section, we present a simulation case study. The ALAADy simulation scenario we consider is: Flying from A to B at a pre-defined altitude with different wind conditions per flight.

In order to reduce ground and air risk, the trajectory should avoid urban areas and manned airspace. As mentioned, geofencing with correct calculated buffers is a means to avoid entering such areas.

Unfortunately, not only adjusting buffer distances, but also finding a suitable geofence for flights over hundreds of kilometers in the first place, which e.g. respects population density, is a difficult task which needs support. Therefore, we present the tool *FencyCreator* which was motivated by Stevens et al. (Stevens and Atkins 2018). FencyCreator offers a graphical user interface for automatically scaling polygonal chains inwards as well as outwards. The polygonal chain can be manually defined directly at the border of a city, using e.g. population density maps. Then, this polygonal chain can be given to the tool for scaling. The result can be manually inspected and adapted. Afterwards, as a key feature, the tool generates directly a formal specification for geofencing the respective polygonal chain. Based on this formal specification, a monitor is automatically generated which can be used in flight.

In the following, we present the tool FencyCreator in more details; we explain the concept of runtime verification, and finally show experimental result of a simulation case study.

4.1 FencyCreator—a Tool for Scaling Geofences and Generating Geofencing Specifications

The input to the tool is a polygonal chain given as KML-file which is an XML notation for expressing geodata used in many Earth browsers, e.g. Google Earth. An example is depicted in Fig. 4 where red lines represent the area the UAS should remain within. In the following, we describe the tool steps until the geofence specification

Fig. 4 Polygonal chain given as KML-file and displayed in GoogleEarth

is generated. First, the KML-file needs to be read. Second, the initial position (lat, lon, alt) and the maximal height have to be set. Third, the geofence can be scaled. Figure 5 shows the result of FencyCreator after using risk buffer scaling of -150 and a contingency buffer scaling of -170. Buffer values are given absolute to the original polygonal chain and negative values indicate inwards scaling. Therefore, in this example, we consider a risk buffer of 150 m and a contingency buffer of 20 m. Note that the tool can also be used to identify minimal buffer distances. In Fig. 6, a contingency buffer distance of 250 m was used. The resulting geofence shows intersections due to scaling. Therefore, a valid continuous geofence does not exist, only parts of the geofence are considered valid. As a result, a safe flight would not be possible with the given geofence buffer. By that, the tool enables to easily check if a geofence exists which allows a safe flight respecting the required contingency and safety buffers. Next, when a valid geofence was found, the fence can be exported to a KML-file for a manual inspection and to the formal Lola specification. Lola allows generating monitors automatically with formal guarantees, e.g. bounded memory consumption, for the corresponding specification. Lola is presented in more detail in the next subsection. FencyCreator supports the core Lola, Real-Time Lola, and FPGA Real-Time Lola. This facilitates different hardware solutions, which is an important feature since different use-cases may require this flexibility.

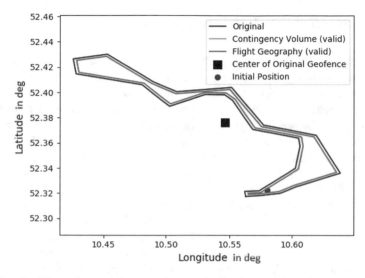

Fig. 5 Result of FencyCreator using a contingency buffer distance of −170 m and a risk buffer distance of −150 m. Negative values indicate inwards scaling. A single valid scaled geofence was identified

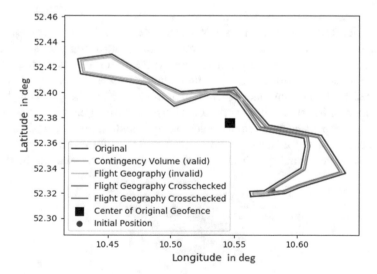

Fig. 6 Result of FencyCreator using a contingency buffer distance of −250 m and a risk buffer distance of −150 m. Negative values indicate inwards scaling. No single valid scaled geofence is identified

4.2   Formal Runtime Monitoring with Lola

Runtime Verification (Bartocci et al. 2018) is a lightweight formal method used in many domains ranging from banking (Colombo et al. 2018) to aviation (Adolf et al. 2017). Based on a formal specification of desired or undesired system properties, a monitor is generated checking the correctness of the events of an instrumented system towards the specification during execution. The monitor outputs a verdict upon which the system can react to. The approach is depicted in Fig. 7.

Lola is such a formal specification language for monitoring (D'Angelo et al. D'Angelo et al. 2005). The declarative specification mechanism is based on stream equations and allows both: specification of correctness as well as statistical properties of the system under observation. Declarative specifications focus on *what* properties should be monitored and do not focus on *how*. This directly avoids implementations error known from general-purpose programming languages, e.g. memory leaks, since *how* the properties are actually implemented is handled by Lola safely. Exemplary Lola specifications are given below to better understand the general idea. When considering the examples, respective references are given in which more detailed information can be found. The order of the examples reflects the evolution of Lola and shows the trend towards using runtime monitoring to supervise complex automated systems.

First, Lola was developed for monitoring circuits using synchronous streams. A Lola specification consists of a set of stream equations. Each stream equation is either an input or output stream representing incoming data or computations based on other stream values, respectively. All streams evolve uniformly based on a common synchronous clock. Streams are typed including Integer and Float. Computations are defined by stream expressions including arithmetic expressions, which allow the specification of incrementally computable statistics. Also, stream expressions have access to past, present, and future values of input and output streams. In case of future accesses, stream evaluations are delayed until the referred future event occurs. By analyzing the formal specification, unbounded future accesses can be detected

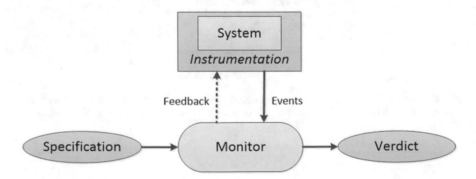

Fig. 7 Overview of the formal method runtime verification. Based on a formal specification, a monitor with formal guarantees is generated which checks the validity of system events to the system specification

and bounds on memory consumption given. Consider Example 1, Lola is used to compute the average velocity. The current velocity is given as sensor reading, i.e. input stream. The output stream *count* represents the number of received velocity values. It accesses the past *count* value by the offset operator $[-1, 0]$, where -1 refers to the previous value and 0 is used as out of bounds value, i.e. when there is no previous value which is the case at the start of the execution. Other values are possible, e.g. -10 and $+10$ which would refer to the tenth value in the past and the future, respectively. The output stream *sum_velocity* aggregates the velocity values by accessing the current *velocity* and the previous *sum_velocity* value. Finally, avg_velocity takes the average by combining both output streams.

```
input double velocity
output double count := count[-1,0] + 1
output double sum_velocity := velocity + sum_velocity[-1,0]
output double avg_velocity := sum_velocity / count
```

Example 1: Example Lola specification, calculating the average velocity

As next step, Lola was extended by parametric streams for network monitoring (Faymonville et al. 2016) and is referred to as Lola 2.0. Instances of stream templates are invoked, extended, and terminated at runtime. Example 2 shows an example specification checking whether a *requestID* is answered with a corresponding *grantID*. When a yet untracked *requestID* arrives, a new instance of the template stream *trackOpenRequests* is invoked (inv) using the parameter **id** to refer to the current *requestID*. The instance remains open until its terminate condition (ter) evaluates to true, i.e. the current *grantID* matches the **id**. The trigger notifies the user if more than ten instances of *trackOpenRequests* exist. The extend condition (ext) is a Boolean expression which evaluates whether a new computation step of the instance needs to be carried out. Hence, only the instances for which the current input is of relevance are considered. All others remain unchanged.

```
input int requestID,   grantID
output bool trackOpenRequests<id : int>;
    inv: requestID;    ext: true;    ter: id = grantID    := true
trigger count(trackOpenRequests) > 10
```

Example 2: Example parametrized Lola specification for a request/grant specification

More recently, asynchronously arriving input data can be handled due to new real-time features (Faymonville et al. 2019) which is an important and convenient language feature when considering embedded systems with multiple sensor sources. In RTLola, which stands for real-time Lola, each stream can individually evolve at arbitrary time. An output stream can be either evaluated event-based and periodically. Event-based outputs are computed each time a new and relevant input/output stream value occurs. To handle inputs which are susceptible for bursts of new data, periodic outputs were introduced. These outputs are evaluated at a fixed pace periodically and sliding windows efficiently aggregate arriving data allowing to still guarantee bounds on the memory consumption. Example 3 gives a specification which computes each second the average velocity over the last five seconds. The periodic stream *avg_velocity* uses aggregation with sliding windows to specific the system property in a concise and readable way.

---

input *velocity* : Float64
output *avg_velocity* :Float64 @1Hz := *velocity*.aggregate(over: 5s, using: avg)
trigger *avg_velocity* > 5 "WARNING: average velocity exceeds threshold"

Example 3: Example RTLola specification, calculating the average velocity

Lola meets the requirements of operation-level monitoring. The presented geofencing algorithm can be encoded directly in Lola and mentioned geofencing extensions incorporating risk and position prediction can be efficiently expressed with the latest Lola language features. Further, the declarative nature promises less error-prone monitors and the modular structure given by the possibility to interconnect streams improves maintainability and keeps specification concise and, therefore, easier to understand. Additionally, proven guarantees on the generated monitors, e.g. bounds on memory consumption, are an important aspect when monitors are intended to be integrated in safety-critical systems like unmanned aircraft.

4.3 Experimental Results

The case-study revolves around validating the risk buffer, i.e. the buffer whose violation leads to terminating the mission. Figure 8 illustrates the risk buffer in red. In white, some of the flown simulation trajectories are shown. The simulation allows varying both: wind conditions and wind directions. As it can be seen, a few trajectories exceed the risk buffer area, reason being strong wind conditions with a wind direction counteracting the passive rotation of the gyrocopter. The simulation results show that either a larger buffer is required or the vehicle is not allowed to fly under strong wind conditions. In general, simulation make it possible to draw safety-critical conclusions before the flight.

In the following, we explain the simulation setup and scenario. Also, we show experimental results and conclude.

Simulation Setup
The experiments are conducted in a MATLAB Simulink environment. The simulation modules are depicted in Fig. 9. The focus of the validation is on the *Geofencing Monitor*. The monitor was generated based on a Lola specification for geofencing. Input to the monitor is the estimated position coming from a *Sensor*

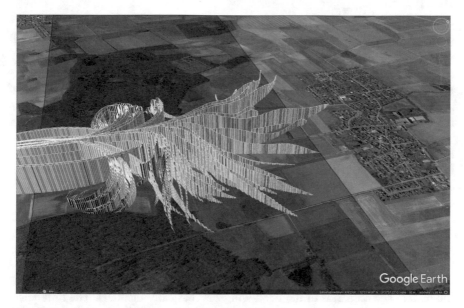

Fig. 8 Visualization of the gyrocopter termination scenario parameter study

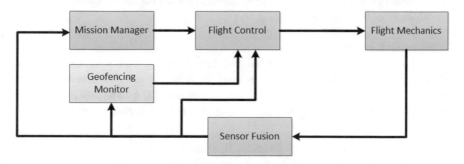

Fig. 9 Simulation setup using a Lola based monitor for geofencing

Fusion module. The trajectory is executed by the *Mission Manager* which keeps track of the current flight phase. Nominally, the *Flight Control* maps commands from the mission manager to actuator commands. In case of a geofence violation, the monitor commands a termination behavior which the flight controller prioritizes. The ALAADy simulation supports three different *Flight Mechanics*: a gyrocopter, a box wing, and a twin boom model. As termination behavior for the gyrocopter, a predefined actuator position is approached and the power supply is cut such that the vehicle spirals down smoothly. For the fixed-wings, parachutes are the chosen means as termination behavior. A distance of 1200 m is used as a risk buffer for the gyrocopter and a distance of 400 m is used for the fixed-wings. The simulation incorporates CS-AWO (All Weather Operations) compliant Environmental Models for wind and turbulences.

Scenario

For the validation of the risk buffer, we provoke a geofence violation in different aircraft states and different wind conditions. We differentiate between four wind conditions based on the Beaufort scale: *none* (0 m/s), *light-breeze* (2.6 m/s), *moderate-breeze* (6.3 m/s), and *strong-breeze* (12.4 m/s). We considered wind from south and south-west which represent the interesting cases for the scenario. Also, the UAS is flying in different attitudes: 60, 100, and 140 m.

In Fig. 10, the scenarios for the gyrocopter are depicted from a top view. As can be seen, all scenarios start at the bottom left corner. The objective of the scenarios is to trigger the flight termination at the same predefined point of the geofence under different attitudes of the aircraft. Therefore, three trajectories are chosen. One triggers the termination when the UAS is on a left turn, one when the UAS is on a right turn, and the last one triggers the UAS when it is flying in a straight line towards the fence. The effects of the different wind condition can be seen in the different simulation runs. In simulation runs under heavier wind conditions, the UAS is dragged further away from the planned trajectory.

For the following, note that the figures of the results focus on the relevant part of the buffer. The scenarios of the fixed-wings are basically identical. The difference being that they use different flight mechanics and rely on a parachute model for the termination behavior. This means, that the termination trajectory changes, but not the principle of terminating the flight.

Gyrocopter Results

Figure 11 depicts the gyrocopter simulation results. It starts ten seconds before the geofence violation and lasts until the ground is reached. The *buffer* distance, i.e. the distance from the geofence to the buffer line, is around 1200 m. The plot indicates the impact of the attitude of the gyrocopter and wind. If the termination is started during a

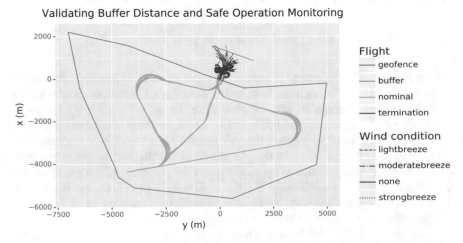

Fig. 10 Top view of the gyrocopter termination simulations. Termination is activated at a similar point of the fence but under different vehicle attitudes (left turn, right turn, straight towards) and wind conditions

left/right turn, the maneuver distance is smaller since the desired spiral behavior of the gyrocopter can be easier reached. Also, the figure shows that heavier wind conditions complicate reaching such desired spiral behavior. Especially when the vehicle is flying in a straight line towards the geofence, wind can even prevent reaching the desired behavior. As can be seen that buffer violations occur, i.e. the orange line is crossed. A possible reason is the underrating of the wind effects on the termination distance. There are two possible mitigations for this. First, the buffer is increased to capture underrated effects and a new geofencing specification is generated using FencyCreator. This results in an earlier flight termination within bounds triggered by the monitor, when running the same simulation. Second, the concept of operations, which describes in detail the operation and is required for the specific operation risk assessment (SORA), is updated and operations under strong-breeze wind conditions are not allowed.

Box Wing Results
The simulation results of the box wing are given in Fig. 12. As before, ten seconds before the geofence violation until ground contact is shown in the plot. The used *buffer* distance is around 400 m. As can be seen, the attitude of the vehicle as well as wind conditions barely influences the termination behavior due to the parachute. In all cases, the buffer line was sufficiently far away and no violations occurred.

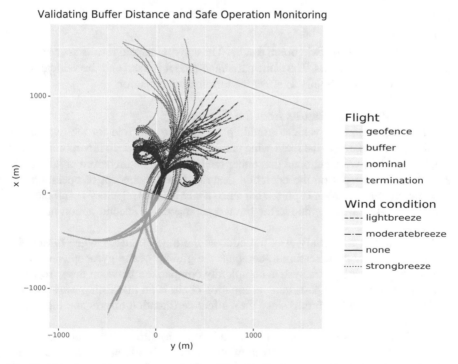

Fig. 11 Top view of gyrocopter termination. The visualization starts 10 s before the termination and lasts until vehicle ground contact

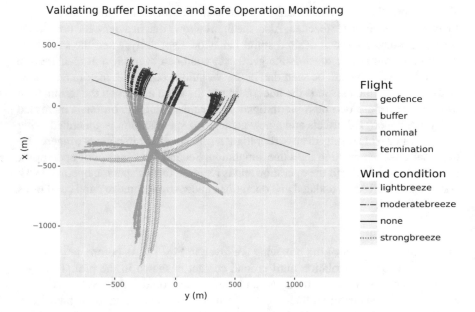

Fig. 12 Top view of box wing termination. The visualization starts 10 s before the termination and lasts until vehicle ground contact

Twin Boom Results

Figure 13 illustrates the twin boom results. The constraints and results are similar to previous box wing results. The same parachute model is used and the vehicle flight mechanics differ too slightly to affect the termination behavior.

Summary of Selected Results from the Simulation

Wind condition and the vehicle attitude play an important role for the gyrocopter termination compared to the fixed-wing parachute termination. To incorporate heavy wind conditions, we mentioned that either the buffer can be increased or additional constraints can be put on the operation described in the concept of operations. If possible, the first solution is preferable since a single safe trajectory is sufficient. If there is no safe trajectory due to the limited air space, the second option may be a solution.

Note that, the case study does not address the impact energy when hitting the ground. It would be interesting to compare the gliding of the gyrocopter with the use of a parachute. Also, a system complexity comparison between the gyrocopter termination chain and a parachute termination chain is an interesting aspect which should be addressed in future since they affect certification efforts and ultimately vehicle cost.

The described simulation is not tool qualified, i.e. the simulation results have to be manually checked. In general, each simulation module has to be an accurate model of the actual system, especially the flight mechanics, which is a challenging task when the buffer validation should be solely based on simulation results. However,

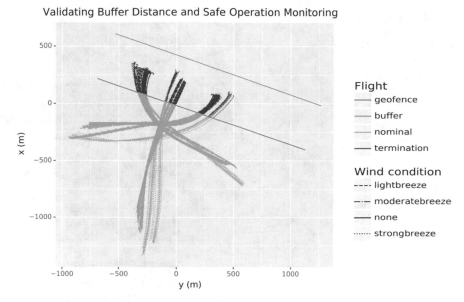

Fig. 13 Top view of twin boom termination. The visualization starts 10 s prior to termination and last until vehicle ground contact

in general. Simulations offer a methodical way to fine tune operational parameters. Often these operational parameters need to be derived since they are unknown a priori. In this paper, we derived a first set of buffer distance values and identified that these distances were under approximating environmental effects. Based on these observations the runtime monitors are refined to avoid buffer breaches due to these effects.

5 Lola Geofencing on FPGA

When integrating the safe operation monitoring architecture into an aircraft, different setups are possible. Generally, a Lola specification can be interpreted by the existing language engine, or the specification can be compiled to a standalone software implementation. Such realizations can run on compatible hardware architecture, an embedded PC, or even a microcontroller. Additionally, it is also possible to compile from RTLola specification into a hardware description language to realize an FPGA. The compilation translates the specification into Very High Speed Integrated Circuit Hardware Description Language (VHDL), which is later used to synthesize and configure an FPGA board. The approach is depicted in Fig. 14. VHDL is a common hardware description language to describe digital and mixed-signal circuits. Further, in contrast to ASICs, FPGAs offer the flexibility to program and reconfigure an integrated circuit.

Fig. 14 Approach from
Lola specification over
VHDL code to a realization
on an FPGA board

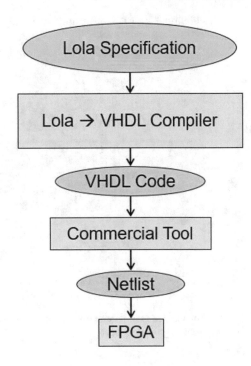

Each of the possible realizations of safe operation monitoring has different characteristics. Important integration aspects are:

- efforts to integrate the safe operation monitoring approach into the aircraft
- unobtrusiveness of the integration
- computing power
- power consumption
- certification considerations

One advantage of the FPGA implementation is that a hardware circuit implicitly gives runtime guarantees. The runtime of the computation can be computed by analyzing the depth of the embedded circuits of the FPGA. Additionally, FPGAs are small and have very low power consumption, since the hardware is tailored to the specific computation. Furthermore, hardware enables the possibility for highly parallel computations. For the geofence algorithm, this can be utilized to independently compute geofence border intersections. Mathematically, the geofence can be described as a set of straight lines, defined by linear equations. The intersections of these lines with the flown UAS trajectory make up the geofence violations. The FPGA implementation could enable to compute violations of all linear equations in parallel. This enables an extremely fast computation of the safety supervision in real time. As a result, the performance of an FPGA can be a multitude faster compared with microprocessors.

Since the safe operation monitoring is at the core of the operational safety, the certification considerations are extremely important for the realization of this methodology. As a result, the design assurance level (DAL) required for the safe operation monitoring often requires the highest criticality. In many cases, this would result in DAL A/B requirements.

Developing a certifiable FPGA implementation is generally possible, but requires a lot of effort, especially for DAL A. Guidance for developing safety critical complex electronic hardware is given by the standard document "DO-254 Design Assurance Guidance For Airborne Electronic Hardware". The standard DO-254 is process oriented. That means it describes the requirements for the processes for the planning, design, verification, and validation of hardware components. In the context of the standard, an FPGA is considered as complex electronic hardware. In contrast to simple electronic hardware, complex electronic hardware requires additional verification efforts. However, if the monitoring can be realized completely on the FPGA, the certification of the FPGA does not require certification of software regarding DO-178C or even the underlying operating system. Contrary, the certification of any software implementation does require the certification of the underlying operating system (if existent), as well as the hardware that runs the software.

In (Baumeister et al. 2019), a translation from a RTLola specification into synthesizable VHDL code is presented, but its limits concerning board utilization in terms of specification complexity remain unknown so far.

As FPGA board, the ZC702 with around 85,000 LUTs on board of the Xilinx Zynq-7000 System-on-a-Chip Evaluation Board is used. The Zynq-7000 supports Ethernet, CAN, USB, and other interfaces. Also, it comes with a dual ARM Cortex-A9 core processor and 4 GB DDR3 RAM. The integration of the FPGA into the system is simplified due to these components. Figure 15 shows the board attached to the payload rail of one of our UAS testbeds. We are preparing first simulation in order to actually test the system in-flight. In future, we plan to report our findings.

Fig. 15 Zynq-7000 System-on-Chip Evaluation Board attached to the payload rail of one of our UAS experimental testbeds

6 Conclusion

With the introduction of the "specific" UAS category by EASA, operational aspects can be incorporated for flight approval. The operational aspects are described in the concept of operation (ConOps) which is input for the specific operation risk assessment (SORA). During operation, these operational limits need to be supervised such that the assessed operational risk remains valid and prepared mitigations can be initiated.

Here, we introduced the notion of safe operation monitoring when referring to the supervision of operational constraints. We identified important properties about mission and system, software and hardware, and environmental limitations. Exemplarily, geofencing was picked as one of the most important mission properties for the specific category, to showcase a supervision property in more details. Geofencing is a technique to virtually limit the freedom of the UAS's movements due to virtual fences. An algorithm was presented in which a geofence is described as polygonal chain and intersections between the polygonal chain and the UAS's movement is calculated. Additionally, we showed important aspect when calculating a (risk) buffer, e.g. around critical infrastructures. For example, a risk buffer can be used to identify the point in space when a UAS needs to initiate its safe landing. Important aspects for the buffer we came up with are: position uncertainty, reaction time, wind, maximal height difference buffer, safety margin, and maneuver.

Next, we presented a simulation case study about geofencing. We motivated the usage of a tool such as FencyCreator to come up with valid polygonal chains for the geofences incorporating buffer distances. Further, we promoted the use of a formal specification language for monitoring in general, but also for geofencing specifically. A formal specification language facilitates the automatic generation of monitors which check the adherence of the system to its specification. Advantages of such an approach are: formal guarantees on monitors due to analysis of the specification, improved maintainability due to brief and concise specifications, and declaring system properties using an higher level of abstraction due to specification language features which support the implementation of complex system properties and are less error-prone compared to handwriting monitor code. The scenario of the case-study revolves around validating the risk buffer, i.e. the buffer whose violation leads to terminating the mission. Different aircraft configurations were considered: a gyrocopter and two fixed-wings (twin boom, box wing) with their individual termination maneuvers autorotation and parachute, respectively. The simulation results showed the impact of identified parameters for the buffer calculation. Especially for the gyrocopter autorotation, wind played the most significant role, since under strong breezes the desired spiral behavior of the gyrocopter could not be entered which finally resulted in geofence violations. Two mitigations could be identified: either the risk buffer has to be increased or flying under strong breezes has to be forbidden by the ConOps.

A realization of the geofence algorithm in hardware, e.g. FPGA, was discussed which could greatly benefit from the inherent parallel nature of such a hardware board.

In future, experimental result using an FPGA will show how well the assumptions on the parallelism could be met. Further, limits of a pure virtual validation of buffer distances in simulations remain an important open research question which should be addressed.

In future, we plan to include the monitor more in the control loop of the UAS. Using RTLola monitors, system health capabilities shall be extended to increase the situational awareness in flight. This enables the precise use of counter-measures and, further, improves automation function, e.g. trajectory planning can tailor its solution to the current system state. Further, monitoring other operational limits on the system in context of SORA is interesting and could enable flying in sparsely populated and populated areas where a safe flight termination of the UAS solely on geofence violations is no option anymore. Similar to this paper, results from simulations of the flights provide leading indicators of the applicability. In the future, it will be necessary to expand the simulation capabilities in order to examine and validate new scenarios.

Acknowledgements We would like to thank the Reactive System Group lead by Prof. Bernd Finkbeiner, Ph.D., for the close collaboration on runtime monitoring using the formal specification language Lola.

References

Adolf FM, Faymonville P, Finkbeiner B, Schirmer S, Torens C (2017) Stream runtime monitoring on UAS. In: Lahiri S, Reger G (eds) Runtime verification. RV 2017. Lecture Notes in Computer Science, vol 10548. Springer, Cham. https://doi.org/10.1007/978-3-319-67531-2_3

ASTM (2017) F3269-17 Standard Practice for Methods to Safely Bound Flight Behavior of Unmanned Aircraft Systems Containing Complex Functions. ASTM

Bartocci E, Falcone Y, Francalanza A, Reger G (2018) Introduction to runtime verification. In: Bartocci E, Falcone Y (eds) Lectures on runtime verification. Lecture Notes in Computer Science, vol 10457. Springer, Cham. https://doi.org/10.1007/978-3-319-75632-5_1

Baumeister J, Finkbeiner B, Schwenger M, Torfah H (2019) FPGA stream-monitoring of real-time properties. In: International conference on embedded software

Colombo C, Pace GJ (2018) Industrial experiences with runtime verification of financial transaction systems: lessons learnt and standing challenges. In: Bartocci E, Falcone Y (eds) Lectures on runtime verification. Lecture Notes in Computer Science, vol 10457. Springer, Cham. https://doi. org/10.1007/978-3-319-75632-5_7

D'Angelo B, Sankaranarayanan S, Sánchez C, Robinson W, Finkbeiner B, Sipma H, et al (2005) Lola: runtime monitoring of synchronous systems. In: 12th international symposium on temporal representation and reasoning, June 2005

Dauer JC, Dittrich JS (2021) Automated cargo delivery in low altitudes: concepts and research questions of an operational-risk-based approach. In: Dauer JC (ed) Automated low-altitude air delivery - towards autonomous cargo transportation with drones. Springer, Heidelberg

EASA (2015) Introduction of a regulatory framework for the operation of unmanned aircraft. Technical opinion. https://www.easa.europa.eu/sites/default/files/dfu/Introduction%20of%20a% 20regulatory%20framework%20for%20the%20operation%20of%20unmanned%20aircraft.pdf. Accessed 08 Oct 2020

EASA (2015-10) Introduction of a regulatory framework for the operation of drones. Advance Notice of Proposed Amendment. https://www.easa.europa.eu/sites/default/files/dfu/A-NPA%202 015-10.pdf. Accessed 08 Oct 2020

EASA (2017) Introduction of a regulatory framework for the operation of drones. Advance Notice of Proposed Amendment. https://www.easa.europa.eu/sites/default/files/dfu/NPA%202017-05% 20(B).pdf. Accessed 08 Oct 2020

Ellis K, Krois P, Davirs MD, Koelling J (2019) In-Time System-Wide Safety Assurance (ISSA) Concept of Operations. NASA Technical Reports. https://ntrs.nasa.gov/citatio6ns/20190032480. Accessed 08 Oct 2020

Faymonville P, Finkbeiner B, Schirmer S, Torfah H (2016) A stream-based specification language for network monitoring. In: Falcone Y, Sánchez C (eds) Runtime verification. RV 2016. Lecture Notes in Computer Science, vol 10012. Springer, Cham. https://doi.org/10.1007/978-3-319-46982-9_10

Faymonville P et al. (2019) StreamLAB: stream-based monitoring of cyber-physical systems. In: Dillig I, Tasiran S (eds) Computer aided verification. CAV 2019. Lecture Notes in Computer Science, vol 11561. Springer, Cham. https://doi.org/10.1007/978-3-030-25540-4_24

Geist J, Rozier KY, Schumann J (2014) Runtime observer pairs and bayesian network reasoners on-board FPGAs: flight-certifiable system health management for embedded systems. In: Bonakdar-pour B, Smolka SA (eds) Runtime verification. RV 2014. Lecture Notes in Computer Science, vol 8734. Springer, Cham. https://doi.org/10.1007/978-3-319-11164-3_18

Gilabert RV, Dill ET, Hayhurst KJ, Young SD (2017) SAFEGUARD: progress and test results for a reliable independent on-board safety net for UAS. In: IEEE/AIAA 36th digital avionics systems conference (DASC), St. Petersburg, FL, pp 1–9. https://doi.org/10.1109/DASC.2017.8102087

JARUS (2016) Guidelines on Specific Operations Risk Assessment (SORA). Draft for public consultation

NASA (2017) Reliable Geo-Limitation System for Unmanned Aircraft - An Assured Safety Net Technology for UAS. Patent reference. https://ntts-prod.s3.amazonaws.com/t2p/prod/t2media/ tops/pdf/LAR-TOPS-244.pdf. Accessed 08 Oct 2020

Nguyen T, Bartocci E, Ničković D, Grosu R, Jaksic S, Selyunin K (2016) The HARMONIA project: hardware monitoring for automotive systems-of-systems. In: Margaria T, Steffen B (eds) Leveraging applications of formal methods, verification and validation: discussion, dissemination, applications. ISoLA 2016. Lecture Notes in Computer Science, vol 9953. Springer, Cham. https:// doi.org/10.1007/978-3-319-47169-3_28

Nikodem F, Rothe D, Dittrich JS (2021) Operations risk based concept for specific cargo drone oper-ation in low altitudes. In: Dauer JC (ed) Automated low-altitude air delivery - towards autonomous cargo transportation with drones. Springer, Heidelberg

Pellizzoni R, Meredith P, Caccamo M, Rosu G (2008) Hardware runtime monitoring for dependable COTS-based real-time embedded systems. In: Real-time systems symposium, Barcelona, 2008, pp 481–491. https://doi.org/10.1109/RTSS.2008.43

Rozier K (2017) On the evaluation and comparison of runtime verification tools for hardware and cyber-physical systems. In: Reger G, Havelund K (eds). RV-CuBES 2017. An international workshop on competitions, usability, benchmarks, evaluation, and standardisation for runtime verification tools, vol 3, pp 123–137

Russell J (2018) Flugmechanische Untersuchungen zu Flugabbruchsystemen von unbemannten Frachtflugzeugen. Master Thesis at RWTH Aachen University

S.A.E. (2010) Guidelines for development of civil aircraft and systems, ARP4754A. SAE International. https://www.sae.org/standards/content/arp4754a/

Sachs F (2021) Configurational aspects and vehicle specific investigations for future unmanned cargo aircraft. In: Dauer JC (ed) Automated low-altitude air delivery - towards autonomous cargo transportation with drones. Springer, Heidelberg

Schirmer S, Torens C, Adolf FM (2018) Formal monitoring of risk-based geo-fences. In: AIAA information systems-AIAA infotech @ aerospace, Florida, Kissimmee, USA. https://doi.org/10. 2514/6.2018-1986

Schopferer S, Donkels A (2021) Trajectory risk modelling and planning for unmanned cargo aircraft. In: Dauer JC (ed) Automated low-altitude air delivery - towards autonomous cargo transportation with drones. Springer, Heidelberg

Skoog MA, Hook LR, Ryan W (2020) Leveraging ASTM industry standard F3269-17 for providing safe operations of a highly autonomous aircraft. In: IEEE aerospace conference, Big Sky, MT, USA, 2020, pp 1–7. https://doi.org/10.1109/AERO47225.2020.9172434

Solet D, Béchennec JL, Briday M, Faucou S, Pillement S (2016) Hardware runtime verification of embedded software in SoPC. In: 11th IEEE symposium on industrial embedded systems (SIES), pp 1–6

Stamenkovich J, Maalolan L, Patterson C (2019) Formal assurances for autonomous systems without verifying application software. In: Workshop on research, education and development of unmanned aerial systems (RED UAS), Cranfield, United Kingdom, 2019, pp 60–69. https://doi.org/10.1109/REDUAS47371.2019.8999690

Stevens MN, Rastgoftar H, Atkins EM (2019) Geofence boundary violation detection in 3D using triangle weight characterization with adjacency. J Intell Robot Syst 95:239–250. https://doi.org/10.1007/s10846-018-0930-5

Stevens M, Atkins E (2018) Layered geofences in complex airspace environments. In: Aviation technology, integration, and operations conference, Georgia, Atlanta, USA. https://doi.org/10.2514/6.2018-3348

Torens C, Durak U, Nikodem F, Schirmer S (2019) Formally bounding UAS behavior to concept of operation with operation-specific scenario description language. In: AIAA scitech forum, California, San Diego, USA. https://doi.org/10.2514/6.2019-1975

Torens C, Durak U, Nikodem F, Dauer JC, Adolf FM, Dittrich, JS (2018) Adapting scenario definition language for formalizing UAS concept of operations. In: AIAA modeling and simulation technologies (MST) conference, Florida, Kissimmee, USA. https://doi.org/10.2514/6.2018-0127

Part IV: Verification, Validation and Discussion

A Multi-disciplinary Scenario Simulation for Low-Altitude Unmanned Air Delivery

Simon Schopferer, Alexander Donkels, Sebastian Schirmer, and Johann C. Dauer

Abstract The design and development of unmanned aircraft is a complex multi-disciplinary task. In the context of risk-based and operation-centric assurance and certification, operational constraints play a key role. Consequently, technologies that impact operational risks such as flight termination, command-and-control links, and onboard autonomy have to be addressed holistically and within the context of the operation. A scenario simulation combines these various aspects originating from different disciplines into one versatile tool. It allows making informed design trade-offs and validating operational requirements at early development stages. In this work, we give insight into the scenario simulation framework developed within the research project ALAADy (Automated Low-Altitude Air Delivery) which addresses unmanned freight operations at low-altitudes within the context of operation-centric certification as introduced by EASA with the so-called specific category. We describe use-cases and requirements derived from research objectives. Furthermore, we present the simulation architecture as well as details on implementation and modelling. Results from an exemplary simulation study regarding safe flight termination in the event of an operational constraint violation are shown to demonstrate the applicability and usefulness of the simulation framework as a tool for holistic unmanned aircraft design.

Keywords Simulation · UAS · Unmanned air delivery · Aircraft design · Parameter study · Risk-based operation · Specific-category

S. Schopferer (✉) · A. Donkels · S. Schirmer · J. C. Dauer
German Aerospace Center (DLR), Institute of Flight Systems, Lilienthalplatz 7 ,
38108 Braunschweig, Germany
e-mail: simon.schopferer@dlr.de

A. Donkels
e-mail: alexander.donkels@dlr.de

S. Schirmer
e-mail: sebastian.schirmer@dlr.de

J. C. Dauer
e-mail: johann.dauer@dlr.de

© Deutsches Zentrum für Luft- und Raumfahrt e. V. (DLR) 2022
J. C. Dauer (ed.), *Automated Low-Altitude Air Delivery*, Research Topics in Aerospace,
https://doi.org/10.1007/978-3-030-83144-8_17

1   Introduction

The foundation of manned aviation lies in the high and assured safety of all systems involved, especially aircraft but also including all systems that are part of the aviation infrastructure. This high degree of safety in technical systems allows for quite general approvals of aircraft and operations restricted only by airport and airspace requirements, pilot training and mandatory procedures. When designing unmanned aircraft in the context of operations within EASA's specific category (EASA 2015, 2017), the same safety targets apply. However, the specific category allows for achieving the safety targets by considering specific operational limitations such as restricted flight corridors, times and durations of flights, or changes to the aircraft itself (e.g. additional safety equipment). Consequently, these operation-specific aspects should be considered during design and development. For example, a small and lightweight unmanned aircraft may be generally approved to fly in rural areas, whereas for flying in urban environments, it may have to be equipped with means to reduce its impact energy and a transponder to increase situational awareness of other airspace users. Also, flights may have to be restricted to night time when there is less traffic. In consequence to these requirements, a parachute system, a transponder and adequate lights could be integrated into the system. It is easy to see that this would have a significant impact on the aircraft's performance and the considered use-case. Possibly, a new aircraft would have to be developed to meet the requirements for urban flights in this example as opposed to using the same aircraft operated in rural areas. This operation-centric approach is in contrast to the vehicle-centric design and development of manned aircraft and poses new challenges and opportunities for unmanned aviation.

In order to meet these new challenges of designing unmanned aircraft, simulations including all aspects of unmanned aircraft operations are required, allowing to analyze the impact of operational constraints and to validate operation-specific requirements. In this work, we present a scenario simulation framework as an example of a multi-disciplinary and holistic validation tool for the operation-centric design and development of unmanned aircraft. The simulation framework was developed within the ALAADy (Automated Low-Altitude Air Delivery) research project (Dauer and Dittrich 2021). It is based on the framework presented in Dauer and Lorenz (2013). The framework has been extended with new modules to simulate long-range and low-altitude missions for three cargo aircraft configuration concepts that were developed and analyzed within the ALAADy project. Furthermore, new automation and configuration facilities were implemented enabling comprehensive and versatile simulation studies. An overview over the simulation framework architecture including its most important modules, variants thereof and signals between is depicted in Fig. 1.

In the following section, an overview of related work on simulation frameworks for unmanned aircraft is given. In Sect. 3, we describe the use-cases of the presented simulation framework, i.e. the design trade-offs and validation tasks for which this scenario simulation has been designed for. Also, we give an overview of general

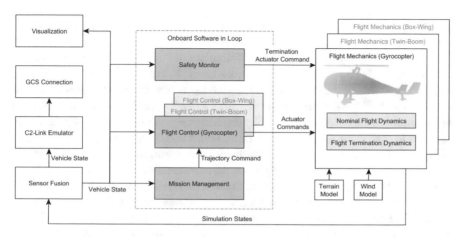

Fig. 1 Overview of the simulation architecture with modules, module variants and signal flow

aspects of simulation development. In the subsequent section, details on the simulation framework architecture are presented with a focus on new modules developed in the ALAADy project. Also, new automation and configuration facilities are described in detail. We show results from an exemplary simulation scenario study addressing the safety of flight termination in varying wind conditions. In Sect. 4, we discuss challenges and lessons learned in regards to the development of the presented simulation framework. Finally, in Sects. 5 and 6, a short summary of the work is given along with an outlook for future work towards multi-disciplinary scenario simulations for unmanned aircraft operations.

2 Related Work of UAS Simulation Environments

Simulation is often used as a means of verification and validation of algorithms and concepts, but also for the analysis of interactions and the influence of parameters and variables. Dynamic simulations are used for the simulation of unmanned aircraft. The term dynamic refers to the necessity of time-wise integration of signals to determine the model behavior.

Although serving similar goals, in comparison to formal methods such as mathematical proofs, simulations have to overcome certain challenges but also provide advantages. Since they cannot be fully represented in written publications like any software technical implementation, simulation as proof is often considered less powerful. However, formal proofs quickly fail for more complex scenarios, they are often more difficult to apply and are not as flexibly adaptable as simulations. A comparison of formal methods and simulation proofs for unmanned aerial vehicles can be found in Osborne et al. (2019). Simulations are often the only available tool to analyze and validate systems exceeding certain levels of complexity, but they lack the credibility of a formal proof.

One way to increase the significance of simulations is standardization. For manned aviation, DO-331 (RTCA 2011) represents a baseline, but is limited to development requirements. Dauer and Lorenz (2013) provide taxonomy for the abstraction of simulations for UAS. Different user roles for a simulation are introduced. From such a role definition the necessity to formally describe a simulation scenario arises. Such a description can be based on parameter trees, which is closely connected to the efforts of Jafer et al. (2016) to define a standardized formulation for simulation scenarios.

To provide the complex behavior of the flight mechanics of manned aircraft in the form of a standardized test environment, a model for a transport aircraft was presented in Hueschen (2011). Fezans and Deiler (2019) progress these concepts and create an open-source project for exactly that purpose, providing a common simulated aircraft for testing purpose. For unmanned aircraft these forms of standardization do not yet exist. However, there is a large number of powerful open-source tools to provide at least comparable simulation environments. These include projects like AirSim, Gazebo, Morse, jMAVSim, Paparazzi Simulator, or HackflightSim, cf. Ebeit et al. (2018). Alternatively, different open-source projects are integrated to provide simulations with a particular focus, e.g. considering environmental sensors (Jerath and Langelaan 2016). Due to their focus on small UAS, these simulation environments are not directly applicable for the conceptual study of the large scale aircraft of ALAADy. Additional simulation projects from the operator's perspective can be found in Osborne et al. (2019).

If certain components of the simulation are exchanged by real-world samples, different types of simulations are distinguished. Software-in-the-loop (SIL) simulations are cases in which concrete implementations of a software component are integrated in the simulation. As an example, this can be an algorithm for navigation or other autonomy functions. Implementations are e.g. autopilot-related simulation projects collected in Ebeit et al. (2018).

Hardware-In-the-loop (HIL) simulations go one step further and integrate complete hardware components and replace physical models with real-world components, see e.g. Kim (2013). Especially for the development of small unmanned aircraft, such simulation systems need to be as flexible and cost-effective as possible, see Pizetta et al. (2016) for such a low-cost HIL environment. Depending on the perspective of the argumentation, for verification purposes these systems can also be referred to as simulation-in-the-loop test environments, cf. Day (2015). A special case of HIL and SIL combined are so-called computer-in-the-loop simulations, see e.g. Dauer and Lorenz (2013) and Xu et al. (2015). In this case, the software implementation is integrated with the target computing unit into the simulation.

Simulations that include people are also widely used (human-in-the-loop simulation). These can be pilot-in-the-loop simulations, for example for educational purposes (Roberts and Beck 2017), preparation and training of flight experiments (Lorenz et al. 2021) as well as identifying the capabilities and impact of the pilot in the experimental setup (Donkels and Voigt 2019). Especially the simulation of air spaces requires such human-in-the-loop environments; see e.g. Mühlhausen and

Peinecke (2021). Another form of combining real-world elements and simulations are so-called mixed reality simulations, cf. Selecký et al. (2019). In this case, the complete experiment is carried out by a real UAS, but environments and interactions are simulated virtually. These forms are suitable for preparing experiments risky for the unmanned aircraft, such as obstacle avoidance or coordination of several vehicles in close proximity.

The simulation of airspaces and integration of UAS into existing structures touches an active field of development. Weinert and Underhill (2018), for example, present a concept for generalizable test trajectories for Monte Carlo simulations of airspace situations. Al-Mousa et al. (2019) presents a simulation concept that covers aspects of airspace integration of small UAS such as Detect and Avoid, communication protocols and management of the UAS. For the simulation of Detect and Avoid of larger configurations like the ALAADy concepts, see Schalk and Peinecke (2021).

3 Use-Cases for the ALAADy Simulation

The purpose of the ALAADy scenario simulation is to serve as a validation tool at different stages and steps of the design and development process of an unmanned transportation aircraft which is to be operated at low-altitudes and certified based on an operation-specific, risk-based set of rules. The aircraft is envisioned to fly autonomously while being supervised and controlled from a control station with the use of a command-and-control (C2) datalink. A key concept of ALAADy is the use of an onboard safety operation monitor to detect violations of any operational constraints and to mitigate the resulting risks by triggering a safe flight termination (Schirmer and Torens 2021).

In order to simulate such a complex operation, various systems must be modeled:

- aircraft flight dynamics,
- aircraft sensor system providing a state estimation,
- flight control system providing stabilization and envelope protection,
- mission management system providing flight guidance and C2 communication,
- flight termination system,
- safety operation monitor,
- C2 datalink.

Furthermore, aspects of the environment must be modeled to increase the degree of realism:

- Terrain defines the vertical profile of the low-altitude trajectory and may affect the C2 datalink availability.
- Wind affects the aerodynamics of the aircraft both in nominal flight and during flight termination.
- Air traffic, manned or unmanned, may have to be taken into account to maintain safety of the operation.

A holistic scenario simulation that models all of the aspects mentioned above may serve for a broad set of use-cases which can be categorized into validation tasks and design trade-offs.

3.1 Validation Tasks

The following validation tasks are exemplary use-cases for a holistic scenario simulation.

Safe Autonomous Flight Under Nominal Operating Conditions
Under nominal operating conditions, the unmanned aircraft shall fly autonomously using its onboard sensors, the flight control system and the mission management system. The interplay between these systems must be validated in simulations covering all automated procedures and phases of automated flight as well as the range of environmental conditions considered to be normal operating condition. Also, procedures on the ground, i.e. monitoring and guiding the autonomous vehicle at the ground control station, can be validated and trained using a scenario simulation.

Safe Flight Termination
A key concept of ALAADy is that flight over sparsely populated areas may be safely terminated at any time using a flight termination system in order to mitigate the risk of violations of the operational constraints. This is accomplished with a safe operation monitor that oversees the system's internal states and activates the flight termination system in case of an imminent violation of any operation constraint. For example, if the aircraft deviates too far from the nominal trajectory, the safety monitor must terminate the flight early enough for the aircraft to touch down within the designated safety buffer zone. A set of rules is implemented in terms of a formal specification which the safety monitor uses to make its decision on continuing or terminating the flight.

Depending on the considered set of parameters, such as wind conditions, aircraft position, speed, attitude and states of the aircraft's systems, it may not be feasible to simulate every combination of parameters for situations that may lead to a violation of operational constraints. However, a scenario simulation may serve as a tool to test a broad range of situations with representative parameter coverage in order to increase confidence in the flight termination system.

C2-Link Loss Procedures
The availability of the C2-link is expected to be mandatory for nominal operation of an ALAADy aircraft. However, due to technical and economic reasons, the datalink must be considered unreliable to a degree at which the safety of the operation cannot rely on the uninterrupted availability of the datalink. Therefore, technical mechanism and procedures will have to be developed to mitigate the risks emerging from a C2-link loss. For example, this could involve the onboard mission management system

initiating a maneuver to increase the probability of quick C2-link recovery and the safety monitor triggering a countdown to flight termination in case the link cannot be recovered. Upon link recovery, measures could be taken at the ground control station to increase availability of the C2-link, e.g. by commanding an alternative route. A scenario simulation which models C2-link properties and availability can be used to validate and to train such complex procedures.

Airspace Integration
Although manned and unmanned air traffic is separated whenever possible for safety reasons, the integration of both manned and unmanned aircraft into the same airspace may be required in particular situations. In the context of ALAADy, such a situation could be landing an unmanned cargo aircraft at a regional airport or a small airfield. Depending on the type of airspace and traffic to be encountered, the unmanned aircraft may have to be guided in a way not to interfere with the other traffic as shown in Mühlhausen and Peinecke (2021). Also, airspace requirements such as a transponder and a voice communication relay to the ground control station may have to be fulfilled. To develop and validate procedures of interaction between the unmanned aircraft and air traffic control in an integrated airspace is an exemplary use-case for a scenario simulation.

3.2   Design Trade-Offs

Besides the aforementioned validation tasks, the following design trade-offs are another set of exemplary use-cases for a holistic scenario simulation.

Vehicle Configuration for Specific Mission Scenarios
For most manned aircraft, the design goal is to find a trade-off between performance, efficiency and safety for a vehicle to transport passengers or cargo at a target range and capacity. The life cycle of manned aircraft usually spans multiple decades, which must be taken into account in the design process. In contrast, unmanned aircraft are developed at a rapid pace for very specific and diverse mission scenarios. The diversity of mission scenarios also results in a diversity of vehicle configurations. This leads to the challenge of finding the optimal configuration for a specific mission scenario. For manned aircraft, performance criteria to analyze and compare vehicle configurations are well established and universal to a certain degree. This is not the case for unmanned aircraft. Instead, the complex interaction of the vehicle, its onboard and ground-based systems and any applicable operational constraints must be considered to analyze the performance of unmanned aircraft configurations for a specific mission scenario. A scenario simulation allows for considering these various aspects when comparing the performance of vehicle configurations.

Safety Features and Mission Performance

A key aspect of the operation-specific certification approach is that operational constraints are taken into account in the safety assessment. This allows to make trade-offs between operational constraints and remaining safety risks and corresponding mitigation measures. For example, by restricting the flight corridor to avoid any infrastructure such as roads, the result may be that a parachute system is not required as opposed to allowing flight over roads with low traffic volume where a parachute might be required as a means to reduce impact energy in case of a flight termination. In order to assess the trade-off, the cost of the parachute system and its impact on the overall system performance must be held against the impact of the additional flight corridor constraint which may result in longer flight times. A scenario simulation may be used to quantitatively assess such complex interactions between mission performance, safety and mitigation features and operational constraints.

Level of Autonomy

For unmanned flight operations, the level of autonomy plays a central role in development of the aircraft and its systems, the safety assessment and certification process and the development of operational procedures. Generally, an increasing level of autonomy reduces the reliance on the command-and-control datalink and the ground crew. At the same time it increases requirements on the performance and reliability of the onboard flight management and control system. Depending on the level of autonomy, safety-relevant events such as C2 link loss, loss of separation or system degradation may have to be handled differently. Both technical solutions as well as operational procedures can be tested and validated with a holistic scenario simulation.

3.3 Aspects of Simulation Development

The diversity of validation tasks and the use of the simulation as a tool to evaluate the conceptual design of the unmanned aircraft and operational procedures create a demand for a high level of flexibility and adaptability of the simulation. Here, flexibility refers to a minimal timely effort to adapt the simulation to changing configurations, operational parameters, or environmental factors. The flexibility can be supported for example if core parts of the simulation can be reused for different use-cases and scenarios. The adaptability means the accessibility to every part of the simulation to change functionalities or parameters which allows direct transferring of design alterations into simulated content. Also, changes at runtime (e.g. the weather or wind situation) may be required. To support the design process, new functions need to be added and integrated into existing parts of the simulation in an easy manner. The extension with new functions must be accompanied by test procedures to avoid jeopardizing the correctness and accuracy of the simulation.

The traceability of user input data to the simulation's output data plays a key role for the evaluation process of different designs and operations and needs to be supported by the simulation infrastructure. Traceability focuses on the relationship between use-case requirements and the simulation implementations. Also, in the context of comparability between investigated configurations, it must be possible to uniquely identify and store the configuration of a considered scenario. A configuration, once stored, allows reviewing correct configuration assembly, restoring results, or deriving new configurations. The comparability is improved by a deterministic behavior, especially of naturally random occurring phenomena like turbulences or gusts.

The evaluation process also requires obtaining a complete set of output data which allows retracing states, decisions, and behavior of all simulated modules of the system. This aspect requires functions to collect and store data which need to adapt to the configuration.

If user interaction is required, the simulation must be real-time capable, i.e. the computation of a single simulation timestep must be less than or equal to the duration of this timestep. The synchronization of simulation time and real time can then be achieved by slowing down the simulation if and as required. With increasing complexity of the simulation framework, the required runtime per timestep can be expected to grow. Furthermore, when conducting simulation experiments covering entire missions and many parameter variations, the simulation runtime becomes an important factor to consider. In this case, it is desirable to speed up the simulation beyond real-time in order to complete as many experiments as possible within a given time frame. The duration of a timestep for such a fast-time simulation is the sum of all runtimes of the simulation modules executed during the timestep. Thus, each module should be optimized in terms of runtime and measures to decrease the overall simulation runtime should be considered when developing the simulation framework.

4 Simulation Framework

This section provides information about the implemented simulation framework. Further, its actual execution is described.

4.1 Simulation Architecture

The simulation is based on a modular Matlab/Simulink framework first presented in Dauer and Lorenz (2013). It features an automated configuration management, which is able to assemble library models and their parameters to an executable simulation experiment. A simulation experiment, called a *scenario*, is the highest level of the simulation architecture. A scenario consists of an environment model and at least one system model representing the unmanned aircraft with all subsystems relevant to the simulation experiment. The system model contains the system-under-test, i.e. the software components that are also deployed to the actual aircraft's components. As typical for a software-in-the-loop simulation the software components are wrapped into simulation models managing incoming and outgoing signals. In contrast to this, the environment model contains physical models or emulated system components which mimic the behavior of subsystems that are not included as part of the system model. For example, the flight dynamics model is a physical model that simulates the aircraft's motion and generates inputs for the navigation sensor emulation.

Each simulation scenario has a specified system model configuration and environment model configuration. The modular structure of the framework supports the easy assembly and exchange of the system model configuration featuring a rapid alteration of scenarios. This ability is a key aspect needed for the investigation of a broad variety of system configurations. Also, new modules representing novel hardware or software parts, like flight termination systems or datalinks, can be easily integrated into the scenario in the form of new modules or module variants.

System and environment model consist of different modules, from which the most important are introduced in Sect. 4.3 and Sect. 4.4. The number and type of modules as well as their relation to each other has to be defined manually by the user. For each module, one variant from the framework can be selected or a new one can be created. From this point, the automated configuration management can take over: according to the chosen variant, a Simulink model for the module is automatically obtained from the library and the required model parameters are set. The selection of modules, each module's variant, and the numerical value of each module parameter form the configuration of the system and environment. The right side of Fig. 2 displays the system model hierarchy as an example. Note that modules may also contain several stages of submodules which also have several variants with different parameters.

4.2 Automated Simulation Build Process

For each scenario, the user can define and alter the configuration using descriptions (see left side of Fig. 2). Definitions are added as options to the description's content. If no definitions were made, pre-defined default configurations are used. As descriptions exist on system, module and submodule level, it is possible that variations of modules are set multiple times starting with default configurations which are overwritten by user definitions made on system, module, and submodule level.

When the simulation build process is started, the available descriptions are automatically collected and evaluated by a startup script within the description phase. The startup script stores options for scenario (simulation time, step size, etc.), system, and modules in a structure which can be interpreted as a pre-configuration. Within the following build phase, the initialization scripts for each scenario, system, and module are automatically called with the pre-configuration as input. The output containing the parameters and Simulink model of each module is then assembled to the system and environment configuration. The assembler conducts automatic consistency checks which ensure that a variant is available and defined for each model. Otherwise, the assembly process is aborted with a notice to define or create a module variant.

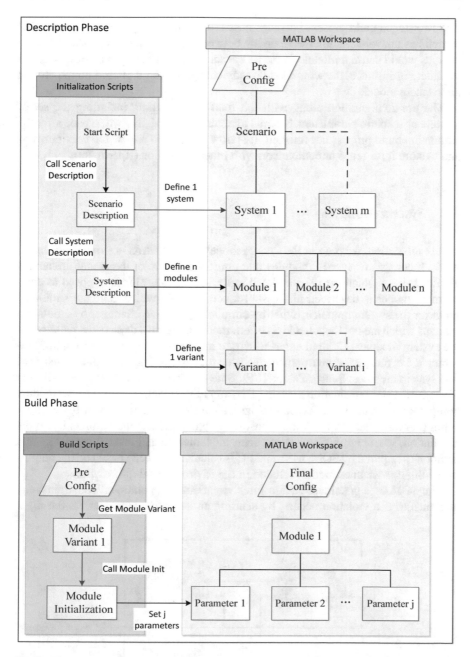

Fig. 2 The automated simulation build process consists of two stages, description phase, which creates the Pre-Configuration structure within the MATLAB workspace and build phase, which uses the Pre-Config. To build the final configuration by setting all module parameters

The two staged build process becomes necessary as there might be undesirable but sometimes unavoidable interdependency between different modules. These dependencies would cause a module to set its variant based on the options of a different module. For instance, the wind module may change its variant based on the chosen atmosphere model.

The pre-configuration along with the final configuration, including the set of module or scenario parameters, become available in the Matlab workspace after the simulation build process has finished. It is now possible to review the configuration and to store it for reuse or comparison with other simulation experiments.

4.3   System Modules

In this subsection, we present the system model of the ALAADy simulation framework. It consists of system modules that contain software of the actual unmanned aircraft system which is to be validated in the simulation. This is opposed to environment modules, described in the subsequent section, which generate simulated input for the system modules, either by simulating physical effects or by emulating systems which are not included in a system module. Figure 3 depicts the dataflow of the system modules. Colored edges illustrate interfaces between System and Environment, i.e. orange for environment and blue for system. The simulation consists of five system modules: Sensor Fusion (SF), Safe Operation Monitor (SOM), Mission Manager (MM), Termination System (TS), and Flight Control (FC). The SF module computes the navigation solution. For the real-world system, the sensor fusion algorithm processes data from various sensors. In order to limit the complexity of the simulation, sensor emulation has not been implemented and instead the navigation solution as required by SOM, MM, and FC is calculated directly based on the outputs of the flight mechanics model included in the environmental information.

The SOM uses positional data to infer violations on operational constrains, e.g. geofencing. If a violation occurs, by sending an activation signal to TS the flight

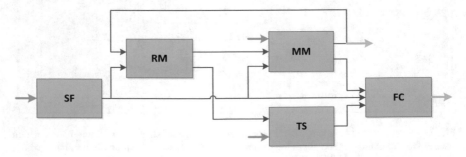

Fig. 3   System modules of the ALAADy simulation. Environment model output and system model output depicted in orange and blue, respectively

termination is initiated. If no violation occurs, the MM executes the trajectory by consecutively sending trajectory tracking commands to FC. Finally, FC translates target positions to actuator commands in order to follow the executed trajectory.

The following tables give more precise information on the inputs, outputs, and module parameter of mentioned system modules. Also, for each module a list of tasks for which it is responsible is given. Further, the inputs and outputs are itemized with the corresponding port numbers enumerated from top to bottom as seen in Fig. 3. Meaning that e.g. at port 3 the mission manager receives the navigation solution (Table 1).

4.4 Environment Modules

In the previous section, modules of the system model were described. In this section, we complement the picture by presenting the modules of the environment model. Figure 4 shows the Flight Mechanics (FM) of the aircraft, a module which simulates wind (WIND), and a map module which incorporates map information (MI), e.g. terrain.

Figure 5 depicts emulated system components that are part of the environment model: C2 datalink emulation to/from the Ground Control Station (DL2GCS, DL2MM) and the Ground Control Station itself. The diagram also shows the dataflow for different communication failure effects when sending data to the Ground Control Station (DL2GCS → GCS) or receiving data from the Ground Control Station (GCS → DL2MM). Again, colored edges illustrate interfaces between System and Environment. In Table 2, the mentioned environment modules are described in detail.

4.5 Automated Execution and Parameter Variations

In the following sections, the automated execution of the simulation framework is described. Especially in the context of parameter studies, where many simulation experiments have to be run with altered configuration and/or parameter settings, an automated execution is indispensable.

Configuration Management
One use-case for the ALAADy simulation is the investigation of different flight situations for different vehicle configurations. Such an investigation requires changes in every level of the simulation's hierarchy. A set of combinations of variations of vehicle, modules, or parameters shall be investigated. A variation means the exchange, addition, or reduction of modules of a vehicle, changes of a module's variant, or alteration of module parameters. For the three investigated vehicle types (Sachs 2021) the variations v per module or parameter result in the amount of N permutations of vehicle configurations, which need to be simulated. The number of

Table 1 Overview of the simulation system modules

Module	Sensor fusion (SF)
Inputs	1. Environment information
Outputs	1. Navigation solution
Tasks	• Receive information • Emulate sensors • Compute solution
Parameter	/
Module	Safe operation monitor (SOM)
Inputs	1. Mission status 2. Navigation solution
Outputs	1. System health 2. Termination signal on board
Tasks	• Checks integrity of navigation solution i.e. system health • Supervises mission
Parameter	/
Module	Mission manager (mm)
Inputs	1. Mission profile 2. System health 3. Navigation solution
Outputs	1. Mission status 2. Vector to next waypoint
Tasks	• Executes trajectory • Leads back to trajectory • Communication with ground control station • Switches to updated trajectory
Parameter	/
Module	Flight controller (FC)
Inputs	1. Vector to next waypoint 2. Navigation solution 3. Termination activated
Outputs	1. Actuator control signals
Tasks	• Translate vector to actuator controls • Activate termination procedure i.e. either activate parachute or pre-defined actuator setting
Parameter	/
Module	Termination system (TS)
Inputs	1. Termination signal on board 2. Termination signal on ground
Outputs	1. Termination mode
Tasks	• Communicates over termination link with people on ground • Switches to termination mode
Parameter	/

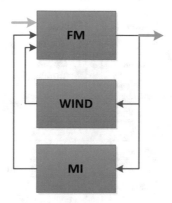

Fig. 4 Environment feedback on wind and map information, e.g. terrain. Environment model output and system model output depicted in orange and blue, respectively

Fig. 5 Datalink emulator. Environment model output and system model output depicted in orange and blue, respectively

permutations N for three vehicles, m varied modules, and p varied parameters can be calculated using:

$$N = 3 \times \prod_{i=1}^{m} v_i \times \prod_{j=1}^{p} v_j. \tag{1}$$

The chain of multiplications in (1) indicates that even few variations of few parameters lead to a significant amount of simulations. The changes to the configuration after each simulation are not reasonable to be conducted manually. Thus, the already automated simulation build process is wrapped by an additional test sequence script, which implements a sequence of simulation experiments and also automates the perturbation process. This script is described in the following.

The test sequence script is able to automatically manipulate the configuration. For each vehicle, module, or parameter, which are generalized to entities in the following, the user may define a list of variations and starts the test sequence script. An exemplary test sequence is shown in Table 3.

When changing vehicle or module variants, the build process has to be restarted as such changes may require to load new modules from the module library and resolve interdependencies between modules. On the other hand, changing parameter values is possible without restarting the build process as all required modules are already

Table 2 Overview of the simulation environment modules

Module	Ground control station (GCS)
Inputs	1. Mission status with errors
Outputs	1. Updated mission profile 2. Termination signal on ground
Tasks	• Human mission supervision
Parameter	/
Module	Datalink (DL2GCS)
Inputs	1. Mission status
Outputs	1. Mission status with errors
Tasks	• Datalink failure model: fading, packet loss, packet delay
Parameter	Fading effects models: • FSPL (Free-Space Path Loss) • FE2R (Flat Earth Two Ray) CE2R(Curved Earth Two Ray)
Module	Datalink (DL2MM)
Inputs	1. Updated mission profile
Outputs	1. Mission profile
Tasks	• Datalink failure model: Fading, packet loss, packet delay
Parameter	Fading effects models: • FSPL (Free-Space Path Loss) • FE2R (Flat Earth Two Ray) CE2R(Curved Earth Two Ray)
Module	Flight mechanics (FM)
Inputs	1. Actuator control signals 2. Map data 3. Wind information
Outputs	1. Vehicle state
Tasks	• Model of aircraft • Termination procedure
Parameter	Choice between three aircraft models: • Box Wing (parachute) • Twin Boom (parachute) Gyrocopter (pre-defined actuator positions)
Module	Wind model (WIND)
Inputs	1. Vehicle state
Outputs	1. Wind information
asks	• Model for wind • Model for turbulences • CS-AWO all weather operations compatible

(continued)

Table 2 (continued)

Module	Wind model (WIND)
Parameter	• Turbulence intensity (none, light, medium, heavy) • Mean wind speed in 20ft above ground level (Beaufort scale) • Wind direction (north, north-east, …, west, north-west)
Module	Map information (MI)
Inputs	1. Vehicle state
Outputs	1. Map data
Tasks	• Connection to map database: obstacles, terrain height
Parameter	/

Table 3 Example list of user defined variations for vehicle, modules, and parameters

Layer	Entity	Variations
Vehicle	Vehicle	Box_Wing, Gyrocopter, Fixed_Wing
Module	Turbulence module	Turbulence_1, Turbulence_2
Module	Wind profile	Wind_Profile_1, Wind_Profile_2, Wind_Profile_3
Parameter	Wind_Velocity	1, 2, 3, 4, 5, 10, 15, 20 (m/s)
Parameter	Wind_Direction	0, 90, 180°

present and loaded. Thus, the variation list from above is split into two lists: a list of vehicle and module permutation and a list with parameter permutations. This way significantly reduces the amount of build processes and shortens the computation times for the test sequence.

An algorithm is used to generate the two lists by collecting all possible permutations of the variations. Given Table 3, the two generated lists are shown in Table 4.

The test sequence script builds the first simulation experiment using the first sequence of vehicle and module variations. The changes are made during the description phase. At the end of the execution phase, the sequence of parameter variants is applied. The experiment is automatically started and terminated according to a specified termination event. This event can be the end of a fixed simulation time or vehicle status variables reaching a defined threshold. For the simulation of flight termination (see Schirmer and Torens 2021) the simulation is terminated as the vehicle reached the ground.

After termination of the simulation experiment, a directory with the name of the simulated variations is automatically created and the results are collected and stored within the directory. Results of experiments are log files which are generated by the used modules. Then, the next parameter sequence is applied, until the last parameter sequence is reached. Finally, the next vehicle and module sequence is used within

Table 4 Example lists of vehicle and module permutations and of parameter permutations

Entity	Sequence					
	1	2	3	4	...	18
Vehicle	Box_Wing	Box_Wing	Box_Wing	Box_Wing	...	Fixed_Wing
Turbulence Module	Turbulence_1	Turbulence_1	Turbulence_1	Turbulence_2	...	Turbulence_2
Wind Profile	Wind_Profile_1	Wind_Profile_2	Wind_Profile_3	Wind_Profile_1	...	Wind_Profile_3
Parameter	Sequence					
	1	2	3	4	...	24
Wind_Velocity (m/s)	1	1	1	2	...	20
Wind_Direction	0°	90°	180°	0°	...	180°

the next simulation build process. In the end, all vehicle, module, and parameter combinations were executed and all logfiles for subsequent analysis were generated.

Example Simulation Scenario Execution

In order to provide an illustrative example, we consider the conducted experiments for flight termination using the presented simulation framework. In the context of ALAADy, the flight termination of the vehicle is activated whenever a critical violation of given operational constrains is detected. The reason for this is that the safety of the operation and, thus, the safety of people on the ground can only be guaranteed if all operational constraints are met. In this section, we focus on the flight termination due to spatial violation. The technique to detect spatial violations is called geofencing. In geofencing, virtual barriers in time and space are defined which the aircraft is not allowed to cross. If one of these virtual barriers is crossed an operational violation is detected, followed by the activation of the flight termination. Experimental results on the actual geofencing implementation can be found in Schirmer and Torens (2021). Figure 6 shows a visualization of the simulated termination trajectories obtained in a simulation study for testing the geo-fencing functionality in different wind conditions.

The following describes how the implemented configuration management could support the execution of simulations. The initial question was under which circumstances a flight termination can occur. The derived circumstances comprise: wind, turbulences, and different initial turn rates. As assumption, a steady vehicle speed and a constant height above ground during flight were given. The configuration setting is shown in Table 5. As can be seen, all vehicles were considered since each of them

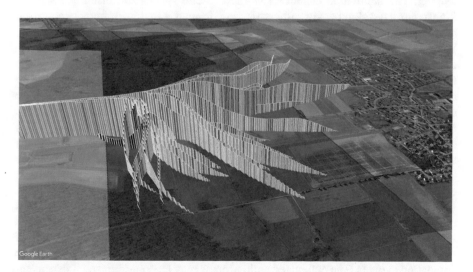

Fig. 6 Simulated flights with different wind conditions and termination activated when entering the risk buffer (red area): for multiple cases, the termination trajectory exceeds the risk buffer in this example

Table 5 Configuration setting for flight termination

Layer	Entity	Variations
Vehicle	Vehicle	Box_Wing, Gyrocopter, Fixed_Wing
Module	Turbulence module	None, light, medium
Module	Wind profile	CS-AWO conform
Parameter	Wind_Speed	0, 2.6, 6.3, 12.4 (m/s)
Parameter	Wind_Direction	North, North-East, East, …, North-West
Parameter	Start height	60, 100, 140 (m)
Parameter	Trajectories	leftTurnInFence, straightAheadInFence, rightTurnInFence

behaves differently under termination. Next, different turbulences modules were defined for different levels. Operational constrains allowed flying up to even medium grade turbulences. Further, a wind module which facilitates required changes in wind speed and wind direction was used. The initial height of the aircraft was defined and different trajectories were used leading to different roll angles of the aircraft at the moment when the geofence violation occurs.

5   Discussion

The creation of one tool to serve the various tasks presented in Sect. 3 is very challenging. The simulation cannot be tailored to one specialized discipline such as flight control development or software testing. Instead, a multi-disciplinary approach addressing aircraft and system design as well as operational aspects must be taken. Still, the simulation should be manageable and maintainable with low cost for development. The choice of a modular framework solved many issues arising from the great number of different configurations and experiments which are designed to answer the questions arising from research, design and validation. Modules are composable such that functions can be re-used in multiple simulation experiments. New functions can either be added as new modules or new module variants which allow a straight-forward extension of the simulation framework while maintaining a synoptic structure. However, the integration of modules and module variants draws high effort for interface design which has to ensure coherent connections for data exchange between the modules while enabling independent exchange of modules. Interfaces not only exist to other modules but also to the automated simulation build process as well as to data collection services. The automated build process, which initializes the modules one by one, specifies a rigid structure for implementation of modules and variants. The structure's complexity grows as the complexity of the simulation evolves. The automated build process may be compromised if dependencies between modules occur. Dependencies occurring between different hierarchy levels of modules intensify this problem. Figure 7 shows the initialization process for three example modules. Part a) shows no dependencies and the order of initialization

Progess of
Initialization

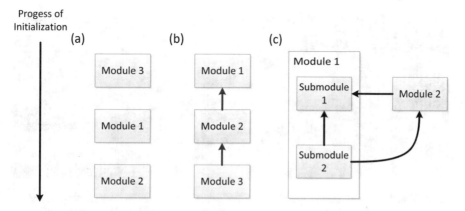

Fig. 7 Initialization of **a** modules without dependencies, **b** modules with dependencies marked by the arrows, and **c** circular dependencies between modules, which disables automatic simulation build processes

can be arbitrarily performed. Part b) shows a sequential dependency of the modules which can be solved by implementing a specific initialization sequence. Part c) shows a circular dependency occurring between two modules within a hierarchical structure. Here, Module 1 is initialized before Module 2, as the initialization of Model 2 depends on Submodule 1. Meanwhile, Submodule 2 of Module 1 depends on Module 2 and thus Module 1 cannot be fully initialized. This conflict cannot be solved by the automatic simulation build process. Possible ways to overcome the problem are to either change the module dependencies or to upgrade Submodule 2 to a new Module 3 which enables sequential initialization according to Part b). Upgrading Submodule 2 to Module 3 results in a diversion between the simulation build architecture and the model architecture: Module 3 will still remain a submodule of Module 1 within the Simulink model.

The dependency problem illustrates the relatively large effort for module integration. However, this is a one-time effort which allows a rapid exchange of modules and a flexible assembly of following simulation experiments. The hierarchical structure enables a broad adaptability of modules but also results in a trade-off between simulation completeness and complexity.

During simulation development, integrity was ensured by a software testing framework which was built in parallel to the modules and adapted to the modular architecture enabling testing on module level. The ability to store the scenario configuration allows to review the simulation build process and to trace the requirements for each use-case.

To enable deterministic behavior for all simulation executions without human interaction, external inputs such as mission tasks defining the system's flight path are loaded from files or tables. Furthermore, stochastic processes and parameters such as amplitude and frequency of gusts are created in a deterministic and reproducible manner using pseudorandom number generators. Also, the simulation bases on sets

Fig. 8 Visualization of interchangeable cargo drone configurations as an example of vehicle simulation modules with dedicated flight mechanics as developed in the ALAADy research project (left to right): box wing, autogiro, fixed wing, fixed wing parachute termination

of deterministic, non-stochastic, connected, and uniquely defined equations, such as kinematical equations, which ensure that deterministic behavior is maintained.

Principally, the presented simulation framework supports all of the use-cases described in Sect. 3. Within the ALAADy project, it was used to validate specific aspects of unmanned air delivery operations. First of all, it served as a testing and proof-of-concept environment for vehicle configurations and systems such as automated flight control and flight management as well as safe operation monitor, flight termination systems, C2-link communication and integration with the ground control station. As an example for the versatility of this modular approach, Fig. 8 depicts a set of vehicle configurations implemented in the ALAADy simulation framework. Also, the flight termination of a fixed-wing drone using parachutes is shown, which is implemented as a dedicated flight mechanics module for the corresponding vehicle. Low-altitude flights over distances of several hundred kilometers were simulated under nominal conditions to assess C2 link coverage and reliability (Schalk and Becker 2021). This also served as a validation of automatically planned trajectories as described in Schopferer and Donkels (2021). Also, a thorough validation of the flight termination, as briefly described in Sect. 4.5, was conducted by varying input parameters such as wind speed and direction. Details on this validation can be found in Schirmer and Torens (2021). These exemplary use-cases demonstrate the applicability and usefulness of the presented holistic scenario simulation.

6   Future Perspective

For future work, the achievable simulation runtime, real-time and fast-time aspects and resulting effects on the simulation quality could be analyzed in depth. In practice, the desirable simulation runtime depends on the use-case. If procedures with human interaction are to be assessed or trained, the simulation timing must be as close to real-time as possible. If, on the other hand, a great number of experiments should be conducted, the simulation should run as fast as possible, preferably much faster than real-time. However, some simulation modules such as the C2 link emulation may require complex calculations that may significantly slow down the simulation. A trade-off between simulation fidelity and simulation runtime must be made in this case. The presented simulation architecture supports switching between variants of simulation modules. In future work, this could be used to investigate the impact of this trade-off for specific use-cases. For example, it would be interesting to analyze how low and high fidelity variants of the simulation can be efficiently combined to achieve both a good coverage of the input parameter space and a high significance and trustworthiness of the simulation results.

Another aspect to consider in future work is the integration of the simulation with an air traffic simulation in order to support assessing airspace integration and airspace management procedures. In the ALAADy project, detect and avoid aspects as well as airspace integration aspects were considered using separate simulations described in Schalk and Peinecke (2021) and Mühlhausen and Peinecke (2021). Integrating the two simulations with each other would allow assessing the impact of nominal and non-nominal unmanned operations to other airspace user and the air traffic system. Also, the integration with simulated UTM (Unmanned Traffic Management) services could provide a powerful tool for assessing the impact of such services on manned and unmanned operations.

7 Summary

In this work, we present a multi-disciplinary scenario simulation for unmanned cargo operations which was developed within the scope of the ALAADy research project. We discuss potential use-cases and highlight the need for holistic scenario simulations in the context of operation-centric design and certification approaches for unmanned aircraft. The simulation framework's architecture and implementation details such as automation facilities for parameter studies are described in detail. Also, we show results from an exemplary simulation study on the safety of the flight termination behavior of a cargo aircraft in the context of safe-operation monitoring applied for geo-fencing.

The presented simulation framework has been used for design and validation of various aspects of the unmanned freight operations concept within the ALAADy research project. For design and validation, a modular framework has been chosen which needs higher effort for module integration or extension with new functionalities. In return, the framework can be used as one tool within different stages of vehicle and operation development as well as to answer diverse research questions. Considering these aspects, we share the experiences and lessons learned from simulation design and development and discuss issues of software complexity.

Also, we highlight the benefits and drawbacks of the presented simulation architecture. These insights may serve as a reference and example for future research on multi-disciplinary simulations for complex unmanned aircraft operations.

References

Al-Mousa A, Sababha BH, Al-Madi N, Barghouthi A, Younisse R (2019) UTSim: a framework and simulator for UAV air traffic integration, control, and communication. Int J Adv Rob Syst 16(5):1–19. https://doi.org/10.1177/1729881419870937

Dauer JC, Lorenz S (2013) Modular simulation framework for unmanned aircraft systems. In: AIAA modeling and simulation technologies (MST) conference, Boston, MA, 19–22 August 2013. AIAA, pp 4974–4986. https://doi.org/10.2514/6.2013-4974

Dauer JC, Dittrich JS (2021) Automated cargo delivery in low altitudes: concepts and research questions of an operational-risk-based approach. In: Dauer JC (ed) Automated low-altitude air delivery - towards autonomous cargo transportation with drones. Springer, Heidelberg

Day MA, Clement MR, Russo JD, Davis D, Chung TH (2015) Multi-UAV software systems and simulation architecture. In: International conference on unmanned aircraft systems (ICUAS), Denver, CO, 9–12 June 2015, p. 426–435. https://doi.org/10.1109/ICUAS.2015.7152319

Donkels A, Voigt A (2019) Helicopter Formation control algorithm for manned-unmanned teaming. In: Vertical flight society's 75th annual forum, Philadelphia, PA, 13–16 May 2019

EASA (2015) Introduction of a regulatory framework for the operation of unmanned aircraft. Technical opinion

EASA (2017) Introduction of a regulatory framework for the operation of drones. Advance notice of proposed amendment

Ebeid E, Skriver M, Terkildsen KH, Jensen K, Schultz UP (2018) A survey of open-source UAV flight controllers and flight simulators. Microprocess Microsyst 61:11–20. https://doi.org/10.1016/j.micpro.2018.05.002

Fezans N, Deiler C (2019) Inside the virtual test aircraft (VIRTTAC) benchmark model: simulation architecture. In: ASIM workshop 2019, Braunschweig, Germany, 21–22 Feb. 2019

Hueschen RM (2011) Development of the transport class model (TCM) aircraft simulation from a sub-scale generic transport model (GTM) simulation. NASA/TM-2011-217169. National Aeronautics and Space Administration, Hampton, Virginia, USA. http://hdl.handle.net/2060/201100 14509. Accessed 30 Jun 2020

Jafer S, Chhaya B, Durak U, Gerlach T (2016) Formal scenario definition language for aviation: aircraft landing case study. In: AIAA modeling and simulation technologies conference, Washington, DC, 13–17 June 2016, p 3521. https://doi.org/10.2514/6.2016-3521

Jerath K, Langelaan JW (2016) Simulation framework for incorporating sensor systems in UAS conceptual design. In: AIAA modeling and simulation technologies conference, San Diego, CA, 4–8 January 2016. AIAA. https://doi.org/10.2514/6.2016-1186

Kim M, Kang S, Kim W, Chun I (2013) Human-interactive hardware-in-the-loop simulation framework for cyber-physical systems. In: 2nd international conference on informatics & applications (ICIA), Lodz, Poland, 23–25. September 2013, p. 198–202. https://doi.org/10.1109/ICoIA.2013.6650255

Lorenz S, Benders S, Goormann L, Bornscheuer T, Laubner M, Pruter I, Dauer JC (2021) Design and flight testing of a gyrocopter drone technology demonstrator. In: Dauer JC (ed) Automated low-altitude air delivery - towards autonomous cargo transportation with drones. Springer, Heidelberg

Mühlhausen T, Peinecke N (2021) Capacity and workload effects of integrating a cargo drone in the airport approach. In: Dauer JC (ed) Automated low-altitude air delivery - towards autonomous cargo transportation with drones. Springer, Heidelberg

Osborne M et al (2019) UAS operators safety and reliability survey: emerging technologies towards the certification of autonomous UAS. In: 2019 4th international conference on system reliability and safety (ICSRS), Rome, Italy, 2019, pp 203–212. https://doi.org/10.1109/ICSRS48664.2019.8987692

Pizetta IHB, Brandão AS, Sarcinelli-Filho MA (2016) Hardware-in-the-loop platform for rotary-wing unmanned aerial vehicles. J Intell Robot Syst 84:725–743. https://doi.org/10.1007/s10846-016-0357-9

Roberts E, Beck A (2017) Management and training programs of military drone small unmanned aircraft systems. Proc Hum Factors Ergon Soc Ann Meet 61(1):1131–1135. https://doi.org/10.1177/1541931213601767

RTCA (2011) DO-331: model-based development and verification supplement to DO-178C and DO-278A. RTCA Inc., Washington, USA

Sachs F (2021) Configurational aspects and vehicle specific investigations for future unmanned cargo aircraft. In: Dauer JC (ed) Automated low-altitude air delivery - towards autonomous cargo transportation with drones. Springer, Heidelberg

Schalk LM, Becker D (2021) Data link concept for unmanned aircraft in the context of operational risk. In: Dauer JC (ed) Automated low-altitude air delivery - towards autonomous cargo transportation with drones. Springer, Heidelberg

Schalk LM, Peinecke N (2021) Detect and avoid for unmanned aircraft in very low level airspace. In: Dauer JC (ed) Automated low-altitude air delivery - towards autonomous cargo transportation with drones. Springer, Heidelberg

Schirmer S, Torens C (2021) Safe Operation monitoring for specific category unmanned aircraft. In: Dauer JC (ed) Automated low-altitude air delivery - towards autonomous cargo transportation with drones. Springer, Heidelberg

Schopferer S, Donkels A (2021) Trajectory risk modelling and planning for unmanned cargo aircraft. In: Dauer JC (ed) Automated low-altitude air delivery - towards autonomous cargo transportation with drones. Springer, Heidelberg

Selecký M, Faigl J, Rollo M (2019) Analysis of using mixed reality simulations for incremental development of multi-UAV systems. J Intell Robot Syst 95:211–222. https://doi.org/10.1007/s10846-018-0875-8

Weinert A, Underhill N, (2018) Generating representative small UAS trajectories using open source data. In: 2018 IEEE/AIAA 37th digital avionics systems conference (DASC), London, 23–27 September 2018, pp 1–10. https://doi.org/10.1109/DASC.2018.8569745

Xu GY, Ren LM, Zhang HM, Tong H (2015) Implementation of a rapid prototyping aircraft simulation system based on iHawk platform. In: International conference on virtual reality and visualization (ICVRV), Xiamen, China, 17–18 October 2015, pp 290–294. https://doi.org/10.1109/ICVRV.2015.50

Capacity and Workload Effects of Integrating a Cargo Drone in the Airport Approach

Thorsten Mühlhausen and Niklas Peinecke

Abstract Within the project ALAADy (Automated Low Altitude Air Delivery) several options for start and destination of a cargo transport mission are discussed. One option in the concept study is the integration of larger transport drones in regular airport operations. In this case, integration of the vehicle in air space class B, C, or D (depending on the country) is required. The feasibility of this option was evaluated at the Air Traffic Validation Center at DLR (German Aerospace Center). As an example, the approach sector of Dusseldorf airport (EDDL) was simulated with different arrival traffic flows with and without a drone included. In a Human in the Loop (HITL) simulation the test persons controlled the air traffic in the sector according to ICAO rules assuring separation between all aircraft (including the drone) at any time. The resulting traffic flow with and without drone was compared to analyze the capacity effects of that vehicle. In addition, the test persons performed a questionnaire asking for workload and situational awareness. Based on the information gained by the analysis and the questionnaire, possible operational improvements were developed.

Keywords Unmanned aircraft airport integration · Human in the loop · Air traffic control validation · Air traffic simulation

1 Introduction

To allow smooth integration of the ALAADy (Automated Low Altitutude Air Delivery) vehicle with other traffic means, especially with manned air traffic, operations at regular airports are preferred. This would allow change of cargo in a hub-and-spoke manner. Large aircraft can bring in cargo and multiple cargo drones can distribute it to the recipients. Integrating the drone into the regular airport operation

T. Mühlhausen (✉) · N. Peinecke
German Aerospace Center (DLR), Institute of Flight Guidance, Lilienthalplatz 7, 38108 Braunschweig, Germany
e-mail: thorsten.muehlhausen@dlr.de

N. Peinecke
e-mail: niklas.peinecke@dlr.de

© Deutsches Zentrum für Luft- und Raumfahrt e. V. (DLR) 2022
J. C. Dauer (ed.), *Automated Low-Altitude Air Delivery*, Research Topics in Aerospace,
https://doi.org/10.1007/978-3-030-83144-8_18

would have the advantage that no additional infrastructure needs to be build. Instead, existing airports could be used.

On the other side, integrating a fully automated vehicle into manned air traffic poses a challenge in itself. All existing safety regulations that apply at an airport need to be adhered to. Further, there is the danger of losing efficiency due to the uncommon operation of an unmanned vehicle in regular airspace.

For the integration, different models can be implemented: Aside from treating the UA (Unmanned Aircraft) as a regular aircraft, another option could also be to block the assigned runway for regular traffic for the time it takes to land the UA. As a last option there is the possibility to have a dedicated UA hub or runway for unmanned traffic exclusively. This, of course, would mean to sacrifice some of the advantages in using an existing airfield.

The feasibility of these approaches was evaluated in real time simulations at the DLR Air Traffic Validation Center at the Institute of Flight Guidance. For the study Dusseldorf airport (EDDL) was selected as example. It is equipped with a parallel runway system in direction 05 resp. 23. For the trials, arrivals at runway 23R with varying traffic flows were simulated. Additionally, one arriving ALAADy cargo drone was injected to the arrival flow.

Section 2 describes the settings of the simulation environment and the implemented scenario. In Sect. 3 the conducted simulation runs as well as the results and derived recommendations are depicted. Section 4 closes the chapter about airport approach integration with final remarks and an outlook on the next steps for integrating low flying unmanned cargo aircraft in airport operations.

2 Related Work

Concepts and simulation studies to integrate UA into controlled airspace exist since at least 2005 (Theunissen 2005) (Korn 2006). The possibility of an integration of an unmanned aircraft into regular airport operations was outlined in Geister (2013) and then demonstrated by DLR (Geister 2017). In this paper the concept of a dedicated workplace for a pilot or controller in the vicinity of the airport assisting especially the approach and landing was suggested. We adopt this concept calling it Service Remote Pilot (SRP). Further, in this work a real landing on an actual airport was performed.

A more complete evaluation of the concepts of airport integration was given in Helm (2018). Aside from an overview of existing airspace integration concepts an extensive simulation study is conducted with multitude options of integrating the UA into the airport approach, including the integration in regular operations, segregated approach routes or corridors, and autonomous Detect and Avoid maneuvers.

3 Simulation Environment

3.1 Facility

The simulation environment was set up at the Air Traffic Validation Center at the DLR Institute of Flight Guidance in Braunschweig. The center offers a large variety of simulation tools to analyze air traffic management and has successfully been used to evaluate airspace integration concepts for UA (Schmitt 2008). This includes fast time and real time simulation tools and facilities. For the intended feasibility study about including a low flying unmanned cargo aircraft in the regular air traffic, a Human in the Loop (HITL) simulation of an approach sector is best suited. Therefore, the ATMOS (Air Traffic Management and Operations Simulator) is configured accordingly.

The simulator is based on the NLRs Air Traffic and Research Simulator (NARSIM). It consists of up to five controller working positions (CWP). These can be configured freely to cover the simulation needs. In addition, several pseudo pilot positions exist. These pseudo pilots control depending on the complexity of the scenario up to twelve aircraft. They get their advisories from the controller via simulated radio connection, read them back according to ICAO phraseology, and put them into the simulation system. This way the controller acts in a very realistic environment. A supervisor controls the sequence of the simulation as well as the data recording. Figure 1 depicts the different workplaces in the ATMOS.

Fig. 1 ATMOS (top-Supervisor, left-Controller, right-Pseudo pilots)

3.2 Scenarios

For this research topic, an existing model of Dusseldorf airport (EDDL) was used. The simulated approach sector modeled the arriving aircraft as well as the integrated ALAADy cargo drone, in case it was part of the scenario. All traffic landed on runway 23R. If not present at the start of the simulation, new arrivals were injected 10 NM ahead of one of the Initial Approach Fixes (IAF) TEBRO, XAMOD, DOMUX, or BIKMU in flight level 110 and a calibrated air speed of 250 kts. Figure 2 shows the layout of the approach used.

For the approach an open trombone was used, so that the air traffic controllers had to explicitly instruct aircraft to turn onto the final approach line. If not, the aircraft remained on the opposite approach and extended them beyond waypoints DL410 respectively DL430.

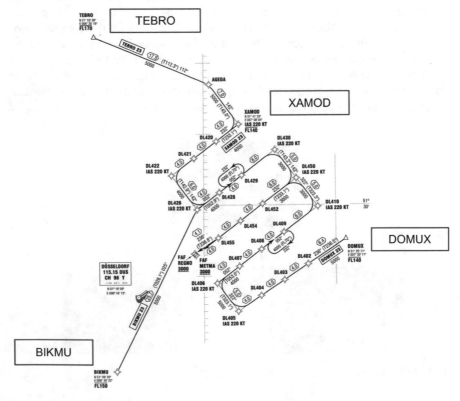

Fig. 2 Approach chart Dusseldorf airport (EDDL) (based on (DFS 2018))

Table 1 Advantages and disadvantages of simulation approaches

	Option 1: connect simulators	Option 2: implement BADA type
Advantages	• Exact model • Extending DLR evaluation competence	• Easy implementation based on BADA standard • Usage of one simulator only
Disadvantages	• High effort to define and implement interfaces and network infrastructure • High effort during simulation due to operating two different simulators	• Performance parameters of vehicle need to be approximated • No HMI evaluation for the ALAADy vehicle possible

3.3 Vehicle Model

Two options were possible to implement the ALAADy cargo drone into the ATMOS simulation: Either the existing vehicle simulation developed in Schopferer et al. (2021) could be used. In this case all interfaces between the two simulators would have to be bridged or adapted. A second possibility would be to devise a BADA (BADA 2014) model and implement this into the aircraft database of ATMOS. Table 1 summarizes advantages and disadvantages of both approaches.

Since both negative arguments for option 2 are not relevant for the selected questions, this option was chosen for the project course. On the one hand, a technology demonstrator is already being created in the project (Lorenz et al. 2021) so that the HMI evaluation can be carried out later using the actual aircraft. On the other hand, approximated performance parameters are sufficient here, since at the moment of executing the study only the target parameters existed. There was no airworthy ALAADy aircraft existing at that time on which the exact parameters could be measured.

The gyrocopter configuration is considered for the further course of the evaluation. Since the desired performance parameters of the three developed ALAADy vehicle variants (Hasan and Sachs 2021) have been fixed in the project and do not differ significantly in terms of BADA (Base of aircraft data) (BADA 2014) parameters, the result of the evaluation can be used for all variants.

The implementation in the NARSIM aircraft database is carried out by adding the aircraft-specific parameters to the database for aircraft published by Eurocontrol and used in NARSIM (BADA). The aircraft performance is calculated on the basis of the total energy model.

3.4 Airspace Environment

An operation start of the cargo drone near Dortmund was planned for Düsseldorf Airport (EDDL). To adhere to the minimum risk approach inhabited areas have to be avoided during the approach. Thus, effectively geofencing was carried out for planning the actual approach path (Peinecke and Mühlhausen 2021). This procedure

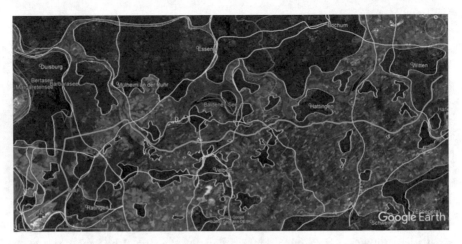

Fig. 3 Approach path to EDDL avoiding inhabited areas

describes flight path planning using software that identifies areas such as residential areas that have to be avoided for a possible flight path. Figure 3 shows red polygons that represent the inhabited area around EDDL. The main roads marked by yellow lines further limit the area of a possible flight path. It cannot be avoided that main roads are crossed or parts of less populated areas are cut. The final planned flight path of the cargo drone from Dortmund to the EDDL is shown as a green line. This path does not interfere with the SIDs and go-around procedures at Düsseldorf Airport.

For the integration point of the cargo drone in the final approach of the airport, a final UA integration point is introduced, the UA Approach Fix (UAF) (see Fig. 3). The idea is to include the UA in the glide path as late as possible to minimize the impact on the remaining air traffic. The rest of the landing approach is then designed in a straight line so that the aircraft is brought down to the ground in a stable manner.

Finally, the flight behavior of the specified aircraft and the simulation environment were verified by checking the following points: On the one hand, it was checked whether the flight characteristics, such as flight speed, climb rate and curve radius, correspond to the specifications and, on the other hand, whether the air traffic controller work station, the pseudo pilot work stations and the flight movements generated by the traffic generator meet the requirements.

4 Experimental Procedure

The tests were carried out with three different test subjects serving as Approach air traffic controller (ATCO) on one day each. Each subject underwent exactly seven scenarios in a randomized order. The scenarios differ in terms of traffic volume and the type of integration of the ALAADy drone, see Table 2.

Table 2 Test scenarios overview

Traffic volume (IFR Flights per hour)	Scenario		
	Baseline without ALAADy drone	A: With ALAADy drone, without Service Remote Pilot	AS: With ALAADy drone, with Service Remote Pilot
Low: 25	–	4	1
Medium: 35	–	5	2
High: 45	7	6	3

The seven scenarios were classified into three different flight plans in terms of traffic volume (low traffic, medium traffic and high traffic). The medium traffic scenario was chosen with a traffic flow of 35 arriving flights per hour. The low and high traffic scenarios each had a traffic flow reduced or increased by ten arriving flights per hour.

Furthermore, the simulation scenarios were classified into three scenario groups. Figure 4 illustrates the types of airport integration: The "H"-symbol marks an optional dedicated UAS hub. The red area symbolizes a possible existing zone that needs to be avoided, for example, an area restricted for reasons of safety or noise emissions.

On one side, the controller had a Service Remote Pilot (SRP) available to pilot the ALAADy drone in three scenarios (see Fig. 4, top). The workplace of an SRP was suggested in Geister (2017). The SRP serves to operate the ALAADy cargo drone, in case the speed needs to be changed or the drone has to be sent to a locally defined holding pattern. The vicinity to the airport allows for a close interaction with ATC.

On the other hand, there were three other scenarios with the same traffic volume only without the possibility of an SRP (see Fig. 4, bottom). In all of these scenarios, the aircraft flew the flight path and the landing completely autonomously. In addition, one of these scenarios was selected at random, in which the cargo drone was terminated shortly before reaching the final approach, i.e. close to the airport.

The seventh scenario had a high volume of traffic, but was simulated without the use of a cargo drone. This scenario served as a reference scenario (see Table 2).

Each trial day was divided into a morning and an afternoon part. Each subject underwent all seven simulated scenarios in one day. In addition to three simulation scenarios, the morning part also included a briefing and a training scenario before the actual simulation runs. In the afternoon part, four simulation scenarios were carried out. At the end there was a debriefing. Here, the air traffic controllers involved were also asked for their personal assessments.

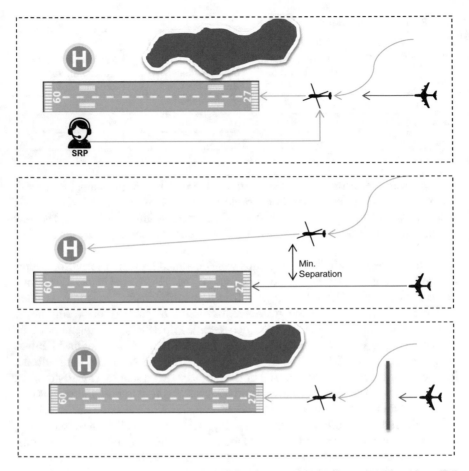

Fig. 4 Options for airport integration: with SRP (top), using dedicated hub (middle), without SRP (bottom)

5 Evaluation

5.1 Evaluation Criteria

The evaluation of the campaign took place through the collection of system-side data (separation performance, actual flight distance), with the help of questionnaires on the workload, the scalability and the situation awareness, as well as individual interviews with the air traffic controllers.

Table 3 ISA rating scale

Level	Work load	Free capacity	Description
1	Underutilized	Extensively	Little or nothing to do, rather boring
2	Relaxed	Plenty	More time than necessary to get the job done. Time passes slowly
3	Good	Some	There is enough to do to stay focused. All tasks are under control
4	High	Little	Nonessential tasks are postponed. The controller is working at the limit and could not hold this level for long. Time flies by
G	Extreme	None	Some tasks cannot be finished. The controller is overloaded and does not feel safe

Instantaneous self-assessment (ISA) is one of the most commonly used subjective measurement methods for mental workloads in real-time simulation. It was developed for the assessment of the workload of air traffic controllers (Kirwan et al. 1997). The operator, in this case the controller, estimates his perceived workload retrospectively. The measurement method uses a five-point rating scale (Table 3).

The radio signals recorded during the simulation runs are also analyzed with regard to the frequency assignment. By comparing the frequency allocation and other key performance indicators, a possible correlation between the integration of an ALAADy drone and a possibly increasing simulated voice radio can be assessed. This procedure provides indirect information about how busy the controller is. Few, long instructions indicate direct vectoring instructions, whereas many short instructions indicate a division of work steps. The percentage frequency assignment is considered. Finally, the frequency assignment should be used to assess the effects of integrating a drone with or without a Service Remote Pilot.

In addition to the ISA questionnaire and frequency allocation, the separation of the aircraft and the average flight distance are also analyzed. If there are separation underruns in the simulations, this can indicate that the controller is under too high work load and the requirements have been violated. The average flight distance is an indicator of delays and thus problems on the final approach or during integration into the airport operations. It is analyzed how much the routes have to be expanded due to the integration so that no separation requirements are violated.

The separation in the airport approach is 3 NM lateral separation and 1000 ft vertical separation for aircraft in category medium with a smaller aircraft is flying ahead of a larger aircraft (ICAO 2001).

The test controllers are also interviewed after their simulation runs. This is done by means of open questions about the general feasibility of the project, the developed integration concept and general abnormalities during the simulation.

5.2 *Results*

Figure 5 summarizes the results of the ISA of the three test subjects. Green bars show the averaged workloads in the different scenarios over time (in minutes), while blue bars indicate the overall average workload in the entire scenario. The self-assessment of the workload of each subject shows a tendency to increase with increasing traffic flow, but shows no tendency with regard to the integration of the cargo drone. However, the only two extreme load values occurred in the scenario with drone without Service Remote Pilot.

The frequency allocation shows the expected increase with increasing traffic volume. However, the additional ALAADy cargo drone in the high-load scenarios (45A and 45AS) does not seem to have any influence on the radio usage in comparison to the high-load scenario without drone (45). Here, a higher number of tests would be necessary in order to be able to analyze any existing dependencies using statistical methods.

The temporal approach separation, which describes the distance to the previous aircraft at the time of landing (see Fig. 6), shows that regular traffic is separated in the range of 60 to 160 s. The separation of the cargo drone (first column) to a previous landing varies vastly. The data reveals that the two lowest values are connected to a separation violation. It should be noted, however, that the two separations of the regular commercial aircraft above 180 s occurred for the flight following the separation violations of the drone. Two main results can be derived from this:

1. The integration of a cargo drone in the approach flow of a commercial airport has a significant impact on capacity. The large separation ahead of the vehicle reduces the amount of traffic that can be handled on the runway.
2. The flight behavior of the cargo drone, which differs significantly from regular commercial aircraft, presents controllers with an additional challenge. This can be reduced using suitable controller support tools such as ghosting (Oberheid et al. 2009) in order to avoid undesired control deviations—such as separation violations.

A further effect of integrating the cargo drone into the traffic flow is an extension of the flight path length. This particularly affects the aircraft following the drone.

In the questionnaires the test subjects themselves did not rate the integration of the cargo drone as a significant increase in workload. However, the uncommon behavior of the vehicle was rated as a potential challenge. The curved flight path and the large difference in flight speeds between manned aircraft and the UA are the greatest difficulties. Assistance systems such as a ghost target, which indicates when the aircraft can turn to the final without risk, were judged to be helpful.

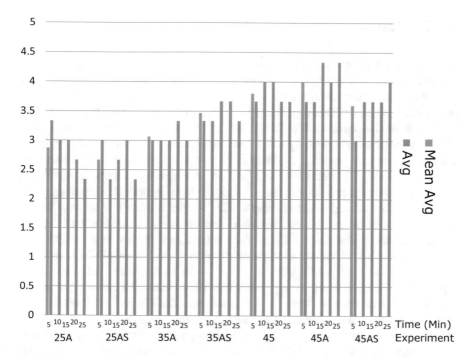

Fig. 5 Results of the ISA

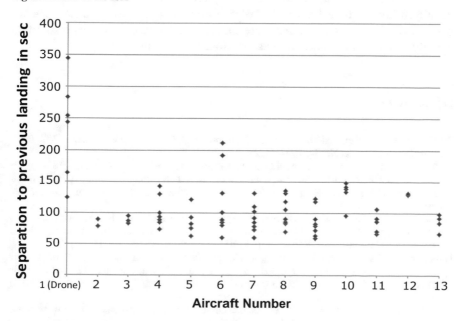

Fig. 6 Separation of A/C from previous landing in seconds

6 Conclusion

We showed the integration of a simulated ALAADy cargo drone based on a gyro-copter into the existing airspace of an air traffic management simulation. However, since the desired performance parameters of the three developed ALAADy vehicle configurations do not differ significantly in terms of the simulation, the result of the evaluation can be considered representative for all ALAADy configurations.

Under the selected boundary conditions (approach of the drone and regular air traffic to a common runway), it has been shown that the cargo drone can be integrated into the regular ATCO workflow. However, the impact on other traffic is significant. It is therefore recommended that the cargo drone lands on a dedicated runway near the airport that is independent of other traffic. Such a runway can be created with relatively little effort, since both length and load are considerable smaller than for a regular runway. It would also have the advantage that if there was the unlikely but possible case of an emergency termination (Schirmer and Torens 2021) immediately before or during other traffic landing, there would be no further influence to the regular operation. However, a runway that is outside the normal operating areas and their safety area can lead to longer turn-around times, for example, in case of the use of the cargo drone as a distributor for disaster relief.

Another result of the study is that additional support systems for controllers have great potential to make air traffic with unmanned cargo drones safer and more effi-cient. For example, traditional ATC support systems do not yet offer the possibility to identify, differentiate, and give dedicated vectoring commands to unmanned vehicles in cases of mixed operations. In this area, there is still a need for further research to identify, develop and validate the best possible tools and processes.

References

BADA (2014) User manual for the base of aircraft data (BADA) revision 3.12. EEC Tech-nical/Scientific Report No. 14/04/24-44, August 2014

DFS (2018) Luftfahrthandbuch Deutschland, DFS Deutsche Flugsicherung GmbH, 2018

Geister D, Geister RM (2013) Integrating unmanned aircraft efficiently into hub airport approach procedures. Navig J Inst Navig Seiten 235–247. ISSN 0028-1522. https://doi.org/10.1002/nav i.40

Geister D, Schwoch G, Geister RM, Korn B (2017) Integrating RPAS into existing ATM structures – published approach procedures vs. local arrangements. In: Integrated communications, navigation and surveillance conference, ICNS 2017, 18–20 April 2017, Herndon, VA, USA

Hasan YJ, Sachs F (2021) Performance-based preliminary design and selection of aircraft configurations for unmanned cargo operations. In: Dauer JC (ed) Automated low-altitude air delivery—towards autonomous cargo transportation with drones. Springer, New York, Berlin, Heidelberg

Helm SM, Temme A, Gerdes I (2018) Unmanned cargo operations at airports: influencing factors, potential use cases and first possibilities for implementation. In: 22nd ATRS world conference, 02-05 Jul 2018, Seoul, South Korea

ICAO (2001) International Civil Aviation Organization, Air traffic management—procedures for air navigation services, 14th edn. Doc 4444 ATM/501

Kirwan B, Evans A, Donohoe L, Kilner A, Lamoureux T, Atkinson T, MacKendrick H (1997) Human factors in the ATM system design life cycle. In: FAA/Eurocontrol ATM R&D seminar, 16–20 June, Paris, Frankreich, pp 16–20

Korn B, Udovic A (2006) "File and Fly"–procedures and techniques for integration of UAVs in controlled airspace. In: 25th ICAS conference, 3–8 September 2006, Hamburg, Germany

Lorenz S, Benders S, Goormann L, Bornscheuer T, Laubner M, Pruter I, Dauer JC (2021) Design and flight testing of a gyrocopter drone technology demonstrator. In: Dauer JC (ed) Automated low-altitude air delivery—towards autonomous cargo transportation with drones. Springer, New York, Berlin, Heidelberg

Oberheid H, Weber B, Temme MM, Kuenz A (2009) Visual assistance to sup-port late merging operations in 4D trajectory-based arrival management. In: 28th digital avionics systems conference (DASC), 23–29 October, Orlando, Florida, USA

Peinecke N, Mühlhausen T (2021) Cargo drone airspace integration in very low level altitude. In: Dauer JC (ed) Automated low-altitude air delivery—towards autonomous cargo transportation with drones. Springer, New York, Berlin, Heidelberg

Schirmer S, Torens C (2021) Safe operation monitoring for specific category unmanned aircraft. In: Dauer JC (ed) Automated low-altitude air delivery—towards autonomous cargo transportation with drones. Springer, New York, Berlin, Heidelberg

Schmitt DR, Kaltenhäuser S, Keck B (2008) Real time simulation of integration of UAVs into Airspace. In: 26th congress of international council of the aeronautical sciences, 14–19 September 2008, Anchorage, Alaska, USA

Schopferer S, Donkels A, Schirmer S, Dauer JC (2021) A multi-disciplinary scenario simulation for low-altitude unmanned air delivery. In: Dauer JC (ed) Automated low-altitude air delivery - towards autonomous cargo transportation with drones. Springer, New York

Theunissen E, Goossens A, Bleeker O, Koeners J (2005) UAV mission management functions to support integration in a strategic and tactical ATC and C2 environment. In: AIAA modeling and simulation technologies conference and exhibit, August 2005, p 6310

Design and Flight Testing of a Gyrocopter Drone Technology Demonstrator

Sven Lorenz, Sebastian Benders, Lukas Goormann, Thorben Bornscheuer, Martin Laubner, Insa Pruter, and Johann C. Dauer

Abstract An unmanned gyrocopter is selected as a technology demonstrator to check the feasibility of the ideas discussed in the automated low altitude air delivery (ALAADy) project in 2016. A gyrocopter airframe is chosen, based on experience in the operation of unmanned aircraft systems (UAS) and manned light aircraft gyrocopters. The specific operations risk assessment (SORA) draft by the joint authorities for rulemaking on unmanned systems (JARUS) served as a basis and guideline for the operational concept. A key aspect of the concept is the choice of flying over sparsely populated areas. Therefor the capability for safe termination of the flight in case of an unexpected failure or other unsafe condition is considered as the main risk mitigation. The development of the demonstrator is characterized by the balance between cost-efficiency, functionality and reliability of hardware and software components. A rapid prototyping approach allows the aircraft to take off as early as possible. Findings from flight tests are leading to the development of several additional features and functions, which illustrates the importance of an agile development. The demonstrator proves to be a capable platform for research on unmanned freight transportation and one of the largest transport drones in civil operation at the time.

Keywords Unmanned freight transportation · Unmanned cargo aircraft · Technology demonstration · Gyrocopter · Prototype · Flight-testing · Safety concept · SORA · UAS

Abbreviations

ACT	– Actuator
ADS-B	– Automatic Dependent Surveillance – Broadcast
AGL	– Above Ground Level
ALAADy	– Automated Low Altitude Air Delivery

S. Lorenz (✉) · S. Benders · L. Goormann · T. Bornscheuer · M. Laubner · I. Pruter · J. C. Dauer
DLR, German Aerospace Center, Institute of Flight Systems, Unmanned Aircraft Department and Flight Dynamics and Simulation Department, Lilienthalplatz 7, 38108 Braunschweig, Germany
e-mail: sven.lorenz@dlr.de

© Deutsches Zentrum für Luft- und Raumfahrt e. V. (DLR) 2022
J. C. Dauer (ed.), *Automated Low-Altitude Air Delivery*, Research Topics in Aerospace,
https://doi.org/10.1007/978-3-030-83144-8_19

ALT   – Altimeter
ARC   – Air Risk Class
BBOO   – Bulletin Board Object Oriented
C2   – Command & Control
CIC   – Core Interface Computer
ConOps   – CONcept of OPerationS
CRM   – Crew Resource Management
ERP   – Emergency Response Plan
FCC   – Flight Control Computer
GCS   – Ground Control Station
GRC   – Ground Risk Class
GRP   – Glass Fiber Reinforced Plastic
HIL   – Hardware-in-the-loop
INS   – Inertial Navigation System
JARUS   – Joint Authorities for Rulemaking on Unmanned Systems
NOTAM   – Notice(s) to Airmen
OSO   – Operational Safety Objectives
OST   – Overall System Tests
PIC   – Pilot In Command
POH   – Pilot Operating Handbook
PSA   – Passive Spiral Autorotation
PUU   – Power Utility Unit
RC   – Remote Control
RECV   – Receiver
RTOS   – Real Time Operating System
SORA   – Specific Operations Risk Assessment
TERM   – Termination
UAS   – Unmanned Aircraft System
UPS   – Uninterruptable Power Supply
VIS   – Vehicle Interface System
VLOS   – Visual Line of Sight
VMC   – Visual Meteorological Conditions

1 Introduction

Within the previous chapters of this book an in-depth look at ALAADy is given. Several aspects are discussed for utilizing unmanned systems for air delivery at low altitudes and challenges are identified. Special attention is paid to safety aspects, the implementation of components of unmanned cargo systems, operating conditions and system architectures.

Strongly connected to the concepts introduced in the previous chapters, the development, integration and operation of a first prototype of a demonstrator aircraft

Fig. 1 Gyrocopter-based unmanned aircraft system, a technology demonstrator for air cargo transportation (*Source* DLR)

(Fig. 1) is presented in this chapter. The demonstrator will enable an early validation and demonstration of concepts for the supply of air cargo at low altitude in terms of feasibility and operational capability. The intended scenario for the demonstrator is derived from the use cases presented in Pak (2021). However, the requirements are scaled down to enable development and operation within the research environment. Therefore, the objective is the transport of payloads up to 200 kg, including the integration into a simple logistic chain, operational risk assessment, automatic flight and adaptation to the arising legal framework.

The development of the demonstrator aircraft is aligned to upcoming innovative rules for certification of UAS for operation in civil airspace (Nikodem et al. 2021). Hereby, the safety assessment is based on a mission centric approach called SORA, which focusses on the risks resulting from the operational scenario, to determine the necessary safety assurance level. Put simply, the higher the risk of fatalities or significant environmental damage in the event of loss of control over the operation, the higher will be the strictness of the safety assurance.

Three different configurations for an unmanned cargo aircraft are proposed in Sachs (2021). Out of these, the gyrocopter is selected as the most promising due to its anticipated inherent safety characteristics without the need for an additional parachute. Since several years, gyrocopters are in the focus of the DLR Institute of Flight Systems (Duda and Seewald 2016; Sachs et al. 2016; Duda et al. 2015, Grima Galisteu et al. 2015; Sachs et al. 2014; Duda et al. 2013a; 2013b; Duda and Pruter 2012; Duda et al. 2011; Pruter and Duda 2011).

Moreover, the arising activities related to the SORA focus on fixed wing aircraft and multirotor configurations only. Thus, a gyrocopter configuration will raise further questions. Its capability of autorotation without further control or stabilization is of particular interest. Inherent stability, very short take-off and landing capabilities combined with a simple mechanical system are unique features of gyrocopters. Following these considerations, a light sport aircraft built by the German company AutoGyro was chosen as the base airframe for the technology demonstrator.

In summary:

- The demonstrator is a prototype for validation and demonstration of the feasibility and operability of automated low altitude air delivery,
- Upcoming innovative rules for certification of UAS operation (SORA) are addressed,
- Requirements of the authorities are collected and discussed,
- A gyrocopter is selected due to its inherent safety.

2 Motivation and Problem Statement

To operate an aircraft in order to validate the theoretical concepts and considerations of the ALAADy concept and support its ideas by practical evaluations are the basic motivations of the demonstrator. Taking the leap from theory and simulation to real flight-testing allows to identify challenges in the practical implementation of conceptual work that are not considered in advance. The development of the demonstrator is a proof of concept for the technical and operational aspects of low altitude air delivery and one of the first applications of the SORA for an unmanned aircraft of this size.

The base airframe, a manned light sport gyrocopter, is converted to an unmanned aircraft capable of transporting cargo. The demonstrator should be able to fly, take-off and land automatically within the given legal framework. Safety for the crew and third parties has the highest priority for operations with the demonstrator, and especially performing flight tests. Therefore, it is necessary to apply mitigations to all relevant loss-of-control scenarios.

The development of the demonstrator includes interdisciplinary tasks from work on the concept of operation, the SORA, equipment of the manned gyrocopter for unmanned and automatic flight, including hardware and software development and modifications, flight mechanical evaluations of the modified aircraft, testing and team training, flight test and the evaluation of flight test results. All these disciplines and tasks are managed in an agile manner, like in various software development approaches. This enables the development team to react on changing requirements from the outcome of the ALAADy concept developed further, upcoming problems within the development of the demonstrator itself, or changing legal regulations, especially since the legal framework for drones is not finalized yet.

For operation of a novel UAS with a take-off mass of up to 500 kg, development processes and operational procedures have to be defined and tested. Since a ready-to-use flight controller for this type of gyrocopter does not exist, a remote pilot with manual control authority is an efficient way in order to perform flight tests as early as possible.

On the other hand, the system architecture is partly influenced by a predefined choice of technology. Electromechanical actuators delivered by a supplier well-known from former projects are selected based on simulation predicted forces and moments in relation to desired actuator deflection speed.

Since the main goal is gaining experiences within the practical realization of the ALAADy concept, the demonstrator presented in this chapter is not supposed to be a complete or product level design, but rather a research aircraft optimized for scientific purposes.

In summary, the motivations and main tasks are:

- Demonstration of aspects of the ALAADy concept,
- Validation of concepts for low altitude cargo transportation,
- Conversion of a manned aircraft to an unmanned aircraft,
- Interdisciplinary agile management for complex system's integration,
- Identification of challenges within the practical realization.

3 Airframe

The light sport gyrocopter MTOfree by AutoGyro GmbH shown in Fig. 2(a) is used as base frame for the unmanned demonstrator. The decision for such a system as a starting point was made initially and as the fundament of the project. The rationales behind this decision were the simple structure, the inexpensive availability and the local support. The airframe has a nose wheel gear chassis and a main landing gear with glass fiber reinforced plastic (GRP) spring spar, and hydraulic disc brakes.

(a)   **(b)**

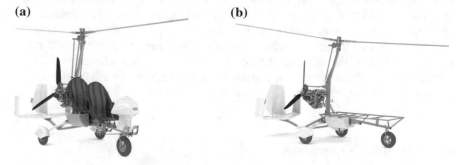

Fig. 2 **a** Gyrocopter MTOfree (with permission of AutoGyro). **b** Initial Unmanned Airframe (with permission of AutoGyro)

Table 1 Dimensions of the MTOfree

Length	5.08 m
Width	1.88 m
Height	2.71 m
Rotor diameter	8.4 m

The framework is manufactured from inert gas welded stainless steel tubes. The teetering main rotor is made from extruded aluminum. Connecting rods control the rotor head and cables control the rudder. Rudder and stabilizer surfaces are made of GRP (or carbon fiber). The gyrocopter is powered by a Rotax 912 ULS engine with a maximum power of 100 HP in pusher configuration with a fixed pitch propeller. This allows for a cruise speed of about 100 km/h and a best climb rate of 3.4 m/s. The operational range is 400 km. Table 1 shows the dimensions of the gyrocopter. Further technical information on the MTOfree is given in the Pilot Operating Handbook (POH) provided by AutoGyro GmbH (AutoGyro 2014).

Compared to the manned version of the MTOfree airframe, modifications to the unmanned airframe were made by the manufacturer (Fig. 2(b)). The unmanned version does not require the cockpit, the seats, sticks, or pedals. In order to get a simple structure to integrate payload section as well as avionic components, a rectangular frame and several lugs were welded to the airframe. The unmanned base airframe has a weight of about 220 kg. With a desirable payload of 200 and 45 kg of fuel, 35 kg would be left for avionics components such as sensors, actuators, computers and housing, if the desired maximum take-off weight of 500 kg were fully utilized. For simplicity, weight optimization is neglected and actuator power supply is based on additional batteries, as presented later. Therefore, the take-off weight of the demonstrator within the context of this publication ends up being up to 420 kg including 45 kg fuel.

The airframe was delivered by AutoGyro GmbH in 2016 and has been further modified by the DLR afterwards. These modifications are described in detail in Sect. 8. The fairing shell manufactured by a third-party company on DLRs request was specifically designed for the demonstrator. It covers the experimental avionic systems and reduces drag during flight. It allows easy access to maintenance interfaces and batteries as well as modifications of the prototypic avionics system.

Alongside with the modifications of the original airframe, the original maintenance manual of the MTOfree (AutoGyro 2014) has been adapted. Additional installation and maintenance instructions have been added for modified and new components, keeping the original guidelines wherever applicable.

4 Concept of Operation

The concept of operations (ConOps) is a document providing information on the planned operation of the UAS. It is part of the SORA (Sect. 6) and a legal requirement to get a permission to fly from a competent authority. The ConOps describes the nature of the intended operation and the roles of the involved personnel in detail. In this ConOps, the term operation is defined from the moment onwards when the unmanned aircraft starts taxiing. The period before is considered as flight preparation. The safety concept is also part of the ConOps, but due to the importance of the topic, it will be covered separately in Sect. 5. The first draft of the ConOps for the demonstrator has been written before the detailed design of the actual vehicle started. Therefore, it can be considered as a design basis for the demonstrator.

4.1 Manual Control Strategy for Initial Flight Testing

An essential question in the development of a suitable ConOps is how the aircraft can be controlled during the initial flight tests. The demonstrator is based on a manned aircraft; hence, one possibility would be developing an optionally piloted vehicle (OPV). OPVs have the advantage that the pilot has an excellent situational awareness and can react very quickly to malfunctions of the automatic flight control system. This option was not chosen because the requirements for safety (e.g. the pilot must be able to overrule actuators at any time) would have been much higher. Furthermore, the available payload would be less, due to the weight of the pilot and required onboard instruments and controls. In addition, a modification of a light sports aircraft would require gaining a complete airworthiness approval. The selection of removing the pilot and his support systems from the aircraft enables the operation with the SORA context and results in a suitable amount of payload.

Another possibility is the use of a remote first-person-view (FPV) system, but the development of such a system is challenging itself because it has to feature low latency and high reliability. To keep the operation and integration as simple and safe as possible, FPV is considered not ideal for the initial phase of flight-testing due to the lack of prevalent reliable systems, which lie in the project budget.

Nevertheless, the demonstrator is intended to be controlled directly by a human pilot for the initial flight-testing phase. Later in the experiments, the manual control serves as a backup interface within the flight test experiments. The pilot flying is able to steer the demonstrator via a remote-control interface. He can always take over control of the aircraft if a controller malfunction occurs. This setup has proven to be very efficient. In addition, the gyrocopter is well suited for this kind of control because it is relatively large with a comparably slow airspeed. On the other hand, the silhouette of a gyrocopter makes observing the attitude more difficult than the silhouette of a similar-sized fixed wing aircraft.

Still, the cues of the pilot flying are limited due to the lack of motion feedback, in particular forces and rotations. Visual impressions and display indicators may have to compensate the missing feedback. However, in addition to tracking the aircraft in the sky, the pilot flying is not able to monitor any kind of instruments, when steering by the remote-control device. Therefore, a pilot monitoring assists the pilot flying by communicating the most important data, such as e.g. airspeed, and supports in estimation of aircraft conditions during abnormal operation.

4.2 Normal Operation

The demonstrator is operated only in daylight under visual meteorological conditions (VMC) with calm wind. In general, the operation follows the standard operating procedures of the MTOfree POH wherever applicable (AutoGyro 2014). Additional procedures are established as required by the remote-controlled operation. All flight maneuvers of the demonstrator can be performed remotely by the flying pilot, automation is in principle not mandatory.

The normal operation consists of a takeoff maneuver as shown in the MTOfree POH (AutoGyro 2014) and a climb to a height of 100 to 200 m AGL (above ground level). Afterwards one or more traffic circuits are flown at constant altitude. Meanwhile system tests or tests of automatic flight control functions are performed. The pilot flying activates automated functions, maneuvers and mission control according to the experimental goals. The ground control station is the interface to interact with the autopilot by sending commands or assigning missions. In automatic flight mode, the pilot flying will be informed about the activities of the UAS and observes the flight passively as long as nominal conditions are maintained. Those nominal conditions are expressed in e.g. attitude, altitude, flight speed, flight direction and position. During assisted flight modes, a flight control system supports the pilot and eases the steering of the aircraft.

The flight maneuvers are based on predefined and practiced procedures; the pilot flying may decide on variations, e.g. to perform a go around instead of a landing. Several sources of information are available to the pilot flying in order to judge the current flight conditions, e.g. altitude and speed. The operation ends by landing according to the operation instructions written in the MTOfree POH (AutoGyro 2014).

If any unexpected event or failure occurs, e.g. the weather conditions change rapidly or the pilot flying is not in full control of the UAS for any other reason, the operation is considered abnormal. In this case, the crew tries to regain full control over the UAS or to land it immediately within the nominal area. More detailed information on abnormal operation can be found in Sect. 5.

4.3 Flight Test Area

The flight test area is chosen among different aerodromes and airfields based on several parameters including size, availability, accessibility and infrastructure. For future operation, a sparsely populated environment is of major interest. Since the experiments are partly manual, remote-controlled flights within the line of sight of at least the controlling pilot, who is standing at a fixed position on the airfield, hilly terrain or visual restrictions by buildings or vegetation are not desired. More detailed information on the flight test area used can be found in Sect. 11.1.

4.4 Crew Team Roles

During flight tests of several unmanned aircraft systems by DLR in the recent years, specific roles and responsibilities have been proven to be efficient and reasonable. Eight roles are identified for this ConOps, as shown with their nominal communication flow in Fig. 3. The roles are documented explicitly within the team to avoid inefficiencies or misunderstanding. During less complex experiments, one person might occupy some of these roles. However, experience shows that all roles should be assigned separately as often as possible to decrease workload and thus the probability of errors.

The **Flight Test Lead** coordinates the experiment procedures and supervises the team interaction. He supports the team by taking care of a smooth execution of the flight test procedures and that every person for each responsibility receives the information required. The flight test lead ensures the safety of the crew and observers, initiates the experiment and may stop it at any time. He takes care that every station is

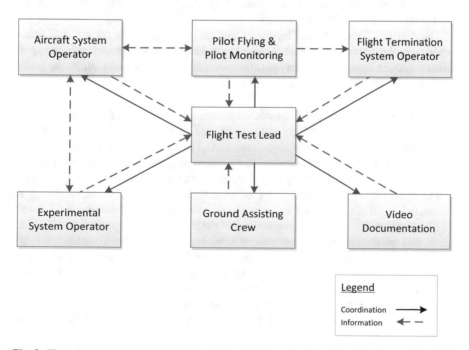

Fig. 3 The roles in flight test with coordination and information flow (based on Dauer et al. 2015)

ready when necessary. The flight test lead is a mandatory role and shall be occupied by an experienced researcher who has participated in flight tests for years.

The **Flight Termination System Operator** observes the flight experiment passively and in nominal operation only reports to verify the presence of the safety feature. By dedicated information sources, he observes the aircraft position and its flight direction, and indicates when the aircraft hits predefined flight area boundaries. Beyond those warnings, he is empowered to trigger the termination signal and therefore terminate the flight if the flight area violation or any other abnormal situation is not mitigated otherwise. Based on information from traditional air traffic control infrastructure like ADS-B or FLARM, he is able to monitor traffic outside the flight area of the demonstrator, and thus takes care of the isolation of the air segments resulting in limiting the risk for uninvolved parties.

The **Aircraft System Operator** is in charge of supervising the state of the system and sends commands, e.g. for mode selection or mission control to the vehicle if necessary. In contrast to manned flight experiments, the aircraft system operator is actively involved in the flight test, which increases workload compared to a solely observing role. In flight tests that require low-latency direct communication with the pilot flying, the aircraft system operator is in direct contact with the pilot flying and vice versa. He informs the flight test lead about problems, delays and readiness of the aircraft system. He communicates within the team about tests, inspections and final adjustments of the vehicle's hardware. In addition, he communicates with the operator of the experimental system about its status and the synchronization of the experiment with the flight experiment.

The optional **Experimental System Operator** is responsible for any additional experimental systems or payloads during flight. He controls an experiment or experimental system, synchronizes its operation with the flight test and gathers the information needed for the goal of the experiment. In case of a failure, he directly informs the flight test lead.

A **Ground Assisting Crew** enables the nominal operation of the flight experiment. In preparation of the flight test, the ground assisting crew cares for payload, experimental systems and support equipment. During flight test, a variable number of persons support the operation by visual inspection of ground and airspace. The tasks of the ground assisting crew may vary depending on the experiments and operations.

The **Pilot Flying** is responsible for safe operation of the aircraft. He observes an automated flight of the aircraft and is ready to take over manual control at all times. He also inspects the aircraft in preparation of the flight. He is empowered to abort an automated mission or experiment without confirmation of the flight test lead. It is his responsibility, that the vehicle is not operational while any other person is within its vicinity. If an unexpected event occurs, either caused by the aircraft, the environment or flight test crew he immediately takes over manual control and returns the vehicle to a predefined safety position if possible. The pilot flying acts as pilot in command (PIC) of the flight, which implies that he also carries the legal responsibility of the aircraft.

Due to high workload and in order to be supported with information about the current aircraft state (airspeed, altitude, engine rpm) the pilot flying is supported by a **Pilot Monitoring**. The pilot flying is not able to neither spend time in observing the environment completely nor watch for instruments, when the aircraft is airborne. The pilot monitoring assists the pilot flying closely by reporting flight conditions, watching for surrounding events, and pointing out flight condition limits.

The **Video Documentation** includes on-ground and on-board cameras which are recording different aspects of the flight test. It is helpful for analyzing complex tests, and is an essential element of any flight test. The camera operators are likely to be subject to higher risks than other ground crew, since they are often positioned for their field of view to cover the most important and thus most risky part of the experiment. These risks have to be explicitly addressed in the flight-planning phase and during mission briefing. Additionally, the persons operating cameras are in continuous contact with the flight test lead who can adjust safety measures if necessary.

In summary, the operation is characterized by:

- Specific roles and responsibilities
- The pilot flying controls the demonstrator manually via remote control
- Operation takes place following defined standard procedures
- The operation takes place during daylight with calm weather conditions
- The aircraft systems state is monitored by the pilot monitoring, by the aircraft system operator and by the flight termination system operator with a dedicated monitoring system.

5 Safety Concept

Safety by reducing the risk of damage to third parties to a minimum is one of the top priorities in the development and operation of the demonstrator (Sect. 2). The safety concept for the operation is a part of the ConOps (Sect. 4). Additionally, the safety concept also incorporates safety measures taken during development and testing.

The airframe of the unmanned demonstrator is based on light aircraft equipment. It is developed and built under consideration of the regulations for ultralight aircraft. Where possible, these regulations were also considered on all modifications done after delivery. All design and verification documents are prepared and checked with regard for the principle of dual control (two-man rule), a standard practice in aviation.

Flight operations take place above the controlled ground area at a closed airport. The ground area is therefore free of third parties, and air traffic is informed about UAS operations by active Notice(s) to Airmen (NOTAM). All flights being performed are in separated airspace approved by the responsible authority. In case of a malfunction of the system, an emergency landing is performed within the restricted ground area if possible. Therefore, the area of operation is divided in different three-dimensional areas: a nominal flight area, a containment area and a buffer area (Fig. 4).

The size of the flight zones is determined by considering the worst-case glide ratio, which is decisive for dimensioning of the buffer area. If a flight is interrupted by shutting down the engine at the outermost top position of the contingency volume,

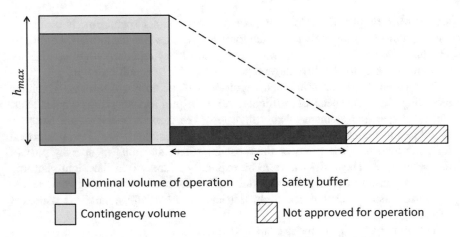

Fig. 4 Permissible operational volumes in dependence of the gliding ratio (DLR)

at best glide ratio and reasonable tail wind the outermost limit of the buffer to the non-approved operating area can be drawn. As a tailwind can cause longer glide distances, it is accounted for by considering the maximum wind speeds the operation is limited to.

All flight altitudes below the maximum lead to a shorter range in terms of distance and therefore leaving the buffer zone can be excluded. The maximum permissible flight altitude h_{max} is calculated from glide ratio n and the safety distance s:

$$n = \frac{h_{max}}{s} \equiv h_{max} = \frac{s}{n}$$

The capability for a safe flight termination is supported by the nature of the gyrocopter itself. Due to the auto-rotating rotor, once the engine is switched of in flight, the aircraft will behave as it is hanging on a parachute. In addition to switching the engine off, when the termination is activated, the actuators for the main rotor and rudder will go into a termination position. The termination position is pitch up and slightly left for the main rotor, and fully left for the rudder. With these control inputs a slow, shallow spiral is anticipated, further reducing the possible impact area. However, since this capability has never been proven in flight yet, the safety buffer for the flight tests is still based on the worst-case scenario using the best glide ratio.

Due to the criticality of the termination functionality, a minimum of two independent radio data links with different characteristics and redundant functionality are implemented. On the ground segment, flight termination can be triggered either by the pilot flying via remote control, the flight termination system operator by high power termination transmitter, or the aircraft system operator by the GCS interface (Fig. 5). The termination command is sent to a relay circuit that is built up of aerospace parts with high reliability. Due to the special importance of the Termination Control Electronic, only components from the military or aviation sector are used and a failure hazard analysis is conducted.

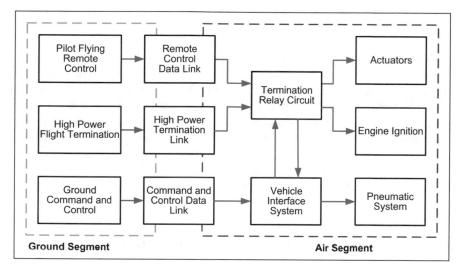

Fig. 5 Connection of ground and air segment by three data links for flight termination (DLR)

In case of termination, the relay circuit (OR logic):

- switches off engine ignition,
- sets hardware flag on actuators to move flight controls in termination position,
- freezes pneumatic trim and pre-rotation system in flight mode.

The safety concept is supported by a dedicated emergency response plan (ERP). The ERP includes information of the rescue services about the planned flight activities as well as the definition of the rescue chain.

Beside safety during the flight test, safety during system tests on ground needs other precautions. For those tests, different color-coded safety plugs are used to protect crew members during the tests (Sect. 8.1). The safety plugs disable different functions of the system, e.g. engine ignition, actuators, or pneumatic system.

In summary, safety during flight operation is achieved by:

- controlled ground area
- a-typical airspace
- flight termination that leads to emergency landing,
- division into three-dimensional flight zones,
- worst case glide ratio that defines buffer area,
- spiral autorotation that reduces the required impact area,
- emergency response plan is in place.

Other safety measures and precautions:

- airframe of the unmanned demonstrator is based on air sports equipment built to the standards of manned aviation,
- rules for design and verification are in place,
- safety plugs reduce risk for personnel during testing.

6 Specific Operations Risk Assessment

In Nikodem et al. (2021) the legal framework of the ALAADy project is described. The process of specific operation risk assessment (SORA) is introduced. Although the EASA regulation of unmanned aircraft systems has not come into effect for the first flight operations, a SORA is conducted for the demonstrator to derive requirements for research operation and in preparation for future flight permissions. This analysis is based on the ConOps (Sect. 4) and the safety concept (Sect. 5) introduced above. Thereof the Ground Risk Class (GRC) and Air Risk Class (ARC) are determined, from which the Specific Assurance and Integrity Level (SAIL) is derived. This assessment is described in detail in the following sections.

6.1 Determination of the Intrinsic Ground Risk Class

The aircraft's dimension and its kinetic energy, as well as the characteristics of the operation determine an Intrinsic GRC. Based on risk classes that are unique to the SORA, the evaluation of the intrinsic (unmitigated) ground risk is performed. Characteristic measures for the size of the aircraft in relation to its typical energy are used as metrics (Fig. 6). The basis for selecting the appropriate operational scenario is the ConOps including the safety concept (Sects. 4 and 5). A controlled ground area is used for the operation of the demonstrator.

Intrinsic UAS ground risk class				
Max UAS characteristics dimension	1 m / approx. 3 ft	3 m / approx. 10 ft	8 m / approx. 25 ft	>8 m / approx. 25 ft
Typical kinetic energy expected	< 700 J (approx. 529 ft lb)	< 34 kJ (approx. 25 000 ft lb)	< 1 084 kJ (approx. 800 000 ft lb)	> 1 084 kJ (approx. 800 000 ft lb)
Operational scenarios				
VLOS/BVLOS over a controlled ground area[6]	1	2	3	4
VLOS in a sparsely populated environment	2	3	4	5
BVLOS in a sparsely populated environment	3	4	5	6
VLOS in a populated environment	4	5	6	8
BVLOS in a populated environment	TBD[7]	TBD[7]	TBD[7]	TBD[7]
VLOS over an assembly of people	7			
BVLOS over an assembly of people	TBD[7]			

Fig. 6 Determination of the intrinsic GRC (EASA 2019)

To tackle different safety characteristics of different classes of aircraft (e.g. fixed-wing and rotorcraft) the reference metrics change while the risk classes remain the same. For fixed-wing aircraft, the measures are wing-span and cruise speed; for rotorcraft, on the other hand, rotor diameter and terminal velocity is used. These metric variations are similarly applied in different other publications, (Dalamagkidis et al. 2008; EASA 2005; JAA 2004).

By these two metrics, the hazardous nature of an uncontrolled aircraft is represented and abstractly matched to ascending risk classes. This process is a simple solution for risk classification that otherwise would need assessment of complex relations between failure modes, crash trajectories, splash patterns and assessment of harm to the environment.

Challenges of this simplification arise, if the safety characteristics of the used aircraft do not correspond to either of the two aircraft classes considered in the SORA, which is the case for the gyrocopter used. The SORA itself implies to individually agree with the Civil Aviation Authority (CAA) on a risk classification in such cases, including the announcement that the gyrocopter is such an exceptional case (EASA 2019). In contrast to other rotorcraft, the rotor of the gyrocopter is in auto rotation state permanently, which is also stable in case of a failure. Nevertheless, in contrast to a fixed wing aircraft the gyrocopter can store energy in its rotor system.

To clarify the ambiguity of risk classification, a comparison of three metrics of energy is used. First the fixed-wing metric using cruise speed, second worst-case for rotorcraft, using an estimate of the terminal velocity assuming high-altitude flight, and third a metric considering a realistic worst-case estimate is calculated. The latter is based on estimating the total energy contained in the aircraft during operation. It is determined using the potential energy, the kinetic energy based on cruise speed and the energy stored in the main rotor.

Fixed-wing Energy Calculation Method of the SORA
The kinetic energy based on mission cruise speed is calculated to

$$E_{kin,cruise} = \frac{1}{2}mV^2_{cruise} \approx 162 \text{ kJ}$$

by using vehicle mass $m = 420$ kg and mission cruise speed $V_{cruise} = 27.8\frac{m}{s}$.

Rotory Wing Energy Calculation Method of the SORA
The terminal velocity can be calculated by the knowledge of the specific drag coefficient and the aerodynamic reference area. This represents the worst case and maximum energy that can arise if the rotor is stalled and produces zero lift. By using an estimate of the drag coefficient of $C_d = 1,0$ and an estimate of the reference area of $A = 1$ m^2, both estimates by Duda and Seewald (2016), the terminal velocity results in

$$V_{term} = \sqrt{\frac{2mg}{C_d \rho A}} \approx 82\frac{m}{s}.$$

with vehicle mass $m = 420$ kg, gravity constant $g = 9.81\frac{m^2}{s}$ and density of air $\rho = 1.225\frac{kg}{m^2}$.

The resulting kinetic energy is calculated by

$$E_{kin,term} = \frac{1}{2}mV_{term}^2 \approx 1413 \text{ kJ}.$$

The terminal velocity will be reached only after sufficient time of free fall.

Proposed Maximum Total Energy Estimation
A flight altitude mainly below 150 m above ground characterizes the operation. Therefore, the terminal velocity clearly overestimates the typical kinetic energy to be expected. The maximum potential energy, which can be gained from this flight altitude, is calculated to

$$E_{pot} = mgh \approx 618 \text{ kJ}$$

by using vehicle mass $m = 420$ kg, typical mission height AGL $h = 150$ m and gravity constant: $g = 9.81\frac{m^2}{s}$.

The rotating rotor will store additional energy during flight. The energy is calculated by Duda and Seewald (2016) to

$$E_{rotor} = \frac{1}{2} \cdot I_R \cdot \Omega_R^2 \approx 109 \text{ kJ}$$

by using inertia: $I_R = 189$ kgm^2 and rotational frequency: $\Omega_R = 34\frac{rad}{s}$.

The total energy calculated as sum of

$$E = E_{kin,cruise} + E_{pot} + E_{rotor} \approx 889 \text{ kJ}.$$

Discussion
From these values it is evident, that for an operation within line of sight over controlled area, located inside a sparsely populated environment:

- the fixed-wing classification with the typical kinetic energy of 158 kJ would lead just by its typical energy to an initial GRC of 3,
- the rotorcraft classification resulting in 1413 kJ by using the terminal velocity to an initial GRC of 4,
- the energy that is calculated as sum of kinetic, potential and rotor energy for the gyrocopter at 150 m flight altitude is 889 kJ. which is the energy for a typical GRC 3 classification.

For the gyrocopter, the characteristic dimension is given by the rotor diameter of 8.4 m, which leads to a GRC of 4. However, the typical energy following from fixed wing energy calculation and extended by the total potential and the rotor's kinetic energy is nearly half of the rotorcraft energy.

Taking autorotation of the gyrocopter into account for the case of a vertical autorotation, the typical energy can be reduced to the sum of kinetic and rotor energy, with

$$E_{kin,sink} = \frac{1}{2} m V_{sink}^2 \approx 13.4 \text{ kJ}$$

by using vehicle mass $m = 420$ kg, typical sink rate at vertical autorotation $V_{sink} \approx 8 \frac{m}{s}$, (Duda and Seewald 2016)

$$E_{kin,sink} + E_{rotor} \approx 125.4 \text{ kJ}.$$

This calculation shows that the total (real) energy of a gyrocopter in vertical autorotation is less than for the fixed wing calculation based on mission speed.

Similar to fixed-wing aircraft certain flight conditions can lead to an uncontrolled unstable crash. In the case of a gyrocopter, the assumption that the autorotation state is maintained may no longer apply. Further investigations in the future will allow to discuss the typical energy generated in such situations. For the moment and with the realistic worst-case estimate presented earlier, it is appropriate to assume the lower risk class.

Nevertheless, for regular and market ready operation, the open question of how to classify nonconventional aircraft in respect to the SORA risk classes needs to be validated. This concrete example shows how the definition of the transition from quantitative risk assessment (size and energy of the aircraft) to qualitative (risk classes) shapes and influences the way how a particular aircraft is operated. Safety and cost of the operation are weighed against each other. From the current point of view, this is a critical point in the applicability of the SORA process, as it will render the outcomes overly conservative and expensive or risky and overly cost efficient. Providing information about the rationale behind or guidance material on the classification is thus needed to handle nonconventional aircraft configurations.

6.2 Final Ground Risk Class Determination

The intrinsic ground risk class is reduced by operational specific mitigations, which reduce the likelihood or the severity of harm from the aircraft. The SORA process proposes three types of mitigation, while each of it is further classified by three levels of integrity and assurance. In the following it is explained how each type of these mitigations is applied to the demonstrator.

Mitigation M1—Strategic mitigations for ground risk enables a reduction of people at risk by the use of a ground risk buffer. The operation of the unmanned gyrocopter is limited to a restricted ground area, and therefore minimal. As technical containment, a method of geofencing is used (Sect. 5). However, the GRC cannot be reduced to a value lower than the lowest value in the applicable column in Fig. 6; the GRC is therefore not modified.

Mitigation M2—Effects of ground impact is reduced in theory by the passive spiral autorotation (PSA), which the demonstrator is designed to perform after a termination. Gyrocopters are able to perform a relatively slow low impact landing via autorotation. This is used as method to reduce the impact energy and the size of the area affected by the impact in case of termination. However, this maneuver is not proven to be robust and possible at any initial condition. Therefore, a low robustness is proposed at this time and the GRC is not modified.

Mitigation M3—an Emergency Response Plan (ERP) is in place. The ERP for the operation is based on the pilot's handbook supplemented for the unmanned operation. The ERP is known to the flight test crew and initially trained before the mission. However, the ERP is for the time being not validated by independent parties and therefore a medium robustness is proposed; the intrinsic ground risk is not modified.

The final GRC corresponds to the initial value of 3.

6.3 Determination of the Initial Air Risk Class

Typical flight tests are conducted in restricted airspace up to 200 m (about 656 ft) AGL. Takeoff and landing are done inside the restricted area. According to the classification of the airspace by JARUS: "Operations in Atypical Airspace"—the initial ARC is ARC-a. As additional tactical mitigation the operation is conducted in visual line of sight (VLOS). Multiple airspace observers assist the pilot flying. In case of incoming traffic, the remote pilot performs either avoiding maneuvers, lands immediately or terminates the UAS. The pilot flying is trained to judge which option is the safest in the situation.

6.4 Specific Assurance and Integrity Level Determination and Operational Safety Objectives

The final determination of the SAIL, which is the basis to assign the Operational Safety Objectives (OSOs), is derived by combining a GRC of 3 with an "ARC-a". The resulting SAIL is SAIL II. In Table 2, the consolidated list of OSOs for a SAIL II is presented, which has to be conformed with the ConOps (Sect. 4).

Table 2 Operational safety objectives for SAIL II (EASA 2019)

OSO Number	Description	Robustness Level
	Technical issue with the UAS	
OSO#01	Ensure the UAS operator is competent and/or proven	Low (L)
OSO#03	UAS maintained by competent and/or proven entity	L
OSO#06	C3 link performance is appropriate for the operation	L
OSO#07	Inspection of the UAS (product inspection) to ensure consistency to the ConOps	L
OSO#08	Operational procedures are defined, validated and adhered to	Medium (M)
OSO#09	Remote crew trained and current and able to control the abnormal situation	L
OSO#10	Safe recovery from technical issue	L
	Deterioration of external systems supporting UAS operation	
OSO#11	Procedures are in-place to handle the deterioration of external systems supporting UAS operation	M
OSO#12	The UAS is designed to manage the deterioration of external systems supporting UAS operation	L
OSO#13	External services supporting UAS operations are adequate to the operation	L
	Human errors	
OSO#14	Operational procedures are defined, validated and adhered to	M
OSO#15	Remote crew trained and current and able to control the abnormal situation	L
OSO#16	Multi crew coordination	L
OSO#17	Remote crew is fit to operate	L
	Adverse operating conditions	
OSO#20	A human factors evaluation has been performed and the human machine interface (HMI) found appropriate for the mission	L
OSO#21	Operational procedures are defined, validated and adhered to	L
OSO#22	The remote crew is trained to identify critical environmental conditions and to avoid them	L
OSO#23	Environmental conditions for safe operations defined, measurable and adhered to	L

6.5   Adjacent Area and Airspace Considerations

To address the risk posed by a loss of control of the operation, the adjacent airspace has to be considered. The adjacent airspace to the area of operation has an uncontrolled status. The safety requirement, that no probable failure of the UAS or any external system shall lead to operation outside the operational volume (JARUS 2019), is tackled by the concept of operation (Sect. 4) and the safety concept (Sect. 5). The containment is achieved by the use of an air risk and ground risk buffer between containment area and adjacent airspace (Fig. 28). The air risk buffer takes best glide ratio and wind speed into account.

In summary:

- The aircraft used in the operation is a modified gyrocopter (with the ability to perform a passive spiral autorotation (PSA) in case of emergency)
- The flight is performed over temporarily restricted area, both in air and on ground
- A geofence is in place to assure the unmanned aircraft does not leave the approved airspace / area
- With the final GRC of 3 and ARC-a the resulting SAIL becomes SAIL II.

7   Design and Development Strategies

In order to quickly obtain a flying demonstrator, several development strategies are pursued. Reliable commercial of the shelf (COTS) hardware are used as much as feasible including the airframe, actuators, computing units, etc. Wherever COTS hardware is not available modifications to existing COTS-components are preferred over new developments from scratch. Existing expertise and knowledge in management, design and operation of manned and unmanned aircraft shall be utilized by choosing compatible components, tools and processes.

Although the demonstrator will be a prototype which is inherently more unreliable than a product-level aircraft, the costs for damage or loss of such a larger unmanned vehicle are still considerable. Additional to safety, which was already discussed in Sect. 5, reliability of the demonstrator is therefore also a major design goal.

Nevertheless, the development and testing effort of components to ensure robustness and reliability significantly increases with the systems complexity. Thus, it is beneficial to isolate safety critical modules from non-critical and more experimental as much as possible. The design should minimize the number of safety critical components in the system, minimize the complexity of critical components and shift complex operations and functionalities to less critical parts. Modules, functions and components that are not used in every mission are implemented in such a way that they can be deactivated in order to test individual components in isolation.

To train flight test procedures, gain system knowledge, test components and algorithms in controlled, isolated environments on the dedicated hardware simulations are essential tools. Therefore, it is recommended to design components with well-defined interfaces, to ease integration and creating modularized models for hardware- and software-in-the-loop simulations. The simulations shall be part of a consistent development, training and release process for the vehicle.

For the prototypical development of the demonstrator it is an integral part to find, test and refine requirements and constraints of the system within an iterative, agile development process, the spiral model (Boehm 1988) is fitting best. To support this agile approach, the system architecture shall be modular and extendible. Furthermore, a thorough communication and documentation of design decisions, any changes to the system and version control of the incorporated documents is necessary. The configuration management is performed by following the same philosophy as with other aircraft operated by DLR, i.e. all the add-ins and the changes are monitored.

One of the goals of a demonstrator is to gather new information and insights based on the data collected in experiments. Experience shows that it is crucial to gather as much information as possible in a consistent time frame, to be able to interpret them in relation to occurring events. The integrated components shall provide either interfaces for logging or store data themselves.

Summarized the system-design of the demonstrator shall provide:

- cost efficiency
- modularity
- high flexibility
- quick iterative results
- traceable configuration management
- time synchronized storage of all available data
- separation of critical from experimental components.

8 System Concept

The demonstrator is based on the gyrocopter airframe described in Sect. 3. In order to control the aircraft by electrical commands, the manual-mechanical-flight controls are replaced by electromechanical actuators. Onboard electrical power system, pneumatics, sensors, flight control computers, a flight termination system and multiple data links are integrated to enable manual remote controlled and automated flight. The system concept includes a core system and an experimental subsystem. The basic system ensures that functionalities for manual flight via remote control meet requirements concerning safety and reliability, whereas the experimental uses interfaces of the basic system for automated control of the demonstrator. Regardless of the flight controls, the safety concept (Sect. 5) requires a highly reliable independent means to terminate the flight of the demonstrator. The software is also an important part of the system concept. Due to the scope of the topic, software is covered in a separate Section (Sect. 9).

Table 3 lists the built-in avionic components and Fig. 7 gives an overview of

Table 3 List of onboard avionic components

Component	Abbreviation	Remark
Core interface computer	CIC	Connection of remote-control receiver and actuators
Flight control computer	FCC	Experimental/automatic flight control
LAN switch		UAS internal network
Control surface and nose wheel actuators	ACT 1-4	Main steering devices
Throttle, choke, wheel brake actuators	ACT 5-7	Engine and brake controls
CAN repeater		Enables cut-off for ACT 1-4 in case of flight termination
Vehicle interface system	VIS	Replaces former pilot instrument panel and buttons
28 V electrical power supply	PUU	Avionics power supply
12 V electrical power supply		Base aircraft power
Remote control receiver	RC 1/ 2	2.4 GHz dual diversity receiver
Termination system receiver	TERM RECV	Independent safety device
Termination system relays		Activation by 3 data links and in case of a loss of the power supply
Command & control data link	C2	Telemetry and tele-command data link
Inertial navigation system	INS	GNSS-IMU system
Air data measurement system	AIR DATA	5-hole-probe system
Radar altimeter	RADAR ALT	Height above ground sensor
Combined FLARM and ADS-B receiver	FLARM/ADS-B	Enables independent position and air traffic monitoring

the avionics system architecture. The avionic system contains no redundancies apart from the datalinks to trigger the flight termination system.

The overview of the avionic system architecture (Fig. 7) consists of two subsystems that are connected via an Ethernet connection. The basic fly-by-wire aircraft system with the central CIC (core interface computer) relies on inputs of the two remote control receivers whose signals are interpreted. The flight control sub systems (actuators ACT 1-7, vehicle interface system VIS, power utility unit PUU) are commanded by the CIC. Depending on manual or automatic flight mode, the CIC switches between the flight control command source, either from the remote-control input or the inputs from the FCC (flight control computer) of the experimental system. The CIC validates and transforms received commands before sending them to the actuators via the CAN busses. Major system components like the VIS (Sect. 8.1) and the PUU (Sect. 8.3) are needed for fulfilling the basic system tasks and therefore connected to the CIC.

The FCC is connected to multiple flight state estimation sensors (inertial navigation system INS, radar altimeter RADAR ALT, air data sensor AIR DATA) as the

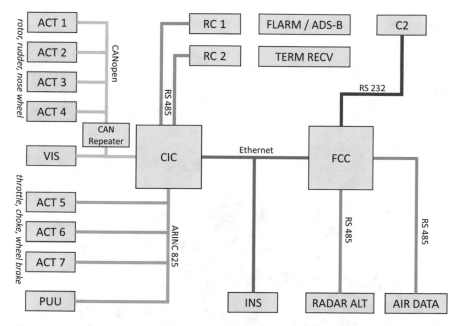

Fig. 7 Overview avionic system (Benders et al. 2018)

automated flight system requires more information about the current flight state of the demonstrator. Since the communication to the ground control station is not used for manual control of the aircraft, it is not considered as a basic system function and therefore performed by the FCC, too.

Independent of the other subsystems a FLARM/ADS-B receiver and the termination system are installed. The termination system can be triggered via multiple data links and is directly connected to a secondary input of major system components (e.g. actuators). The commands send by the termination system follow the procedures required be the safety concept.

8.1 Vehicle Interface System

The vehicle interface system (VIS) is the interface to the former instrument panel of the manned gyrocopter. Electrical and pneumatic functions as well as instruments of the basic aircraft are interfaced via a CAN bus and a WAGO fieldbus system. The VIS detects analogue and digital sensors and publishes the sensor information (e.g. rotor speed, engine data) via a CANopen protocol. In return, it converts commands including the starter of the engine, the pneumatic pump and the engine's ignition magnetos from the CAN bus to digital outputs. Additionally, the VIS provides the

Table 4 Safety level modes implemented by the safety-plugs

Function	Flight	Red	Yellow	Green
Starter	✓	✗	✗	✗
Ignition	✓	✗	✗	✗
Pneumatics	✓	✓	✓	✗
Actuators enabled	✓	✓	✗	✗
Nose wheel angle steering limit	nominal	extended	-	-

Fig. 8 Unique safety plugs (DLR)

possibility to trigger the independent flight termination system by onboard software components. The VIS is powered by 12 V DC of the base aircraft's battery.

For testing, a series of safety levels are defined (Table 4), which disable or enable pre-defined functions of the aircraft. These safety levels can be activated by connecting different unique safety plugs (Fig. 8) to the VIS.

The shown safety plugs are mapped to unique color codes that represent the grade of freedom of the actuators and the engine from green (everything deactivated) to flight (everything activated). Dependent on the functions that shall be tested the appropriate safety plug is selected. Only the flight safety plug enables all functions of the system, while all ground tests can be done without modification of the software.

The CIC interacts with the VIS to adopt parameter based on the selected safety plug. In flight, especially for landing for example, the nose wheel steering angles are limited to prevent a rollover. However, under testing and maintenance conditions (red safety plug) large angles shall be allowed to increase the maneuverability of the demonstrator.

Fig. 9 Rotor actuators and actuator control units as mounted in the demonstrator (DLR)

8.2 Electro-mechanical Actuators

The automation of the aircraft system required to install electro-mechanical actuators for primary flight controls, engine controls and wheel brakes. COTS actuators[1] with a DLR in house control unit are used to actuate throttle, choke, and the pneumatic wheel brake. The communication to the controller is implemented via a CAN interface and the ARINC 825 protocol. Position commands are commanded to the actuators; internal states, and sensor information of the controller and actuators are provided.

COTS actuators[2] in combination with DLR in-house developed actuator control units[3] are used to actuate the rotor (two actuators for controlling pitch and roll, Fig. 9), rudder, nose wheel steering. The communication to the controllers is implemented via a CAN interface and the CANopen protocol. The actuators can be initialized and commanded and internal states and sensor information can be accessed. Further information on the design considerations, like functional and performance requirements, tests, and general information can be found in (Bierig et al. 2018).

8.3 Electrical Power Supply

The electrical power supply is divided into two independent (but not redundant) networks with 12 and 28 V nominal voltage. The VIS, the pneumatic system and the engine starter is powered by a 12 V battery, which is charged by the ROTAX engine's

---

[1] Volz DA 26 actuators.

[2] Harmonic Drive CHA-17A actuators.

[3] Based on ELMO Gold Solo Hornet controllers.

generator. The rest of the system, including actuators, flight computers, flight state estimation sensors, flight termination and telemetry modules is powered by eight 28 V grids.

The 28 V grids are supplied by a Li-Ion rechargeable battery system. The power is distributed and controlled by a power distribution unit and the power utility unit (PUU) with integrated electrical controlled circuit protection devices. The Li-ion batteries consist of eight battery blocks, which can easily be exchanged after usage.

8.4 Sensors

The demonstrator is equipped with sensors for system surveillance, flight state estimation and airspace observation. The sensors are partially components of the manned base system and partially equipment used for the automated flight.

The VIS provides sensor information of the originally manned system, which are engine data, cooling system data, pneumatic system data, rotor speed and temperatures. A COTS air traffic and collision avoidance system[4] with FLARM broadcast/receive and automatic dependent surveillance—broadcast (ADS-B) receiving capabilities (Fig. 10(a)) is used to observe the surrounding airspace. The airspace surveillance is not yet integrated into the onboard automation system, but used for observing the airspace for intruder aircraft and as an independent resource for monitoring the demonstrator's position from a ground control station. In addition, the demonstrator is equipped with three sensors for flight state estimation:

- Inertial navigation system (INS) with integrated differential GPS (two onboard antennas) and 3 axis accelerometers (Fig. 10(d))
- Air data system with nose boom (Fig. 10(b))
- Radar altimeter (Fig. 10(c)).

All sensors for flight state estimation have built-in filters and provide meaningful data for unchanged use in the automation system. The automation system uses the three sensors for flight state estimation as well as the sensors for engine and rotor data. All sensor data are logged onboard and are transmitted to the ground control station via a C2 data link.

8.5 Flight Computers

The cores of the onboard avionic system are industry-hardened embedded computers (Fig. 11) especially designed and constructed for the demonstrator. Two of these computers (core interface computer, CIC; flight control computer, FCC) are installed on the demonstrator and are used for the fly-by-wire system (CIC) and the automated

---

[4] PowerFLARM Core.

(a) Combined FLARM and ADS-B module for air traffic surveillance. (https://flarm.com/de/medien/bild-und-logomaterial/)

(b) Air Data System (DLR)

(c) Radar Altimeter (DLR)

(d) Inertial Navigation System for flight state estimation (DLR)

Fig. 10 Sensors of the demonstrator

Fig. 11 Core interface computer (DLR)

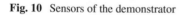

Table 5 Flight computer specifications

Property	Description
Power supply	7–30 V DC, max. 150 W
Form factor	PC/104
Processor	Celeron 2000E 2 × 2.2 GHz
Random-access memory	4 GB
Hard disk drive	256 GB SSD
I/O interfaces	6× serial (RS232, RS485 or RS422), 2× CAN, 2× Ethernet, USB, GPIO, VGA
Operating system	QNX 6.6.0
Notes	Passively cooled, vibration damped, rugged housing, screw-mountable connectors

flight control and experimental system (FCC). The two computers do not serve for redundancy even though the hardware is identical. Both computers run a QNX real time operating system. More detailed specifications of the computers and interfaces are given in Table 5:

As mentioned the computers fulfill two different functions. On the one hand the CIC handles the basic system tasks of the fly-by-wire system, which can be subdivided into the following tasks:

- receive remote control commands
- receive automated flight control commands
- command actuators, VIS and electrical power system
- receive sensor data from actuators, VIS and electrical power system
- log data.

On the other hand, the FCC is mainly used for the automation system. The subdivided tasks and additional functions are:

- receive flight states estimation sensor data
- generate automated flight control commands
- log data
- establish the C2 communication to ground control stations
- execute experimental software.

Note that the aircraft can be controlled via remote control with only the CIC in the loop without the FCC. Specifically, the stability and robustness of the CIC and its software components is critical for flight. Failure of the CIC or its critical software components in flight will result in a loss of the aircraft.

The FCC and its software components are necessary for the ongoing flight test as experimental software is running on it. Additionally, the FCC in connected to the C2 data link and acts as relays for data send by and towards the CIC.

The flight computers are integrated in the avionic system architecture as depicted in Fig. 7. For communication between the CIC and FCC an Ethernet interface is used. Due to this distributed setup, communication latencies are induced. This needs to be addressed in the software design, especially in real-time-critical tasks like the automated flight control loop.

8.6 Radio Communication

Various radio communications are required during operation of the demonstrator. Before going into detail, the communication can be classified into three major groups. Communication between the demonstrator and ground control station, demonstrator and infrastructure/environment and internal ground control station components. Figure 12 visualizes an overview of the radio communication during flight test.

The selection of data links for controlling the aircraft and other purposes in flight tests has to take into consideration the possible interference between different links. Exact internal documents and the following Table 6 show the used data links and their frequencies.

The most safety critical data link is the flight termination data link. The flight termination data link is a 400 MHz signal and is tested prior to each flight. The termination system, is primary developed as termination system for rocket flight experiments, but also successfully used for the operation of unmanned aircraft.

The ground control station (GCS) command & control (C2) data link uses 900 MHz. Status information such as engine or battery states are transmitted and automatic flight modes are commanded via the C2 data link. The received information are forwarded to multiple other GCS via a 5.8 GHz WLAN connection locally on ground (e.g. to the pilot monitoring).

Independent from the C2 data link the position of the demonstrator is monitored by a FLARM-receiver. FLARM is a non-certified collision avoidance system mainly used by gliders and slow powered aircraft and allows not only the GCS, but also possible other unintentional users of the airspace to discover the demonstrator.

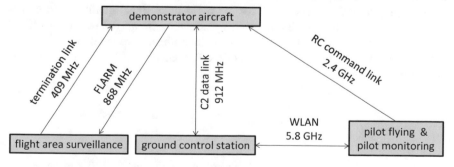

Fig. 12 Radio communication between the demonstrator and the ground control station (DLR)

Table 6 Radio communication during flight test

Description	Aircraft communication		Frequency	Devices	
	Sender	Receiver		On-board	Ground
Flight termination		X	400 MHz	1 (2 antennas)	1
C2 link	X	X	900 MHz	1	1
FLARM	X	X	868 MHz	1	1
ADS-B		X	1090 MHz	1	1
GNSS		X	GPS, GALILEO, GLONASS	2 (3 antennas)	1
Pilot flying remote control	X	X	2,4 GHz	2	1
WLAN	GCS internal comm.		5,8 GHz		4
Flight test crew radio comm.	GCS internal comm.		ISM band		8–10
Airport radio communication	GCS internal comm.		VHF		3
Airband radio communication	GCS		VHF		1

For manual control of the flight the demonstrator is equipped with a remote control (RC) receiver for the pilot flying command link. A standard commercial 2.4 GHz RC link is used to transmit the pilot remote control commands.

The antenna installation positions of (not only, but especially) the RC receivers are crucial for the signal quality and range of the connection. The two RC receiver antennas are mounted sidewise on the demonstrator. In case one antenna is shadowed by the fuselage, the other receiver has direct visual contact to the pilot flying remote control. In order to test the data links the critical data links flight termination, pilot flying remote control and C2 data link are tested during the flight preparation tests, following a pre-defined test procedure. Nevertheless, these tests, with the vehicle on the ground, give only a benchmark for the radio links signal qualities and ranges during flight. Flight tests with a manned gyrocopter with an RC receiver installed at various positions showed that RC receivers on both sides of the demonstrator are needed and that the ground-to-air range of the RC receivers is five times larger than the ground-to-ground range.

8.7 Payload

The payload area of the aircraft is designed in such a way that it can be loaded from the side using a forklift. The payload itself persists of a base plate and is securely attached with four bolts on top of the corner fixtures connected to the

Fig. 13 Payload area and corner fixture (DLR)

airframe. Figure 13 shows the payload position on the airframe and details of the corner fixtures. The size and shape of the cargo are to be evaluated, but the airframe and corner fixtures are already designed to carry a payload of more than 200 kg. The side loader design is a compromise compared to the designs within the front loader design in the ALAADy design study (Sachs 2021) and the possibilities of the modified demonstrator airframe.

8.8   Ground Support Equipment

For operating the demonstrator additional ground infrastructure is needed. In this section the relevant ground infrastructure equipment to command, control and observe the demonstrator is described. Most relevant are the ground control station (GCS), the pilot flying remote control and the flight termination system.

The ground support equipment is integrated into two existing ground control station vehicles, which are equipped with basic systems (e.g. power supply, network devices) installed and are extended with multiple features for the demonstrator flight tests.

Ground Control Station Vehicles
The GCS equipment is build-in into two vehicles. The vehicles (Fig. 14) allow the

Fig. 14 Demonstrator with ground control station (DLR)

transport of the flight test personnel and equipment to the flight test sites and are equipped with an uninterruptable power supply (UPS).

Network routers with LAN and WLAN interfaces are installed inside of the vehicles and interconnect the GCS computers. The build-in desks are ergonomic workplaces and protect the GCS operator from weather and especially difficult light conditions that disturb the visibility of displays. Additionally, the vehicles offer standardized mounting positions for antennas and weather monitors.

Electrical Power Supply

At flight test sites, the availability of electrical infrastructure cannot be assumed. Furthermore, the electrical power supply must be redundant for a high reliability. Therefore, generators and back up batteries are part of the flight test equipment. The electrical power supply is guaranteed by a 230 V generator and a UPS per vehicle. Batteries inside the devices (e.g. laptops and flight termination) complete the redundant power management concept of the GCSs.

Several devices, like the pilot flying remote control, the pilot monitoring tablets, or voice radio communication devices rely on a single build in battery and cannot be equipped with a redundant power supply. These devices are equipped with warning mechanism and the battery states of charge (SOC) are checked during the pre-flight checks in particular. For the case of a loss of electrical power supply emergency procedures are defined and trained.

Flight Test Control and Command Devices

The demonstrator onboard systems and flight states are monitored in the ground control station. All data are received by the C2 ground segment is distributed by an Ethernet router via LAN and WLAN to the ground control laptops and tablets.

The tablets allow the pilot monitoring fast lookups of elementary flight state data required for direct support of the pilot flying and are positioned multiple meters away of the vehicles (see also the flight test setup in Sect. 11.1). An extended system status monitoring and automatic flight control command & control is given by the laptops positioned inside of the vehicles.

The software and hardware setup of the ground segment is designed modular in order to quickly adapt to changing requirements to displays and ground control station setups. Further information about the software and intercommunication can be found in Sect. 9.6.

Flight Termination System and Airspace Surveillance

The flight termination ground segment is split from the vehicle function centric ground control segment. The segment is composed of two redundant top-views to determine whether the vehicle leaves the pre-defined flight test area and the flight termination needs to be triggered. Figure 15 shows the FLARM display for monitoring the position of the demonstrator independent from the experimental systems. Together with the flight termination system the FLARM display is supplied with redundant power. In the flight tests surrounding airspace was observed by an additional observer with use of the given airport tower infrastructure.

Pilot Flying Remote Control

The demonstrator is designed to be controllable by a remote control. The remote control is depicted in Fig. 16. The pilot flying has full manual and direct control of the vehicle with the remote control and can switch between automatic and manual flight mode. The pilot flying can, in addition to the already mentioned systems, terminate the vehicle in case of an emergency via the remote control.

A pilot monitoring supports the pilot flying. The pilot monitoring is using a tablet, which presents most relevant system data and flight states in a clearly arranged view. Based on a top-view of the flight test area and the current position of the vehicle, he can give advice to the pilot in command, when critical conditions appear.

Fig. 15 FLARM display for airspace monitoring (DLR)

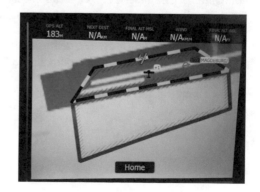

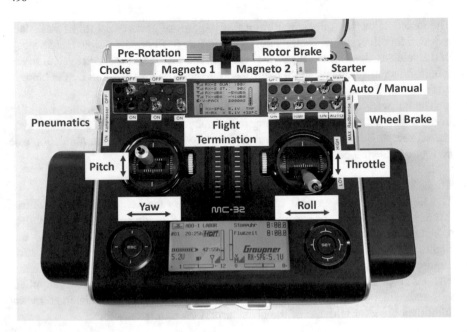

Fig. 16 Pilot flying remote control (DLR)

9 Software

Beside the modifications of the airframe and the hardware of the avionics system, a number of software modules are required to operate the demonstrator. Most of the used avionics components contain a software component, called firmware, which is taken as COTS if available. Almost any avionic subsystem (e.g. actuator, air-data system) is driven by firmware. In-house developed avionic components like the PUU uses firmware implemented by DLR. Some firmware contains persistent configurable parameters, e.g. calibration factors of sensors, actuator speed setting, etc. Like all software, firmware versions and their configurations are versioned as part of a configuration management system.

In addition to the firmware, software is executed on the two onboard flight computers and several GCS computers. The following section focuses on the flight computer software architecture, the software modules, and the GCS software. Further details about the software design and development can also be found in (Benders et al. 2018).

9.1 Software Architecture

The software architecture is following the design strategies of the demonstrator, as presented in Sect. 7, which will be briefly recalled. In particular "modularity" and the "isolation of critical and experimental components" are major design objectives. For more experimental parts of the software it is desired to get "quick iterative results". The "time synchronized storage of all available data" also applies to all software modules handling such data from external inputs, which does not store the data themselves.

More critical components in terms of reliability are implemented as simple as possible by design and in a way that they are never blocked, interrupted or restrained from using shared resources like memory or computing power. The software architecture is scalable to be able to include further computing hardware onboard, if it would become necessary due to increasing computational requirements in the itcrative development process.

The software architecture of the onboard and ground control station software is shown in Fig. 17. Two similar computers are installed in the demonstrator which are called CIC and FCC (Sect. 8.5). The CIC handles the basic system tasks like sending commands to the actuators etc. according to the current inputs of the pilot flying remote control or respectively the FCC. It also receives their responses and states, records and forwards them to FCC and GCS computers. The basic software system of the CIC consists of a fly-by-wire application and a communication and logging application. The CIC software interacts with the FCC and its datalink to the ground by using the in-house development of a communication middleware (Sect. 9.4), which is used onboard and on the ground for inter-process communication.

The FCC is relying on receiving and recording data of sensors (Sect. 8.4) to calculate automatic flight control commands, run experimental software, and handles the communication over the C2 data link. It runs different software modules essential for automated flight and experimental mission execution.

The GCS software on ground shows the current system state and flight conditions for the ground crew, and allows high level commands like switching flight controller modes, select the level of automation, change flight path, or enable experimental functions. The ground control system consists of several devices running different configurations of the GCS software components.

9.2 Core Interface Computer Software

To separate two key functions with respect to their intended reliability, software on the CIC is divided in two main applications: the fly-by-wire application and the communication and logging application. An additional application is used to initialize the system at startup once.

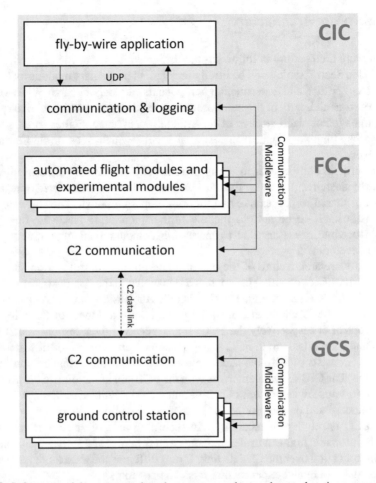

Fig. 17 Software modules on two onboard computers and ground control stations, connected by communication middleware and the datalink (DLR)

The fly-by-wire application enables manual controlled flights by forwarding remote control commands of the pilot flying to the actuators controlling the aircraft's motion, and other subsystems like the vehicle interface system to control the engine. The fly-by-wire application is also able to forward automated flight control commands from the FCC when desired. The source of control is selected by the pilot flying only.

In addition to these two modes a third mode for flight termination is implemented as illustrated in Fig. 18. In this case the whole system including the CIC is set into a mode, where emergency routines are enabled. Due to its important significance in the operating concept this mode can be triggered by three sources: from pilot, from ground control station software, or via a dedicated datalink connection independent from software applications.

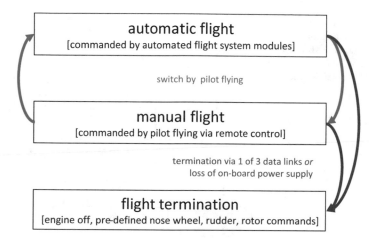

Fig. 18 Fundamental modes of the core interface computer (DLR)

Since a working and responding fly-by-wire function is mandatory for reliable control, it is not only designed and tested with special care, but also supervised by a watchdog system. The watchdog is partly a component of the operating system and will restart the fly-by-wire application within several milliseconds in the unlikely case that the application is stuck or aborted. To achieve arbitrary and instant restarts, the fly-by-wire application is implemented stateless, which means, it is generating its outputs solely based on the inputs of the current program cycle. A clear application design and minimal logic decisions lead to an application design, which can be easily developed and reviewed.

The flight control commands in manual flight mode from the pilot flying are received by two remote control receivers. Inputs from both remote-control receivers are available during operation. In order to vote between the receivers a checksum of the serial data and a flag, which reports the remote-control receiver's internal heath monitoring are used for further processing. If the commands of both receivers are valid, the command of the last used receiver is used as long as this receiver becomes invalid. Then, the other receiver is evaluated. If no valid commands are received at all, the latest valid commands are hold and the throttle setting is set to idle. As part of the risk mitigation, this behavior enables the crew to regain the control without increasing the risk of an uncontrolled fly-away in manual flight mode.

Within Fig. 19 the interaction of the modes and signals are shown. The remote-control receivers are "voted" by validity. The mode selected by the pilot flying determines the active control command source, which is forwarded to the aircraft system.

To separate logging and communication from the essential flight control functions without possibly interrupting them, the second application is executed with lower priority and privileges. Relevant data processed by the fly-by-wire application is forwarded to the communication and logging application and logged onto the built-in SSD. This data contains the remote-control commands, received flight

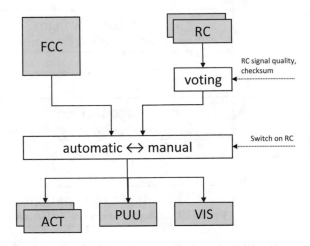

Fig. 19 Signal routing and switch between automatic and manual commands in the application of the Core Interface Computer (DLR)

control commands of the automatic modules, sensor and system state feedback from actuators, VIS, and PUU. The data is also published via the middleware to the FCC and GCS if requested. This partitioning supports software robustness requirements by separating resource access for logging from the functionality of processing and forwarding flight control commands. The logging is described in Sect. 9.5 in detail.

The system startup procedure is outsourced into a separate initialization application, to achieve that the main fly-by-wire application is stateless. The initialization program reads the configuration of the safety-plug first and initializes the system according to the safety level modes as presented in Table 4. Afterwards, the selected electrical power supply channels are activated as well as actuator configuration is performed. The actuators are moved to their initial position at slow speed, in order to avoid harm to people or the system by sudden step inputs to the actuators. Additionally, basic checks of the connected avionic sub-systems are performed before the application reports completion and starts up the fly-by-wire and communication applications.

9.3   Software Modules of the Flight Control Computer

The demonstrator will support experiments for fully automated transport missions. A mission includes at least the main phases: taxi, take-off, enroute flight and landing. In addition, actions such as engine start, system check and engine shutdown will be required for a fully automated mission execution. For this purpose, automated flight software is implemented on the flight control computer FCC. The automated flight software consists of three modules: sensor fusion, mission management and flight control system. Figure 20 shows the architecture of the automated flight software.

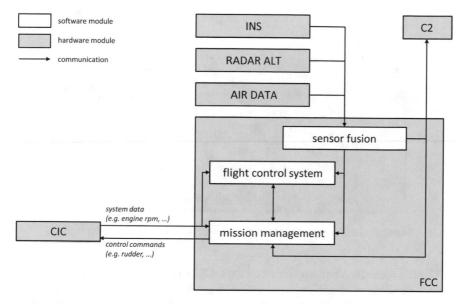

Fig. 20 Module architecture of the automated flight software (DLR)

The sensor fusion software receives sensor data from the inertial navigation system, radar altimeter, and air data measurement system, stores them in data log files and provides the data to other software modules and the GCS via the middleware (Sect. 9.4) with a data rate of 50, respectively 100 Hz, depending on the sensors output rate. Additional aircraft system condition data, e.g. engine status, or actuator conditions are provided to the controller and mission management modules by the CIC.

The mission management module sets the controller modes according to the current mission task and forwards control commands of the flight control system to the fly-by-wire application module on the CIC. The mission management is implemented as a hierarchical event driven finite state machine. Event driven means, the mission manager only reacts on incoming data and does not run at a pre-defined framerate. Figure 21 illustrates the state machine for manual and automatic mode.

In general, the mission manager differs between manual and automated flight. Automated flight can be assisted (semi-automated) or fully automated flight modes. In manual flight the mission manager is in a stand-by condition, namely idle. If the automated flight mode is enabled via GCS and the pilot flying activates the automated flight mode, the mission manager transits into the preselected mode, either assisted or taxi.

In several assisted modes, the pilot flying remains in partial manual control, while individual controls are commanded by the flight control system. These modes are used for parameter tuning and as support modes for manual flights and are selected via GCS. In contrast to a pure manual control, all control commands are routed through the FCC and send back to the CIC.

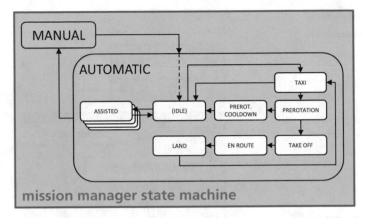

Fig. 21 Transitions between flight modes and sub-states (DLR)

Control modes can be set up either by the GCS on request, or are initiated automatically if pre-defined conditions are satisfied. If a mode is activated, which corresponds to a certain control mode of the controller, the controller mode is enabled and flight control systems' commands are calculated and subsequently routed to the fly-by-wire application on the CIC.

The basic initial mission to be performed fully automatically consists of a prerotation, followed by a takeoff, climb, level flight in a traffic pattern, descend and landing (Fig. 22). After touch down and deceleration, the main rotor is slowed down and the engine is turned off.

Controlling the automatic flight comprises the control of throttle, pitch and bank angle of the rotor head, nose wheel steering, rudder deflection, wheel brake and operation of the pre-rotation unit. The flight control system for this demonstration mission is divided in eight different modes, which are individually adapted to the flight phases. In detail, the modes specified in Table 7 are executed during a mission.

Fig. 22 Demonstration mission: automatic traffic pattern (DLR)

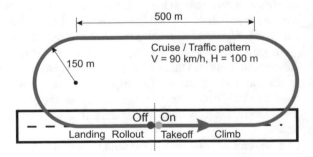

Table 7 Control modes for the automatic traffic pattern

Control mode	Description
Initialization	Initial values are set and actuators are held at specified positions
Pre-rotation	Accelerating rotor to starting speed
Takeoff	Accelerating on runway and lift-off
Climb	Climbing with full throttle in take-off direction
Cruise	Maintaining altitude and speed, right/left turn, descent
Landing	Performing flare, decrab and touchdown
Roll out	Decelerating on runway
Stop	Stopping on the runway

9.4 Middleware for Communication

The applications running on the flight control computers as well as on ground control stations are interacting by sharing data with each other. For the communication onboard, especially the flight control loop, low latencies are crucial. An in-house developed communication middleware BBOO is used for data exchange within a single computer as well as for communication between network participants. Figure 18 on illustrates the participating software components, which are interacting by the communication middleware. BBOO is an acronym for a decentralized ad hoc and Bulletin Board Object Oriented concept.

Major design objectives of BBOO are robustness, flexibility, low latency and scalability. BBOO is mainly based on UDP, but also acts as gateway for communication channels other than IP/ UDP network interfaces. The C2 datalink of the demonstrator is integrated in the network by using BBOO as gateway. BBOO is capable to route packages automatically via those gateways and therefore connects multiple distributed networks e.g. via a serial data interface. Multiple instances of participating applications can run simultaneously on any connected device in the network. If an application of the network fails or is not responding, it will only affect the communication with that particular participant. It enables point-to-point messages as well as communication in the publish-subscribe-pattern. To address the varying bandwidth capabilities of different datalinks BBOO offers a message priority mechanism.

The application of BBOO enables the flexible software module architecture, supports simulation setups and easy extension of the system. To gain this flexibility BBOO is also the standard communication module used for inter-process communication (IPC) on any single computer. However, for the IPC between the fly-by-wire application and the communication and logging application on the CIC another lightweight IPC implementation is chosen. Main reason for this design decision is to encapsulate the fly-by-wire application from connections and interference with any other program and reduce its complexity.

To support the encapsulation and avoid blocking of the fly-by-wire application, the communication shall follow the "fire-and-forget"-message-pattern, with no queueing and no need to clean up or doing extra administrative work in case the receiver is not listening. This is sufficient since for the data exchanged—like actuator commands and system states—it is more important to exchange the latest consistent data record than to process every single data point. Therefore, the connection-less non-blocking UDP protocol was used for the IPC as it offers a low management and data overhead, and met the requirements when used with some static configurations.

9.5   Flight Data Logging

Data acquired during tests and experiments is recorded for further findings and analysis in post-processing. Generally, the partitioning of access to resources on the flight computers is desired. Usually this is enabled by decoupling applications from using hardware resources simultaneously when possible. The hard disk access turned out to be one of the resources which did not meet these requirements by using the interfaces of the operating system. If several applications have hard run-time requirements but require a single resource access, the priority mechanism and scheduler used to balance its time frames will be overruled. Disk access is therefore managed centrally by one application only, in order to run the recording of data into files, afterwards called logging.

Logging applications run on all computers, on-board and on-ground. All logging on the CIC is performed by the "communication and logging" application, all FCC logging is performed by the "logging" application (Fig. 18). Applications send data to be logged via UDP respectively the communication middleware BBOO to the logging applications. A logging service handles several different data structures to support multiple sources and multiple targeted files on a Solid-State Drive.

To increase the responsiveness of the logging service and therefore minimize influence on accessing programs, the receiving of data and the access to the filesystem is separated into multiple threads. One central thread handles the access to the filesystem interface of the operating system. Several threads are receiving data and hand them to a "frontend" of the logging service. The frontend writes the data into a lock-free, thread-safe buffer. The central logging thread processes the data following a predefined priority and by an at the runtime calculated schedule.

The central logging thread periodically checks and sorts the known logging channels depending on their absolute priority, defined a priori, and the current size of the data buffer. These capabilities of the logging service allow the developer to prioritize log data, while the logging service has the flexibility to handle dynamic changes and upsets at runtime. As the logging service has never the highest priority process on the system, interruptions of the process and especially the single logging thread occur regularly.

The buffer, acting between the applications and the logging to the filesystem, could therefore grow rapidly, if the hardware is blocked or too busy. The logging service mitigates these situations by controlling the size of the buffer. When a critical buffer size is exceeded the central thread can discard data. This guarantees that at least the latest data is written to the hard disk. Furthermore, data might arrive at a data rate that cannot be handled by the central thread. For this situation an emergency routine exists that discards the data before writing it into the buffer. This prevents an overload of the Random-Access Memory (RAM) to no interfere with other processes running on the same system. The data is logged in general in a human-readable tabular format into multiple files. The average amount of recorded data is about 2 Mbit/s for CIC and FCC respectively.

9.6 Ground Control Station Software

In flight tests the aircraft system operators and the pilots need information about the internal conditions of the demonstrator. The aircraft system operator needs detailed information about the flight condition, system state and details about the flight control and experimental systems. In addition, he needs an interface to preselect the mode of the aircraft as well as to define the flight trajectory. The pilot flying supported by the pilot monitoring need an intuitive and well-structured overview of the aircraft flight conditions in relation to their limits. Both types of the ground control station software are described in the following.

Pilot Monitoring Display: For the pilot flying it is hard to switch his visual focus between the demonstrator and the GCS display. The reorientation would generate time gaps, which increase the risk of loss of control remarkably. Therefore, the pilot monitoring communicates relevant information via voice radio. Figure 23 shows the used display—containing traditional flight instruments combined with a flight area top view. Graphical interface is based on the DLR's U-Fly GCS software (Friedrich et al. 2021). Basic system status as the engine speed and rotor speed is shown, as well as relevant flight states like the airspeed, altitude and position of the demonstrator and the flight test area boundaries. Due to the relevance of that data in terms of the operation and therefore related to the risk of damage of the aircraft, the pilot monitoring can switch to a second device in case of a malfunction of the first tablet. Both tablets are connected to the aircraft system operator GCS by a WLAN connection.

Aircraft System Operator Displays: This ground control station interface is designed with a tradeoff between well-structured visual design, the ability to show a large set of information and the flexibility to frequent changes in the implementation requirements. In order to highlight important information, colored fonts are used, especially if system statuses are out of range. A console-based application enables deeper insights and is combined with graphical user interfaces, like those used by the pilots, but also enables mode selection and trajectory planning and upload.

Communication with the GCS via the C2 data link is managed by a software application on an aircraft system operator notebook. Data from the demonstrator is

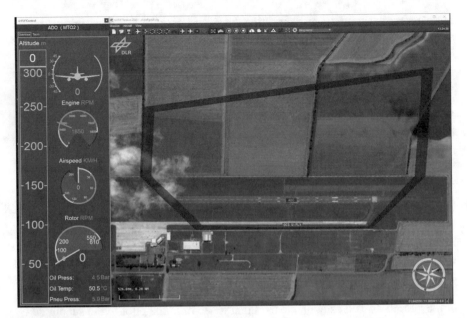

Fig. 23 Flight conditions and test area display used by the pilot monitoring and the aircraft system operator (DLR)

forwarded to other GCS software modules via LAN or WLAN using the BBOO middleware.

10 Development Processes, Releasing and Testing

The development is roughly split in hardware and software development by the composition of the developer teams. Nevertheless, the development cycles are harmonized and the teams work tightly coupled with a quick and simple communication, which is especially required during the system integration phases.

Since the demonstrator is operated at SAIL II of the SORA (Sect. 6), it is not mandatory to use dedicated development standards. For reasons of cost, the hardware and software are expressly not developed according to standards such as DO-178C for software development or DO-254 for avionics components. Instead, internal processes based on DO-254 and DO-178C are followed. The development, release and test process are depicted in Fig. 24.

The development process is based on the development strategies presented in Sect. 7 utilizing user stories and observed issues. User stories are known from agile project management approaches and basically describe features, without specifying a solution explicitly. Instead, desired properties and usability is described. Software modifications during development are continuously checked on correct building,

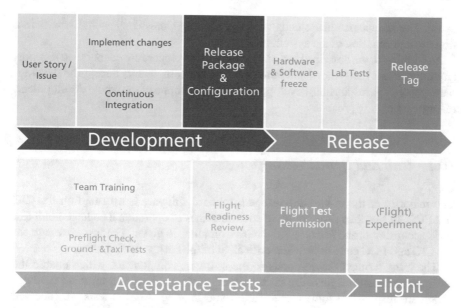

Fig. 24 Development, release and test process (DLR)

execution and interaction with other software components in an automated contin-uous integration process. Additional integration tests, especially on test rigs with hardware-in-the-loop are conducted manually. If the changes of a development cycle pass the continuous integration tests, a release package and configuration can be achieved. In case the release package and configuration consist of all user stories and resolved issues required for a flight test the release process can be started.

For the release process, all current developments for the relevant hardware and software components of the demonstrator are frozen temporarily. This means that change of the demonstrators' configuration is no longer allowed. The frozen software version is transferred to a software release candidate and is by that decoupled from further ongoing software developments. The release candidate software and hardware configuration are now subject of predefined release tests in a laboratory environment. During the release phase minor changes of the software are allowed and are merged into the development branch at any point in time. If all release tests are passed, the demonstrator configuration is released for the acceptance tests.

During the acceptance test phase, the flight test team trains in a hardware and software-in-the-loop simulation environment for the next flight test according to new features or modifications within the aircraft system. These tests follow the objectives of a team training and further test the involved hard- and software under more realistic conditions than in unit tests or the laboratory. In parallel, the demonstrator undergoes a maintenance and overall system test procedure. Afterwards ground and taxi tests are conducted. These tests allow observing and comparing of long-term effects under reproducible conditions and tasks. During the acceptance phase no changes to the software are allowed. If these are required, the system under test must be released again. The acceptance phase ends with the flight readiness review and results in a flight test permission of the system under test.

In addition to the described test procedures all developers are encouraged to report any unexpected behavior of software or hardware components during the development process. A responsible development is the backbone of the whole development, release and testing process.

10.1   Unit and Bench Testing

The majority of the in-house developed onboard software is executed on the CIC and FCC. In Fig. 25 a distribution of software lines of operational code and unit test code according to the executing onboard computer is given. The test coverage for the CIC and FCC code reflects the criticality of the corresponding code.

The term "critical" corresponds to the software components, which enable the fly-by-wire functionalities. The term "other" corresponds to the software modules which are responsible for communication and logging. All unit tests are executed automatically by the continuous integration process. Reports of the unit tests are directly incorporated into the ongoing development.

Automated flight software modules and experimental software modules running on the FCC highly depend on in-house developed legacy libraries, external libraries and auto-generated code from MATLAB/Simulink models. Functional testing of these software parts is either done in the library projects or in software-in-the-loop tests. The unit testing of these modules is subject of interface tests, plausibility analysis and extensive usage during the development, release and acceptance phase.

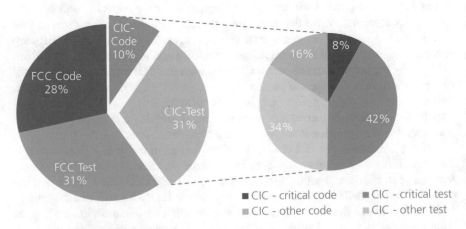

Fig. 25 Onboard software number of lines of operational code and unit test code, according to the executing computer (DLR)

The amount of internally developed operating code is similar to the amount of test code for the FCC software.

Malfunction or crash of the fly-by-wire application, running on the CIC will result in a loss of control of the demonstrator. For these critical software modules, the software code is as lean and independent from external code base as possible. This results in fewer lines of codes which are easy to review and as well black-box and white-box testable. The CIC software is therefore subject to extensive testing in unit tests. The tests cover aspects of interface testing, functional testing, and software stability tests within the environment of the real time operating system, including long time tests. The resulting number of lines of unit test code is five times higher than the number of lines of operational code.

All in-house developed avionic components are tested on test benches prior to the integration in the demonstrator. The tests are partly conducted manually; partly automated test procedures are executed. During the integration of onboard computer with other avionics, for example actuators or sensors, the test benches were very useful laboratory environments for integration tests. Detailed information about the actuator testing as one example can be found e.g. in (Bierig et al. 2018).

10.2   Pilot Training

Despite the aim to demonstrate a fully automatic gyrocopter suitable for cargo transport, a pilot flying is necessary during the test flights, so that all safety measures and system components can be tested in flight. The task of the pilot flying in testing the air vehicle and the flight controller in particular is to steer the demonstrator to the desired aircraft attitude via a remote-control device on ground. Further tasks are to intervene in critical situations and stabilize the flight if necessary. Even an experienced UAS test pilot must be adequately trained for the vehicles specific flight characteristics and the particular experimental missions in order to control the demonstrator safely. The quality and fidelity of the simulation used for the training should be high enough to ensure realistic real-time feedback of the simulated aircraft.

Therefor a gyrocopter flight simulator developed at DLR is used, which has been adapted to the flight characteristics of the modified MTOfree, for more details see Pruter et al. (2019). For the training of the test pilots the angle of vision of the pilot flying is configured as a fixed position relative to the runway in the visual system.

The training comprises the entire flight. This includes the take-off procedure as well as crew resource management (CRM) in flight and the landing procedure. The proper performance of the take-off procedure could already be significantly improved during the first training units. CRM is also significantly improved during the first training units. The pilot flying requires very little information about flight conditions in most phases of flight, which allows him to concentrate much better on the essential tasks. The outermost challenge during training was found to be the landing procedure. Again, the great benefit of the pilot monitoring was shown. Especially during landings, a remarkable learning effect can be observed (Fig. 26).

Fig. 26 Qualitative
evaluation of landings in the
simulator (DLR)

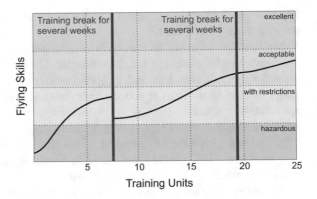

At the beginning of the training, the quality of the landings could only be evaluated with restrictions. If the safety pilot trains continuously over a longer period of time in the simulator, he is able to land the demonstrator reliably and precisely. The landings are consistent within the acceptable range. After longer breaks from training, a certain number of units is necessary to reach the previous training level. However, with an increasing number of training units, a better repertoire of skills becomes apparent. Within the 25 training units, a total of 384 landings were performed in the simulator. This number is of the same order of magnitude as a student pilot needs to be able to land safely.

After the flight tests the pilot flying reported only small differences in the behavior of the demonstrator compared to the simulation. Just the very small permissible flight range causes difficulties, and visibility decreased more than expected at distances above 700 m. This shows the benefit of indicating the position of the gyrocopter within the permissible flight range, which improves situational awareness and ensures that the permissible range is maintained.

10.3   Hardware-in-the-Loop Simulation

Hardware-in-the-loop (HIL) tests are needed in order to test the flight control functions on the onboard computer. HIL means that an identical copy of the hardware of the flying system with software also identical to the flying system is used in a simulation environment in such a way that it behaves exactly like the system under real conditions.

Depending on what components are represented virtually respectively real, the fidelity of a HIL simulation differs. For the demonstrator HIL simulation, CIC, FCC, RC receivers, and remote-control device as well as GCSs are real components. In contrary, all actuators and sensors as well as the VIS and the PUU are simulated on a dSpace-System. The flight dynamics are simulated on a second computer that runs a real time operating system (RTOS) and is connected to a visualization system (Duda et al. 2013a; 2013b).

Figure 27 shows the interaction of devices. A dSpace-System is used to execute code, which is generated by MATLAB/Simulink. A dedicated simulator infrastructure of a manned gyrocopter is interfaced to run the flight mechanics and a spherical visualization (Sect. 10.3). The dSpace-System emulates the interfaces of the gyrocopter that cannot be integrated in a laboratory environment. Sensor values and actuator monitoring feedback is generated based on states of the flight mechanics simulator. Connected via original interfaces (e.g. CAN, RS422 and LAN) the flight computers are connected to the dSpace-System. The flight computers execute the flight software and are interacting by wireless connections with the ground control station and the remote-control unit equivalent to real flight operation.

The actuation system emulation works as follows: The target position for the actuator is commanded from CIC or remote control via their specific interface. The actuator model simulates the dynamics of the real actuator. It reports about the simulated current position as well as about internal measurements to the CIC, and forwards the actual actuator position to the flight mechanics simulation to affect the simulated aircraft motion.

For simulated sensor measurements the motion of the center of gravity is send from the flight mechanics simulation system to the dSpace-System. On the dSpace system, the sensor characteristics like internal dynamics, noise and limits are added.

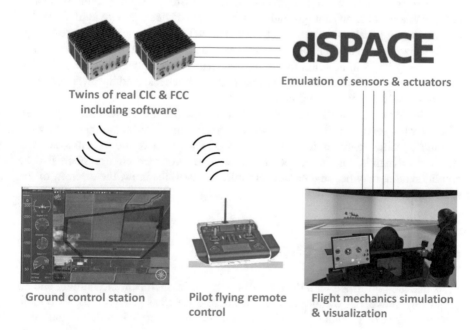

Fig. 27 Hardware in the loop simulation architecture (DLR)

Based on those values, the sensors data stream is generated before it is forwarded via the specific interface to the CIC or FCC.

The simulation of the VIS works in a similar way. The VIS is used e.g. to control the simulated engine ignition. A command from the CIC to switch the ignition on is processed by the VIS-emulation on the dSpace and forwarded to the flight mechanical model to control the simulated engine modes. Feedback of the VIS is sent to the CIC in a similar way as the real aircraft would respond.

HIL simulation is especially useful for the flight controller development, because it is the only way to test the controller properly in closed-loop conditions similar to a flight test. The flight controller is developed as a MATLAB/Simulink model. The resulting block-model is than converted into C++ code using the automatic code generation feature. This process naturally incorporates risks regarding the interface definitions and real-time requirements. Additionally, to interface tests that are part of the overall system test (OST), HIL simulation allows for much more extensive testing of the flight controller, because states can be simulated that are risky to set up in a real flight test, but may occur occasionally.

10.4   Maintenance and Overall System Tests

Maintenance includes the regular checks like the airframe, engine, avionic components, cable connections and ground control system devices. Where applicable, the maintenance manuals of the manufacturer are used. Due to frequent hardware configuration changes and comparably low flight time a maintenance routine is performed before every flight in addition to the regular maintenance intervals of the aircraft. The maintenance check is part of the flight readiness checks and hardware changes are not allowed afterwards, before flight.

While the maintenance focusses on the hardware of the aircraft system, the purpose of an overall system tests (OST) is testing the system functions including the software within a specific configuration. Parts of the OST are tests of the overall system in static and dynamic condition, but in a controlled, protected environment. Failure conditions that may become obvious should not harm the crew, the aircraft, or the environment.

10.5   Ground and Taxi Tests

Tests that are performed with the aircraft on the airfield with engine running and all sensors operative, but without taking off, are referred to as ground tests. Two main different types of ground tests exist: tests where the aircraft is tethered and tests where it is not.

Ground tests with the tethered aircraft can take place in principle at any time during the development phase. They are typically engine run-up tests or pre-rotation tests. A ground test where the aircraft is not tethered is called a taxi test. This test is performed at least after all system tests including HIL-simulation and OST are completed. Taxi testing involves testing the maximum operating time as an endurance test and shall ensure that all systems are working properly and reliably.

Taxi tests are significant because they provide the opportunity to test the entire system in a complete setup identical to a flight test. Unlike HIL simulation, there are no more simulated components. The ground control station and the monitoring of the flight area are involved, so that a taxi test is basically like a flight test without flight. Therefore, in addition to the main purpose of testing the aircraft system, communication and crew coordination are also trained in a taxi test.

The aircraft is tethered for the initial engine start-up and main rotor pre-rotation tests in a development cycle. Because of that it is very unlikely to move, even if there is any malfunction. A start-up test can be performed both with the main rotor installed or uninstalled. The aim is to achieve a certain operating time of the engine in order to exclude malfunctions or anomalies.

A special case of such a ground test is a pre-rotation test. In a pre-rotation test the main rotor is sped up by the engine. When the rotor has reached the rated speed of 200 rpm, the engine is decoupled and the rotor brake is activated. The test is finished when the rotor comes to a standstill. With this test, the technical and procedural readiness of the pre-rotation function is tested, either in automatic or manual operating mode. Automatic pre-rotation is part of the automatic takeoff procedure. The flight control system is able to automatically perform the correct procedure.

Automatic taxying along a predefined path is necessary within different phases of an automatic flight. After engine startup, the taxi to the take-off point has to be performed. After pre-rotation, during acceleration to take off speed, automatic steering is required. After touch down, the aircraft must hold its direction on the runway, followed by leaving the runway if desired.

After all testing is successfully completed, the final flight readiness review takes place. All stakeholders and developers meet and review at least the following items:

- Mission objectives and flight test plan
- Risk analysis
- Risk of mission failure
- Modifications compared to last flight
- Aircraft configuration
- Aircraft condition
- Regulatory requirements
- Flight test crew staffing
- Schedule.

On the basis of the flight readiness check, the decision to carry out the flight test is taken. A flight order document ensures the knowledge and approval of the superiors.

11 Flight Test Operation

After approval of flight readiness and issued flight order, the demonstrator is cleared for takeoff. Depending on the flight test area selected, the aircraft is already moved to the airfield and tested on site. Often, the flight test takes place at the National Facility for Unmanned Aircraft Testing at the Cochstedt Airport (Germany). The following sections describe the current typical flight test operation that has been conducted several times.

11.1 Flight Test Area

The former public airport has a runway with a length of 2500 m and a width of 45 m. There are no settlements in the vicinity of the airport which makes it ideal for flight testing of large unmanned aircraft. The flight test volume used for the demonstrator is depicted in Fig. 28. The different colored areas correspond to safety boundaries presented in Sect. 5. The green colored area represents the nominal flight volume; yellow marks the containment area and red the buffer zone. The permissible flight altitude was lowered over the airport to limit the buffer zone according to the available distance in relation to the street in the north of the airport. In the south, the red area ranges into the fields adjacent to the airport. Paths within the range of the possible flight area are blocked to satisfy the condition of controlled ground area (Sect. 6).

11.2 Flight Test Schedule

A well-defined schedule enables a smooth and efficient workflow during flight test. A typical schedule for a flight test is presented within this section. Based on experiences, the preparation of a flight test starts by preparation of the flight approval (Table 8).

Compared to smaller unmanned aircraft, this type of UAS requires this rather extensive effort for a flight test. The reasons can be deduced from the scientific context of the operation of the vehicle. The procedures for operating such an experimental system are based on recommendations and best-practices and change continuously over time. The tasks are usually performed by limited available personnel. Most team members are active in several roles (e.g. as developer and member of the flight test team).

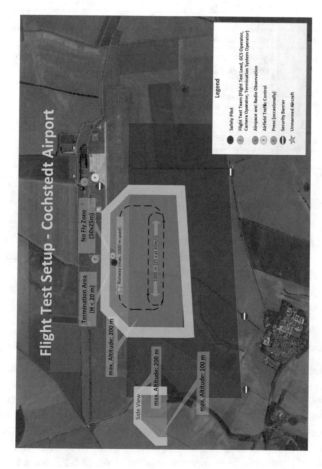

Fig. 28 Flight test area at the unmanned aircraft test range Cochstedt, side view (left) indicates height profile (green: operational volume, yellow: contingency volume, red: volume for ground and air risk mitigation, DLR)

Table 8 Typical flight test schedule

Timeline	Event
1 month before flight test	Flight approval
1 week before flight test	Pilot and team training in the simulator, information for airport personnel, farmers, and hunters within flight area
2 days before flight test	Maintenance and overall system test, flight readiness-review passed
1 day before flight test	Charge battery pack
2 h before flight test	Preparation of ground equipment, fuel
1:30 h before flight test	Final safety briefing and flight test briefing
1:15 h before flight test	System start, system pre-flight check, taxi to take-off location, prepare ground equipment to designated location
0:30 h before flight test	Final check of readiness to take-off by all flight test team members
0:15 h before flight test	Test pre-rotation and aborted take-off
0:00 h	Take-off
After each flight	Post-flight briefing (if required: briefing for next flight)
After flight test	Data evaluation, feedback evaluation

11.3 Atmospheric Conditions

Cochstedt Airport is located in a temperate zone. It has a humid continental climate bordering on an oceanic climate. The prevailing wind direction is west, reflected by the runway heading 255/75°.

For the first flight tests, the operation highly depends on weather conditions. The main restricting factors are:

- Precipitation: electronics are not water-resistant,
- Temperature: reduced concentration and sensitivity of the pilot flying, and Wind.

The manned version of the gyrocopter can be operated in wind conditions up to 40 kts with a maximum demonstrated crosswind component of 20 kts for takeoff and landing (AutoGyro 2014). Flight tests have shown that the pilot flying has difficulties to align the demonstrator with the runway heading during landing because of his perspective. As a consequence, the demonstrator tends to touch down at a slight pushing angle, which carries the risk of rolling over. To reduce this risk, the maximum permissible crosswind component is initially set to 6 kts for a flight test. The permissible mean wind is limited to 12 kts with low gust intensity.

For risk mitigation, flight tests are scheduled in the early morning or late evening when the wind is reasonable lower. However, the typical orientation of the sun during morning and evening flights can lead to visual disturbances for the pilots.

11.4 Flight Tests

Between 2018 and 2020 several flight tests were conducted at Cochstedt airport. Within the first flight tests, the demonstrator was flown in manual flight mode. Based on this manual flight test campaigns a lot of knowledge about the system, flight mechanics and proceedings were obtained. In 2019 test campaigns for testing first controller functions in flight started. A significant milestone was reached in August 2020: the first automatic racetrack was flown, giving full authority to the flight controller for the first time in flight.

The purposes of the first flight tests ranged from training of the pilot flying and flight test team, data acquisition, flight mechanical evaluations and system tests to evaluate the applicability of the given legal framework. In addition to simulation and ground testing, manual flights facilitate the adjustment of several parameters. In flight, noise and loads are evaluated; dynamics and sensor measurements are validated. Furthermore, flight test data allows for simulation model validation and improvement. Collected results of the first manual flights include:

- Sensors work within their expected accuracy. In flight tests, air and radar altimeter data were validated with data from the inertial navigation system in post-processing.
- Initialization procedures for the navigation system are established and refined.
- Mathematical model of the flight dynamics of the unmanned aircraft system is improved by system and parameter estimation methods (e.g. damping coefficients).
- Crosswind impacts the safety during landing phase. Consequently, the permissible atmospheric conditions for flights have been limited.
- Visual perception of the flight attitude and orientation in turns is difficult for the pilot flying. This was improved by installing landing lights to the nose wheel and colored labels to the vertical stabilizers.
- The pilot flying needs precise advisory to stay within the flight area. This was solved by providing a tablet PC with top-view of the flight test area to the pilot monitoring.

Subsequent to manual flight tests, the level of automation is increased step by step. Different automatic flight modes are implemented and tested in order to give the flight control system authority of single or multiple control axes. During these tests, the pilot flying trims the aircraft to a steady state flight in sufficient height and then enables an automatic flight mode. In case of any unforeseen event, the pilot flying is always able to take over manual control.

12 Summary and Concluding Remarks

The technology demonstrator presented in this work is designed to test prototypical aspects regarding the real operation of the ALAADy concept. As the most suitable option the gyrocopter configuration was selected. The development process gained from experiences in the operation of UAS as well as experience with the handling, flight mechanics and simulation of gyrocopters existed within the research organization as a basis for the development of the demonstrator. Basic components are a MTOfree airframe, electro-mechanical actuators, sensors, onboard-computers, battery power-supply and a remote control system for manual operation. In this research context the development team also performed the flight operation. This shared responsibility did prove to be beneficial, as it enabled the optimal use of the team's system knowledge. However due to the timely effort of flight testing it also slowed down the development pace considerably. Nevertheless, the chosen rapid prototyping approach allowed flight operations with the system in early stages of the project. This work focuses on the specific operation risk assessment, the technical aspects of system integration and the continuous development and release process.

A specific operation risk assessment using the JARUS-draft was conducted for the initial flight testing of the demonstrator. The SORA was used as a basis and guideline for the operation. As a result, a SAIL II operation was enabled with the constraints to operate in VLOS and to restrict third party access to the ground area of operation. A concept of operation was elaborated according to SORA. The capability for a safe, redundant termination of the flight in case of any unexpected failure or other unsafe condition was factored in the ConOps. By activating the flight termination, the engine's ignition is switched off and the actuators move to a position leading to a passive spiral autorotation of the aircraft. Due to the glide ratio of a gyrocopter, switching off the ignition is sufficient to ensure that the demonstrator does not leave the buffer area. Safety—for uninvolved persons as well as for operating crew— was considered highest priority during development. To lift safety to an adequate level, many different measures have been taken, including procedural and technical precautions, establishing rules for design and verification, as well as considering the environmental conditions.

The development of the demonstrator was characterized by the trade-off between cost-efficiency (resulting in the renunciation of using dedicated development standards) and achieving a sufficient level of assurance for the functionality and reliability of hardware and software components. The separation of flight-critical and experimental subsystems is an example of these considerations. With the gyrocopter airframe's cost efficiency, robustness and modifiability it proves to be a well-suited platform for the UAS-conversion.

A continuous, reproducible development release process with a hierarchical test structure was established. This release process is completed before each flight test but is also streamlined enough to be feasible in a research environment. In-depth testing and hardware-in-the-loop simulation have been conducted specifically to test automated flight functions. These functions are developed and tested in an agile step-by-step approach to minimize risks of unexpected events.

The agile development process also allowed to quickly adapt to findings from flight tests, which led to the development of several unplanned additional features and functions. These additional features include a yaw controller for landing assistance, a pilot monitoring display, lights and visibility improvements and software modifications to improve the manual flight characteristics. Despite the fact that the demonstrator is an unmanned and automated aircraft system, a remarkable number of challenges and adaptions were related to human performance limitations.

However, it is reasonable to conclude that the major design choices and decisions were correct and resulted in all flight tests being successful. Today the demonstrator is a capable platform for research on unmanned freight transportation.

13 Future Perspective

The ALAADy demonstrator (Fig. 29) is in active flight operation. Nevertheless, the development of the demonstrator is an ongoing process. To fulfill the goal of validating the different concepts of ALAADy, such as the integration into civil airspace like of U-space, further continuous modifications and improvements of the systems are necessary.

The short-term goal for the operation of the demonstrator is a fully autonomous, but highly monitored BVLOS freight transport demonstration mission. To achieve this, further advancements regarding the SORA and technical solutions to fulfill its requirements are planned. In a next step, the impact of longer flight times and

Fig. 29 Onboard-view of the demonstrator during approach (DLR)

distances on the system design and operation will be explored. Therefore, new long-range datalinks, e.g. by satellite, need to be integrated and evaluated. These datalinks need to be either redundant or the demonstrator needs to be able to make decisions autonomously. With longer distances additional challenges arise, like handovers of multiple ground control stations, system integrity monitoring and containment procedures, airspace integration and detect and avoid capability. Adaptation to dynamically changing environmental conditions, e.g. caused by dynamic U-space geofencing, rules and concepts need to be designed. This and further research will address additional aspects of the ALAADy concept study, like the implementation of a representative logistic chain.

Acknowledgements The realization of a project as complex as this demonstrator requires the expertise, brainpower, craftsmanship and dedication of a motivated team. Credit and thanks go to all members of the team for their efforts and participation, especially Andreas Buschbaum, Jörg Dittrich, Holger Duda, Patrick Gallun, Christian Hoffmann, Ingo Jessen, Michael Kislat-Schmidt, Uwe Klamann, Laura Madero Coronado, Frank Möller, Florian Nikodem, Falk Sachs, Rainer Schmidt, and Jörg Seewald.

References

AutoGyro (2014) MTOfree pilot operating handbook. https://www.auto-gyro.com/en/Services/Downloads/Manuals/. Accessed 14 Sept 2020

Benders S, Goormann L, Lorenz S, Dauer JC (2018) Softwarearchitektur für einen Unbemannten Luftfrachttransportdemonstrator. Publikationen zum Deutschen Luft- und Raumfahrtkongress 2018, Friedrichshafen, Germany

Bierig A, Lorenz S, Rahm M, Gallun P (2018) Design considerations and test of the flight control actuators for a demonstrator for an unmanned freight transportation aircraft. recent advances in aerospace actuation systems and components, Toulouse, France

Boehm BW (1988) A spiral model of software development and enhancement. Computer 21:61–72

Dalamagkidis K, Valavanis KP, Piegl LA (2008) Evaluating the risk of unmanned aircraft ground impacts. In: 16th Mediterranean conference on control and automation, Ajaccio, France

Dauer JC, Adolf FM, Lorenz S (2015) Flight testing of an unmanned aircraft system—a research perspective. In: Systems concepts and integration panel (SCI) symposium, Seiten 4-1. NATO. STO-MP-SCI-269 - Flight testing of unmanned aerial systems (UAS), Ottawa, Canada, 12–13 May 2015. ISBN 978-92-837-2015-7

Duda H, Gerlach T, Advani S, Potter M (2013a) Design of the DLR AVES research flight simulator. In: AIAA MST 2013, Boston, USA, 19–22 August 2013

Duda H, Seewald J (2016) Flugphysik der Tragschrauber. Springer, Heidelberg. ISBN 978-3-662-52833-4

Duda H, Sachs F, Seewald J, Rohardt CH (2015) Effects of rotor contamination on gyroplane flight performance. In: 41st European rotorcraft forum 2015, München, Deutschland

Duda H, Seewald J, Cremer M (2013b) Data gathering for gyroplane flight dynamics and simulation research. In: 24th SFTE-EC symposium, Braunschweig, Germany

Duda H, Pruter I (2012) Flight performance of lightweight gyroplanes. In: ICAS 28th international congress of the aeronautical sciences, Brisbane, Australia, 23–28 September 2012

Duda H, Pruter I, Deiler C, Oertel H, Zach A (2011) Gyroplane longitudinal flight dynamics. In: CEAS 3rd air and space conference, Venice, Italy, 24–28 October 2011

European Aviation Safety Agency - EASA (2005) A-NPA, No. 16/2005, policy for unmanned aerial vehicle (UAV) certification

European Union Aviation Safety Agency (2019) Acceptable means of compliance (AMC) and guidance material (GM) to commission implementing regulation (EU) 2019/947. Last access 04 March 2020 von Official Publication: Acceptable means of compliance (AMC) and guidance material (GM): https://www.easa.europa.eu/official-publication/acceptable-means-of-compliance-and-guidance-materials. Accessed 14 Sept 2020

Friedrich M, Peinecke N, Geister D (2021) Human machine interface aspects of the ground control station for unmanned air transport. In: Dauer JC (ed) Automated low-altitude air delivery - towards autonomous cargo transportation with drones. Springer, Heidelberg

Grima Galisteu D, Adolf FM, Dittrich J, Sachs F, Duda H (2015) Towards autonomous emergency landing for an optionally piloted Autogyro. In: AHS international 71st annual forum, Virginia Beach, VA, USA, , 5–7 May 2015. ISSN 2167-1281

Joint JAA/Eurocontrol Initiative on UAVs (2004) A concept for european regulations for civil unmanned aerial vehicles (UAV). Final report

Joint authorities for rulemaking on unmanned systems (2019) JARUS guidelines on specific operations risk assessment (SORA), edn 2.0

Nikodem F, Rothe D, Dittrich JS (2021) Operations risk based concept for specific cargo drone operation in low altitudes. In: Dauer JC (ed) Automated low-altitude air delivery - towards autonomous cargo transportation with drones. Springer, Heidelberg

Pak H (2021) Use-cases for heavy lift unmanned cargo aircraft. In: Dauer JC (ed) Automated low-altitude air delivery - towards autonomous cargo transportation with drones. Springer, Heidelberg

Pruter I, Benders S, Duda H, Lorenz S, Sachs F (2019) Erprobung eines Flugreglers für einen unbemannten Tragschrauber, Deutscher Luft- und Raumfahrtkongress, Darmstadt, Deutschland

Pruter I, Duda H (2011) A new flight training device for modern lightweight gyroplanes. In: AIAA modeling and simulation technologies conference 2011, Portland, USA, 08–11 August 2011

Sachs F (2021) Configurational aspects and vehicle specific investigations for future unmanned cargo aircraft. In: Dauer JC (ed) Automated low-altitude air delivery - towards autonomous cargo transportation with drones. Springer, Heidelberg

Sachs F, Duda H, Seewald J (2016) Flugerprobung zukünftiger Tragschrauber anhand skalierter Versuchsträger. Braunschweig, Deutschland: Deutscher Luft- und Raumfahrtkongress

Sachs F, Duda H, Seewald J (2014) Leistungssteigerung von Tragschraubern durch Starrflügel. DLRK 2014, Augsburg, 16–18 September 2014

Unmanned Aircraft for Transportation in Low-Level Altitudes: A Systems Perspective on Design and Operation

Johann C. Dauer

Abstract Using Unmanned Aircraft Systems (UAS) for transportation is an increasingly widespread concept. However, the majority of UAS only carry payloads of a few kilograms. This payload limitation will change in the future. New regulatory approaches facilitate the use of larger UAS with significantly increased payload capacities by incorporating the operational risks during certification. To assess the risk, the *Specific Operations Risk Assessment* (SORA) can be used, which defines a method that incorporates all safety affecting factors holistically. However, so far these factors ranging from automation or autonomy, aircraft configurations and design, airspace infrastructure and integration, certification and assurance, to the economics of potential use-cases have been studied in literature independently. The reason for this separation is the complexity of such holistic analysis of the aircraft system and operation. To overcome this complexity, this chapter identifies four segments of particularly strong interdependencies. This segmentation allows to map the air and ground risk classes of the SORA to the different system and operational components. At the same time, it supports balancing the level of autonomy with the reliability of required external services. Furthermore, it allows to study and adjust potential use-cases with the resulting system and operational requirements. The segmentation is discussed in detail based on the results of the research project *Automated Low Altitude Air Delivery (ALAADy)*. The project conceptually studied transport UAS with payload capacities of up to one metric ton. Several technology demonstrations and flight experiments are included within this project. The following sections present a comprehensive view of the project's results and conclude four years of research on unmanned aerial transport at the German Aerospace Center (DLR).

Keywords Automated Low Altitude Air Delivery · Heavy-lift cargo drones · Humanitarian aid · Safe autonomy · Safe operation · Specific Operations Risks Assessment · SORA · Technology demonstration · Unmanned Aircraft System · UAS · Unmanned cargo aircraft

J. C. Dauer (✉)
Unmanned Aircraft Department, Institute of Flight Systems, German Aerospace Center (DLR), Lilienthalplatz 7, 38108 Braunschweig, Germany
e-mail: johann.dauer@dlr.de

© Deutsches Zentrum für Luft- und Raumfahrt e. V. (DLR) 2022
J. C. Dauer (ed.), *Automated Low-Altitude Air Delivery*, Research Topics in Aerospace,
https://doi.org/10.1007/978-3-030-83144-8_20

1 Introduction

Is it possible to realize a safe unmanned aircraft system (UAS) for transportation more cost-efficiently, if the operation is limited to low-level altitudes and above sparsely populated areas? This question is the focus of the research project Automated Low Altitude Air Delivery (ALAADy) at the German Aerospace Center (DLR). The project focuses on payload capacities of the UAS of up to one metric ton. All chapters of this book highlight individual aspects of the realization of such cargo transportation. Market aspects are considered as well as questions on technical realization. Unique to this research project is the hypothesis that a UAS of this size can be operated within the *Specific Category* of the European Union Aviation Safety Agency (EASA) and compliant to the Specific Operations Risk Assessment (SORA).

   The SORA provides the means to interrelate all safety affecting elements and balance operation conditions with the guarantee achievable of the UAS reliability. The project ALAADy is a first step to explore the new possibilities a holistic approach like the SORA provides for realization of UAS and its operation. However, such analysis requires a broad set of aspects to be studied and at the same time cover their interdependencies. While many publications exist that assess specific technical concepts for drones in general or transportation in particular, publications in scientific literature always strive to focus on specific and sharply traced out problems. In contrast, this book integrates a large set of aspects to achieve a systems perspective as comprehensive as currently possible. It lays the foundation of a holistic design of UAS and operation. Within this concluding chapter, a set of particularly strong interrelations are identified and discussed based on the results of the project. The proposed segmentation of interrelations simplifies the analysis and discussion by limiting the interplay to those most significant. At the same time, it allows to discuss and explore the new degrees of freedom that the SORA method enables.

   Three aspects for the aircraft design and its operation are fixed for this study. First, a set of Top-Level Aircraft Requirements (TLARs) define the range and size of the aircraft considered. For example, with this TLARs, the payload capacity is fixed. Second, the operational risk is minimized by operating in low-level altitudes and third, operation is restricted to fly above sparsely populated areas only. With these definitions, we challenge the SORA with a comparably large-scale aircraft and minimized operational risks to study the potential to optimize cost-efficiency while maintaining the high level of safety required in aviation. In particular, within the set operational conditions, safely terminating the flight is an appropriate last contingency to ensure safety. This argument is valid from a safety perspective, as no other airspace types need to be crossed when triggering the safe termination and, choosing adequate flight routes, no infrastructure or people on ground are endangered. This concept is not meant to accept regular safety landings as they can imply financial loss and environmental hazards. Rather, within this book we focus on the questions whether the immense development costs due to certification can be reduced and still result in an overall safe operation of a UAS.

The following sections are structured as follows: First, the overall study of ALAADy is recapitulated and four major interrelations of UAS operations are identified. These four aspects are (a) Use-Case & Reliability, (b) Aircraft Configuration & Contingency Concept, (c) Airspace and Detect and Avoid (DAA) and (d) Operability & Safe Autonomy. These dependencies then form the layout of the proceeding sections connecting the results of the previous chapters in this book. Afterwards, simulations and three aspects of technology demonstration are discussed. In all sections, open questions and future research perspectives are outlined which directly follow from relating the different topics to each other. The chapter concludes with a condensed form of discussion and future direction of ALAADy.

2 Recapitulation of the Project ALAADy

The first chapter of this book (Dauer and Dittrich 2021) spreads open the different topics and research questions and structurs it in the four discipline-focused parts of the book. The uniting elements across these different topics addressed is a set of TLARs, and the intended application the SORA. The use of the SORA allows for new possibilities to achieve operational safety without necessarily maximizing the aircraft's reliability by certifying the technical realization. Thus, PART I of the book assesses the unique aspects of SORA for the proposed size of aircraft and possible use-cases including their economic aspects. Part II of the book focusses on the aircraft configurations, its propulsion and details of the design.

Part III considers components and aspects having major impact on the safety of the operation. Infrastructure modifications for the airspace in the context of U-space are discussed, as well as command, control and safety datalinks, human machine interface, and DAA. The safe autonomy is advanced by two technologies. First, a route planning is proposed which is based on risk modelling. Such an algorithm can now directly be linked to the SORA by explicitly including its underlying risk model to find routes that minimize the operational risks to the environment. Such an implementation as well as all the other autonomy enabling functions involve a significant amount of software that need to be assured. For this reason, we propose the use of a Safe Operation Monitor (SOM), which is designed to be of minimal complexity and sufficiently independent from the remainder of the avionics and software components. The SOM surveils the safe operational conditions defined in the SORA Concept of Operations (ConOps[1]) and invokes counter measures (contingencies) as soon as these conditions are about to be impeded. The final part of the book focusses on simulations for different aspects of knowledge gain and for validation, and on the development of experimental technology demonstrations.

---

[1]

In this chapter we understand ConOps in the SORA terminology. Its content is explicitly defined in SORA ANNEX A (Murzilli 2019) and is the basis for the safety assessment. This SORA ConOps is not necessarily identical to a general concept of operations, which might have other focus areas, e.g. business and stakeholder requirements.

Throughout this project, many different aspects have been analyzed in detail, and interplay and dependencies of these aspects have been discussed and studied. Initially, the new concept for certification allows for a holistic approach of designing the UAS in combination with its application and interrelation of all components and operational aspects are discussed, see Dauer and Dittrich (2021). However, for this particular project, focusing on heavy-lift transportation, a set of strong dependencies have been identified, see Fig. 1. The figure introduces four circular dependencies that also form the structure of the discussion in the following sections.

The use-cases define the requirements to the UAS as well as the operations' functional requirements. Based on identified use-cases, a rough sketch of the ConOps can be formulated as indicated in Fig. 1(a). The ConOps allows for the discussion of the SORA. In this specific project, the concept of airport integration has a major influence on the resulting overall risk classification. The decision, whether the UAS should be integrated into commercial airports, use separated airports, or alternatively

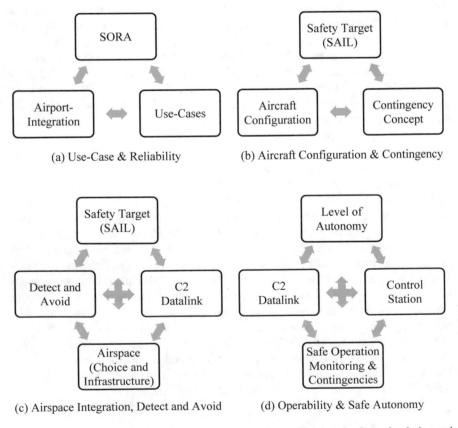

(a) Use-Case & Reliability   (b) Aircraft Configuration & Contingency

(c) Airspace Integration, Detect and Avoid   (d) Operability & Safe Autonomy

Fig. 1 Strong dependencies of system components and operation that dominate the design and safety of the unmanned cargo aircraft

take-off and land off-field, has major influence due to the involved airspaces the UAS needs to be operated in.

The SORA introduces the ground and air risk which are assessed separately. The higher of the two grades is used to determine the overall safety target which defines the level of rigor for all safety related aspects of the UAS. Figures 1(b) and (c) show the conceptual measures to influence the overall safety target. The aircraft configuration in Fig. 1(b) indicates the ground-risk-affecting aspect. The influence of size and impact energy on the risk for ground-based third parties is assessed by application of the SORA process. However, different aircraft configurations have varying potential to take credit of risk mitigations. Selected contingencies influence the affected ground area and impact energy to be considered. Configurations with inherent safety characteristics may surpass conventional and well-known configurations from manned aviation in this context. The second part of the overall risk assessment is the air risk influenced by dependency circle of Fig. 1(c). The services and infrastructure provided for an airspace and its encounter rates define the overall air risk and are strongly intertwined with the datalinks and DAA.

The last dependency cycle of Fig. 1(d) develops around the trade-off between human interaction and autonomy. A certain capability of a UAS can be achieved either by providing an adequate human machine interface (HMI) in most cases, the alternative is to equip the UAS with automatic capabilities. This trade-off, however, has safety implications. The concept of ALAADy is to apply a SOM to assure the capabilities of autonomy during runtime. Nevertheless, this SOM needs to be sufficiently reliable if the UAS level of autonomy is high. In contrast, if the level of autonomy is low, the C2 datalink as well as the control station needs to be sufficiently powerful and reliable. An academic, yet illustrating example of this dependency is a fully manual remote control without any flight control assistance. Each control surface deflection needs to be commanded from the control station, and already the loss of only a few data packages can result in a catastrophic event.

3 Discussion of the Study

The following sections use the previously identified strong dependencies, see Fig. 1, to structure and connect the results of this book's different chapters. Afterwards, major results found during the realization of the technology demonstration are included into the discussion. The goal of the following sections is to achieve a condensed discussion of the many detailed aspects relevant for the transportation concept of this project.

3.1 Interrelation: Use-Case and UAS Reliability

The evaluation of the SORA shows that the operation of the ALAADy concept can span SAIL levels III to VI, see Nikodem et al. (2021). Aiming for SAIL III implies an emergency response plan of high robustness, which, in many situations, does not seem feasible in short term operational contexts, like humanitarian applications or large distance operations involving different emergency services. Thus, the lowest practical SAIL is IV. Initial investigations into the certification effort and system complexity required for the three remaining levels show that there is a significant leap between SAIL IV and V (Rothe and Nikodem 2021). If the operation aims at SAIL IV, this implies a reduced development and certification effort while allowing for operating the aircraft over sparsely populated areas and in low-level altitudes, as initially intended by the project. Aiming for SAIL IV requires the aircraft to be equipped with some means to reduce impact energy (SORA mitigation M2) to medium robustness.

SAIL IV allows for operations in low-level altitudes including uncontrolled airports or heliports. However, it constraints operation outside of airspaces B, C and D, i.e. of many general commercial airports. The results from Mühlhausen and Peinecke (2021) confirm the need to avoid commercial airports also from an economic perspective. The authors use airport approach simulations to determine the impact of an integration of the cargo UAS into regular air traffic. One of the main findings is that such aircraft would block multiple slots at the airport. They suggest to at least introduce separated and dedicated runways for drone operation. The results do not differ for the three ALAADy aircraft configurations as the slot requirements are dominated by the initially defined and uniform TLARs, which already fix the relevant flight performance characteristics for this simulation study. The ALAADy concept hence implies that commercial airports are to be excluded due to the SAIL selection and due to the airport traffic management. As a result, dedicated runways, separated airports, or off-field take-off and landing need to be used, where approvable.

If it is considered to design dedicated landing sites for drone operation, the automation of the ground handling deserves closer attention. At first glance, it might seem obvious that full automation saves a significant amount of cargo handling time and personnel costs. Nevertheless, full automation leads to significant efficiency loss in the event of failures of individual process steps (Meincke 2021). Such dedicated handling lines thus require either redundancies of critical sections or maintenance personnel available at short notice. In the latter case, in particular, it is also conceivable to allow for semi-automatic handling and to have the personnel required for the manual tasks also perform maintenance, if necessary. Meincke examines different variants of aircraft and cargo handling concepts and also shows their impact on the aircraft configuration by the simulation of loading and unloading operations. The orientation of the cargo bay surface or position of the loading flap can hinder or assist automatic cargo handling. Another important aspect, especially for humanitarian aid operations, is the ability to air-drop cargo. This need has been confirmed in simulated humanitarian drone operations (see Sect. 4.3) and should be considered directly in the TLARs in future projects.

Some conditions have now been clarified including the appropriate SAIL, landing sites, and ground handling. However, an important open question remains, which is the identification of promising use-cases. The research conducted early in the project is summarized in Pak (2021). Pak shows that use-cases are promising if the use of UAS provides a beneficial ratio of the value connected to the time reduction opposing the additional costs due to airborne transport. The value of time savings depends on the benefits associated or the harm avoided and corresponds to the time criticality of the delivered goods. Time savings are determined by the available competing ground-based transportation options, the detours required to remain over sparsely populated areas, and the effort needed to integrate the UAS into the complete logistics chain. The additional costs due to air transport are determined by the costs of the drones and their operation, e.g. the increased energy demand due to the airborne transport. As a result, the transport of time-critical spare parts and humanitarian aid appear to be especially suitable for the transport drones considered here.

Liebhardt and Pertz (2021) model and study both applications, time-critical spare parts and humanitarian aid, in more detail. The different elements that influence the costs of the operation are described using a parametric cost model. This model considers depreciation of the UAS, insurances, pilot wages, the fuel (assuming a conventional engine), interests, fees for airport and datalink as well as maintenance costs. A challenge of this work is the lack of available reference values for the disruptive aspects of this transportation concept. References from manned aviation are used to fill in these gaps. For example, the Development and Production Costs for Aircraft (DAPCA) model is used to estimate the maintenance and production costs. Using references from manned aviation sets the limit of the current validity of this modelling. Especially in the context of the SAIL IV as the selected operational risk class, these two factors do not require the same level of rigor as in manned aviation. The maintenance dominates the operating costs in Liebhardt and Pertz (2021), at least in the case of time-critical spare parts logistics. This aspect suggests that the model overestimates the operating costs. However, the overestimation is reduced by two aspects. A use of reference information from manned aviation underestimates the costs of development and production of autonomy related components including avionics and software, which plays a major role for unmanned systems. Additionally, a reduction of maintenance can also lead to a reduction of system reliability and thus shorten the lifetime compared to that used in the suggested cost model.

For the spare parts logistics, the cost model is applied to a specific exemplary company in Europe and a logistics network using the transport drones is designed according to this company's requirements. A comparison with ground-based vans shows that the cost of using drones in a one-way mission leads roughly to doubled prices but halved delivery time. If the application, like the example in Liebhardt and Pertz (2021), requires an empty return of the aircraft, the price ratio worsens, as it is unlikely that vans or trucks would return empty. If non-empty returns of the competitors are considered, the costs ratio worsens accordingly. Nevertheless, situations of poorer ground infrastructure compared to the example, e.g. with a reduced road quality or frequent traffic jams, can influence the choice of the airborne delivery significantly.

An extreme case of a lack of ground competitors are scenarios of humanitarian aid. Here, Liebhardt and Pertz consider a flooding incident in Mozambique caused by the cyclone Idai in March 2019 as a concrete case for the assessment. For this event, a supply chain is modeled describing the mission execution for the last-mile delivery. It becomes evident that for such use-cases, the ability of airdrops is essential, as already suggested in the ground handling analysis. Assuming this capability, Liebhardt and Pertz show that the drone transport concept is able to compete with the helicopter Mil Mi-8, which was used in Mozambique. However, the mission profile differs to some extent. While the helicopter is used for bulky goods and selected destinations, delivery by drone focuses on more targeted supply of smaller cargo. This conclusion considers the fact that the available payload is reduced by additional space required for the dropping equipment.

3.2   Interrelation: Aircraft Configuration and Contingency Concept

At the beginning of the project, specific TLARs were formulated as an initial definition: one metric ton of payload capacity, 600 km range, 200 km/h cruise speed, maximum take-off and landing distance of 400 m, and a cargo bay for two Euro-pallets. These TLARs initiate the following discussion: First, the applicability of a large set of potential aircraft configurations is examined. Subsequently, after a first pre-selection, the flight performance of the candidates is calculated and as a result, three aircraft configuration candidates are selected for the detailed considerations of the project. For these three configurations, the cell structure of the aircraft is designed to specify and evaluate the technical feasibility more reliably. Additionally, the safety characteristics in terms of SORA are considered and appropriate measures are identified to reduce the ground risk.

Initially, Hasan and Sachs (2021) consider a variety of different configurations and select candidates based on flight performance calculations. In addition to conventional fixed-wing and rotary-wing aircraft, more unusual configurations such as box wing aircraft, blended wing body aircraft, tilt-rotors, (hybrid) airships, and paragliders are considered. From the initially large number of different configurations, Hasan and Sachs pre-select fixed-wing aircraft and rotorcraft for more in-depth considerations. In this pre-selection, the requirement of short take-off and landing distances turned out to be particularly selective. In a following assessment based on a qualitative assessment with respect to their suitability for the transportation task, Hasan and Sachs model the flight performance of the pre-selected fixed-wing aircraft and rotorcraft and compare them separately. In a second step, the fixed-wing configurations are additionally equipped with electrically powered propellers only active during take-off and landing. These propellers deliver additional thrust and by exploiting the slip-stream effect also additional lift. So, the wings surface can be reduced to get a slender wing with a high wing load which is favorable for cruise efficiency. The flight performance calculation indeed shows an improvement related to the ALAADy TLARs. For the further investigations in the project, a twin-boom configuration with 16 m wingspan is selected, since it has the lowest fuel consumption. To facilitate ground handling, a box-wing configuration with a wing span of 12 m is proposed as an alternative, since it has efficiency advantages at lower wing spans thanks to its nonplanar wing system.

With regard to rotorcraft, Hasan and Sachs focus on configurations with one main rotor, i.e. helicopters of classical configuration and gyrocopters/gyroplanes. The reasons for this choice are the efficiency of the low-speed rotors and the possibility of using autorotation as a safety measure. A comparison of the flight performances shows that both helicopters and gyrocopters have the potential to fulfill the TLARs of ALAADy. In order to achieve comparable flight performance, the rotor diameter of the gyroplane is chosen to be 14 m; larger than the 10 m of the helicopter. However, the lower aircraft complexity is in favor of the gyroplane. Furthermore, the main rotor of the gyroplane is permanently in autorotation. Consequently, using the autorotation in the event of a system failure is a simpler flight maneuver for the gyrocopter. A helicopter would require active and approved flight control to enable autorotation as a safety measure.

This trade-off is a first argument where the SORA has an impact on the configuration choice. This safety properties can be directly applied to the mitigation M2 of the ground risk. It is this mitigation that was evaluated in the previous section as particularly critical and that is required to have at least medium robustness to enable an operation compliant to SAIL IV. The technical simplicity of the gyroplane also influences acquisition and maintenance costs. The parametric cost models from Liebhardt and Pertz (2021) indicate the significant impact of the maintenance on the overall operating costs. However, the advantage of simplicity is traded for the Vertical Take-Off and Landing (VTOL) capability of the helicopter. For future investigations, it is worthwhile to weigh the VTOL capability imposed by market demands of additional use-cases versus the increased price of the helicopter configuration.

Three aircraft configurations now remain: a comparatively classic fixed-wing configuration with twin tail boom and push propeller, a box-wing, and a gyroplane with additional wings to improve cruise efficiency. What are the advantages and disadvantages of the different configurations in terms of ground risk? This question can be analyzed with the SORA mitigations M1 and M2 considering the restriction of the operational concept to sparsely populated areas. Additionally, the containment from areas adjacent to the operation also has to be ensured (Step #9 of the SORA). A key to answer this question is the choice of technology for safe flight termination. For example, if no additional measures are chosen in the event of a failure except deactivation of the engine, the allowable flight area is significantly reduced due to the required large safety buffers. In such a case, neglecting environmental conditions, the best glide ratio of the aircraft directly defines the necessary special safety buffer. However, if additional means such as parachutes or predefined maneuvers are used, the special safety buffer can be significantly reduced. Furthermore, if measures are found to simultaneously reduce the impact energy in the event of a safety landing or crash, this will significantly reduce the ground risk class. As a consequence, the SORA Operational Safety Objectives (OSOs) robustness are reduced, which implies relaxed reliability requirements on the aircraft, the developer, manufacturer, and operator. The components involved with these mitigations require a high level of rigor for development and implementation. Thus, in this context, we are particularly interested in low-complexity solutions that ideally require only passive control actions. Additionally, the interplay between the nominal aircraft system and the rescue or contingency components needs to be minimized.

With respect to the ALAADy operational concept, the next step suggested in Sachs (2021) is to investigate which safety measures can be used for the three remaining configurations in order to limit the ground risk. Sachs investigated various procedural concepts, such as deliberately initiating a flat-spin situation or a steady sideslip maneuver. Due to their disadvantages, however, these approaches were discarded in the course of the study. Either the necessary flight conditions for these mitigations cannot be reliably achieved from all flight situations, or they impose requirements on the aircraft that are disadvantageous to nominal flight. For this reason, parachutes are selected for the two fixed-wing aircraft. For the gyrocopter, the initial hypothesis was to use a vertical autorotation state so that the aircraft descends moderately with no forward speed. Such a behavior would be the ideal safety characteristic for an aircraft in terms of ground risk according to SORA. However, simulations show that this process must be actively controlled in order to enable a transition from all flight situations. A more promising variant in terms of passive control input is therefore to allow a low forward speed in a descending spiral flight. With the help of a manned microlight gyrocopter, Sachs demonstrated this procedure in flight tests. The descending speed of 8 m/s was confirmed and also the resulting spiral with 45 m diameter is a satisfactory result to achieve containment. However, the flight test showed a distinct forward speed. Future flight tests with an unmanned system will investigate to what extent this forward speed can be reduced to further approach the ideal safety characteristic.

In a next step, supplementary wings are added to the gyrocopter to increase the flight efficiency and its range. Also, the use of two pulling propellers instead of the often-used single pusher version is expected to improve the noise characteristics. In order to evaluate such comparably unusual configuration, a scaled demonstrator was built and studied in detail in wind tunnel and flight tests by Sachs (2021). In this way, various effects were uncovered which are not predictable in a conceptional and preliminary design. In particular, the arrangement and size of the control surfaces were corrected. Sachs concludes with a set of concrete configuration specifications to implement such configurations. This modification of the classic gyrocopter configuration allows for two major advances: the increase of range and of social acceptance by reducing noise emissions. Future work needs to investigate the interaction between the wing modification and ground risk mitigations further.

Hecken et al. (2021) examine the three pre-selected configurations in more detail to determine the influence of the aircraft structure on the mass estimation using Finite Element Methods (FEM). The aircraft configurations including the geometric arrangement of the cargo compartment, avionics, propulsion, and a rescue system are used to create a geometrically simplified representation. This mesh-based representation of the aircraft is enriched with material properties and structure technologies of the airframe. In a next step, the load cases that the structure has to withstand are determined. Normally, these load cases are taken from Certification Specifications (CS). Appropriate CS for SORA based drones were not available at the time of the project. For this reason, Hecken et al. selected the CS-23 (normal, utility, aerobatic and commuter airplanes) for the fixed-wing aircraft and, due to the lack of a suitable CS for gyroplanes, opted for CS-27 (small rotorcraft). For the operational concepts in this project, safely terminating the flight is a special and essential aspect. Since parachutes are installed for the fixed-wing aircraft, the structure must be able to withstand loads when they are deployed. In the case of the gyrocopter, the maneuver concerned with the transition into vertical descent equivalently implies loads on the structure. The structural design is therefore based on load cases, which are determined from gusts, the landing, and the safety termination. The influence of the load cases on the structure is calculated by FEM in an iterative process and the sizing of the structural elements is adjusted until the structure is sufficiently converged. In this way, the design of the aircraft structure is specified and hence, the mass estimates are improved.

In the investigation performed, Hecken et al. show that the empty weight estimates from Hasan and Sachs (2021) need to be increased by 10 to 20%. The gyrocopter performs best in this investigation. The configuration does not require a parachute, due to the main rotor always being in autorotation, saving the weight of an additional emergency landing system. The box wing is heaviest and supersedes the gyrocopter by about 18%. This disadvantage is opposed by the smaller wingspan beneficial for ground handling. A parachute has a significant influence on the structural sizing for the configurations considered in ALAADy. Rescue technologies are increasingly coming into focus for unmanned aircraft and so are Acceptable Means of Compliance (AMC) for parachute verification, see e.g. ASTM F322-18 (2018) for small UAS. The flexibility generated by SORA is limited by the availability of applicable AMC at the current state. Future AMC should not only consider the requirements for the rescue system itself, but also define performance characteristics against which the aircraft need to be designed depending on chosen contingency maneuvers and technologies. Defining these performance characteristics will be especially challenging for unconventional configurations, which may now gain advantages in SORA-driven designs.

For the discussion of the results, it is important to recall that the original TLARs were an assumption formulated at the beginning of the project. A variation of these requirements may well lead to a different pre-selection of aircraft configurations–for example, it has become clear from the previous sections that dropping cargo seems to be more important than the short take-off and landing capability. In order to arrive at a final configuration selection in future work, it is necessary to let the market considerations and options of the aircraft configurations have repercussions on the TLARs themselves. It is also interesting to note that increasing numbers of available small-scale transport drones share the VTOL capability. Here, a combination of multirotor and fixed-wing in a hybrid configuration is increasingly used. To what extent these configurations scale to larger payloads, and benchmarking them to the configurations selected in ALAADy are aspects for future research activities.

In respect of the three configurations analyzed in more detail, the twin-boom configuration provides the highest flight efficiency which is traded for an improved ground-handling in case of the box-wing. Both configurations share the need of additional parachutes to achieve a sufficient reduction of the ground risk class. The drawback of this additional rescue system is not only a system complexity aspect, but it also implies an increase in structural weight to enable the aircraft withstand the loads in a contingency event. The gyrocopter excels in containment and weight characteristics, yet has the lowest cruise flight efficiency. Cruise flight efficiency of the gyrocopter model presented in ALAADy has been increased by additional wings to make the configuration more competitive. The impact on ground risk, however, needs to be thoroughly studied and confirmed in flight testing.

3.3   Interrelation: Operability and Safe Autonomy

The term autonomy is used in many contexts, various technical definitions exist, and it is often subdivided into different levels. There is a long-lasting debate about how to differentiate between autonomy and automation. Here, we use the term autonomy to discuss the extent to which human pilots can or must monitor or influence the flight to ensure the safety of operation. To exert this task, datalinks and a control station are necessary. For the autonomy, software components increasingly take over the pilot's functions and safety tasks. The following section discusses the results of the datalink and control station concepts for the automated transport missions of ALAADy. The achievable reliability of the datalinks directly interacts with the properties to be proven for the autopilot. This fact is discussed with the example of a motion planner. Further, we consider an approach that uses a runtime monitor, so called *Safe Operation Monitors* (SOM), to bridge the gap of software assurance challenges in this context.

Operation well Beyond Visual Line-Of-Sight (BVLOS) excludes purely radio link-based datalinks for Command and Control (C2). Schalk and Becker (2021) focus on cellular network and satellite-based datalinks using commercially available components. The datalink concept provides the complementary use of the two communication technologies. Compatible data channels are established and the one with highest availability is chosen automatically. This approach has the advantage that an increase in reliability over a single datalink can be achieved without having to adapt the communication protocols. If the available communication infrastructure in the area of operation differs from the suggested concept, new communication channels can be added transparently to the user. Considering security aspects, a Virtual Private Network (VPN) between control station and aircraft is used to ensure encrypted and authenticated data transmissions.

The investigations of the datalink concept are carried out using the spare parts logistics example in Europe outlined in Sect. 3.1. Here, publicly available data on 4G cellular ground stations (Long Term Evolution—LTE) is used in a real-time simulation to determine the datalink availability during simulated flights. Radiation patterns, attenuation, and handover of LTE ground stations are included within the simulation. Schalk and Becker show that flying in low-level altitudes poses a particular challenge for the availability of LTE and can lead to dropouts during flight. Beneficial operating conditions for the availability of the datalink would be an increased flight altitude to avoid radiation obstacles by the geography or by buildings. However, Fresnel zones and exact surface structures including buildings and vegetation are not yet considered in the simulation. The coverage shown in the simulation is thus likely to overestimate the actual availability. Therefore, an additional datalink, e.g. satellite-based, are safety critical to support continuous availability.

In general, the C2 datalink enables the pilot to monitor the overall situation of the aircraft. The pilot can initiate contingencies if the operation evolves unexpectedly. For this purpose, the Human Machine Interface (HMI) has to be designed to prevent human errors, and it requires design principles to consistently present information to the remote pilot (Friedrichs et al. 2021). As an alternative, the technological challenge of cost-effective and sufficiently reliable datalinks can be reduced by finding on-board automatic solutions for system monitoring. The need for such runtime monitoring components also arises from the increasing level of autonomy of the unmanned aircraft. The concept to monitor and safeguard complex functions during runtime, and, in case of unexpected behavior, trigger contingencies to bring back the system to a safe state is considered in the ASTM standard F3269-17 (ASTM 2017).

With regard to limiting operational risks of an unamend aircraft, the following conditions need to be supervised:

- operating conditions on aircraft level including the safe flight envelope,
- operating conditions of the safety critical system components,
- functionality and availability of external services,
- human errors,
- external operating conditions (e.g. weather),
- compliance of the operation to the SORA ConOps.

From this list of aspects, it is clear that monitoring spans across different system levels: operation, human interaction, aircraft, and components. Additionally, during the life cycle of the UAS, its missions and its operations may change Thus, the constraints on the operation that are monitored change over time and need to be adapted flexibly within the monitoring implementation. However, if such monitoring systems are to reduce the safety-criticality of the pilot, it must meet strong safety requirements for integrity and robustness, implying a rigorous development process. In conclusion, the complexity and flexibility of the specifications to be monitored partly contradict the integrity and the safety requirements.

A promising method to overcome this challenge is formal languages, which can be used to separate the specification of system requirements that must be monitored from the actual implementation of a monitor. In essence, they distinguish between what shall be monitored and how it should be implemented. Therefore, both can be checked separately for consistency and correctness. Also, the specification can be used to automatically generate a monitor implementation with certain guarantees on correctness and memory demands. In the following, we refer to the presented approach of formal language-based runtime monitoring to supervise the safety conditions of a drone as *Safe Operation Monitor* (SOM). A SOM concept based on the formal monitoring language *Lola* and *RTLola* is presented in Schirmer and Torens (2021).

Schirmer and Torens discuss the SOM using the example *geofencing*, which is directly related to the SORA ConOps. The reliable containment of the unmanned aircraft to a specific operational area is essential for the SORA mitigation M1 to reduce the ground risk and for the separation to adjacent areas and airspaces (SORA Step #9). Schirmer and Torens show that the specification for geofencing can be formalized using Lola. Based on chains of polygons, a software tool is presented that allows to scale geofences based on initial flight regions and desired contingency and safety buffer regions. This tool automatically generates formal specifications for the SOM. The authors show that such a SOM can be realized in software and hardware, e.g. an FPGA (Field Programmable Gate Array), and they demonstrate the functionality in simulation.

A necessary step to unlock the potential of this concept is the development of a qualified tool to translate the formal specification to the implementation, i.e. to synthesize an implementation. The SOM needs to be embedded in a system architecture designed according to the safety requirements of the corresponding SAIL (Rothe and Nikodem 2021).

If software components are too complex, e.g. due to high input dimensions or the algorithmic complexity, static analysis tools often reach their limits. Therefore, other techniques must be used to facilitate the use of these components. One of such a technique is runtime assurance. In runtime assurance, a runtime monitor is used to switch from a statically unverified component to a verified backup component. Hence, the safety of the system is guaranteed. Complex software components, which were previously considered impossible to qualify or could only be qualified with extensive effort, may be integrated in safety-critical environments using this concept. A significant step towards the autonomy of unmanned systems is hereby enabled. An example of such complex software component is path planning, which not only determines the shortest or fastest routes, but also optimizes the flight with respect to the mission goals. In the context of operational risks, in order to maximize the safety margins, it is desirable to include the risk itself during planning. The result of such optimization would be routes where the unmanned aircraft poses the lowest possible risk to its environment. Resulting routes maximize the distance to populated areas and minimize the time spent crossing roads.

Schopferer and Donkels (2021) present a concept enabling such route planning. The authors show that three components are required for such risk-minimizing planner:

1. A semantic geospatial map, which combines information about geography with land use and traffic information
2. A risk model that represents the safety contingencies and includes statistical uncertainties, impact, and probabilities
3. An optimization or search algorithm that can handle a cost function that represents the risk.

The authors present the concept of a database, which can incorporate layered information from different sources, including land cover, elevation maps, and street and building maps. Based on this information, risk models are designed and stored in this database as an additional layer. Methods of modelling the characteristics of the safety termination according to the ALAADy safety concept and the associated descent trajectories with probability propagation are outlined and included into the route planning. The authors use a sampling-based free space representation and a search graph to generate the routes. They show that with these components, flyable trajectories representative of ALAADy transport missions can be determined within a few seconds of computation time. Transport missions from the previously determined spare parts logistics use-case are the basis for a specific case study.

The results show that route planning based on these risk models avoids areas with higher operational risks. Applying this method, routes can be determined even in environments with a complex arrangement of settlements, obstacles, airspaces, and sparsely populated areas. This route planning is an example of complex software components that depend on a variety of external information, such as traffic, weather, but also on the topographic map information itself. An integrity check of the algorithms alone is therefore not sufficient to guarantee the safety of a flight operation.

This is where the SOM unfolds its advantages. From the autonomy perspective, the idea is to receive a general qualification for such software components. This qualification would enable the unmanned aircraft to generate a risk-limited trajectory and react to unforeseen situations during operations. Towards this vision, a step-wise approach to mature the reliability can be taken involving the approval of individual flight routes determined offline as a first step. Next, the formal methods also applied within the SOM can be used to validate the safety of a calculated trajectory automatically before the flight. In this situation, only the SOM inspired evaluation tool needs to be qualified. Approval of the complex route planning is not necessary. Eventually, the use of the SOM will be key to enable in-flight planning by monitoring the safe trajectory generation and its tracking by the aircraft. Nevertheless, a prerequisite for these steps is the availability of sufficiently reliable, up-to-date, and accepted map information including all necessary external information. Another open research question is the relative quantification of operational risks for the different overflown areas. The relative weights currently assumed in the planning algorithm need to be supported by accepted ground risk models. The publication of the SORA Annex F, which is currently pending, may become the key to overcoming this challenge in the near future.

3.4 Interrelation: Airspace Integration, Detect and Avoid

The concept of ALAADy envisions the flight below regular air traffic. An original hypothesis behind this idea suggested that the air risk can be significantly reduced. Although the operation is below the minimum safe altitude of manned aviation, it is still possible that the cargo UAS will encounter other air traffic participants, including general aviation, microlight, gliders, balloons, parachutes, aircraft of search and rescue missions, and other unmanned systems. For this reason, additional infrastructure to provide airspace services and DAA systems are discussed in the following section.

The approach of technologically enabling unmanned aircraft to behave like a manned aircraft from the perspectives of airspace and deconfliction, especially in the context of DAA, has not proven successful over the past decades. For this reason, concepts for establishing digital services for specific airspace volumes to enable UAS operation are currently being planned and gradually implemented. The best-known representatives of these concepts are the NASA/FAA UTM (Unmanned Aircraft System Traffic Management) concept and Single European Sky ATM Research Program (SESAR) U-space. Investigations for airspace design, specifically for the ALAADy project, were performed from 2016 to 2017 (Peinecke et al. 2017). Peinecke et al. propose dedicated airspace volumes with UAS specific enhancements consisting of an information network using ground-based stations (D2X—drone to infrastructure). This network can also be used to convert and interface with transponders established in manned aviation. The authors refer to this extended airspace as G+ into which manned systems may enter only if they have the appropriate equipment to register with the services and to share the airspace cooperatively with the unmanned systems. This concept is challenged most in countries where air traffic is dense and diverse. Germany is an example providing such airspace structure, which was the reason why Peinecke et al. conducted the study here using Germany's airspace topology as an example. U-space and UTM had not yet been conceptualized at the time of this work. In a subsequent analysis, the compatibility of the G+ airspace with U-space is investigated and equivalence to the Concept of Operations for European UTM Systems (CORUS) of U-space is established (Peinecke and Mühlhausen 2021).

Making commercial airports available to the UAS is a challenge also from the point of view of airspace structures. Peinecke and Mühlhausen (2021) propose a separate Service Remote Pilot (SRP) that handles the approach and take-off of a UAS in the airspace D usually present around airports. The results from a simulation study confirm the conclusion made above that dedicated drone ports are preferable against commercial airports for regular transportation. However, in exceptional situations, such as humanitarian disasters, SRPs can provide an alternative option to enable the use of transshipment points of regular airports.

Furthermore, it is essential to evaluate the criticality of a failure of this G+ infrastructure. Peinecke and Mühlhausen (2021) predict the safety termination to be the last measure of the unmanned aircraft in this case. Since, in principle, any number of unmanned aircraft can be present in the airspace segment affected by the failure, the impact of this mitigation is significant. The availability requirement for the airspace services is therefore very high. However, the actual risks depend on whether manned systems are in the direct vicinity of the unmanned aircraft at the time of the service failure. Therefore, it may be worth discussing more moderate mitigations in the future if, for example, the entry of manned aircraft can be prevented when airspace services are offline.

In addition to the design of the airspace, tactical mitigations, i.e. DAA, are equally important to control the air risk. Schalk and Peinecke (2021) investigate two aspects in this context: the system architecture of the necessary DAA components and the evaluation of the sensing range of available transponder technologies. The authors decompose the DAA into three components: sensing, processing, and decision making. They assume that it is possible to distribute these components between the control station and the aircraft. This function's distribution requires the C2 datalink and HMI to be included in the safety evaluation also in the context of DAA. If, due to a low degree of autonomy of the aircraft, the intervention of the remote pilot is necessary to resolve an airspace conflict safely, the required reliability of the datalink and HMI increases. Accordingly, the three functions must be allocated in such a way that the latency caused by data transmission and pilot integration minimizes the impact on the overall deconfliction performance. Equivalently to the discussion of the previous section, increasing functions of autonomy combined with SOM have the potential to relax reliability requirements of the C2 datalinks and HMI.

A simulation study performed by Schalk and Peinecke of possible conflicting airspace participants revealed a minimum sensing range of 4–5 km for collision avoidance and guaranteeing self-separation. While currently many different research efforts approach technical solutions for such systems, existing solutions like FLARM (flight alarm) originally designed for gliders can already provide a feasible technical solution to fulfill these sensing range requirements. Nevertheless, compatibility to the U-space services and scalability depending on expected numbers of unmanned aircraft need to be considered for the DAA technology for future transport operations in low-level altitudes.

4 Experimental Validation

Two levels of validation were used in the course of the project: simulation as well as technology demonstrations at varying technology readiness levels. The following section discusses these aspects in three parts. First, the use of different digital simulations is discussed. Second, the use of cargo UAS technology demonstrators is presented. Last, a commercially available and modified unmanned helicopter is used to evaluate and discuss the results of the use-case study.

4.1 Simulations for Knowledge Gain and Validation

Some of the simulation setups used in the project have been developed specifically to investigate individual research questions, e.g. for ground handling (Meincke 2021), for the investigation of approach properties for airport integration (Mühlhausen and Peinecke 2021), and for the determination of sensor requirements for DAA (Schalk and Peinecke 2021). Beyond these problem-specific simulations, an integrated system simulation was created to study the interactions of the different aspects of the study (Schopferer et al. 2021). Schopferer et al. refer to this concept as scenario simulation. The results of the scenario simulation are used in the individual chapters of this book to discuss the impact on the specific studies; see Schalk and Becker (2021) for the influence of the C2 datalink, Schopferer and Donkels (2021) for risk-based route planning, and Schirmer and Torens (2021) for the SOM along with descent behavior upon safe termination of the different aircraft configurations. In addition, simulation plays a role in crew training and system testing for the technology demonstration in the next section and in Lorenz et al. (2021).

To support the results of this project, two key requirements for the simulation software architecture were identified. First, the simulation components from the different work packages of the ALAADy project should be integrated and combined. These components include, for example, flight dynamics in the regular operation and safe termination cases, datalink characteristics, the ground station, and on-board functions such as risk-based planning, flight control, and SOM. The second requirement is a flexible and efficient variation of simulation components. An exemplary application is studying the impact of the aircraft configurations on the safety in a termination scenario.

The solution presented in Schopferer et al. (2021) is a modularization concept for a software-in-the-loop (SIL) simulation and its separation into interface-compatible module variants. Furthermore, an automation of the execution and evaluation of the simulation trails are presented. The variation studies made possible by this automation belong to two categories:

1. Design variations: Software and simulation components are exchanged to assess the impact of the variation on the overall system behavior. An example is the exchange of the aircraft configurations between gyrocopter, twin-boom, and box-wing configuration.

2. Monte Carlo-like variations: Parameters with statistical properties are sampled and varied to obtain a statistical distribution of the resulting overall system behavior in the simulated scenario. Such simulations cannot be represented in a closed form expression. For this reason, a sampling-based method allows to statistically analyze problems with stochastic characteristics. An example is the variation of the descent trajectories upon safe termination under an expected and variable set of weather conditions.

The simulation scenarios are executed in real- and fast-time. For each of the modules used, a trade-off between efficient computability and detailed implementation was chosen to be real-time capable. Limitations of the simulation results arise exactly from this trade-off. New dynamics can result from the interactions of the individual components. Additionally, all components have their own validated or representative intervals of the input and system parameters and states, which need to be considered during assessment of the resulting simulation data.

The scenario simulation tool is used to test a variety hypotheses. An example is the validation of safety buffers, which are estimated using simple first principal heuristics such as flight altitude and best glide ratio of the aircraft configuration. The influence of varying flight and weather conditions on the compliance with the safety requirements can be determined using the simulation. The groundwork and results from Jann (2015) and Jann et al. (2015) have proven to be particularly valuable in this context, as they provide previously validated components to simulate parachutes and can thus be used to simulation parachute-based termination variants.

For the simulation, there are two options for future development. To enable the application beyond the evaluation of a concept study, the different simulation components have to be validated. This validation ranges from testing the actual datalink availability to validating the numerically determined parameters of the flight dynamics. Since such a validation is rarely possible globally in the sense of parameter and input intervals, the simulation results achieved so far can be used to specify likely operating points and domains. The second aspect of enhancement is a harmonization and shared use of the environment representation. Various components simulate physical properties using these representations including sensors, datalinks, and also software functions for higher autonomy, like risk-based planning and the SOM. In addition, the visualization for the simulation user also requires a representation of the environment, see Schopferer et al. (2021). When using different environment representations, checking consistency of data sources is not easily possible at runtime. A unified environment representation proving environmental information such as land cover and ground characteristics with configurable data properties is an opportunity to save computational resources and to systematically model intentional representation discrepancies. The already mentioned need of trustworthy map sources for trajectory and route planning from Schopferer and Donkels (2021) also need to be considered in this context.

4.2 Technology Demonstration

The limitations of simulation-driven concept studies are the validation of underlying assumptions and the identification of *unknown unknowns*. To overcome these limitations and to increase the maturity of the technology under consideration, technology demonstrators are designed. For a complex concept study as it was undertaken in ALAADy, the realization of such technology demonstrators (of about one metric ton payload capacity) is very cost intensive. For this reason, the underlying comprehensive demonstration goal is divided into different experiments to prove the most significant aspects of the concepts cost-efficiently. Two experimental platforms are used to assess the previously found relationships:

1) Design, realization, and testing of a new scaled demonstrator for the aircraft configuration study (Sachs 2021).
2) Modifying an existing microlight gyrocopter to achieve a technology demonstrator balancing the costs involved with the realization and approaching the conceptual size of the transport UAS (Lorenz, et al. 2021).

In addition, a commercially available UAS was modified to enable the use in humanitarian applications. This third aspect of technology demonstration is discussed in Sect. 4.3.

Sachs (2021) describes the realization of a scaled gyrocopter with additional wings. This aircraft configuration is one of the three most promising options from the preliminary aircraft selection (Hasan and Sachs 2021). In particular, the advantages in terms of expected total mass and presumed characteristics in descent dynamics render this configuration particularly interesting. However, even if the configurations are carefully pre-designed, especially in the case of uncommon configurations, it is hardly possible to identify all revenant physical effects in advance. The demonstrator aims to understand this configuration in more detail. The scaled demonstrator (Air Cargo Gyro 2) allows both wind tunnel tests and flight experiments. Significantly improved design specifications as well as more accurate relations of the aircraft dimensions resulted from the experiments, see Sachs (2021). Combined with the simulation results for the descent trajectories of this configuration in the safe termination case (Schirmer and Torens 2021), it can be concluded that the additional wings increase flight efficiency. However, the wings also seem to reduce the set of flight states from which an ideally contained downwards spiral for safe termination can be assumed. Future work requires more detailed analysis in this respect to find an appropriate trade-off between cruise flight efficiency and descent characteristics.

The second technology demonstrator is based on a manned microlight gyrocopter, see Lorenz et al. (2021). This gyrocopter was converted into a cargo UAS and is currently operated in line-of-sight. The objective of this demonstrator is to test the assumptions regarding assurance requirements related to the operational risk assessment and assess the technical feasibility of such a heavy-lift cargo UAS. Lorenz et al. present results on the following aspects:

- Technical realization of the emergency system and the corresponding safe design
- Flight test role allocation and operational procedures of the flight crew
- Test procedures starting with ground test and iteratively unlocking the flight envelope and automation
- Concrete application of the EASA SORA and realization of a ConOps including a review and exchange with the local aviation authority
- Realization of an avionics concept which allows to perform research validation experiments for autonomy and safety functions

  - Safety mechanisms for experimental operation
  - Design and selection of actuator and avionics components

- Software design and development processes that balance safety and flexibility required by a research environment
- Hardware-in-the-loop simulation

- Concepts and validation of the training of the experimental crew, in particular simulative training of the pilot in control
- Development of different HMI for test operation and empowerment of the different roles in flight operation
- A payload concept for cargo handling with little infrastructure requirements

A microlight gyrocopter well-known in the research environment of the DLR institutes was chosen as the technology demonstrator. Thus, experienced pilots where available whose expertise could be directly considered in the development for the flight test procedures and development of the automation. In addition, an identified flight mechanical model of this gyrocopter already existed at the beginning of the project. This approach eliminated many errors and dead ends in advance. In the general case, for the unconstrained development of new aircraft which are not based on manned systems, such prior knowledge is not available. The use of scaled demonstrators can be an interesting alternative for these situations. However, this approach has limitations in the case of non-scalable flight characteristics, for example the impact of material properties. Another important issue is the initial flight of new aircraft. Here, two options oppose each other with separate challenges. The first option, a manual maiden flight requires an excellent control station to provide sufficient situational awareness and a highly qualified pilot to handle unexpected situations. The second option is a flight control assisted maiden flight, which, however, requires early high-fidelity flight mechanical models and a robust controller design to handle unexpected model discrepancies.

The technology demonstrator is currently operated in visual-line-of-sight with the purpose to test the components that have already been designed conceptually in the project ALAADy. These tests include risk-based planning (Schopferer and Donkels 2021), SOM (Schirmer and Torens 2021), HMI (Friedrichs et al. 2021), airspace integration (Peinecke and Mühlhausen 2021) in the near future. In addition, the aircraft is a concrete benchmark to apply the currently maturing AMCs for airworthiness and safe operation for larger unmanned aircraft. The experience gained by this process facilitates a stepping stone into industrial application. In particular, BVLOS operation is an important challenge, and its impact on system design and operational procedures will be developed and discussed in the coming years.

4.3 Use-Case Validation in an Experimental Context

Unmanned aircraft systems in the scale envisioned in ALAADy were not commercially available at the time of the project. Additionally, it can be assumed that these cargo UAS will have disruptive character on the market. Thus, it is a particular challenge to critically discuss the use-cases identified for these drones. In particular, the use of UAS in the context of humanitarian aid, which is one of the most apt examples for the transport concept considered here, provides many open questions in the scientific context. In order to gain first insights in an early stage, an experiment was designed based on a commercially available UAS which is representative in terms of the operation in low-level altitudes above sparsely populates areas, and applying the SORA process. This experiment consisted of three steps:

1) In a dialogue with Non-Governmental Organizations (NGOs), a technical solution was worked out which had a real potential for humanitarian aid application.
2) The resulting technical solution was implemented and trial operations were prepared. Real disasters served as a template for the experimental operation.
3) In field trial operations were conducted, and jointly with local NGOs the benefit of the UAS technology for the intended humanitarian operation was assessed.

In Step 1, the need to deliver relief supplies such as food by air-dropping became evident. A cooperation with the foundation Wings for Aid (WFA 2014) was therefore key for this experiment. The foundation develops innovative cardboard boxes that are biodegradable and enable payloads of up to 20 kg to be dropped from unmanned aircraft without the need of a parachute. Together with the NGOs and WFA, flooding scenarios in the Dominican Republic were selected as a template. Based on actual incidents, transport missions were developed for the unmanned helicopter shown in Fig. 2 to deliver humanitarian goods to locations that had been cut off from the environment by actual flooding events in the past.

These trials were accompanied by local humanitarian first responders who evaluated the applicability of the drone technology. The operation took place in two regions of the Dominican Republic for two weeks in 2018. First, the salt lake Enriquillo was crossed for transportation from the settlement Jimani to Nuevo Boca de Cachón in BVLOS operations spanning about 7 km distance. The second region in the northeast of the island is affected by the Rio Yuna and is subject to regular flooding due to the topography. The second region is much better connected logistically than the first. Nevertheless, there have been frequent situations in the past where people had to be supplied with food and goods at short notice. With these two scenarios, two classes of applications are identified: one-time critical events (A), and logistically critical situations which are recurring (B).

Fig. 2 Impressions of delivery by dropping 20 kg WFA cargo boxes using DLR's unmanned helicopter *superARTIS* in missions to evaluate UAS-based delivery in humanitarian emergency situations

In general, in this experiments it has been shown that short-term and safe operation is possible using the SORA methodology in humanitarian aid scenarios. However, a sufficiently generalized SORA ConOps enabling an easy transfer to different mission locations is necessary. At the time of the test operation, the individual flight locations with their respective flight and safety areas were individually coordinated with the aviation authorities. In the future, it would be desirable to define boundary conditions that are either technically or procedurally established and that can be approved by the authorities in a general way. Such an approval would significantly reduce the mission overhead for suppling different destinations. On the one hand, however, it must first be ensured that the drone operator reliably complies with these boundary conditions at all times, also considering the local features. On the other hand, the authorities need information about the operating parameters as reference for the approval so that safe operation is guaranteed. SORA does provide the tools for this purpose. However, concrete local features with different forms ConOps still need to be linked. Currently, the only way forward seems to be a step-by-step approach where many civil aviation authorities (CAA) gain experience in concrete applications and exchange on the lessons learned.

In many situations, significant effort is required to establish operations which hinders a fast reaction time. Examples are: non-harmonized rules for unmanned aviation, especially airspace integration, the effort of evaluating a SORA-based application by the CAA, availability of operation-relevant data at the local aviation authorities, dual-use export regulations for drones and transport restrictions for batteries and fuel. Other essential elements for on-site operation are regulations to be observed, such as fuel handling, or the involvement of local landowners. Also, practical aspects need to be considered such as the available logistics for transporting the drone locally. The latter might be eliminated when using larger drones, directly flying to the transshipment points. However, the flexibility of drones with smaller payloads, for which take-off and landing points can be determined more flexibly, is beneficial in many situations.

Two general economic concepts seem appropriate. For short-term disasters (A), a centralized service provider seems to be the best solution. A service provider maintains appropriate qualifications, trained personnel, and equipment to operate drones in difficult conditions. The service provider would travel to the disasters on short notice. In contrast, in the case of a frequent or reoccurring event (B), operators and equipment should be permanently established locally. Knowledge of the local conditions and maintaining permanent contact with the local CAA seems beneficial in order to maintain short-term or permanent operating permits and to increase operational efficiency.

Small airports where available that could serve as take-off and landing sites for the envisioned heavy-lift cargo drones of ALAADy. These airports were sufficiently close to cover the entire country with the original TLARs of ALAADy (Dauer and Dittrich 2021). However, at the intended destinations, landings sites are not available. Finding appropriate take-off and landing sites for the DLR helicopter superARTIS (see Fig. 2) with maximum take-off weight of 90 kg used in these mission simulations proved difficult already. Dropping cargo safely seems to be the only feasible solution to meet sufficient mission conditions.

However, dropping goods implies additional safety requirements. The helicopter used is already significantly larger than widespread commercial drones. Nevertheless, the predominant reaction of third-party observers is curiosity and trust in the technology rather than caution, even while dropping goods. In these simulated missions, the drop site was secured by the operating personnel, and the drop was triggered only with visual clearance of the target area. Since this manual procedure is practicable only in exceptional cases for actual missions, DLR has initiated an additional initiative called Drones4Good in 2021. In Drones4Good, camera-based artificial intelligence (AI) is used to monitor the drone's surroundings for people and crowds. This technology will increase the safety of BVLOS operations, since overflying people can be actively avoided based on real-time information, and the trigger of dropping humanitarian goods can be constrained to safe conditions. The triggering and condition-monitoring can be implemented and safety requirements can be assured using the above proposed SOM. However, the reliability properties of the AI algorithms in conjunction with the sensors that are needed to base a safety argumentation on it, is an open question of research.

5 Conclusion and Future Work

5.1 General Remarks on the Application of the SORA

The SORA was developed to enable scalable certification at a reasonable cost while still ensuring a sufficient level of safety. However, this form of scaling creates additional challenges for the authorities. To evaluate a SORA ConOps and its interpretation of the safety assessment method, a deep understanding of many details of the operation is necessary. This understanding includes the qualification of the operator, qualification of personnel, and of the operation itself, for example standard operating or contingency procedures. For establishing regular operations, the practicability in terms of effort and the transregional harmonizability of this approach remain to be proven in the future. One tool to overcome this challenge is the collection of standard scenarios, which may reduce the variation of ConOps. Nevertheless, non-standard concepts will always require a case-by-case evaluation.

The SORA methodology projects the numerous dimensions of safety and risk factors to a one-dimensional (scalar) metric–the SAIL. The level of rigor of the operational safety objectives scale simultaneously with this SAIL. It is, for example, not possible to claim credit for operating a technically highly reliable (failsafe, autonomous) UAS with little pilot qualification. Nevertheless, this projection and scaling is a significant advancement compared to existing more classical approaches to certification. It still leaves many questions open for the near future. In particular, there is a gap in the AMC for many operational safety objectives of the SORA.

Related to the point just mentioned, two limitations of the SORA have emerged in the course of the ALAADy project. First, the ground risk for low-flying aircraft is overestimated compared to platforms operated in higher altitudes. The reason is that currently, identical energy considerations are used to determine the GRC regardless of flight altitudes. Second, assessing the risks of first flights of newly developed aircraft is not explicitly considered in the process. The very first flight tests of a new aircraft are subject to significantly changing operation conditions. Creating a ConOps for each of these flight tests does not seem practicable. Furthermore, the scalar SAIL implies that safety objectives cannot be adjusted independently. For example, if assistance or autonomy functions have not yet been validated in flight, the datalink has a significantly increased influence on safety compared to later regular operation. The SORA does not capture these particularities. At the moment, however, the SORA is still under continuous development. Hence, it is likely that the consideration of these special cases will be improved in the future.

5.2 *Conclusion*

The original hypothesis of the project ALAADy could be phrased as: *Is it possible to design and build a low-cost and heavy-lift transportation drone by exploiting operational mitigations?* Technically, the answer is positive. The project started before the SORA and U-space as such took shape. The goal was to investigate whether the consideration of operational risks leads to a more cost-efficient, yet safe and ecologically valuable solution. Due to the parallel emergence of the SORA approach, the study could even be taken one step further and was discussed with the future AMC for the EASA Specific Category. During the project, we have shown that there is a possible mid-range risk solution (SAIL IV) for the proposed application. This realization has the potential to save significant effort compared to a fully (classically) certified system.

It was found that for the aircraft of the size considered here, measures to reduce the ground risk are mandatory. These measures can be inherent properties of the system, as the gyroplane might provide, or a complementary parachute as an alternative. A parachute, however, implies increased system complexity and additional weight. Initial experience with authorities shows that the SORA is currently not always literally applied. Diverging interpretations are to be expected in the next few years before a harmonization can be achieved. It is also to be expected that the still outstanding Annex F of the SORA will have a significant influence on this interpretation.

Regarding the different use-cases, it becomes clear that the ALAADy solution is beneficial if there is little competition by ground infrastructure, or if the cargo and its timely delivery justify the additional costs of the airborne transportation. From today's perspective, the most promising use-case seems to be humanitarian aid, more concretely, regular delivery of goods to establish a continuous supply of water, food, medication and larger equipment like shelter. The observation that first heavy-lift transportation drone manufacturers follow exactly this route strengthens this assessment. Nonetheless, it is required to add the cargo dropping capability to the TLARs of ALAADy. Both, service provider or local operators are appropriate means of implementation of such emergency response services. Frequency and severity of the events considered may favor one of the two solutions in specific cases.

Additional infrastructure is required to establish safe and efficient operation. Dedicated take-off and landing sites seem to be the best solution due to the restriction to SAIL IV and due to the aircraft's flight performance. It is also important to enable drone services within the airspace as foreseen by U-space. Especially in the case of humanitarian aid, mobile solutions are possible. Mobile solutions potentially also compensate for the current limitations of datalinks. Additionally, the application of SOM also has the potential to compensate for a loss in datalink availability and, in the future, it becomes a tool to qualify more complex on-board software and hardware at lower cost.

During the first years of ALAADy, three technology demonstrations were set up with different objectives. A scaled demonstrator was used to investigate aspects of the aircraft configuration. A system demonstrator based on a modified microlight gyrocopter was the basis to investigate the system concepts in a concrete realization and to test them in flight operation. The use of a smaller and commercially available unmanned helicopter allowed experiments to gain evidence of the applicability to humanitarian help. Each of these three demonstrations also revealed questions and future research directions.

5.3 Future Work

So far, ALAADy has focused on the technical feasibility and benefits of considering operational risk for certification. The restriction of the flight to altitudes below manned air traffic was intended to minimize possible interference. One of the original reasons for introducing a minimum flight altitude for manned aviation is noise reduction. This aspect has to be examined more closely also for drone operations in the future. This research is especially important, since the gyrocopter, which was evaluated particularly positive, has disadvantages in terms of noise. Acceptance is an essential aspect for establishing such transportation solutions. Especially, when night flights are considered to reduce the overflown population density, noise characteristics gain importance.

Furthermore, as many results preliminarily confirm the gyrocopter configuration's inherent safety characteristics, these need to be proven and assessed in flight tests. Long-term operations and the accumulation of flight hours will generate research questions and will increase the technology readiness level. This increasing maturity will not only focus on the technology, but also the concepts of operational risk assessment. The influence of the maturing AMC for airworthiness of the flight system must be further investigated with the goal to enable BVLOS flight operations.

From a technical and functional perspective, the development of the airspace infrastructure is an additional issue to handle the air risk. The ability to detect people and crowds in the vicinity of the drone operation in real-time will significantly reduce ground risk and enable capabilities such as automatic and safe airdrop of humanitarian goods.

Acknowledgements The Author would like thank the entire team that contributed to the ALAADy project during the last years. Special thanks also go to Barry Koperberg, Alexis Roseillier, Eva Spoor, David Guerin, Oleg Aleksandrov, William Vigil, and Elisabet Fadul, for valuable discussions and hard work to advance this technology one step further. The entire team will also remember forever Deyvy Roa, who paved the way in many situations in the Dominican Republic. Many thanks!

References

ASTM F3322-18 (2018) Standard specification for small unmanned aircraft system (sUAS) parachutes, ASTM International, West Conshohocken, PA. https://doi.org/10.1520/F3322-18

ASTM F3269-17 (2017) Standard practice for methods to safely bound flight behavior of unmanned aircraft systems containing complex functions. ASTM International, West Conshohocken, PA. https://doi.org/10.1520/F3269-17

Dauer JC, Dittrich JS (2021) Automated cargo delivery in low altitudes: concepts and research questions of an operational-risk-based approach. In: Dauer JC (ed) Automated low-altitude air delivery - towards autonomous cargo transportation with drones. Springer, Heidelberg

Friedrich M, Peinecke N, Geister D (2021) Human machine interface aspects of the ground control station for unmanned air transport. In: Dauer JC (ed) Automated low-altitude air delivery - towards autonomous cargo transportation with drones. Springer, Heidelberg

Hasan YJ, Sachs F (2021) Performance-based preliminary design and selection of aircraft configurations for unmanned cargo operations. In: Dauer JC (ed) Automated low-altitude air delivery - towards autonomous cargo transportation with drones. Springer, Heidelberg

Hecken T, Cumnuantip S (2021) Structural design of heavy-lift unmanned cargo drones in low altitudes. In: Dauer JC (ed) Automated low-altitude air delivery - towards autonomous cargo transportation with drones. Springer, Heidelberg

Jann T (2015) Implementation of a flight dynamic simulation for cargo airdrop with complex parachute deployment sequences. In: AIAA 2015-2144, 23rd AIAA aerodynamic decelerator systems technology conference, Daytona Beach, FL, 30 March–2 April

Jann T, Geisbauer S, et al (2015) Multi-fidelity simulation of cargo airdrop: from the payload bay to the ground. In: AIAA 2015-2654, AIAA modeling and simulation technologies conference, Dallas, TX, 22–26 June

Liebhardt B, Pertz J (2021) Automated cargo delivery in low altitudes: business cases and operating models. In: Dauer JC (ed) Automated low-altitude air delivery - towards autonomous cargo transportation with drones. Springer, Heidelberg

Lorenz S, Benders S, Goormann L, Bornscheuer T, Laubner M, Pruter I, Dauer JC (2021) Design and flight testing of a gyrocopter drone technology demonstrator. In: Dauer JC (ed) Automated low-altitude air delivery - towards autonomous cargo transportation with drones. Springer, Heidelberg

Meincke P (2021) Cargo handling, transport and logistics processes in the context of drone operation. In: Dauer JC (ed) Automated low-altitude air delivery - towards autonomous cargo transportation with drones. Springer, Heidelberg

Mühlhausen T, Peinecke N (2021) Capacity and workload effects of integrating a cargo drone in the airport approach. In: Dauer JC (ed) Automated low-altitude air delivery – towards

Murzilli L et al. (2019) JARUS guidelines on specific operations risk assessment (SORA). Joint authorities for rulemaking on unmanned systems. Available via JARUS. http://jarus-rpas.org/sites/jarus-rpas.org/files/jar_doc_06_jarus_sora_v2.0.pdf. Accessed 7 Jun 2020

Nikodem F, Rothe D, Dittrich JS (2021) Operations risk based concept for specific cargo drone operation in low altitudes. In: Dauer JC (ed) Automated low-altitude air delivery - towards autonomous cargo transportation with drones. Springer, Heidelberg

Pak H (2021) Use-cases for heavy lift unmanned cargo aircraft. In: Dauer JC (ed) Automated low-altitude air delivery - towards autonomous cargo transportation with drones. Springer, Heidelberg

Peinecke N, Mühlhausen T (2021) Cargo drone airspace integration in very low level altitude. In: Dauer JC (ed) Automated low-altitude air delivery - towards autonomous cargo transportation with drones. Springer, Heidelberg

Rothe D, Nikodem F (2021) System architectures and its development efforts based on different risk classifications. In: Dauer JC (ed) Automated low-altitude air delivery - towards autonomous cargo transportation with drones. Springer, Heidelberg

Sachs F (2021) Configurational aspects and vehicle specific investigations for future unmanned cargo aircraft. In: Dauer JC (ed) Automated low-altitude air delivery - towards autonomous cargo transportation with drones. Springer, Heidelberg

Schalk LM, Becker D (2021) Datalink concept for unmanned aircraft in the context of operational risk. In: Dauer JC (ed) Automated low-altitude air delivery - towards autonomous cargo transportation with drones. Springer, Heidelberg

Schalk LM, Peinecke N (2021) Detect and avoid for unmanned aircraft in very low level airspace. In: Dauer JC (ed) Automated low-altitude air delivery - towards autonomous cargo transportation with drones. Springer, Heidelberg

Schirmer S, Torens C (2021) Safe operation monitoring for specific category unmanned aircraft. In: Dauer JC (ed) Automated low-altitude air delivery - towards autonomous cargo transportation with drones. Springer, Heidelberg

Schopferer S, Donkels A (2021) Trajectory risk modelling and planning for unmanned cargo aircraft. In: Dauer JC (ed) Automated low-altitude air delivery - towards autonomous cargo transportation with drones. Springer, Heidelberg

Schopferer S, Donkels A, Schirmer S, Dauer JC (2021) A multi-disciplinary scenario simulation for low-altitude unmanned air delivery. In: Dauer JC (ed) Automated low-altitude air delivery - towards autonomous cargo transportation with drones. Springer, Heidelberg

Wings for Aid (2014) The wings for aid foundation. https://www.wingsforaid.org. Accessed 07 Jun 2020

Peinecke N, Volkert A, Korn B (2017) Minimum Risk Low Altitude Airspace Integration for Larger Cargo UAS. In: 17th Integrated Communications, Navigation and Surveillance Systems Conference, ICNS 2017. IEEE Press. Integrated Communications Navigation and Surveillance Conference (ICNS 2017), 18.-20. April 2017, Washington DC, USA. DOI: 10.1109/ICNSURV.2017.8011946

Printed in the United States
by Baker & Taylor Publisher Services